Smart Sensors, Measurement and Instrumentation

Volume 29

Series editor

Subhas Chandra Mukhopadhyay
Department of Engineering, Faculty of Science and Engineering
Macquarie University
Sydney, NSW
Australia
e-mail: subhas.mukhopadhyay@mq.edu.au

More information about this series at http://www.springer.com/series/10617

Subhas Chandra Mukhopadhyay
Krishanthi P. Jayasundera
Octavian Adrian Postolache
Editors

Modern Sensing Technologies

Springer

Editors
Subhas Chandra Mukhopadhyay
Macquarie University
Sydney, NSW, Australia

Krishanthi P. Jayasundera
Macquarie University
Sydney, NSW, Australia

Octavian Adrian Postolache
Instituto de Telecomunicações and ISCTE-IUL
Lisbon, Portugal

ISSN 2194-8402 ISSN 2194-8410 (electronic)
Smart Sensors, Measurement and Instrumentation
ISBN 978-3-030-07607-8 ISBN 978-3-319-99540-3 (eBook)
https://doi.org/10.1007/978-3-319-99540-3

This Springer imprint is published by the registered company Springer Nature Switzerland AG
The registered company address is: Gewerbestrasse 11, 6330 Cham, Switzerland

Preface

The advancement in material sciences, electronics, embedded processing, and communication devices as well as the increased demand for small size, affordable sensors for accurate and reliable data recording, processing, storage, and communication are visible in the domain of modern sensing technologies. Sensors are extremely important in our everyday life. The sensors are used to gather data from environment, and information on weather, traffic congestion, air pollution, water pollution, etc., is obtained; they gather data on human body, and information on health, treatment, or therapy outcomes is obtained; they are also used to measure data on objects, and information for monitoring and control of these objects is obtained. For instance, the weather information is used to choose adequate clothes, the battery-level sensor permits smartphone power management optimization, and the level of blood glucose allows better healthcare management. Data collected through these modern sensors enhance our lives and our connections to each other and with our environment, allow real-time monitoring of many phenomenon around us, provide information about quality of products and services, and improve the performance of equipment based on sensor information.

The book on modern sensing technologies consists of 20 chapters specially selected from the papers presented at the Eleventh International Conference on Sensing Technology (ICST 2017), held at Macquarie University, Sydney, Australia, from December 3 to 6, 2017. The selected authors have extended the paper with more research results to be included in the book.

The first few chapters are on the area of human health monitoring. Chapter "Non-invasive Monitoring of Glycogen in Real-Time Using an Electromagnetic Sensor" presents a novel noninvasive electromagnetic sensor operating at microwave frequencies for real-time monitoring of glycogen in vitro, developed for forthcoming human trails. Skeletal muscle glycogen stores are a key indicator of athletic performance in activities requiring high levels of energy supply. Real-time monitoring of glycogen stores would allow the optimization of nutritional strategies (mainly CHO intake) to maintain energy supply and high performance. The electromagnetic sensor used in this study swept frequencies between 10 MHz and 4

GHz, an ideal range to locate any possible frequencies that match glycogens electromagnetic footprint.

Implantable devices are used extensively in the medical field for treatment and rehabilitation. However, one key issue with the use of implantable medical devices is the absence of a safe, reliable method of evaluating the effectiveness of the device during the period of implantation. The concept of using sensors to keep track of implants is found in various applications, but the challenges in designing the ideal sensor have yet to be solved. These problems are the disruption of transmission by the tissue barrier, the longevity and safety of the battery source, and the danger of infection of artificial components such as wires. Chapter "Sensors for Implants: Real-Time Failure Detection on the Arabin Pessary" has introduced a novel design of implant sensor that addressing the problems and describes its theoretical application on the Arabin Pessary implant.

Chapter "Development and Application of an Orthodontic Photometer and Thermometer to Monitor the Effect of Near Infrared Light on Root Resorption and Orthodontic Tooth Movement" describes the development, calibration, and testing of a custom-made photometric–thermometric sensor used to measure the penetration of NIR radiation from an intraoral therapeutic LED source, and the resultant temperature increase in tooth sockets. The use of NIR light to aid with bone resorption and tooth movement is a relatively new technique used to hasten tooth realignment after an orthodontic procedure.

Chapter "Wide Band Antennae System for Remote Vital Signs Detecting Doppler Radar Sensor" presents a novel antennae system for human vital signs detection. In this work, system working principles and different types of patch antennae are introduced, along with the measurement setup of the vital signs detecting radar sensor system. A wideband (from 900 MHz to 12 GHz) patch antennae system with beam-enhanced capacity is developed in FR4 substrate. This substrate has dielectric constant 4.4 and 1.2 mm of height. To reduce the size of the antennae system, a 3D-orthogonal structure was utilized to design the transmitting and receiving antennae. The multi-patch element transmitting antenna was placed orthogonally with the receiving antenna to decrease the size of the antennae system. Moreover, the bandwidth and the directional capacity of the antennae were put in high priority to identify the human's chest displacement at different frequencies, from L band to the X band. The measurement outcome shows that human vital signs could be revealed by the proposed 3D antennae system.

Chapter "Smart Sensing and Biofeedback for Vertical Jump in Sports" has reported a comprehensive system architecture of wearable embedded devices for Drop Vertical Jump (DVJ) and Vertical Jump Height (VJH) measurements toward monitoring performance of an athlete during sports activities and for biofeedback during DVJ and VJH.

The development of a facial recognition program tailored to evaluating facial behavior for real-time application has been reported in Chapter "Spontaneous Facial Expression Analysis Using Optical Flow Technique". An exploratory analysis of optical flow data was conducted with an aim to detect patterns and trends to differentiate between the emotional facial expressions: amusement, sadness, and fear from the frontal and profile facial orientations. Analysis was in the form of emotion maps constructed from feature vectors obtained by using the Lucas–Kanade implementation of optical flow.

Detection of abnormalities in heart sound to predict the abnormal condition of the human heart using the acoustic stethoscope is difficult if the signal intensity is low. This problem is solved using a simple developed electronic stethoscope in MATLAB environment with the real-time approach. Chapter "Heart Sound: Detection and Analytical Approach Towards Diseases" reviewed physiological aspects of the heart sound, evolution of heart sound detection methods, and analytical techniques to extract heart sound features. The Kalman filter response has been studied for normal and abnormal heart sounds to detect the cardiac murmur.

Chapter "Serious Games Based on Kinect and Leap Motion Controller for Upper Limbs Physical Rehabilitation" presents smart physical therapy architectures that combine multimodal sensing and virtual reality scenarios. The developed VR serious games that are used for objective evaluation of physical rehabilitation are based on multi-sensing force platform and Kinect V2 optical that provide information about the upper limb or lower limb motion detection as so as the user balance during the training session. Elements about the IoT compatibility of the implemented platform are also considered.

The next few chapters are on the monitoring of environment conditions especially in a harsh and challenging conditions. Chapter "Microwaves and Functional Materials: A Novel Method to Continuously Detect Metal Ions in Water" presents a feasibility study using unique functionalized electromagnetic (EM) sensors for continuous monitoring of zinc in water. The reaction between Zn and a Bi_2O_3-based thick film that is screen-printed onto a planar interdigitated electrode (IDE) sensors starts within 30 s, and the adsorption equilibrium was attained within 10 min. The response is faster during the initial stage and slows as equilibrium is reached. Results show good linear correlations between C (capacitance), S_{11} (reflection coefficient), and Zn concentration. The recovery time of sensors has been evaluated to be 100–150 s which demonstrates the reusability and potential application for continuous monitoring.

Chapter "Reusable Surface Acoustic Wave Immunosensor for Monitoring of Mite Allergens" has described a reusable immunosensor for monitoring of a HDM allergen—*Dermatophagoides farinae* group 1 (*Der f* 1). The immunosensor was fabricated using a surface acoustic wave (SAW) device and pH-tolerant protein scaffold (ORLA85 protein). Capture antibodies were immobilized on the ORLA85 protein-modified SAW device surface, and *Der f* 1 was measured by detecting viscoelastic change induced by sandwich assay. Differential method was employed to shorten the measurement time. It utilized a slope of the signal change as a sensor output instead of a signal shift that was used conventionally, and the measurement time was shortened by 10 min from 30 min while maintaining the sensitivity.

Chapter "Performance Enhancement of Polypyrrole Based Nano-Biosensors by Different Enzyme Deposition Techniques" has presented a comparison of results from a range of experiments carried out to investigate the performance dependency of polypyrrole-based nano-biosensors on fabrication and enzyme immobilization techniques. The methods compared are drop casting, co-entrapment, and electrophoretic enzyme deposition techniques. Templated polypyrrole nanotube array sensors provided high sensitivities and quick response times. The size of the template pore diameter plays a vital role in enzyme immobilizing in terms of loading capacity.

Chapter "Piezoresistive Pressure Sensors for Applications in Harsh Environments—A Roadmap" has discussed and reviewed materials (silicon, pSOI, SOI, 3C SiC, and 4H/6H SiC) and technologies, which are applicable to realize MEMS piezoresistive mechanical sensor elements for applications at elevated temperatures. Existing semiconductor devices based on silicon are limited to operating temperatures below 150 °C, as thermal generation of charge carriers severely degrades device operation at higher temperatures. The development of silicon on insulator (SOI) technology helped to extend device's operating temperatures to approximately 400 °C. However, at temperatures over 400 °C, the material silicon reaches its physical limits as plastic deformation starts to occur when mechanical stress is applied. Silicon carbide is considered to be the most promising semiconductor for future high-temperature and harsh-environment applications as it features a unique combination of favorable physical, electrical, mechanical, and chemical properties.

A new noncontact method to measure both surface and internal temperatures of a heated cylindrical rod has been presented in Chapter "Noncontact Temperature Sensing of Heated Cylindrical Rod by Laser-Ultrasonic Method". In the method, a laser ultrasonic technique that provides noncontact measurements of ultrasonic waves in such heated rod is effectively employed. To quantitatively determine both the surface and internal temperatures near the rod end, an ultrasonic thermometry that is a technique for measuring temperature by ultrasound has been developed by considering the direction and path of the ultrasonic waves propagating in the rod. The thermometry is basically a combined method consisting of ultrasonic wave velocity measurements based on pitch–catch configurations and a one-dimensional finite difference calculation for unsteady heat conduction. A laser ultrasonic system consisting of a pulsed laser generator (Nd:YAG, 1064 nm, 180 mJ) and a laser Doppler vibrometer (He-Ne, 633 nm, <1 mW) is used for making noncontact measurements of shear and longitudinal waves propagating near the rod end during the gas burner heating.

In the research as reported in Chapter "Development of a PVDF Multi-resonance Vibration Sensor with a Wide Range of Frequency Resolution", an attempt has been made to develop an electromechanical concept to study dynamics of piezoelectric poly(vinylidene fluoride) (PVDF) films as employed to develop a multi-resonance vibration sensor. The real-time sensor has the capability of adjusting its resonant frequency to the ambient vibration frequency. The system is able to measure the vibration at a wide range of frequencies and overcome many sources of error associated with traditional sensors such as hysteresis, variation in tension and geometry with time, and influence of temperature on electromechanical properties using only one sensor.

The following three chapters dealt with wireless sensor networks and Internet of Things. A new sensing architecture inspired from nature's sensing systems has been reported in Chapter "Unique and Unclonable Capacitive Sensors for Nature-Inspired Secure Sensing" which can prevent direct attacks on sensor network. The idea is to develop sensors with unique and unclonable characteristics and enroll the unique sensor characteristic into the verifier, for instance a cryptography module. As the verifier is tuned to the specific characteristic of a particular sensor, direct attacks are difficult. In order to develop such highly secure sensing modules, unique and unclonable sensors that are similar to sensors found in nature are required. Unique and unclonable sensors respond to specific mechanical parameter, and each sensor output is different from others. Random structural variations present in the sensor structures are sources of the uniqueness. As random structural variation is difficult to measure, model, or duplicate, such sensors are unclonable. Such sensors can be identified from their output. Design, modeling, simulation, fabrication, testing, and application of such unique and unclonable sensors are discussed in this chapter.

Chapter "Internet of Things Scalability: Communications and Data Management" has introduced a novel scheduling algorithm called Long Hop (LH) first to optimize energy usage on a wireless sensor network (WSN) that enables IoT system. The selected algorithm proposes an optimized solution to the energy efficient for scalable IoT networks. LH technique assigns high priority for packets coming with more hops and longer distances to be served first at the cluster head (CH) nodes. However, these packets require more links and nodes (thus increased energy and bandwidth usage) to reach the ultimate destination if not prioritized. The proposed technique reduces the overall energy usage and minimizes the total number of packets re-transmission and the effective data transmission distances.

Chapter "Internet of Things: Vision, Future Directions and Opportunities" looks into providing a concise review of the concepts on IoT and applications describing the main features, vision, and future directions. Furthermore, open issues and challenges that need addressing by the research community and some potential solutions are discussed. The idea of IoT has enticed significant research attentions since the massive connectivity brings varieties of challenges and obstacles, including heterogeneity, scalability, security, big data, energy requirements.

The next two Chapters "Internet of Things: Vision, Future Directions and Opportunities" and "A Novel Unobtrusive Vibration Sensing System for Machine Inspection" are on the sensing system to monitor the health of structures. Design of an affordable, unobtrusive, and autonomous vibration sensing system, called "ShakeMeter," where we combine two cost-effective measurement principles, namely optical stroboscope with a standard low-frame rate camera and a Doppler sensor has been reported in Chapter "A Novel Unobtrusive Vibration Sensing System for Machine Inspection". The optical stroboscope detects the vibration frequency with a high degree of precision by capturing the modulo (of division) between the difference between the object's vibration frequency and the optical frequency. The vibration signal consists of multiple independent signals. For incorporating such effects, multiple mutually co-prime strobing frequencies are generated and utilized to efficiently estimate the multiple frequency components via Chinese Remainder Theorem (CRT). Experimental results show that our proposed technique can effectively estimate multiple frequencies with a frequency detection error of 0.5% for vibrations occurring up to 1 kHz.

Chapter "Compensation Techniques for Vibration Sensors with Application in Structural Health Monitoring" describes a new modification in the vibration sensors with applications structural health monitoring (SHM). The frequency response of the mass–spring velocity meters (geophone) which is one of the commonly used sensors in SHM application has been addressed and modified in this chapter. The modification includes several methods with the concentration on the analog signal conditioning of the sensors.

The last chapter is on sensors for precision agriculture. Accurate sensor technologies able to give real-time information on the nutritional status of crops are needed to realize precision agriculture. The chapter has introduced different types of chemometric analyses on spectra generated by laser-induced breakdown spectroscopy (LIBS) to measure micro- and macronutrients in pasture under laboratory conditions.

We would like to express our appreciation to all authors of the chapters whose expertise and professionalism have certainly contributed significantly to this book.

We are very happy to be able to offer the readers such diverse sensors technologies and applications. We hope that this book can shed light on various technological aspect-related sensors for healthcare context and stimulate further research in this field.

Sydney, NSW, Australia Subhas Chandra Mukhopadhyay
Sydney, NSW, Australia Krishanthi P. Jayasundera
Lisbon, Portugal Octavian Adrian Postolache

Contents

About the Editors

Prof. Subhas Chandra Mukhopadhyay holds a B.E.E. (gold medalist), M.E.E., Ph.D. (India) and Doctor of Engineering (Japan). He has over 29 years of teaching, industrial, and research experience. Currently, he is working as a Professor of Mechanical/ Electronics Engineering, Macquarie University, Australia, and is the Discipline Leader of the Mechatronics Engineering Degree Program. His fields of interest include smart sensors and sensing technology, instrumentation techniques, wireless sensors and network (WSN), Internet of Things (IoT), numerical field calculation, electromagnetics. He has supervised over 40 postgraduate students and over 100 Honors students. He has examined over 50 postgraduate theses. He has published over 400 papers in different international journals and conference proceedings, written six books and forty chapters and edited sixteen conference proceedings. He has also edited thirty books with Springer-Verlag and twenty journal special issues. He has organized over 20 international conferences as either General Chair/Co-chair or Technical Program Chair. He has delivered 309 presentations including keynote, invited, tutorial, and special lectures. He is a Fellow of IEEE (USA), a Fellow of IET (UK), a Fellow of IETE (India), a Topical Editor of IEEE Sensors journal, and an associate editor of IEEE Transactions on Instrumentation and Measurements. He is a Distinguished Lecturer of the IEEE Sensors Council from

2017 to 2019. He chairs the IEEE IMS Technical Committee 18 on Environmental Measurements. He is the Founding Chair of the IEEE IMS New South Wales Chapter.

More details can be available at http://web.science.mq.edu.au/directory/listing/person.htm?id=smukhopa.

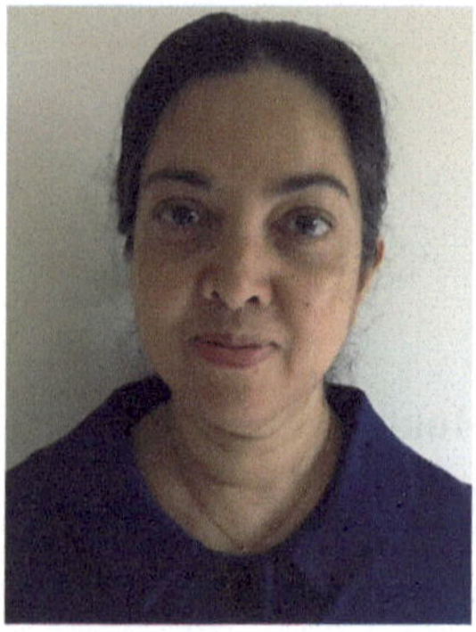

Dr. Krishanthi P. Jayasundera graduated from University of Peradeniya, Sri Lanka, with Honors degree in chemistry. She obtained her both master and Ph.D. in organic chemistry from Kanazawa University, Japan. She worked as postdoctoral researcher at Massey University nearly 14 years in New Zealand involving various projects focused on the chemical synthesis of architecturally interesting molecules which have biological, environmental, and/or medicinal significance. Currently, she is honorary researcher at Macquarie University, Sydney. She specializes in organic chemistry, biosciences, sensitivity analysis using NMR, HPLC, SPR, and so on. She has published over 30 papers in different international journals and conference proceedings.

Dr. Octavian Adrian Postolache graduated in Electrical Engineering at Gheorghe Asachi Technical University of Iasi, Romania, in 1992 and he received the Ph.D. in 1999 from the same university. The university habilitation degree was obtained in 2016 from Instituto Superior Tecnico, Universidade de Lisboa, Portugal. In 2000, he became principal researcher of Instituto de Telecomunicações where he is now Senior Researcher. He joined the Instituto Universitário de Lisboa/ISCTE-IUL Lisbon in 2011 where he is director of doctoral program on Science and Information Technology and director of research center IT-IUL. His fields of interests are smart sensors for biomedical and environmental applications, IoT for smart healthcare, wireless sensor networks, signal processing for biomedical applications, and computational intelligence with application in automated measurement systems. He is leader and active member of national and international research teams involved in Portuguese and International projects such as Tailored

Environments for Smart Physiotherapy. Dr. Postolache is author and co-author of 10 patents, 10 books, 18 chapters, 75 papers in international journals with peer review, more than 250 papers in proceedings of international conferences. He is IEEE Senior Member I&M Society, Distinguished Lecturer of IEEE IMS, chair of IEEE IMS TC-13 Wireless and Telecommunications in Measurements and chair of IEEE IMS Portugal Chapter. He is Associate Editor of IEEE Sensors Journal and Sensors MDPI open journal. He served as general chair of IEEE MeMeA 2014, and TPC chair of ICST 2014, Liverpool and ICST 2017 in Sydney and he is the general chair of ISSI 2018, Shanghai, China. He received IEEE best reviewer and the best associate editor in 2011, 2013 and 2017, and other awards related to his research activity in the field of smart sensing and applications.

Non-invasive Monitoring of Glycogen in Real-Time Using an Electromagnetic Sensor

Jacob Greene, Badr Abdullah, Jeff Cullen, Olga Korostynska, Julien Louis and Alex Mason

Abstract This work presents a novel non-invasive electromagnetic sensor operating at microwave frequencies for real-time monitoring of glycogen in-vitro, developed for forthcoming human trails. Skeletal muscle glycogen stores are a key indicator of athletic performance in activities requiring high levels of energy supply. However, glycogen stores are limited, and depletion can lead to fatigue and reductions in exercise intensity. Thus, athletes need to ensure optimal glycogen synthesis through adapted carbohydrate intake. Real-time monitoring of glycogen stores would allow the optimisation of nutritional strategies (mainly CHO intake) to maintain energy supply and high performance. However, invasive muscle biopsies remain the gold standard method of analysis, only providing a snapshot in time. Glycogen from oyster mixed into a water solution was used to manipulate concentrations observed in healthy humans ranging from 0–400 mmol/L. The electromagnetic sensor used in this study swept frequencies between 10 MHz and 4 GHz, an ideal range to locate any possible frequencies that match glycogens electromagnetic footprint. Data produced from the scattering parameter S_{11} identified a strong linear correlation between glycogen (mmol/L) and S_{11} (dBm), $r = 0.9$, $p = \leq 0.002$, with a $R^2 = 0.87$ at 2.11 GHz. This book chapter reports the first significant data that an electromagnetic sensor can successfully monitor change in glycogen concentration. This provides an encouraging basis for future work, as practical non-invasive method for in vivo monitoring of glycogen in human skeletal muscle. The progression of this research will be to analyse the sensor during different glycogen depleting exercise trials in human subjects.

J. Greene · B. Abdullah · J. Cullen · O. Korostynska
Faculty of Engineering and Technology,
Liverpool John Moores University, Liverpool L3 3AF, UK
e-mail: J.Greene2013@ljmu.ac.uk

J. Louis
School of Sports and Exercise Science,
Liverpool John Moores University, Liverpool L3 3AF, UK

A. Mason (✉)
Animalia AS, Norwegian Meat, and Poultry Research Centre, Oslo, Norway
e-mail: alex.mason@animalia.no

S. C. Mukhopadhyay et al. (eds.), *Modern Sensing Technologies*,
Smart Sensors, Measurement and Instrumentation 29,
https://doi.org/10.1007/978-3-319-99540-3_1

Keywords Glycogen · Non-invasive measurement · Electromagnetic wave sensors · Microwave spectroscopy

Introduction and Literature Review

In the last decade, there has been an emergence of wearable technology ranging from physical activity trackers [1, 2], through to medical-grade devices [3]. In recent years, advances in technology have allowed athletes, sports teams, and physicians to monitor performance workload indicators [4], rudimentary biological markers [5] and examine athlete movement in meticulous detail [6]. In team sports, there has been a growing need to monitor athletes every movement during training and competition, wearable technology that incorporates an array of sensors and advanced GPS trackers have become the norm at the elite level [6]. This data provides sports scientists, practitioners and coaches with a better insight into the physiological demands of their athletes, allowing for bespoke training protocols, thus minimising injury and improving performance [7, 8]. Athletic capabilities translate into many other areas outside of sport, especially where optimal human performance is not only a fundamental requirement but also a necessity. Military, aerospace, aviation, deep-sea diving, exploration and many other professionals need to be able to monitor vital signs to remain at peak physical and mental performance to ensure safe and effective outcomes [9–12]. Wearable sensor technology that continuously monitors changes in key physiological markers throughout demanding tasks and situations would ensure risks are limited and strategies are in place for whatever scenario occurs. Monitoring biomarkers provides a challenge, often requiring invasive producers in a clinical setting. Skeletal muscle glycogen is a key indicator of athletic performance and a predictor of fatigue. Despite the advances in modern technology, there remains a need for a device that is non-invasive, continuous and accurate. The aim of this study was to analyse the potential of a novel real-time electromagnetic sensor to detect concentration changes of glycogen in a water solution, thus providing the necessary preliminary data needed to advance to human trails.

Glycogen and Current Methods of Analysis

Carbohydrates (CHO) are an important source of energy for intense and continued bouts of exercise. CHO contain carbon, hydrogen, and oxygen atoms. The simplest carbohydrate is glucose, which is a monosaccharide, with a molecular formula of $C_6H_{12}O_6$. Glycogen is a polysaccharide containing hundreds of glucose molecules, equating to the storage form of carbohydrates within animals (Fig. 1) [13]. In humans, only the liver (~100 g) and skeletal muscle (~400 g) cells can store significant quantities of glycogen [14].

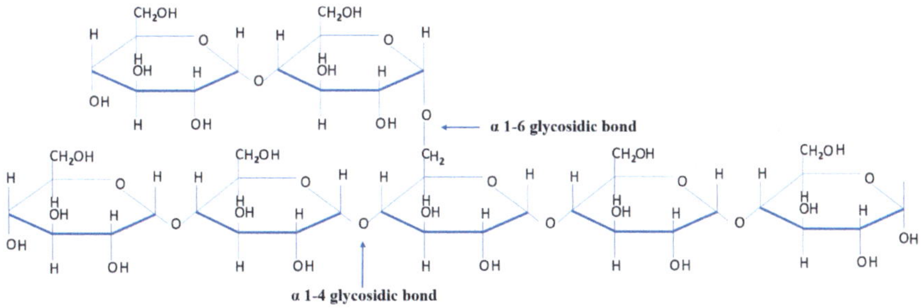

Fig. 1 A section of a glycogen molecule illustrating individual glucosyl units. It shows the two different types of glycosidic bonds used to make up glycogen

Skeletal muscle glycogen stores are a key indicator of athletic performance in activities requiring high levels of energy supply. Glycogen is one of the main energy sources for adenosine triphosphate (ATP) production, enabling muscle contraction during a wide variety of exercises, from short high-intensity exercises to longer endurance type exercises [15]. However, glycogen stores are limited and depletion can lead to fatigue and reductions in exercise intensity. Thus, athletes need to ensure optimal glycogen synthesis through adapted carbohydrate intake [16, 17]. Professional athletes and coaches are in a constant effort to maximise performance and aid recovery through improving training protocols and ensuring optimal nutritional strategies. It is widely recognised within the athletic population that ensuring sufficient CHO availability before, during and after exercise is a fundamental nutritional strategy. Thus, athletes need to ensure optimal glycogen synthesis through adapted carbohydrate intake. Exercise intensity (energy expenditure) contributes significantly to the contribution of CHO metabolism for ATP production. Naturally, as exercise intensity increases from moderate (65% VO_{2max}) to high intensity (85% VO_{2max}) CHO metabolism predominates [18]. Therefore, to reach peak performances during endurance based activities, specific preparation is required ensuring CHO consumption can match and sustain the high demands of energy expenditure during exercise [19, 20]. Glycogen storage in humans is dependent upon daily CHO consumption and activity status [13], resting muscle glycogen content of an untrained person consuming a mixed diet is around ~80–85 mmol/kg of muscle wet wt and higher for individuals who undertake regular endurance-type exercise training at around ~120 mmol/kg wet wt [13]. Increasing CHO consumption in the days before a major endurance competition is known as "CHO loading", this super-compensates muscle and liver glycogen stores and has been common practice since the 1960s [15]. Athletes can reach muscle glycogen levels of around ~200 mmol/kg wet wt with CHO loading [13]. Additionally, during competition athletes will continue to consume CHO rich drinks, bars and gels to maintain a high glycogen availability.

In recent years, there has been emerging acceptance that CHO availability needs to be tailored to meet the athletes individual energy requirements rather than a one size fits all approach [21]. Additionally, there is new understanding of the role CHO availability as a regulator of training adaptations and how different nutritional strategies may influence performance [21]. Many scientists and coaches now are beginning to see the benefits of day-by-day, meal-by-meal management of CHO availability. Knowledge of an athlete's exact glycogen levels during exercise would have many benefits throughout the sporting world, specifically in elite cycling. For example, athletes having to retrieve CHO sources from support staff/vehicles is time-consuming, energy inefficient, and comes with a high degree of risk, these all can alter the final positioning in a race where the margin for error is so small [22]. Monitoring CHO availably during competition would, in theory, improve the timing of CHO intake during key parts of the race. However, for practitioners without access to laboratories and equipment that can monitor glycogen levels non-invasively and regularly, reliance on meticulous calculation of dietary intake and energy expenditure will have to be sufficient.

Muscle Biopsies

Since the 1960s, invasive muscle biopsies have remained the elusive gold standard method for the analysis of skeletal muscle glycogen in humans [23]. However, even though biopsy equipment has improved throughout the years using modern equipment, this procedure does have its limitations. Administration by a trained practitioner within a controlled sterile laboratory environment is needed, meaning glycogen can only be assessed before and after exercise protocols (Fig. 2). Athlete willingness to undergo muscle biopsies before competition and throughout the season is often hindered due to its uncomfortable invasive nature and direct aftercare required, it usually takes between 5–7 days for soreness and swelling to fully dissipate and can potentially leave small scars. Muscle biopsies in real-world settings are limited and this technique is not used to regularly monitor glycogen content. Nevertheless, muscle biopsies remain the chosen method of glycogen analysis despite its drawbacks, until an alternative device is available which reaches a high standard of accuracy, allowing portable, real-time, and non-invasive assessments.

Magnetic Resonance Spectroscopy

Magnetic resonance spectroscopy (MRS) is the only method that can non-invasively measure skeletal muscle glycogen in vivo accurately to a medical standard. MRS is typically done in conjunction with the more universally used magnetic resonance imaging (MRI) scan that uses a combination of powerful magnets, radio waves and computers to create detailed images. MRS measures the chemical content of MR-visible nuclei, which include the metabolic elements of hydrogen, carbon, and phosphorus [24]. Spectroscopy is added to the MRI scan across specific regions of the body, these series of tests assess chemical metabolism and allow for the

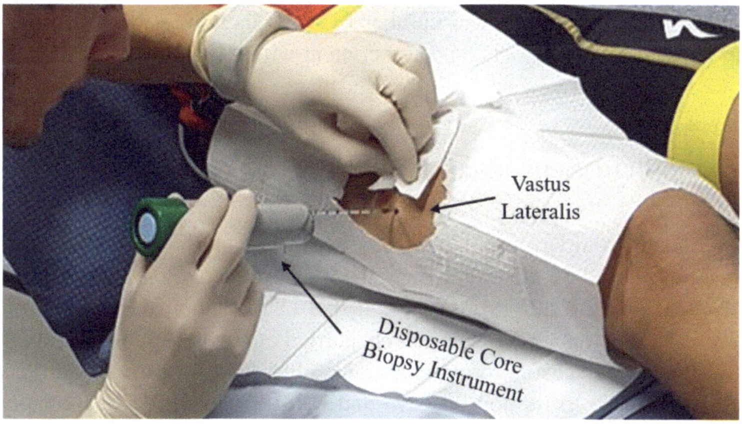

Fig. 2 Muscle biopsy procedure performed by one of the author's highlights invasive nature of assessing skeletal muscle glycogen quantities

detection of glycogen [12]. MRS delivers accurate, fast and non-invasive monitoring of muscle glycogen. However, MRS requires a radiologist to perform the procedure in a specialised MRI suite in a hospital or imaging clinic. This method is not a practical means of assessment for glycogen status in athletes due to the cost, size and stationary setting in a clinical environment. Thus far, MRS focuses on clinical trials rather than athletic performance when conducting an examination into skeletal muscle glycogen [25–27].

Musculoskeletal High-Frequency Ultrasound

Musculoskeletal Ultrasound imaging is commonly used in sports and exercise science and provides many insights such as musculoskeletal pathology diagnosis [28], visualisation of muscle and tendons [29], and determination of hydration status in athletes [30]. Recently, there have been attempts to use musculoskeletal high-frequency ultrasound to detect muscle glycogen non-invasively. Initially, data from these trials showed promising results [29, 31]. The rationale behind the technology was based on the relationship between glycogen and water within skeletal muscle. When glycogen stores are high, so is water content; this produces an ultrasound image that is hypoechoic (dark), to the contrary, when glycogen stores are minimal, there is less water and the ultrasound image is hyperechoic (brighter) [32]. MuscleSound© uses specialised software as a technique to quantify this association and interpret glycogen storage. When the ultrasound probe is placed onto the muscle of interest, the software then uses algorithms and image processing to interpret the level of muscle glycogen content [32]. However, this method was not able to produce accurate data when trialled under varying conditions, with manipulated glycogen levels in a number of ways including the consumption of creatine which directly impacts on water content within the muscle [33].

Electromagnetic Sensors as a Solution

A research team from Liverpool John Moores University have developed a novel sensing system based on electromagnetic (EM) wave technology, this technique is a rapidly developing multipurpose system that can be applied across a range of materials and industries. Previous research, carried out by the authors in this chapter highlighted that EM sensors, provide a viable alternative to current industrial and medical devices due to its non-invasive and real-time nature [34]. Recently, EM sensors have been successful in monitoring parameters in industries such as quality control in food [35–37], environmental issues such as detecting water contaminants [38, 39] and sports and exercise science, such as monitoring metabolites during exercise [40]. The sensors used in this research can typically be characterised as requiring low power (<1 mW) while retaining a good level of penetration depth, ideal for solutions beneath a surface. The principle of microwave spectroscopy is based on the interaction between the EM waves and the molecules of the sample. This interaction with the molecules, bonds etc. (rotation/vibration) allows us measure the samples conductivity and permittivity [40]. To clarify, when the sensor is placed in the vicinity or direct contact with a sample, the change in the reflected (S_{11}) and transmitted (S_{21}) microwave signals can be correlated with the composition and concentration of the material (Fig. 3) [41].

Microwave sensors could provide a solution to real-time non-invasive analysis, which is cost-effective, robust and has many practical applications from a medical and athletic standpoint. Sensors in this paper were developed with in-vivo application in mind, the size and shape of the sensor as shown in Fig. 4 makes it ideal to be placed directly over the skin. Future research will aim to use flexible materials that will enable sensors to be biomechanically positioned, working with the body rather than against it, this limits movement of the sensor during exercise.

Methodology and Measurements

This study was used as a preliminary trial before using human subjects to ensure that the electromagnetic sensor is sensitive to changes in glycogen under in vitro conditions. The study was conducted within a controlled laboratory environment ensuring a true glycogen measurement. To determine repeatability/error, seven different concentrations were used, each measured five times.

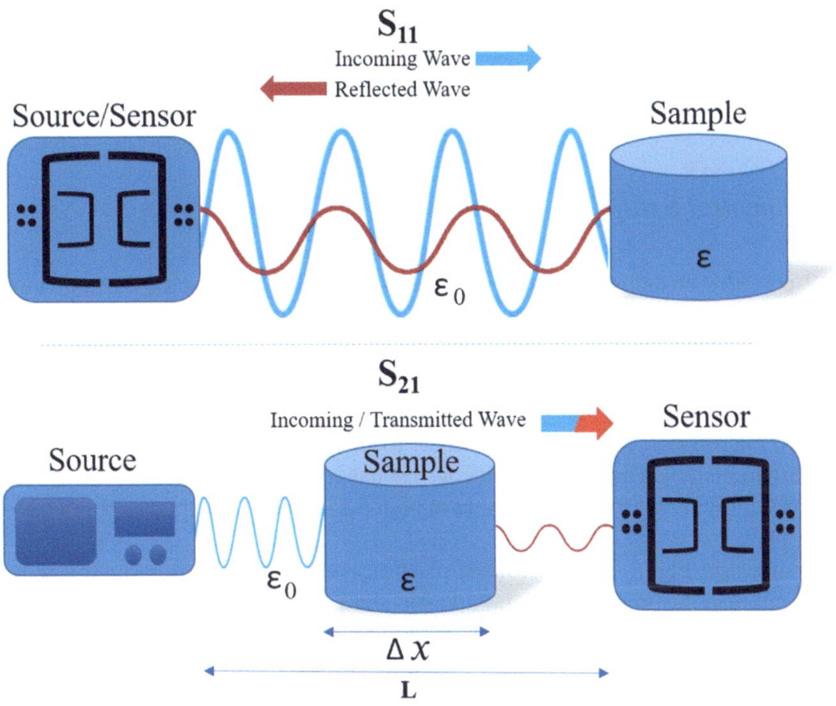

Fig. 3 Schematic showing the change in the reflected (S_{11}) and transmitted (S_{21}) microwave signal, interacting with a sample determining the composition and concentration

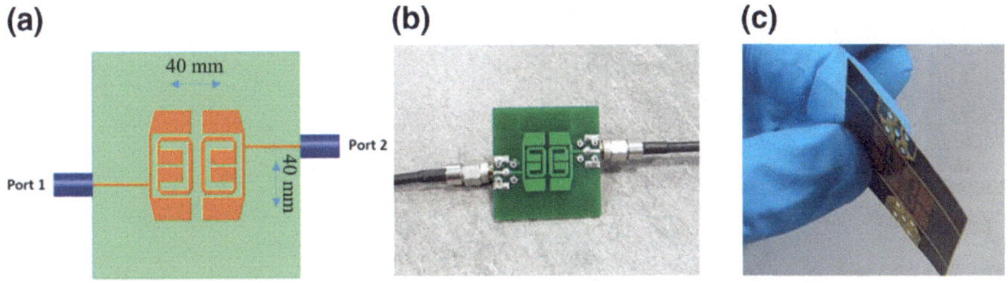

Fig. 4 Schematic of sensor development. **a** HFSS model of a *hairpin resonator* configuration sensor. **b** Working ridged FR4 PCB sensor attached to coaxial (SMA) cables. **c** Flexible polyimide film and copper sensor

Glycogen from Oyster

Glycogen is a branched polymer of glucose synthesised by animal cells for energy storage and release. The glycogen used in this experiment was glycogen (Sigma, G8751) from oyster (Type II). The glycogen was dissolved in water by stirring continuously for 2 h. Glycogen concentration was as follows: 400, 200, 100, 50,

25, 12.5, and 0 mmol/L. The glycogen concentration range was carefully selected to mimic glycogen quantities in average healthy humans under varying nutritional conditions.

Experimental Setup

The 2-port EM sensor used frequencies sweeps between 10 MHz and 4 GHz allowing for an ideal range to locate any possible frequencies that match glycogens electromagnetic signature. For data acquisition, the sensor was connected to a Rohde and Schwarz ZVA24 *Vector Network Analyzer* (VNA) (Fig. 5), calibrated to manufacturer specifications. The multi-parameter nature of microwave analysis allows the captured microwaves to be presented in the form of scattering parameters, referred to as S-parameters. The electromagnetic resonance sensor used for this work has two ports allowing for the measurement of both the reflected S_{11} and transmitted S_{21} (Table 1).

As shown in Fig. 6, a custom-made plastic casing was engineered to ensure no movement once testing had begun. A plastic slot is placed directly over the sensor allowing for the 200 μl of glycogen to be inserted consistently for each administration of the samples. The slot was cleaned with water and dried prior to each new sample.

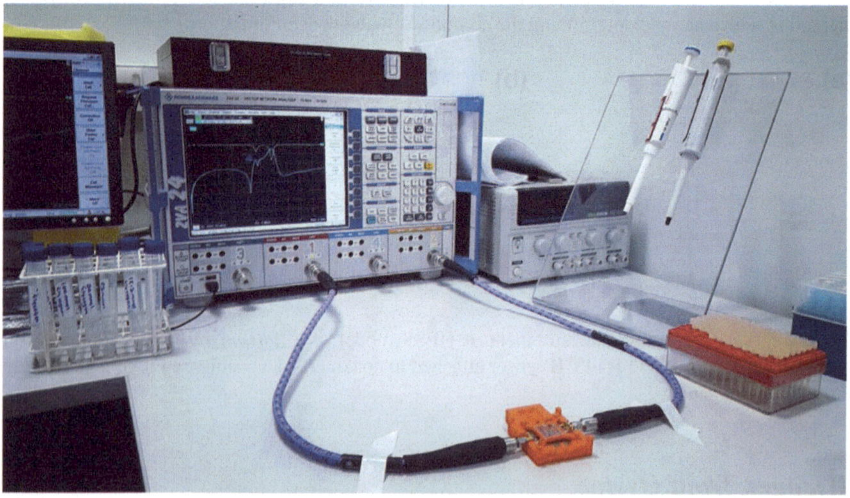

Fig. 5 Measurement Setup showing Rohde and Schwarz ZVA24 and a microwave sensor connected via 2 coaxial cables

Table 1 Measurement specifications/storage conditions

No. of measurements	7 samples × 5 repetitions = 35
EM sensor	2 Ports (S_{11}–S_{21})
Volume of samples	200 ml
Temperature	20 °C
Frequency sweep	10 MHz–4 GHz
Channel base power	0 dBm

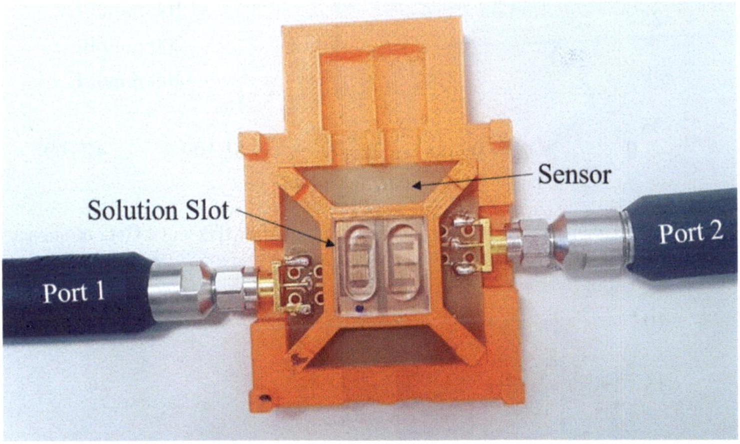

Fig. 6 Illustrates the sensor 2 port setup including plastic slots and custom casting

Statistical Analysis

Enterprise IBM SPSS 22 Statistical analysing package was used to determine the significance of the data. Pearson's correlation was used to report the S-parameters of glycogen and correlation was significant at the 0.01 level. Microsoft Excel 2013 was used for the visual interpretation of the data.

Results

S_{11} Parameter

When observing the S_{11} data obtained from the sensor at 10 MHz to 4 GHz, there were two distinct frequency shifts across this spectrum, namely at 1.6 GHz and 2.1 GHz as shown in Fig. 7.

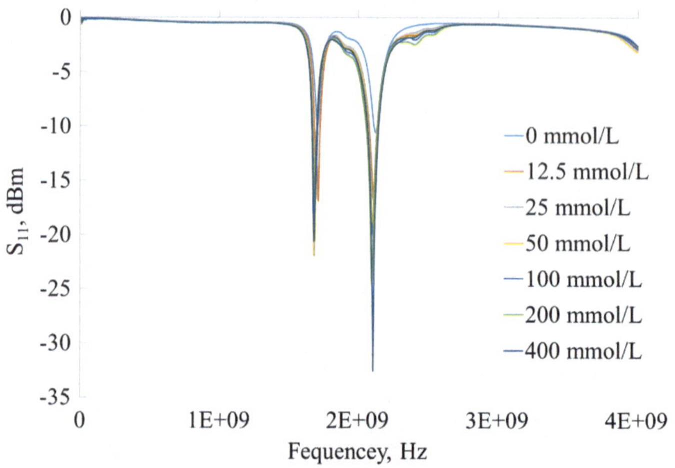

Fig. 7 S_{21} signal distribution of microwave sensor between 10 MHz and 4 GHz frequency ranges under varying concentrations of glycogen in water (mmol/L)

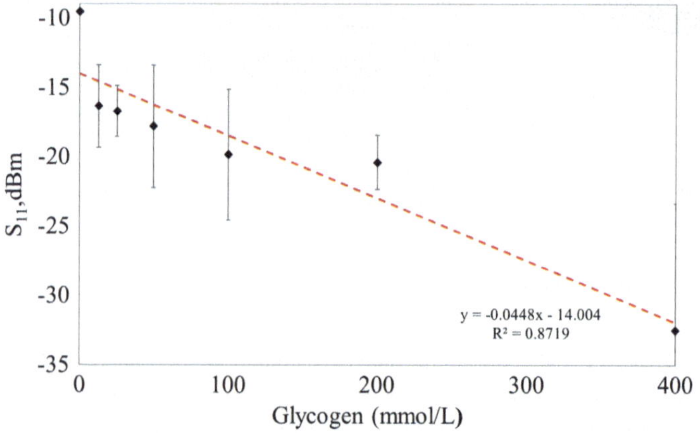

Fig. 8 The linear relationship between S_{11} variations (mean ± SD) and the response to the 7 varying glycogen concentrations at a 2.11 GHz

Further analysis identified that there was a strong linear correlation ($R^2 = 0.87$, $p \leq 0.002$) detected between S_{11} (dB) and glycogen (mmol/L) at 2.11 GHz, shown in Fig. 8.

This correlation can be observed in Fig. 9. There is a clear frequency shift between the ranging concentrations in the order of weakest through to the highest concentration in the correct ascending order.

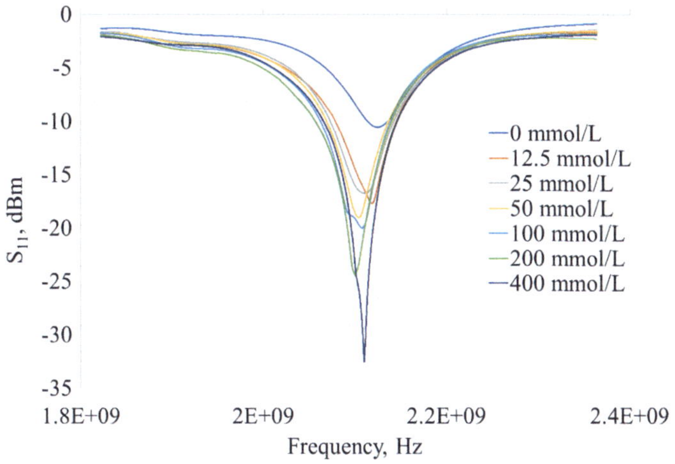

Fig. 9 S_{11} signal distribution of microwave sensor between 1.8 and 2.4 GHz frequency ranges under different concentrations of glycogen in water (mmol/L)

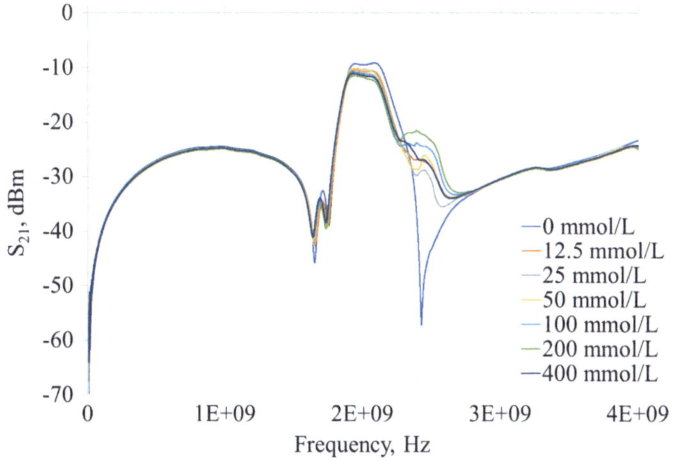

Fig. 10 S_{21} signal distribution of microwave sensor between 10 MHz and 4 GHz frequency ranges under varying concentrations of glycogen in water (mmol/L)

S_{21} Parameter

When observing the S_{21} data obtained from the sensor across 10 MHz to 4 GHz, there were two resonant peaks as seen in Fig. 10. However, further analysis showed correlation with glycogen concentraion was poor ($R^2 < 0.5$).

Discussion

This study was conducted as a preliminary examination, analysing which frequencies in the microwave range best matched the electromagnetic footprint of glycogen in a water solution before advancing to human trials. Identifying the optimum frequencies allows future development of a portable sensor device, eliminating the need for bulky apparatus such as a VNA by incorporating wearable wireless monitoring. A strong linear correlation ($R^2 = 0.87$, $p \leq 0.002$) was found in the S_{11} data collected at approx. 2.1 GHz. However, there is significant variation shown in the data collected, indicating that repeatability is not optimised and should be considered in future work. Nevertheless, since the in vitro measurement of glycogen concentration was to determine whether the sensor could monitor and detect changes, the results are considered sufficient as the impetus for further in vivo study. The EM sensor used in the research demonstrated that S_{11} data at 2.11 GHz is sensitive to changing concentrations of glycogen in water, equivalent to the concentrations naturally observed in healthy humans.

In S_{21} data, variation exists at approx. 1.7 and 2.1–2.6 GHz, but correlation is poor ($R^2 < 0.5$). This outcome is not too disappointing however for two reasons: Firstly, the poor repeatability is likely attributable to the measurement method used in this experiment and secondly, 2-port measurements are viewed as less interesting than 1-port for practical reasons of cost and system complexity in the eventual effort to make the system wearable in some form. Similar results were observed in a study using the same technology detecting blood lactate in humans [6], where S_{11} was the best performing measurement. However, although S_{21}, does not produce a strong correlation to glycogen within water, it is unknown if it will react the same to skeletal muscle so further examination is needed.

A major benefit of a non-invasive continuous device is the potential to be able to monitor live patient information. Current techniques, as noted earlier do not give enough resolution to understand whether a patient's glycogen levels are rising or falling, therefore, providing an intervention strategy can be challenging for coaches and nutritionists. Thus, being able to track the direction of glycogen during exercise is perhaps as important as knowing its absolute value. This study has shown that EM wave sensors, operating at microwave frequencies, are able to detect molecules such as glycogen. Thus, with further research, has the potential of becoming a novel approach as an alternative non-invasive method to monitor changes in glycogen levels in vivo. This technique shows a promising future and currently leads the way in the pursuit of a practical non-invasive method for the continuous monitoring of glycogen. This would solve many of the issues faced by scientists and coaches using invasive equipment, restricted to laboratory environments. However, work in this paper limited to in vitro measurements, additional in vivo studies will be required to enable the technique to have practical application.

Future Progression

Despite many recent technological advances, significant work remains to be accomplished before a reliable smart sensor for in vivo glycogen monitoring is achievable. The progression of this research will be to analyse the sensor during glycogen depletion exercise trials in human subjects. We consider the participation of endurance training athletes undergoing alternating carbohydrate diets (high, medium and low) prior to exercise. Muscle biopsies remain the preferred method of choice to measure muscle glycogen despite its invasive method. This will remain so until an alternative device is available which reaches a high standard of accuracy, allowing for portable, real-time, and non-invasive assessments. Another technical challenge is the personal calibration of wearable devices. Every person is uniquely different, and various factors could affect the EM waves under in vivo conditions. It is unknown how the EM waves will react with glycogen through the skin. Unlike in a water sample, penetrating through muscle sample may present other challenges and variables. Future research will look to incorporate personal calibration and machine learning based analysis of the date, this will ensure a more accurate and individualised method of monitoring glycogen during exercise.

Conclusions

In the pursuit of a non-invasive device to detect skeletal muscle glycogen in athletes, this study provides the first piece of evidence that an electromagnetic sensor can detect and track changes in glycogen under in vitro conditions. With recent studies researching the effects of different levels of CHO from 'real life' perspectives within the world of sport, the need for real-time non-invasive measuring equipment is ever more valuable. This technique can be further optimised with additional support and development of online data collection and processing tools driven by software like LabVIEW to operate in conjunction with the microwave sensors. The data produced from this study allows us now to focus within a specific range for better optimisation of the equipment. Furthermore, this data has now given us a frequency range look for during in vivo trails as a guided estimate.

Acknowledgements The authors would like to acknowledge the continued support of Liverpool John Moores University, Faculty of Engineering and Technology Ph.D. scholarship Programme, enabling this research to happen.

References

1. C.D. Foster, C. Twist, K.L. Lamb, C.W. Nicholas, Heart rate responses to small-sided games among elite junior rugby league players. J. Strength Cond. Res. **24**(4), 906–911 (2010)
2. K.R. Evenson, M.M. Goto, R.D. Furberg, Systematic review of the validity and reliability of consumer-wearable activity trackers. Int. J. Behav. Nutr. Phys. Act. **12**(1), 159 (2015)
3. Z.C. Haberman et al., Wireless smartphone ECG enables large-scale screening in diverse populations. J Cardiovasc. Electrophysiol. **26**(5), 520–526 (2015)
4. M.G. Mooney, J.R. Hunter, B.J. O'Brien, J.T. Berry, W.B. Young, Reliability and validity of a novel intermittent peak running speed test for Australian football. J. Strength Cond. Res. **25**(4), 973–979 (2011)
5. W. Gao et al., Fully integrated wearable sensor arrays for multiplexed in situ perspiration analysis. Nature **529**(7587), 509 (2016)
6. J. Loader, P. Montgomery, M. Williams, C. Lorenzen, J. Kemp, Classifying training drills based on movement demands in Australian football. Int. J. Sports Sci. Coach. **7**(1), 57–68 (2012)
7. J.D. Bartlett, F. O'Connor, N. Pitchford, L. Torres-Ronda, S.J. Robertson, Relationships between internal and external training load in team-sport athletes: evidence for an individualized approach. Int. J. Sports Physiol. Perform. **12**(2), 230–234 (2017)
8. B.M. Jackson, T. Polglaze, B. Dawson, T. King, P. Peeling, Comparing global positioning system (GPS) and Global navigation satellite system (GNSS) measures of team sport movements. Int. J. Sports Physiol. Perform. 1–22 (2018)
9. R. Nayak, L. Wang, R. Padhye, Electronic textiles for military personnel, in *Electronic Textiles* (Elsevier, 2015), pp. 239–256
10. H.C. Koydemir, A. Ozcan, Wearable and implantable sensors for biomedical applications. Ann. Rev. Anal. Chem. (0) (2018)
11. R. Wright, L. Keith, Wearable technology: If the tech fits, wear it. J. Electr. Resour. Med. Libr. **11**(4), 204–216 (2014)
12. J. Greene, J. Louis, O. Korostynska, A. Mason, State-of-the-art methods for skeletal muscle glycogen analysis in athletes—the need for novel non-invasive techniques. Biosensors **7**(1), 11 (2017)
13. L.M. Burke, L.J. van Loon, J.A. Hawley, Postexercise muscle glycogen resynthesis in humans. J Appl. Physiol. **122**(5), 1055–1067 (2016)
14. P. Tiidus, A.R. Tupling, M. Houston, *Biochemistry Primer for Exercise Science* (Human Kinetics, 2012)
15. J. Bergström, L. Hermansen, E. Hultman, B. Saltin, Diet, muscle glycogen and physical performance. Acta Physiol. Scand. **71**(2–3), 140–150 (1967)
16. D.L. Costill, M. Hargreaves, Carbohydrate nutrition and fatigue. Sports Med **13**(2), 86–92 (1992)
17. P. Gollnick, K. Piehl, B. Saltin, Selective glycogen depletion pattern in human muscle fibres after exercise of varying intensity and at varying pedalling rates. J. Physiol. **241**(1), 45 (1974)
18. L.J. Loon, P.L. Greenhaff, D. Constantin-Teodosiu, W.H. Saris, A.J. Wagenmakers, The effects of increasing exercise intensity on muscle fuel utilisation in humans. J. Physiol. **536**(1), 295–304 (2001)
19. J.D. Bartlett, J.A. Hawley, J.P. Morton, Carbohydrate availability and exercise training adaptation: too much of a good thing? Eur. J. Sport Sci. **15**(1), 3–12 (2015)
20. J.P. Morton et al., Reduced carbohydrate availability does not modulate training-induced heat shock protein adaptations but does upregulate oxidative enzyme activity in human skeletal muscle. J. Appl. Physiol. **106**(5), 1513–1521 (2009)
21. S.G. Impey et al., Fuel for the work required: a theoretical framework for carbohydrate periodization and the glycogen threshold hypothesis. Sports Med. 1–18 (2018)
22. P.J. Pinckaers, T.A. Churchward-Venne, D. Bailey, L.J. van Loon, Ketone bodies and exercise performance: the next magic bullet or merely hype? Sports Med. **47**(3), 383–391 (2017)
23. J. Bergström, Percutaneous needle biopsy of skeletal muscle in physiological and clinical research. Scand. J. Clin. Lab. Investig. **35**(7), 609–616 (1975)

24. D.E. Befroy, G.I. Shulman, Magnetic resonance spectroscopy studies of human metabolism. Diabetes **60**(5), 1361–1369 (2011)
25. T.B. Price, Magnetic resonance technology in training and sports. Br. J. Sports Med. **34**(5), 323–324 (2000)
26. C.J. Fuchs et al., Sucrose ingestion after exhaustive exercise accelerates liver, but not muscle glycogen repletion compared with glucose ingestion in trained athletes. J. Appl. Physiol. **120**(11), 1328–1334 (2016)
27. R. Taylor, T.B. Price, D.L. Rothman, R.G. Shulman, G.I. Shulman, Validation of 13C NMR measurement of human skeletal muscle glycogen by direct biochemical assay of needle biopsy samples. Magn. Reson. Med. **27**(1), 13–20 (1992)
28. E.S. Yim, G. Corrado, Ultrasound in athletes: emerging techniques in point-of-care practice. Curr. Sports Med. Rep. **11**(6), 298–303 (2012)
29. S. Sikdar, Q. Wei, N. Cortes, Dynamic ultrasound imaging applications to quantify musculoskeletal function. Exerc. Sport Sci. Rev. **42**(3), 126 (2014)
30. A.C. Utter, S.R. McAnulty, A. Sarvazyan, M.C. Query, M.J. Landram, Evaluation of ultrasound velocity to assess the hydration status of wrestlers. J. Strength Cond. Res. **24**(6), 1451–1457 (2010)
31. E.S. Yim, G. Corrado, Ultrasound in sports medicine. Sports Med. **42**(8), 665–680 (2012)
32. D.C. Nieman, R.A. Shanely, K.A. Zwetsloot, M.P. Meaney, G.E. Farris, Ultrasonic assessment of exercise-induced change in skeletal muscle glycogen content. BMC Sports Sci. Med. Rehabil. **7**(1), 1 (2015)
33. J.L. Bone, M.L. Ross, K.A. Tomcik, N.A. Jeacocke, J.A. Hawley, L.M. Burke, Ultrasound technology fails to provide indirect estimate of muscle glycogen concentration: 1891 Board# 43 June 2, 2: 00 PM–3: 30 PM. Med. Sci. Sports Exerc. **48**(5 Suppl 1), 520 (2016)
34. O. Korostynska, A. Mason, A. Al-Shamma'a, Microwave sensors for the non-invasive monitoring of industrial and medical applications. Sens. Rev. **34**(2), 182–191 (2014)
35. K.H. Joshi, A. Mason, O. Korostynska, A. Al-Shamma'a, Milk quality monitoring using electromagnetic wave sensors, in *Sensors for Everyday Life* (Springer, 2017), pp. 205–227
36. S. Bjarnadottir, K. Lunde, O. Alvseike, A. Mason, A. Al-Shamma'a, Assessing quality parameters in dry-cured ham using microwave spectroscopy. Meat Sci. **108**, 109–114 (2015)
37. R. Blakey, O. Korostynska, A. Mason, A. Al-Shamma'a, Real-time microwave based sensing method for vegetable oil type verification. Proced. Eng. **47**, 623–626 (2012)
38. I. Frau, O. Korostynska, A. Mason, P. Byrne, Comparison of electromagnetic wave sensors with optical and low-frequency spectroscopy methods for real-time monitoring of lead concentrations in mine water. Mine Water Environ.1–8 (2018)
39. O. Korostynska, A. Mason, A. Al-Shamma'a, Flexible electromagnetic wave sensors for real-time assessment of water contaminants, in *Sensing Technology: Current Status and Future Trends I* (Springer, 2014), pp. 99–115
40. A. Mason et al., Non-invasive in-situ measurement of blood lactate using microwave sensors. IEEE Trans. Biomed. Eng. (2017)
41. O. Korostynska, A. Mason, A.I. Al-Shamma'a, Flexible microwave sensors for real-time analysis of water contaminants. J. Electromagn. Waves Appl. **27**(16), 2075–2089 (2013)

Sensors for Implants: Real-Time Failure Detection on the Arabin Pessary

W. Ge, G. Brooker, S. C. McDonald and J. Hyett

Abstract Implantable devices are used extensively in the medical field for treatment and rehabilitation. However, one key issue with the use of implantable medical devices is the absence of a safe, reliable method of evaluating the effectiveness of the device during the period of implantation. Majority of the current monitoring methods rely on the use of imaging techniques and analysis of chemical composition of fluids. The information provided by these methods lack information on the physical performance and condition of the implants. Additionally, the discrete nature of these methods limits the amount of information provided and risks the late discovery of complications. The concept of using sensors to keep track of implants is found in various applications, but the challenges in designing the ideal sensor have yet to be solved. These problems are the disruption of transmission by the tissue barrier, the longevity and safety of the battery source, and the danger of infection of artificial components such as wires. This chapter will introduce a novel design of implant sensor that addresses these problems and describes its theoretical application on the Arabin Pessary implant.

Introduction and Literature Review

In 2010, 14.9 million infants were born preterm with the majority of cases occurring in Southern Asia. For these infants, the risk of mortality can fluctuate between 10–90% depending on the availability and quality of neonatal intensive care facilities in the country [1]. However, even in developed countries like Australia, many regional hospitals do not have the specific support facilities for preterm infants. In the cases where the preterm birth was unpredicted, the infants born at these hospitals must

W. Ge · G. Brooker (✉) · S. C. McDonald · J. Hyett
School of Aerospace, Mechanical and Mechatronic Engineering
University of Sydney, Sydney, Australia
e-mail: gbrooker@acfr.usyd.edu.au

W. Ge
e-mail: wege5641@uni.sydney.edu.au

S. C. Mukhopadhyay et al. (eds.), *Modern Sensing Technologies*,
Smart Sensors, Measurement and Instrumentation 29,
https://doi.org/10.1007/978-3-319-99540-3_2

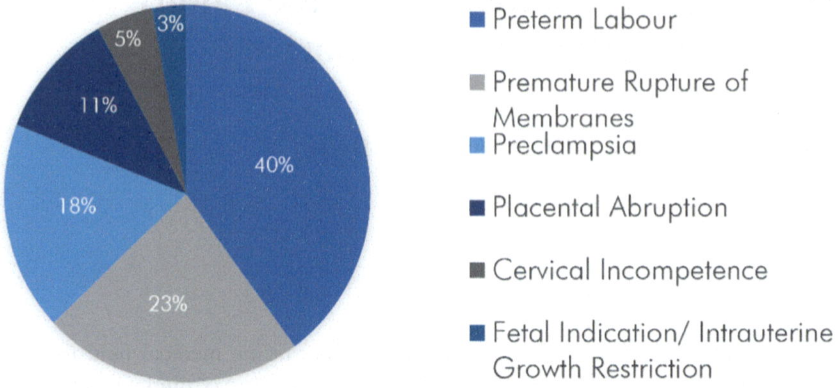

Fig. 1 Complications preceding birth before 28 weeks of gestation [5]

be transported to immediately to intensive care units after birth [2]. Not only is this a risky maneuver for the newborn but also devastating for the mother to be separated so soon after birth. Surviving infants continue to suffer from the risk of neurodevelopmental issues (60%), cerebral palsy (27%), deafness (7%), blindness (11%), motor skill impairment (10%) and epilepsy (1%) [3]. It is these circumstances that make preterm births one of the worst obstetric outcomes for families and the healthcare system.

Despite the extensive research into this field, the complex multi-factorial pathway that leads to preterm birth is still not fully understood by researchers. Its occurrence can, however, usually be linked to the emotional, physical and genetic state of the mother. One of the identified conditions that can lead to preterm birth is cervical insufficiency or cervical incompetence which defined as the painless cervical dilation in the second trimester [4]. It is the cause of 5% or extreme preterm birth cases, but in reality, this may be higher because many preterm birth cases are not assigned a definitive phenotype [5] (Fig. 1).

The diagnosis of cervical insufficiency is reliant on identifying two characteristics ultrasound scans, a short cervical length and cervical funneling. A short cervix is regarded as one of the most reliable predictors of preterm birth and used as an inclusion criterion for many studies, but its predictive value is disputed by some papers [6, 7]. Furthermore, the cut-off length of a short cervix is not well-defined, ranging from 15 to 25 mm [7]. This ambiguity creates problems for the accuracy and consistency of the studies that cite a short cervical length as the inclusion criteria. Cervical funneling is also a phenomenon observed on ultrasound as the change in physical shape of the cervix. The "TYVU" progression, as shown in Fig. 2, is used to describe cervical funneling [8, 9]. In a normal pregnancy, the 'T' shape is observed, and the cervix is fully enclosed. In the case of extreme funneling in the 'U' shape is seen cervix is almost completely exposed to the external os [8].

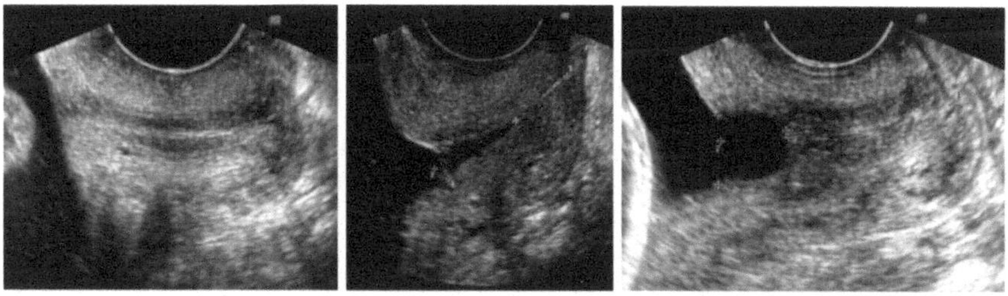

Fig. 2 TVU image of 'T', 'V' and 'U' funneling of the cervix [8]

For women who are predicted to have preterm births based on the above symptoms, the goal is to minimise the negative effects of preterm births by prolonging the pregnancy. Cervical pessaries, progesterone gel and cervical sutures, are some of the methods designed to extend the gestational period. Cervical pessaries and cervical sutures work mechanically to provide support to the weakened cervix [9, 10] while the use of progesterone gel is said to affect the insufficient cervix biochemically by influencing the collagen content [11]. A large number of clinical trials have been conducted on each of these methods [11–25], but the contradictive nature of the results mean that there is no conclusion on which of the method is the most effective in preventing preterm birth. The use of cervical sutures is associated with several negative side effects including infection, bleeding and membrane rupture [26]. In comparison, the side effects of using a cervical pessary such as vaginal discharge are considered relatively harmless and therefore is increasing in popularity [27].

The basic design of the cervical pessary, as shown in Fig. 3a, is a silicone ring-like structure that encloses the cervix and provides mechanical support. The two main types of pessaries used are the Hodge pessary and the Arabin Pessary. The primary difference between these two designs of pessaries is the placement position and the area of pressure application. The Arabin Pessary is placed at the level of the internal orifice such that the inner ring of the pessary applies an inward pressure on the cervix to prevent it from opening, as shown in Fig. 3b, whereas the Hodge pessary is positioned below the cervix to provide an upward lift. Although they are one of the most popular methods of countering preterm birth, the exact mechanism of the cervical pessary is still under speculation. Several theories on how it works have been suggested below [27]:

- It redirects pressure from the cervix to the uterine segment by changing the utero-cervical angle and hence providing structural support for the cervix;
- The pressure of the pessary prevents the degradation of the mucosal plug and prevents infection;
- It prevents the opening of the internal orifice.

The actual effectiveness of the cervical pessary has also become a highly debated topic. Many studies [21–23, 29–31] suggest that the cervical pessary is a highly

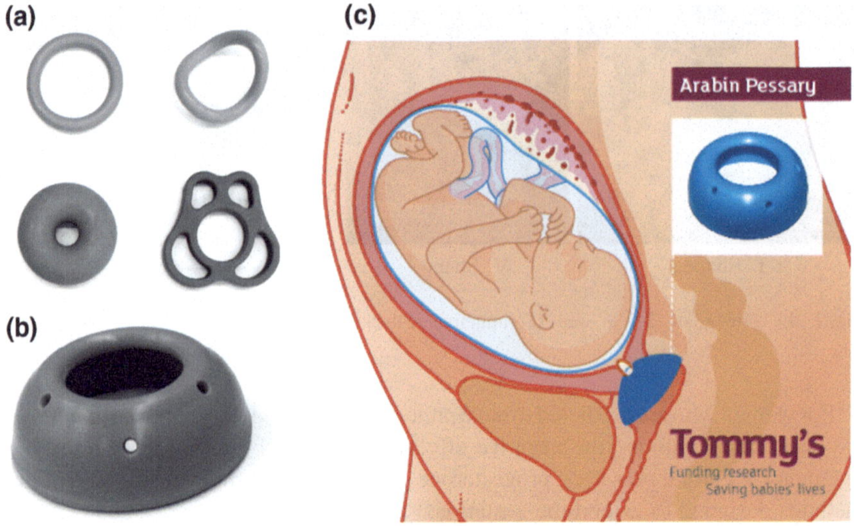

Fig. 3 **a** Ring pessary (top left), hodge pessary (top right), doughnut pessary (bottom left), butterfly-shaped pessary (bottom right); **b** Arabin cervical pessary [10]; **c** position and placement of Arabin Pessary [28]

beneficial instrument in treating preterm birth and can prolong the pregnancy by up to 5 weeks. However, other studies [24, 25, 32] claim it had little to no advantage over the control. The large variations in these controlled trials may be attributed to several factors. Firstly, it was reported that approximately 14–15% of patients required repositioning of the implant [23, 33]. These failures would only be detected in the follow-up scans which would be weeks or months apart which would negatively influence the results. Secondly, the variations in the inclusion criteria for these studies, as previously mentioned, would also cause the results to be skewed. Finally, there are also inconsistencies in the placement and follow-up procedures while conducting the studies [34].

The root of these problems is our gap in knowledge about the cause of preterm birth, but due to the sensitive nature of pregnancies, detailed research in this area is difficult to conduct. In light of this hindrance, this chapter will propose the design of wireless, battery-less sensor that could be attached to the cervical pessary. The device aims to detect both the forces applied by the pessary onto the cervix and the positioning of the pessary internally. As these readings are provided in real-time during the process of treatment, it can act as a failure detection as well as a effective device for research. It can provide a much higher resolution of information than current imaging or testing methods. The hope is that with this data, researchers will be able to pinpoint the cause of preterm birth, determine the actual mechanism of the pessary and improve diagnosis and treatment methods to target the cause.

Real-Time Failure Detection on the Arabin Pessary

Design Concept

The main challenges faced in designing a sensor that collects information in real-time is:

- How will it transmit information?
- How will it be powered?
- How will it detect information?

The transmission of information across the flesh barrier allows two options, data transfer through a wire or wireless communication. Due to the possibility of infection, wireless communication is the preferred choice but face challenges in attenuation and efficiency. Wireless communication with implanted medical devices is a heavily researched area. Currently, the use of radio-frequency, electromagnetic induction and body conductivity have been considered as viable communication options [35–37], but in this chapter, the proposed method will use an inductive magnetic coupling as a platform for communication. To overcome the problem of complicated circuitry being implanted inside the body, the internal sensor is fully powered by the magnetic induction from the external coil. This technology has been designed to utilise the change in resonance frequency and change in the supply current to measure the orientation and the applied force on the Arabin Pessary. The core technology behind the sensor design is the analysis of the change in inductive power transfer from the external monitor coil to the internal sensor coil by varying the orientation of the external coil or the frequency of the oscillator to detect the peak power transfer.

When the resonant frequency of two inductively coupled circuits matches, the power transfer from the source circuit to the load is highly efficient as very little power is lost to surrounding non-resonant objects [37]. The resonant frequency, f_{res} (Hz), of a simple LC-tank circuit can be calculated using (1), and the coupling efficiency of the 2-coil system is dependent only on the coupling coefficient, k_{12}, and the Q-factor of the coils (2) [16]. The maximum Q-factor is achieved by maximising the loop radius, wire-diameter and number of turns.

$$f_{res} = \frac{1}{2\pi\sqrt{LC}} \tag{1}$$

$$\eta = \frac{k_{12}^2 Q_1 Q_2}{\left(1 + \sqrt{1 + k_{12}^2 Q_1 Q_2}\right)^2} \tag{2}$$

where

$$k_{12} = \frac{M_{12}}{\sqrt{L_1 L_2}}$$

The efficiency of the power transfer and the resonant frequency is affected by several factors, namely: stray capacitance, coupling distance, angular misalignment and lateral misalignment. In this instance, the effect of angular misalignment is used to measure the angle of the sensor, but the other factors will limit its accuracy. Stray capacitance is present due to the proximity of the coils to one another and the high permittivity of human tissue [38]. Its presence affects the actual capacitance reading and thus limits the resolution of the data. The tuning capacitance must remain significantly larger than the stray capacitance to minimise its effect on the overall efficiency. The effect of the coupling distance is to limit the coupling factor. As the coupling distance increases, the coupling factor decreases causing the efficiency of the power transfer to also decreases exponentially [39]. Additionally, although low-frequency RF magnetic fields are not highly attenuated with tissue, some power losses may occur due to magnetic-field-induced eddy currents in the tissue [38].

Sensor Design

The proposed sensor design takes into consideration these limiting factors and is composed of two components, the internal sensor and the external monitor. The internal sensor is a simple LC-tank circuit with a variable capacitor which changes concerning the force applied. As shown in Fig. 1 the coil is concentric with the Pessary and coincident to its edges (Fig. 4).

The external circuitry is composed of a modified GDO with the circuitry shown in Fig. 2, and the values for each component is shown in Table 1. The voltage output is approximately 5–7 V measured using a multimeter with an accuracy of 2 decimal places (Fig. 5).

The measurement of the angular orientation of the sensor inside the body is dependent on the effect of angular misalignment between the two coils. When there is an

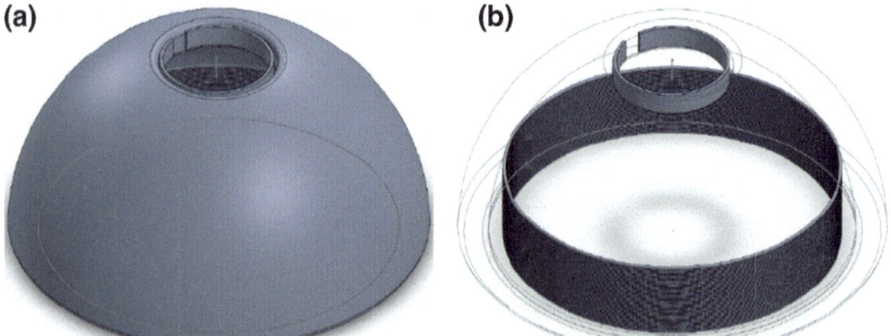

Fig. 4 **a** Model of the Arabin Pessary; **b** Model of the internal sensor component attached to the Arabin Pessary

Table 1 Value of circuitry components

Component	Value
R1	1.2 k
R2	1.2 k
R3	100 k
R4	100 k
R5	680 k
R6	680 k
R7	4.7 k
R8	10 k
C1	10 pF
C2	10 pF
C3	0.1 nF
C4	1 uF
C5	1 uF
L1	0.12 mH
L2	0.14 mH
Offset voltage	5 V
NPN transistor	2N2222
Op-Amp	LM324

angular displacement in two magnetically coupled coils, the coupling coefficient between the inductors decrease which decreases the efficiency of power transfer as shown in (3) [40].

$$\eta = \frac{\mu_0^2 N_{RX}^2 N_{TX}^2 \omega^2 b^4 a^4 \pi^2 \cos\theta^2}{16 R_{TX} R_{RX} \left(a^2 + d^2\right)^3} \tag{3}$$

η Power Transfer Efficiency (PTE)
N Number of turns
b Receiving coil radius (m)
a Transmitting coil radius (m)
θ Angle of coil (°)
d Distance of separation (m)

Assuming that the separation distance of the coil remains constant, the power transfer efficiency is minimum when the two coils are perpendicular to one another. By determining the minimum coupling on the transmission coil, the angular orientation of the receiving coil can be determined.

The capacitive force sensor is composed of an insulating foam layer sandwiched between two conductive layers. The thickness of the insulating foam varies as pressure is applied, which changes the overall capacitance. The change in thickness, in turn, will vary the resonant frequency, f (Hz), of the tank circuit. This altered resonant

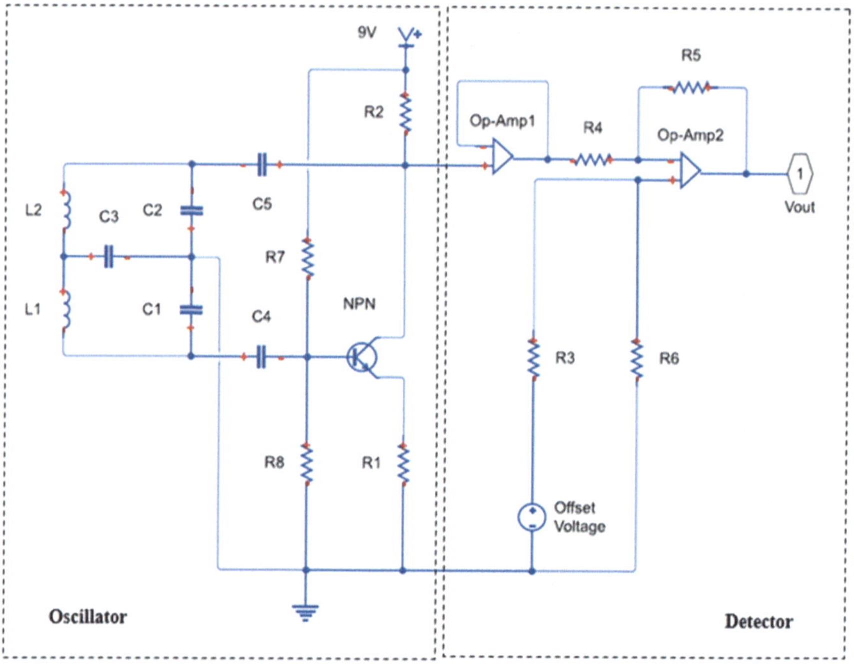

Fig. 5 The circuit design for the external monitor component

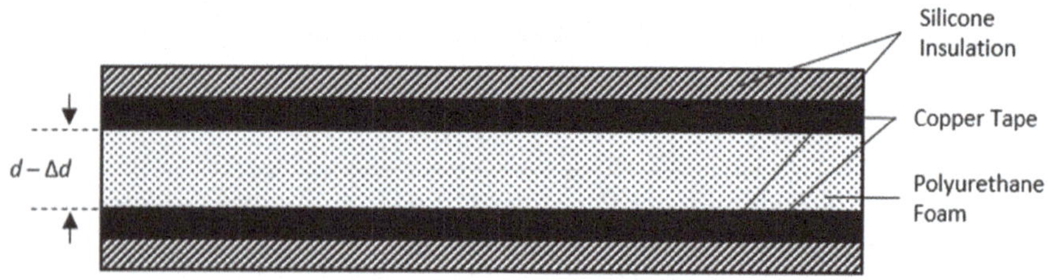

Fig. 6 Layered structure of the capacitive force sensor

frequency is detected by the external monitor and translates into a measure of the applied force (7) (Fig. 6).

$$C = \frac{\varepsilon_r \varepsilon_0 A}{d} \tag{4}$$

$$f = \frac{1}{2\pi\sqrt{LC_0}} \tag{5}$$

$$E = \frac{F d_0}{A \Delta d} \tag{6}$$

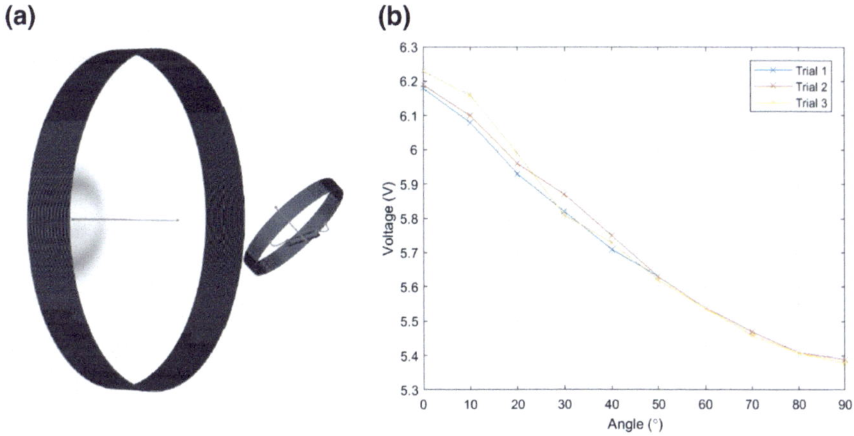

Fig. 7 **a** Experimental set-up; **b** voltage measured on external sensor against orientation of the internal coil

$$f = \frac{1}{2\pi}\sqrt{\frac{(F + EA)d_0}{L\varepsilon_r\varepsilon_0 A^2 E}} \tag{7}$$

Proof-of-Concept

To provide evidence that the concept of the design is valid, a simulation of the internal environment of the cervix is used. A 70 cm thick barrier of water is placed between the external monitor and the internal sensor. As shown in Fig. 7a, the orientation of the sensor coil is varied from 0° to 90°. The capacitive force sensor is placed on a piece animal tissue analogue to emulate the compression of biological tissue. The range of force applied is 0–5 N.

As shown from the three trials in Fig. 7b, the measured voltage follows the trend of $\cos^2\theta$ with small variations of 0.02 V in the 0°–40° range. The slight discrepancies may be due to the change in the lateral displacement between the internal sensor coil and the external monitor coil as the orientation is changed. This would not affect the results in practical applications as using peak detection would automatically calibrate the results (Fig. 8).

The change in force caused a linear change in capacitance for the first 3 N of force before gradually approaches a plateau. The maximum capacitance value reached is 44 pF. This is due to the compression limitations of the polyurethane foam. In the linear region, the approximate sensitivity is 1 pF/N corresponding to approximately 0.77 kHz/N for a system with 575 kHz baseline frequency. Trials conducted with foam of lower porosity yielded better results at higher forces but did not achieve the

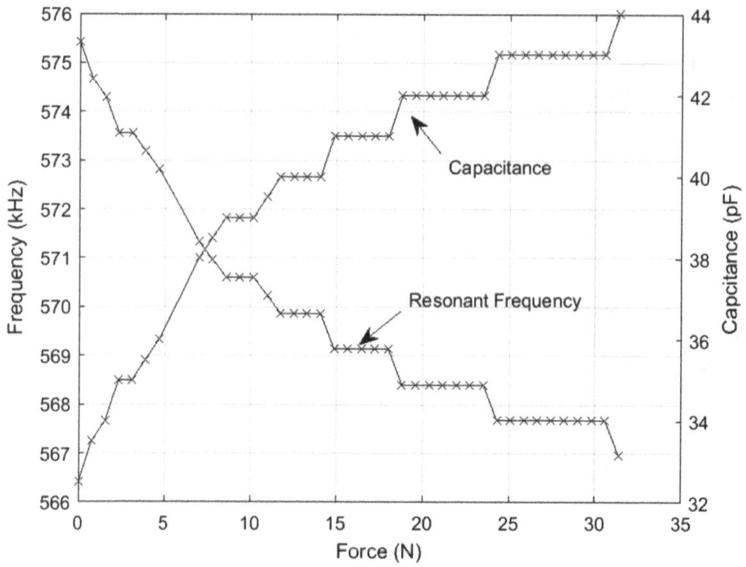

Fig. 8 Change in frequency and capacitance measured due to change in force

same level of sensitivity. Due to the small change measured, the overall capacitance of the sensor circuit would have to be decreased to account for a higher percentage change in the overall resonant frequency. Increasing the frequency of the circuit for the force detection phase to 1.4 MHz did not significantly affect the signal strength received but increased the sensitivity to 11.9 kHz/N.

Conclusions

This sensor has been designed with the intention of detecting the slippage of the pessary implant as well as to provide high-resolution biomechanical information on the preterm birth. The main differentiating factor of this device is its wireless and battery-less design which means that it would pose a minimal health risk to the mother and child. The basic idea behind the technology can be applied to other biomedical devices, and such a tool would allow us to design better implantable devices, improve clinical outcomes and reduce failure rates. This paper provides a method that will allow clinicians to track the performance of implanted medical devices by monitoring, in real time, the placement and positioning of the implantable medical device throughout treatment. The method of sensing and transmitting the data has a negligible impact on the health and the safety of the patient or the functionality of the device itself. Future design aims include increasing the accuracy and sensitivity of the device.

References

1. H. Blencowe et al., National, regional, and worldwide estimates of preterm birth rates in the year 2010 with time trends since 1990 for selected countries: a systematic analysis and implications. Lancet **379**(9832), 2162–2172 (2012)
2. N.H. and M.R. Council, *Clinical Practice Guidelines: Care around Preterm Birth* (1996)
3. M.K. Mwaniki, M. Atieno, J.E. Lawn, C.R.J.C. Newton, Long-term neurodevelopmental outcomes after intrauterine and neonatal insults: a systematic review. Lancet **379**(9814), 445–452 (2012)
4. A. Roman, A. Suhag, Overview of cervical insufficiency: diagnosis, etiologies, and risk factors. Clin. Obstet. Gynecol. **59**(2), 237–240 (2016)
5. T.F. Mcelrath et al., Pregnancy disorders that lead to delivery before the 28th week of gestation: an epidemiologic approach to classification. Am. J. Epidemiol. **168**(9), 980–989 (2008)
6. V. Berghella et al., Prediction of preterm delivery with transvaginal ultrasonography of the cervix in patients with high-risk pregnancies: does cerclage prevent prematurity? Am. J Obstet. Gynecol. **181**(4), 809–815 (1999)
7. A. Conde-Agudelo, R. Romero, Predictive accuracy of changes in transvaginal sonographic cervical length over time for preterm birth: a systematic review and metaanalysis. Am. J. Obstet. Gynecol. **213**(6), 789–801 (2015)
8. M.T. Mella, V. Berghella, Prediction of preterm birth: cervical sonography. Semin. Perinatol. **33**(5), 317–324 (2009)
9. M. House, The cervix as a biomechanical structure. Ultrasound Obs. Gynecol. **28**, 745–749 (2006)
10. B. Arabin, Z. Alfirevic, Cervical pessaries for prevention of spontaneous preterm birth: Past, present and future. Ultrasound Obstet. Gynecol. **42**(4), 390–399 (2013)
11. E.B. Fonseca, E. Celik, M. Parra, M. Singh, K.H. Nicolaides, Progesterone and the risk of preterm birth among women with a short cervix. N. Engl. J. Med. **357**(5), 462–469 (2007)
12. O.A. Rust, R.O. Atlas, K. Jo Jones, B.N. Benham, J. Balducci, A randomized trial of cerclage versus no cerclage among patients with ultrasonographically detected second-trimester preterm dilatation of the internal os. Am. J. Obstet. Gynecol. **183**(4), 830–835 (2000)
13. G. Davis, V. Berghella, M. Talucci, R.J. Wapner, Patients with a prior failed transvaginal cerclage: a comparison of obstetric outcomes with either transabdominal or transvaginal cerclage. Am. J. Obstet. Gynecol. **183**(4), 836–839 (2000)
14. O.A. Rust, R.O. Atlas, J. Reed, J. Van Gaalen, J. Balducci, Revisiting the short cervix detected by transvaginal ultrasound in the second trimester: why cerclage therapy may not help. Am. J. Obstet. Gynecol. **185**(5), 1098–1105 (2001)
15. S.M. Althuisius, G.A. Dekker, H.P. Van Geijn, D.J. Bekedam, P. Hummel, Cervical incompetence prevention randomized cerclage trial (CIPRACT): study design and preliminary results. Am. J. Obstet. Gynecol. **183**(4), 823–829 (2000)
16. V. Berghella, A.O. Odibo, J.E. Tolosa, Cerclage for prevention of preterm birth in women with a short cervix found on transvaginal ultrasound examination: a randomized trial. Am. J. Obstet. Gynecol. **191**(4), 1311–1317 (2004)
17. J. Owen et al., Multicenter randomized trial of cerclage for preterm birth prevention in high-risk women with shortened midtrimester cervical length. Am. J. Obstet. Gynecol. **201**(4), 375.e1–375.e8 (2009)
18. S.S. Hassan et al., Vaginal progesterone reduces the rate of preterm birth in women with a sonographic short cervix: a multicenter, randomized, double-blind, placebo-controlled trial. Ultrasound Obstet. Gynecol. **38**(1), 18–31 (2011)
19. E.A. DeFranco et al., Vaginal progesterone is associated with a decrease in risk for early preterm birth and improved neonatal outcome in women with a short cervix: a secondary analysis from a randomized, double-blind, placebo-controlled trial. Ultrasound Obstet. Gynecol. **30**(5), 697–705 (2007)
20. J.E. Norman et al., Vaginal progesterone prophylaxis for preterm birth (the OPPTIMUM study): a multicentre, randomised, double-blind trial. Lancet **387**(10033), 2106–2116 (2016)

21. B. Arabin, J.R. Halbesma, F. Vork, M. Hübener, J. Van Eyck, Is treatment with vaginal pessaries an option in patients with a sonographically detected short cervix? J. Perinat. Med. **31**(2), 122–133 (2003)
22. P. Sieroszewski, A. Jasiński, M. Perenc, R. Banach, P. Oszukowski, The Arabin Pessary for the treatment of threatened mid-trimester miscarriage or premature labour and miscarriage: a case series. J. Matern. Fetal. Neonatal Med. **22**(6), 469–472 (2009)
23. M. Goya et al., Cervical pessary in pregnant women with a short cervix (PECEP): an open-label randomised controlled trial. Lancet **379**(9828), 1800–1806 (2012)
24. S.A. Hui, C.M. Chor, T.K. Lau, T.T. Lao, T.Y. Leung, Cerclage pessary for preventing preterm birth in women with a singleton pregnancy and a short cervix at 20 to 24 weeks: a randomized controlled trial. Am. J. Perinatol. **30**(4), 283–288 (2013)
25. K.H. Nicolaides et al., A randomized trial of a cervical pessary to prevent preterm singleton birth. N. Engl. J. Med. **374**(11), 1044–1052 (2016)
26. S.L. Wood, J. Owen, Cerclage: Shirodkar, McDonald, and Modifications. Clin. Obstet. Gynecol. **59**(2), 302–310 (2016)
27. L.A. Wetta, J.R. Biggio, Cervical pessary use and preterm birth. Clin. Obstet. Gynecol. **57**(3), 607–615 (2014)
28. Tommy's, Treatment for cervical incompetence or risk of premature birth (Online). https://www.tommys.org/pregnancy-information/pregnancy-complications/premature-birth/reducing-risk-premature-birth/treatment-cervical-incompetence-or-risk-premature-birth. Accessed 02 Oct 2017
29. G. Acharya, B. Eschler, M. Grønberg, M. Hentemann, T. Ottersen, J.M. Maltau, Noninvasive cerclage for the management of cervical incompetence: a prospective study. Arch. Gynecol. Obstet. **273**(5), 283–287 (2006)
30. Y.H. Ting et al., Arabin cerclage pessary in the management of cervical insufficiency. J. Matern. Fetal. Neonatal. Med. **25**(12), 2693–2695 (2012)
31. M. Goya et al., Cervical pessary to prevent preterm birth in women with twin gestation and sonographic short cervix: a multicenter randomized controlled trial (PECEP-Twins). Am. J. Obstet. Gynecol. **214**(2), 145–152 (2016)
32. S.M.S. Liem et al., Economic analysis of use of pessary to prevent preterm birth in women with multiple pregnancy (ProTWIN trial). Ultrasound Obstet. Gynecol. **44**(3), 338–345 (2014)
33. M.M. Cannie et al., Arabin cervical pessary in women at high risk of preterm birth: a magnetic resonance imaging observational follow-up study. Ultrasound Obstet. Gynecol. **42**(4), 426–433 (2013)
34. G. Saccone et al., Cervical pessary for preventing preterm birth in singleton pregnancies with short cervical length: a systematic review and meta-analysis. J. Ultrasound Med. (Cl), 1–9 (2017)
35. H. Aubert, RFID technology for human implant devices. Comptes Rendus Phys. **12**, 675–683 (2011)
36. R. Bashirullah, Wireless implants. IEEE Microw. Mag. **11**(7 Suppl.), 14–23 (2010)
37. J.E. Ferguson, A.D. Redish, Wireless communication with implanted medical devices using the conductive properties of the body. Expert Rev. Med. Devices **8**(4), 427–433 (2011)
38. A.K. Ramrakhyani, S. Mirabbasi, M. Chiao, Design and optimization of resonance-based efficient wireless power delivery systems for biomedical implants. IEEE Trans. Biomed. Circuits Syst. **5**(1), 48–63 (2011)
39. M. Pinuela, D.C. Yates, S. Lucyszyn, P.D. Mitcheson, Maximizing DC-to-load efficiency for inductive power transfer. IEEE Trans. Power Electron **28**(5), 2437–2447 (2013)
40. K. Fotopoulou, B.W. Flynn, Wireless power transfer in loosely coupled links : coil misalignment model. **47**(2), 416–430 (2011)

Development and Application of an Orthodontic Photometer and Thermometer to Monitor the Effect of Near Infrared Light on Root Resorption and Orthodontic Tooth Movement

G. Brooker and T. Tang

Abstract The chapter describes the development, calibration and testing of a custom made photometric-thermometric sensor used to measure the penetration of NIR radiation from an intraoral therapeutic LED source, and the resultant temperature increase in tooth sockets. The use of NIR light to aid with bone resorption and tooth movement is a relatively new technique used to hasten tooth realignment after an orthodontic procedure. One of the unanswered questions in this application is whether there is sufficient light penetration through the tissue and bone into the tooth socket, from an external array of LEDs to have the required effect. The difficulty of conducting these measurements from within a tooth socket is due to the small size of the sensor required. However a judicious choice of probes enabled successful development, calibration and in vivo testing to take place. Measurements on the full cohort of 18 patients (36 measurements in total as each patient had two teeth removed) showed much higher attenuation than expected through to the socket. The attenuation was found to be strongly correlated to the bone thickness with an attenuation coefficient close to that of haemoglobin, implying that the in vivo bone is saturated with blood. Apart from photometric and thermometric data the heart rate of the patients was also apparent, due to swelling of the capillaries in the gum tissue during systole—the contraction phase—of the cardiac cycle.

G. Brooker (✉) · T. Tang
School of Aerospace, Mechanical and Mechatronic Engineering,
University of Sydney, Sydney, Australia
e-mail: gbrooker@acfr.usyd.edu.au

S. C. Mukhopadhyay et al. (eds.), *Modern Sensing Technologies*,
Smart Sensors, Measurement and Instrumentation 29,
https://doi.org/10.1007/978-3-319-99540-3_3

Introduction and Literature Review

Until recently the process of reducing the duration of orthodontic treatment was a traumatic and painful experience. For example, alveolar corticotomies involves cutting deep grooves in the cortical bones. A technique that relies on the theory that trauma applied to the bone tissue will stimulate remodeling and hasten the rate of bone healing. A slightly less invasive method called Piezocision involves smaller vertical incisions below the dental papilla into the alveolar bone, after which an ultrasound device penetrates these incisions to create space for a bone graft [1–4].

Near-Infrared radiation is well documented as a treatment for a number of clinical conditions including disorders of the central nervous system, stroke, macular degeneration [5], neuro-technology research [6] and orthodontic problems [7] amongst many. Radiation between 650 and 1350 nm, in the so called therapeutic or optical window, provides the maximum penetration depth in human (and animal) tissue. This window is bounded by the opacity of haemoglobin at shorter wavelengths and water at longer wavelengths as shown in Fig. 1 using data taken from [8–10].

Bone is mostly a matrix of calcium hydroxyapatite and collagen containing some blood, lipids and proteins and is generally considered to be almost transparent in the therapeutic window, as can be seen from Fig. 1. Haemoglobin is a globular protein made from four polypeptide chains, each containing an iron atom. It has a strong affinity for oxygen molecules with each chain able to bind to one O_2 molecule to form oxy-haemoglobin. This has a high absorption coefficient with a steep cutoff at around 600 nm.

Orthopaedic applications of NIR therapy are primarily aimed at speeding up the process of tooth re-alignment and so rely on good optical penetration into the jaw. A number of laser and LED based devices are on the market to provide illumination for this therapy. However there is little data to support significant penetration into the relevant tissue or whether any therapeutic benefits result from the light stimulus of

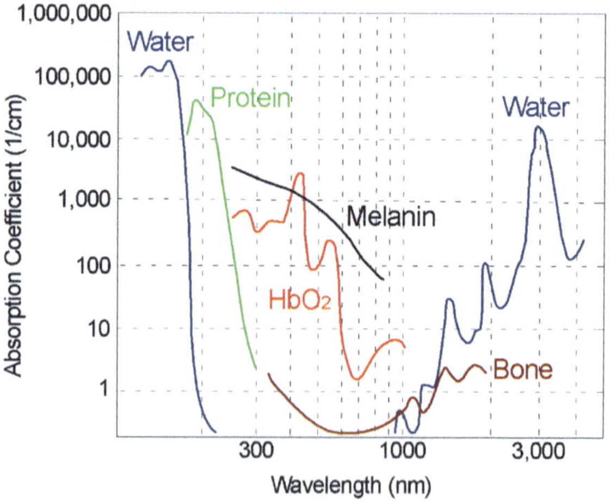

Fig. 1 Optical absorption coefficients of different tissue types

mitochondria or from local heating effects. A study was undertaken at Sydney University Dental Hospital in an attempt to provide some insight into these uncertainties, and some of the results are documented here.

Study Overview

As part of the usual orthodontic procedure performed at the Dental Hospital, patients who had been using an intraoral LED device (OrthoPulse™, Biolux Research Ltd.) [11] had premolars removed bilaterally from their upper jaw, and immediately afterwards had NIR power density and temperature measurements taken by a combined photometer and thermometer inserted into the newly vacated tooth sockets, while the OrthoPulse device was in operation. This procedure is shown graphically in Fig. 2.

To allow these measurements to be made, a custom photometer-thermometer with a probe tip sufficiently small to insert into a tooth socket was developed. This chapter describes the development, calibration and testing of this sensor along with the results obtained for a cohort of 18 patients.

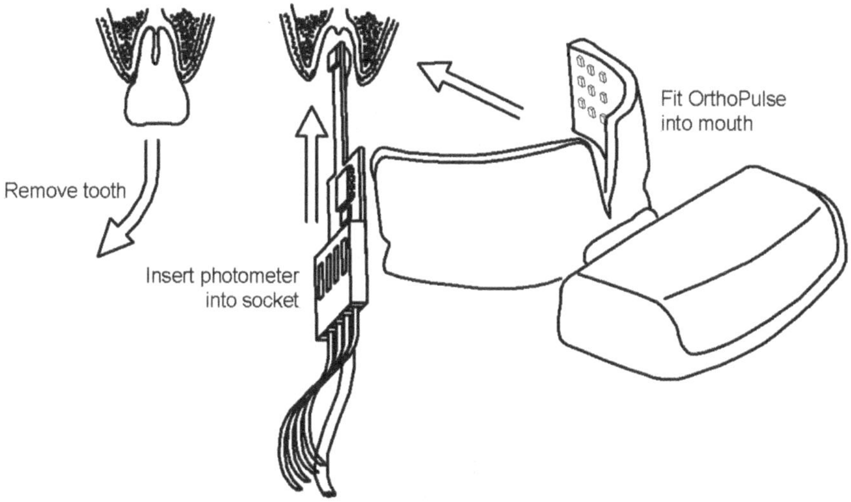

Fig. 2 Test procedure used to measure the penetration of NIR radiation and the temperature in a tooth socket

Sensor Development

Photometer

To allow the sensor to be inserted right into the base of the tooth socket it needed to be less than 4 mm in diameter and 20 mm long. A conventional transimpedance photodiode amplifier is implemented [12], as seen in Fig. 3. In this configuration the photodiode operates in photovoltaic mode with the opamp ensuring that the voltage across the diode is clamped to 0 V.

A small photodiode (Vishay VEMD6010) which has a footprint of only $4 \times 2 \times 1.05$ mm was selected as suitable for the probe tip. Unfortunately its small sensitive area (only 0.85 mm^2) results in a very low sensitivity, typically 9 μA for an illu-

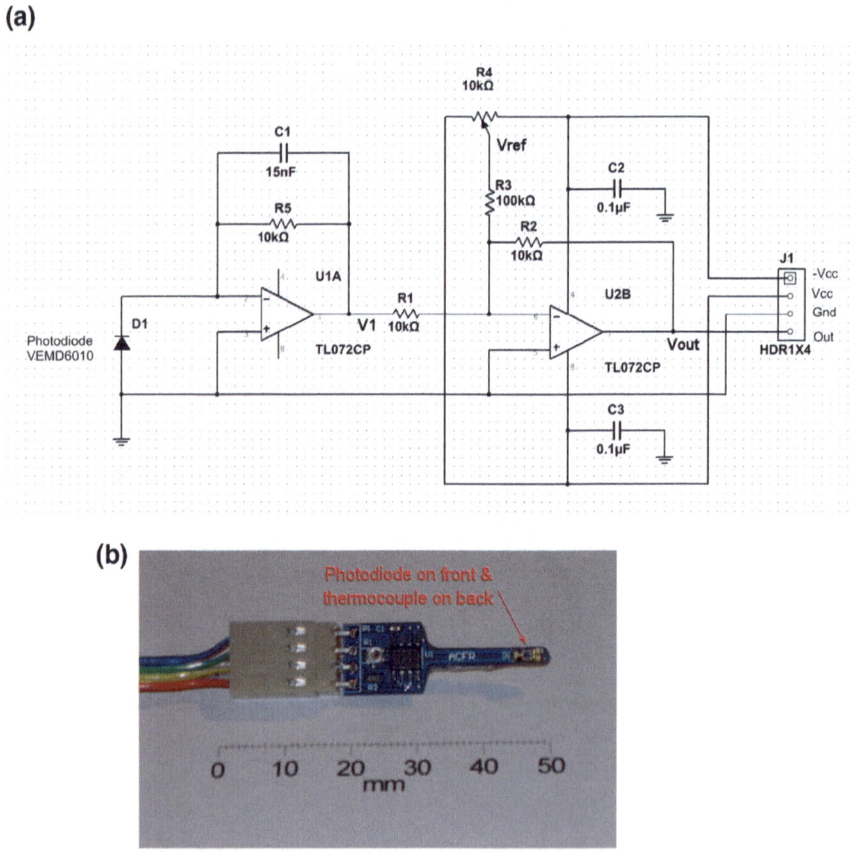

Fig. 3 Schematic diagram (**a**) and a photograph (**b**) of the transimpedance photodiode amplifier

minance of 1 mW/cm^2 at 950 nm. Its relative spectral sensitivity at the OrthoPulse frequency of 850 nm is 0.9.

As a first approximation, the voltage output by the first opamp (U1A) is equal to the product of the current generated by the diode and the feedback resistor value, R_5. A tradeoff between the input offset voltage and the leakage current performance is required when selecting a suitable opamp and the TL072 was found to be satisfactory. Any input offset voltage appears across the photodiode and increases the dark current which further increases the offset error. As regards the leakage current, i_{leak}, it also adds to the offset of the output voltage as defined in Eq. (1).

$$v_1 = -R_5(i_{photo} + i_{dark} + i_{leak}) \tag{1}$$

A capacitor C_1 across the feedback resistor performs a lowpass function with a cutoff frequency of 10 Hz for a 1 MΩ feedback resistor, increasing to 1 kHz for a 10 kΩ resistor, as the transimpedance sensitivity is reduced.

To determine the value required for R_5, the photodiode sensitivity, the illumination power density and the expected attenuation through the jaw into the tooth socket are required, and these are determined later in this chapter.

The second opamp (U1B) functions as a summing junction to allow any offset voltage generated by biases in the first opamp to be trimmed out. The equation which relates the output voltage, v_{out} to the input to that stage, v_1 is

$$v_{out} = -\left(\frac{R_2}{R_3}v_1 + \frac{R_2}{R_1}v_{ref}\right) \tag{2}$$

Light Level Estimation

The OrthoPulse NIR source used in these tests shown in Fig. 4, consists of an array of forty four 850 nm LEDs spaced 5 mm apart in two banks of 3×9 elements embedded within a flexible transparent silicone matrix. This LED array is specified to produce an illuminance (power density) of 60 mW/cm^2. However, as no range is indicated as part of the specification for the device and the illumination source is not continuous being made from discrete “point” sources, it is difficult to reconcile this specification to any meaningful measurements.

As illustrated in Fig. 5, the optical path from the OrthoPulse unit into the tooth socket includes the skin of the gums that may include some melanin and a thin layer of tissue with a combined thickness of 0.5 mm followed by 1–2 mm of bone. Finally the light will pass through 1–3 mm of blood depending on the size of the socket and the position of the sensor within it.

The absorption coefficients for the 850 nm wavelength of the OrthoPulse device are obtained from Fig. 1 and listed in Table 1.

For the overlaying skin-tissue layer, ignoring any melanin, the combined absorption coefficient can be approximated by [9]

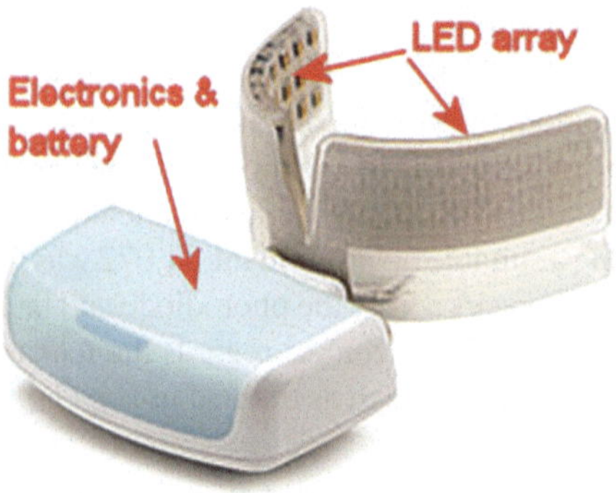

Fig. 4 Photograph of the OrthoPulse unit [11]

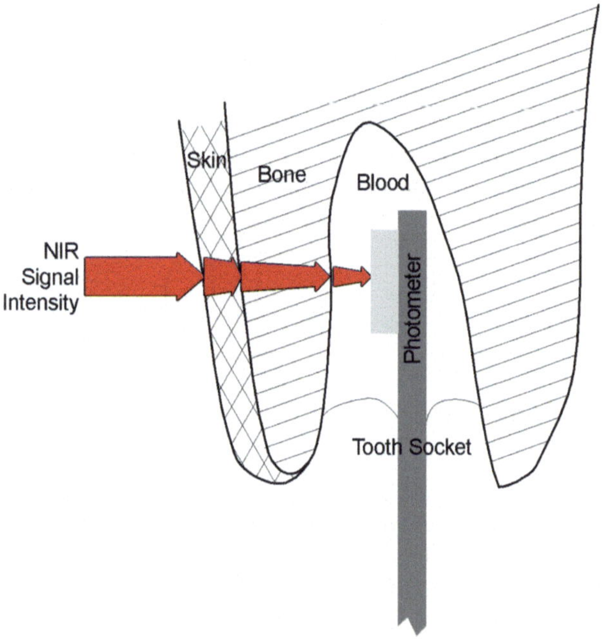

Fig. 5 Drawing showing the relative attenuation of the NIR signal as it travels from the source to the photometer

$$\mu_{askin} = BS\mu_{aoxy} + B(1 - S)\mu_{adeox} + W\mu_{awat} \tag{3}$$

where $B = 0.002$ and $W = 0.65$ are empirical coefficients based on the relative proportions of the tissue types and $S = 0.7$ is the proportion of oxygenated to deoxygenated blood in the gums. Using these coefficients and the thickness of the various components, the expected absorption coefficient will be $\mu_{askin} = 0.039\ \text{cm}^{-1}$.

Table 1 Absorption coefficients of different tissue components at 850 nm

Tissue component	Attenuation coefficient μ_a (1/cm)
Melanin	60
Blood (oxygenated)	5.5
Blood (deoxygenated)	4
Bone	0.2
Water	0.045

Using Beer's law to calculate the fraction of the incident signal absorbed in each case

$$I = I_o e^{-\mu_o x} \tag{4}$$

where I_o is the incident power onto the sample and I is the power after the signal has propagated through x (cm) of the sample with an absorption coefficient μ_o.

The following equation combines these individual components to generate the fraction of the radiated signal reaching the photometer to be

$$I = I_0 e^{-(\mu_{skin} x_1 + \mu_{bone} x_2 + \mu_{blood} x_3)} \tag{5}$$

It is convenient to examine the individual contributions to the total attenuation so that a range of values can be obtained for different thicknesses of the various materials. Ignoring any scattering through the medium, the normalised power is reduced to 0.998 after travelling through 0.5 mm of skin and underlying tissue. Travelling through 2 mm of bone, the power is reduced by a further factor of only 0.976. Finally, because of the high absorption coefficient of oxygenated blood, the factor will be 0.58 for 1 mm down to 0.22 for 3 mm for the blood in the socket. The proportion of the power passing through to the sensor is therefore $0.998 \times 0.976 \times 0.58 = 0.56$ for 1 mm of blood down to 0.21 for 3 mm. For a 60 mW/cm^2 incident power density, the power density at the sensor would be between 34 and 13 mW/cm^2.

It is critical that the sensor circuit doesn't saturate, therefore the gain (volts per amp) is calculated to produce an output voltage of about −3 V for the highest expected intensity level. For the specified sensitivity of the diode of 9 μA at an illuminance of 1 mW/cm^2 at 950 nm, reduced by a factor of 0.9 at 850 nm, a 10 kΩ feedback resistor provides an output voltage of approximately $-9 \times 10^{-6} \times 0.9 \times 10 \times 10^3 \times 34 = -2.75$ V for a power density of 34 mW/cm^2.

Temperature Sensor

Because of the size restriction for the sensor head, it was not possible to use a semiconductor temperature transducer like the LM35. In its place, a much smaller thermocouple is used. The Maxim AD8945 thermocouple amplifier breakout board from Adafruit produces an analog signal, V_{out}, proportional to the temperature, T (°C), using the following conversion

$$T = \frac{(V_{out} - 1.25)}{0.005} \tag{6}$$

Peripheral Circuitry

Peripheral circuitry includes a 9 V battery pack, followed by positive and negative voltage regulators based on the LM317 and LM337 adjustable voltage regulators that drop the 9 to 6 V. This ensures that battery droop has no effect on the accuracy of the sensor. An infrared LED is also incorporated into the circuitry to test the operation of the sensor probe. The voltages output by the photometer and the thermometer are accessible via a connector, and in this chapter were logged using a LabJack U6 data logger with a laptop running LJLog sampling at 10 Hz.

The complete system is built into jiffy box as shown in Fig. 6.

Sensor Calibration

An Ames Photonics 8112 Oculus photometer [13] with specifications as listed in Table 2 and responsivity shown in Fig. 7 was used to calibrate the photometer.

The saturation signal is used as a reference for calibration. This is specified to be 5.0 nJ at a wavelength of 410 nm in Table 2 and must be converted to the saturation

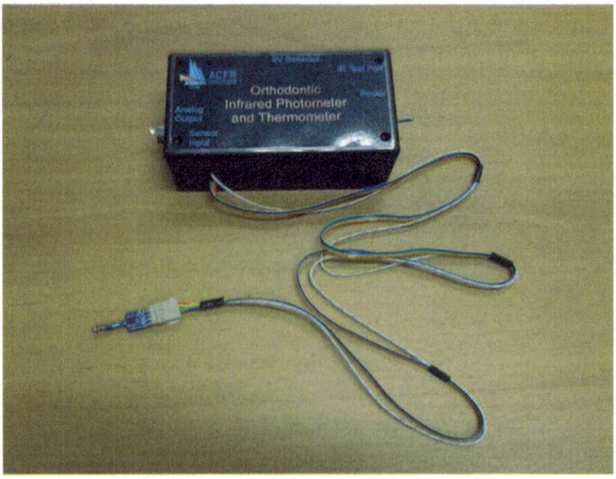

Fig. 6 The complete orthodontic infrared photometer and thermometer

energy at 890 nm for the test LED, 850 nm for the OrthoPulse and 660 nm for the Thor laser probe (used in similar research).

The responsivity of the photometer at 410 nm is specified to be $E_{sat} = 0.24$ A/V, while at 660 nm it is 0.4 A/V. and at 850 nm it is 0.5 A/V and at 890 nm (LED calibration frequency) it is 0.51 A/W. The increased responsivity at the longer wavelengths results in a proportionally lower saturation signal. At 660 nm it is $5.0 \times 0.24/0.4 =$ 3 nJ, at 850 nm the saturation signal is 2.4 nJ, and at 890 nm it is 2.35 nJ.

Unlike conventional photometers which use a transimpedance amplifier (current to voltage converter), the Oculus uses a gated integrator in which a small capacitor is charged by the photodiode current to generate an output voltage. The integration time can be altered to change the sensitivity of the device. In our measurements we set the integration time to $T_{int} = 30$ μs for the lowest sensitivity, so can calculate the incident optical power, P_{sat} (mW), to just saturate the Oculus.

$$P_{sat} = \frac{E_{sat}}{T_{\mathrm{int}}} \tag{7}$$

The power density, S_{sat} (mW/cm^2), can then be determined from the active area of the Oculus sensor. This has a diameter of 2.5 mm, so will have an area of $A_{sen} =$ 0.0491 cm^2.

Table 2 Specification of the 8112 Oculus Photometer

Diameter of active area (mm)	Sensor type	Optimum wavelength λ_o (nm)	Saturation signal at λ_o (nJ)	Minimum resolvable signal at λ_o (pJ)
2.5	Si Blue	410	5.0	3.1

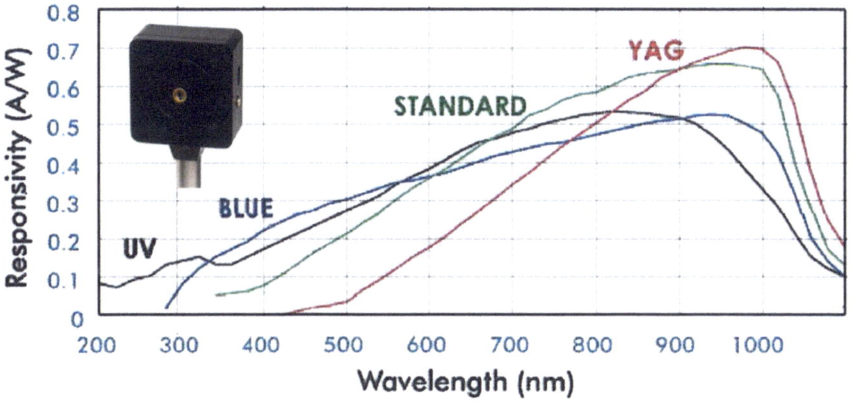

Fig. 7 Responsivity curves for a range of Oculus Photometers with the BLUE curve showing the response for the 8112 model

Table 3 Oculus photometer saturation considerations at different wavelengths

Incident wavelength λ (nm)	Saturation energy E_{sat} (nJ)	Saturation power, P_{sat} for $T_{int} = 30\ \mu s$ (mW)	Saturation power density, S_{sat} (mW/cm^2)
410 nm Reference	5.0	0.167	3.39
660 nm Thor laser	3.0	0.10	2.03
850 nm OrthoPulse	2.4	0.08	1.63
890 nm Test LED	2.35	0.078	1.59

Table 4 Orthodontic optical sensor calibration

Range (cm)	1.8	3.9	4.5	7.6
LED current to saturate Oculus (mA)	2.24	6.63	8.16	22.3
Optical sensor output (mV)	−61.4	−51.6	−45.4	−58.8
Calibration (mW/cm^2 per V)	−25.9	−30.8	−35	−27

$$S_{sat} = \frac{P_{sat}}{A_{sen}} \tag{8}$$

Table 3 provides the calculated power densities to saturate the sensor for the various optical sources.

It is convenient to use an IR LED (TSHF5210) as a reference source to transfer the calibration levels of the Oculus across to the Orthodontic photometer. To do this, the Oculus is first placed in the centre of the LED beam and the LED drive current increased until the Oculus just saturates. This occurs at a power density of 1.59 mW/cm^2 (from Table 3). The Oculus is then replaced by the Orthodontic photometer at the same distance to the LED, and the sensor output voltage is measured. This is repeated at a number of different ranges to ensure consistency, and the relationship between the incident power density, which should remain unchanged, and the measured voltage provides a calibration factor for the sensor as shown in Table 4.

Variations in the calculated calibration values are due to uncertainties with the sequential lateral alignment of the two sensors with respect to the centre of the illumination source at the different ranges. Based on the measured results at the four different ranges, the mean calibration factor is −29.7 mW/cm^2 per V, and this value is used throughout the trial to convert measured voltages out of the sensor to power density levels.

Sensor Linearity

To generate an equivalent irradiance to that provided by the OrthoPulse unit, an array of 8 TSHF5210 IR LEDs is connected to a linear power supply (Parameters P4303) each in series with a 1 kΩ resistor. An ammeter in series with the power supply monitors the current as the voltage is increased. As the optical power output by the LEDs is directly proportional to the current, it can be used to determine the linearity of the Orthodontic photometer response. The photometer is placed about 2 cm from the array at the point where the beams from the LEDs overlap and the sensor output voltage is recorded while the current is varied. It can be seen from Fig. 8 that the response of the photometer is linear until the output is limited by the supply voltage rails.

For expected power densities in the range of 13–34 mW/cm^2, the output voltage would be between −0.44 and −1.15 V. Even in the case where the full 60 mW/cm^2 reaches the sensor, the output voltage will not exceed −2 V. This range is well away from any saturation effects that begin at about −5 V.

The absolute accuracy of the LabJack U6 on the ±10 V range, is 2 mV which corresponds to only 0.06 mW/cm^2 which is insignificant compared to variations seen in the actual measurements made using the orthodontic photometer.

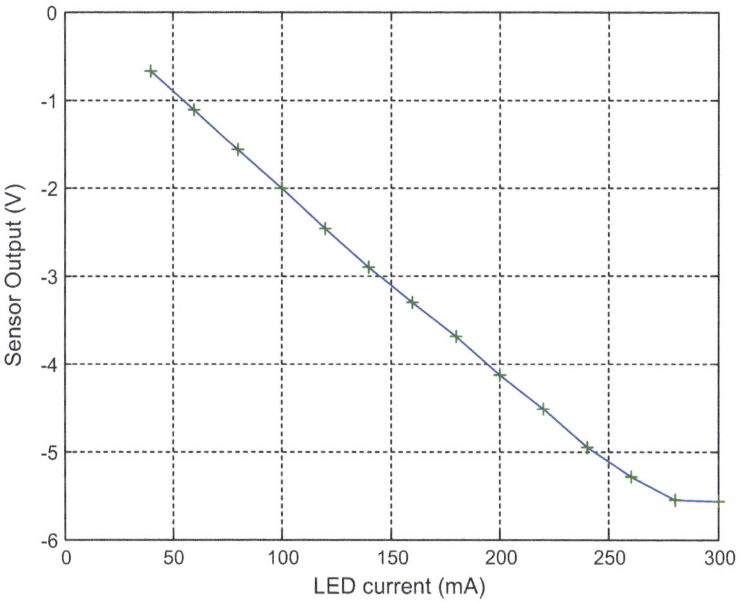

Fig. 8 Measured linearity of the Orthodontic IR Photometer as a function of incident power

Lab Measurements with the OrthoPulse Device

Measurements were made in the lab using the OrthoPulse device as a source, and placing the NIR photometer at different ranges. These were complicated by the shape of the device (see Fig. 4) and an interlock that only allowed it to operate when in the mouth, or when gripped firmly between the fingers (a mechanical clamp was not effective).

Figure 9 shows the measured and theoretical power density from the OrthoPulse device as a function of range. In theory, if it was a point source, the intensity should decrease by 20 dB per decade, as shown in the upper graph. However measurements show that the Intensity tapers off for ranges less than 40 mm. This is because the light outputs of the individual LEDs of the OrthoPulse device only start to overlap at ranges in excess of 30 mm, while at shorter distances only the output from a single LED illuminates the photometer. This relationship is illustrated in Fig. 10.

Note that the intensity for a single LED at a range of 1 cm is about 18.5 dB which equates to 70 mW/cm^2 (very close to the 60 mW/cm^2 specified for a single diode). However, if the output power density of all of the diodes could be combined, at that range, a power density of 631 mW/cm^2 would be expected as shown by the theoretical intensity at a range of 10 mm.

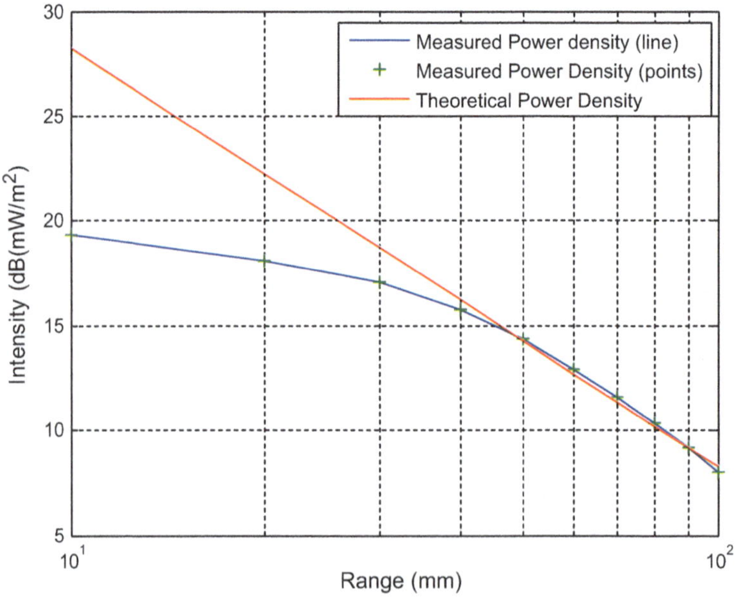

Fig. 9 Measured and theoretical power density as a function of distance from the OrthoPulse device

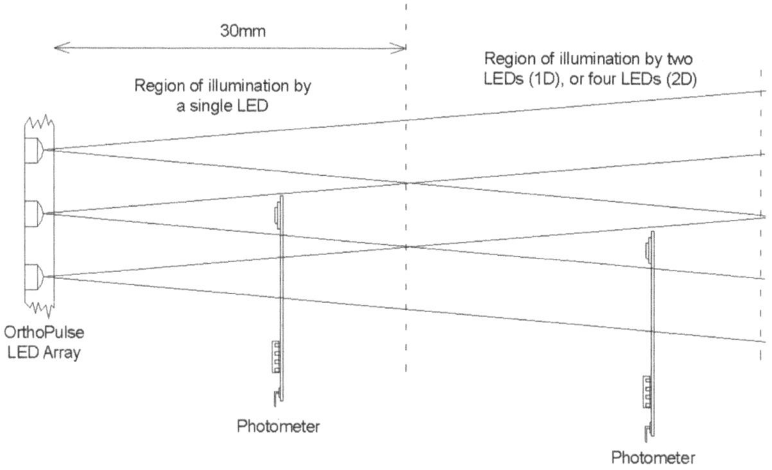

Fig. 10 Diagram showing the relationship between the distance to the photometer and the expected intensity

Results

Optical Attenuation

As discussed in the introduction and illustrated in Fig. 2, a trial using 18 patients was undertaken at the dental school.

The sequence for the test on each participant is as follows:

- A thin sterile transparent plastic sleeve with negligible attenuation in the NIR is placed over the sensor probe.
- Test A: Insert probe into left hand tooth socket for about 60 s with the OrthoPulse device turned off. This allows the socket temperature to be monitored without any effects of the NIR radiation.
- Test B: The probe is removed and then reinserted into the left hand socket for a further 60 s with the OrthoPulse device turned on.
- Test C: The probe is transferred to the right hand tooth socket for a final 60 s before the probe is removed from the mouth.

Figure 11 shows the measured NIR intensity results for one patient using the calculated scale factor of -29.7 mW/cm^2 per V and the temperature scaled using Eq. (6) from the voltage output by the Maxim AD8945 thermocouple amplifier.

It can be seen that more than half of the NIR radiation (given a source power density of 60 mW/cm^2) passes through the gum and the jaw into the tooth socket. However, this varies significantly from patient to patient and appears to be primarily dependent on the size of the socket and the volume of blood between the sensor and the OrthoPulse source. This variation is shown quite clearly in Test C where

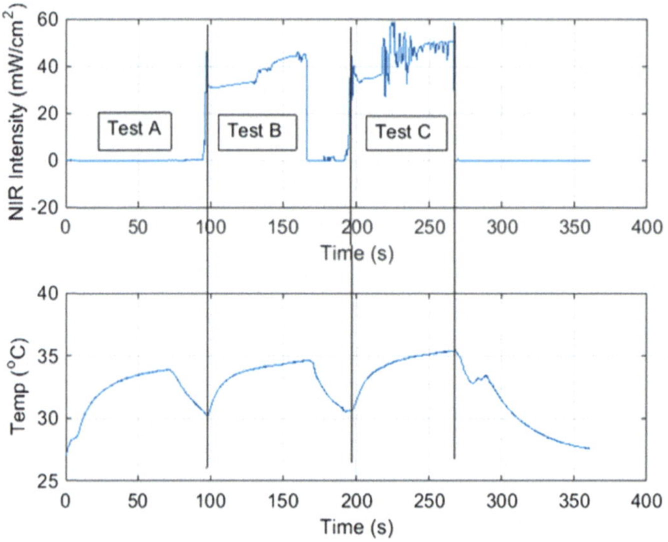

Fig. 11 In vivo data taken from one of the trial patients

the sensor probe is moved back and forth in the socket with the resultant changes in measured power density.

When the measured power density for all of the trial members is analysed, it is much lower on average that would be expected if the model presented at the start of this chapter was accurate. Additionally, if these data are plotted as a function of the measured bone thickness obtained from X-ray images made for each patient, then a strong correlation is found, as seen in Fig. 12.

Firstly, there is one instance, through the thinnest bone section, where the measured power density exceeds the nominal maximum of 60 mW/cm^2 from the source. This is not an outlier, as the same measurement processes were used in all cases, so we speculate that the curvature of the gum and jaw bone sections were able to focus the radiation from a number of LEDs onto the photometer.

It must be assumed that most of the variation in the received power density at a specific bone thicknesses is due to attenuation through different amounts of blood in the individual tooth sockets. Variation within each measurement is shown by the error bars. Additionally, the highest intensity measurements at each thickness can only occur where the blood thickness is a minimum, and that there may be instances where there is no blood in the tooth socket between the source and the photometer. An exponential based on the Beer relationship is then fitted through the highest intensity points, as shown in Fig. 12, and the revised attenuation coefficient for bone is obtained. For the curve shown, the attenuation coefficient for bone $\mu_{bone} =$ 5.36 1/cm. This is a factor of 27 larger than that obtained for the in vitro measurements given in Table 1 and is close to that of oxygenated blood.

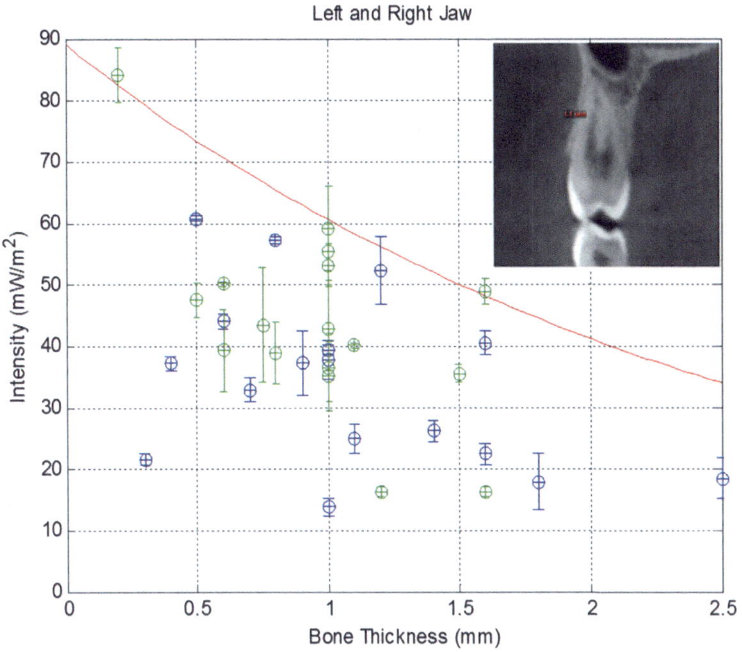

Fig. 12 Graph showing the mean measured power density for all of the trial patients, and an X-ray image showing the measured bone thickness for one of the patients

Pulse Rate Effects

An interesting and somewhat unexpected bonus from the measurements is that the heart rate can be measured using small variations in the intensity of the measured radiation, as shown in an expanded view of Test B in Fig. 13.

It is assumed that the changes occur due to swelling of the capillaries in the gum tissue, and hence increased attenuation, during systole, the contraction phase of the cardiac cycle.

The heartrate of this patient is measured to be 96 beats per minute.

Temperature Effects

As the thermal capacity of the probe tip is quite small, the responsiveness of the temperature sensor is fast. Figure 14a shows the measured temperature compared to that of an exponential function with a time constant of 15 s in the case where the OrthoPulse unit is turned off. This is the expected relationship when heating the thermocouple from a high thermal capacity, constant temperature reservoir.

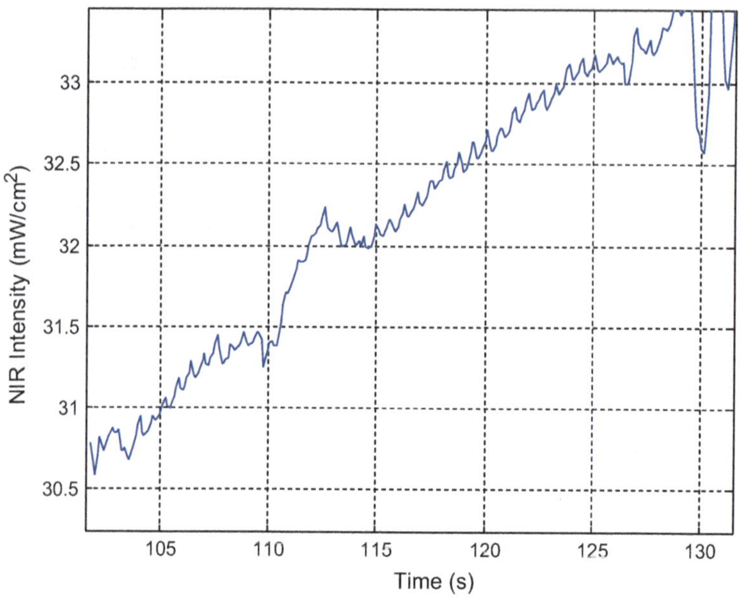

Fig. 13 Expanded view of the NIR intensity showing the patient's heart beat

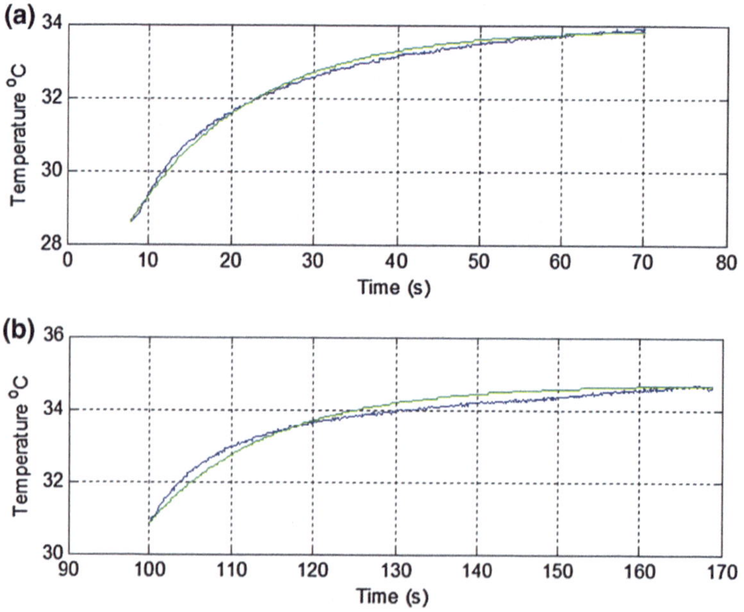

Fig. 14 Comparison between the measured temperature and an exponential function with a time constant of 15 s for **a** test A with the illumination OFF and **b** test B with the illumination ON

The RMS error compared to the exponential for Test A is 0.115 °C with the maximum error peaking at 0.25 °C. In Test B with the OrthoPulse unit on, the RMS error increases to 0.207 °C with the peak difference reaching 0.4 °C. In this case, the overall fit to an exponential is slightly worse indicating that there may be some localised tissue heating caused by the OrthoPulse unit.

Conclusions

This chapter describes the successful development and calibration of a NIR photometer to measure the amount light that propagates through the gum and jaw into a newly vacated tooth socket.

The probe proved to be sufficiently small to enter the tooth socket of even the smallest patient and was sufficiently reliable and accurate to provide good measurements for both sides of the mouth for all 18 patients, providing a total of 36 independent measurements.

As predicted theoretically and confirmed with measurements, the attenuation through the gum is insignificant compared to the attenuation by blood in the socket. However, a strong correlation between the total attenuation and the bone thickness was found, indicating that the in vivo bone provided an attenuation coefficient of 5.36 1/cm which is close to that of oxygenated blood.

A suggestion is that live bone contains much more haemoglobin than dead bone, or that, it removing the teeth, the bone interface was broken up, allowing blood to leak into the porous underlying bone.

From a sensing perspective, it was found that the photometer was sufficiently sensitive to measure the subtle changes in the attenuation over the period of a heartbeat, thus allowing measurements of the patient's heart rate to be made, during the procedure.

Finally, first indications are that in addition to light penetration deep into the tissue, energy from the OrthoPulse unit also resulted in a slightly increased temperature.

Acknowledgements We would like to acknowledge the contribution of Dr. John Sambevski and Professor M. Ali Darendeliler of the University of Sydney Dental School for supporting this research and providing the measurement data based on our sensor. We would also like to thank Ben Stewart and Victor Chan of the ACFR for designing and populating the sensor PCB.

References

1. W. Wilcko, T. Wilcko, J. Bouquot, D. Ferguson, Rapid orthodontics with alveolar reshaping: two case reports on decrowding. Int. J. Periodontics Restor. Dent. **21**, 9–19 (2001)
2. S. Kim, Y. Kook, D. Jeong, W. Lee, K. Chung, G. Nelson, Clinical application of accelerated osteogenic orthodontics and partially osseointegrated mini-implants for minor tooth movement. Am. J. Dentofac. Orthod. **136**, 431–439 (2009)

3. S. Baloul, Mechanism of action and morphological changes in the alveolar bone in response to selective alveolar decortication facilitated tooth movement (abstract), in *American Association of Orthodontists Annual Session—Passion for Excellence*, Washington DC (2010), p. 6
4. T. Vercellotti, A. Podesta, Orthodontic microsurgery: a new surgically guided technique for dental movement. Int. J. Periodontics Restor. Dent. **27**, 325–331 (2007)
5. M. Fitzgerald, S. Hodgetts, C. Van Den Heuvel, R. Natoli, N. Hart, K. Valter et al., Red/near-infrared irradiation therapy for treatment of central nervous system injuries and disorders. Rev. Neurosci. **24**, 205–226 (2013)
6. A. von Luhmann, H. Wabnitz, T. Sander, K. Muller, "M3BA: a mobile, modular, multimodal biosignal acquisition architecture for miniaturized EEG-NIRS-based hybrid BCI and monitoring. IEEE Trans. Biomed. Eng. **64**, 1190–1210 (2017)
7. T. Shaughnessy, A. Kantarci, C.H. Kau, D. Skrenes, S. Skrenes, D. Ma, Intraoral photobiomodulation-induced orthodontic tooth alignment: a preliminary study. BMC Oral Health **16** (2016)
8. A. Bashkatov, E.K. Genina, V, V. Tuchin, Optical properties of human cranial bone in the spectral range from 800 to 2000 nm, in *Optical Technologies in Biophysics and Medicine VII* (2005)
9. S. Jacques, *Generic Tissue Optical Properties* (Oregon Medical Laser Center, 2015)
10. T. Henderson, L. Morries, Near-Infrared photonic energy penetration: can infrared phototherapy effectively reach the human brain? Neuropsychiatr. Dis. Treat. **11**, 2191–2208 (2015)
11. OrthoPulse (June) *OrthoPulse*. http://orthopulse.com/
12. L. Orozco, Optimizing Precision Photodiode Sensor Circuit Design, ed by Analog Devices, vol. Technical Article MS-2624 (2014)
13. Ames Photonics, *Oculus User Manual*, ed byAmes Photonics (2008)

Wide Band Antennae System for Remote Vital Signs Detecting Doppler Radar Sensor

Thi Phuoc Van Nguyen, Liqiong Tang, Duc Minh Nguyen, Faraz Hasan and Subhas Mukhopadhyay

Abstract This chapter presents a novel antennae system for human vital signs detection. In this work, system working principles and different types of patch antennae are introduced, along with the measurement set up of the vital signs detecting radar sensor system. A wide band (from 900 MHz to 12 GHz) patch antennae system with beam-enhanced capacity is developed in FR4 substrate. This substrate has dielectric constant 4.4 and 1.2 mm of height. To reduce the size of the antennae system, a 3D-orthogonal structure was utilized to design the transmitting and receiving antennae. The multi-patch elements transmitting antenna was placed orthogonally with the receiving antenna to decrease the size of the antennae system. Moreover, the bandwidth and the directional capacity of the antennae were put in high priority to identify the human's chest displacement at different frequencies, from L band to the X band. The measurement outcome shows that human vital signs could be revealed by the proposed 3D antennae system.

Keywords Vital signs detection · Wide band antenna · Orthogonal antenna Doppler radar · Micro strip antenna

T. P. V. Nguyen (✉) · L. Tang · F. Hasan
Massey University, Palmerston North, New Zealand
e-mail: V.Nguyen@massey.ac.nz

L. Tang
e-mail: L.Tang@massey.ac.nz

F. Hasan
e-mail: F.Hasan@massey.ac.nz

D. M. Nguyen
Hanoi University of Science and Technology, Hanoi, Vietnam
e-mail: minh.nguyenduc1@hust.edu.vn

S. Mukhopadhyay
Macquarie University, Sydney, Australia
e-mail: subhas.mukhopadhyay@mq.edu.au

S. C. Mukhopadhyay et al. (eds.), *Modern Sensing Technologies*,
Smart Sensors, Measurement and Instrumentation 29,
https://doi.org/10.1007/978-3-319-99540-3_4

Introduction

Non-contact human vital signs detection based on microwave Doppler radar has high potential applications in different areas from health care implementation to surveillance or rescuing areas. There is a lot of work which pays attention to this system [1–9]. It is widely known that the signal to noise ratio (SNR) in the remote vital signs sensing system is quite low. Therefore, extracting the high accuracy vital signs signal from this system is really challenging to researchers. Li et al. [1] proposed an advanced signal processing method to extract vital signs from an ultra-wide band (UWB) radar system. They first applied the Curvelet transform to discard transmitter-receiver direct coupling. The next step was to reduce noise in the receiving signal by using singular value decomposition (SVD). Finally, the vital signs signal was extracted by Fast Fourier Transform (FFT) and Hilbert-Huang transform. Their method can locate the position of people in a complex environment and extract the breathing rate and heartbeat of humans. Kazemi et al. [2] also work on the signal processing issue of the microwave Doppler radar sensor. They investigated a new algorithm based on generalized Warblet transform (GWT) to get a high accuracy of cardio-respiratory signals. After that He et al. [3] developed an innovative technique situated on a double parameter LMS filter which showed better results in terms of breathing rate and heart beat than the band pass filter and wavelet transform (WT) methods. Along with the signal processing methods for the vital signs detecting purposes, the hardware development plays an essential role in improving the performance of the microwave Doppler radar. Different aspects of hardware improvement were discussed in [4, 10–12]. Kao et al. [4] presented their development for the vital signs sensing system in the system and circuit levels. They also proposed a new algorithm to enhance the detecting accuracy of the vital-sign sensing system. Liu and Liu [5] set up the measurement of breathing rate in the system level to detect the respiratory rate for rescue purposes. Their result showed that the breathing rate can be detected in quite a short time (around 5 s). Similarly, the authors in Ref. [10] developed a compact portable microwave life detector system to find people under the debris. Their system can detect humans at a distance of 1.5 m through construction material. In the vital signs sensing radar system, the antennae system plays the role of probes to detect the vital signs. It is obvious that antennae should be small, have high directivity and gain. Authors in Refs. [13, 14] optimized the antennae system to detect vital signs with operating frequencies around 60 GHz. The size of antennae is proportional to the wavelength. Therefore, in such high frequency, the dimensions of antennae system are very small, for rescue application to find the live people under the debris, the lower frequency radio signal can go through the obstacle better than higher frequency signals. When the vital signs detecting radar system has to operate at low frequency, the size of its patch antennae is a significant problem [15]. In addition, to improve the flexibility of the life-detection system in terms of operating frequency, the antennae should have very wide band width. Therefore, in this study, portable multiple bands antennae with a novel design structure is presented. This chapter discuss the investigation the 3D multiple bands (from L-band to X-band) patch antennae for vital Doppler radar detector. The vital signs detecting ability of the system at different operating frequencies is tested through Line-Of-Sign (LOS) and Non-Line-Of-Sign (NLOS) environments. The remainder of this chapter is organized as follows. Section "Antenna Design" states the design of the antennae, underlying the structure, simulation and measurement results. The vital signs detectings using the developed antennae is presented in section "Vital Signs Sensing Using Extra Wide Band 3D Antennae", and the conclusions are given in section "Conclusions".

Antenna Design

There are different types of antennae. The patch antenna is very popular in sensor systems because of its low cost, a reasonable size, light, and easy to connect to the printed circuits. The mircrostrip antenna is composed of three layers. The metallic strip—patch is put on top of the substrate and the third layer is the ground plane [16]. There are many types of microstrip patch elements such as square, rectangular, dipole, circular, elliptical, triangular, disc sector, circular ring and ring sector. To enhance scanning capability and directivity of the antenna, the arrays of microstrip elements were considered.

With respect to the feeding method to the antennae, there are four main ways including coaxial probe, microstrip line, aperture coupling, and proximity coupling [16, 18–21] as can be seen in Fig. 1. The coaxial-line feed (Fig. 1a) has two conductors; the inner conductor is connected to the radiation patch and the other one is soldered to the ground plane. This type of feed is easy to fabricate, match, and has low spurious radiation. However, it has bandwidth limitations and difficulty in modelling. Similar to the advantages of the coaxial-line feed, the microstrip line feed (Fig. 1b) is simple to fabricate, match and model. But, the spurious radiation increases proportionally with the thickness of the substrate, and the bandwidth is quite narrow (around 2–5%).

In the first two methods, coaxial probe and mirostrip line, asymmetries always exist. This problem leads to the high order modes and cross-polarized radiation. These problems could be reduced by the non-contacting aperture-coupling feeds (Fig. 1c, d). The aperture coupled feed (Fig. 1c) has two substrates separated by the ground plane in the middle. The patch is etched on the top substrate while the microstrip feed line is imprinted on the bottom of the lower substrate. The feed line is coupled to the patch by a slot on the ground. Obviously, the parameters of this type of feeding can be chosen separately. The independent choices of substrate dielectric constants, thickness, feed line width, slot size, and position support the optimized antenna design. The most popular advantages of the aperture coupled feed are easy to model, and have low spurious radiation. However, this method is very hard to fabricate and the bandwidth is not very wide. The last feeding method is proximity-coupled feed as shown in Fig. 1d. This feeding consists of two substrates; the upper and the lower ones. The radiation patch is on top of the substrate, while the feeding line is placed

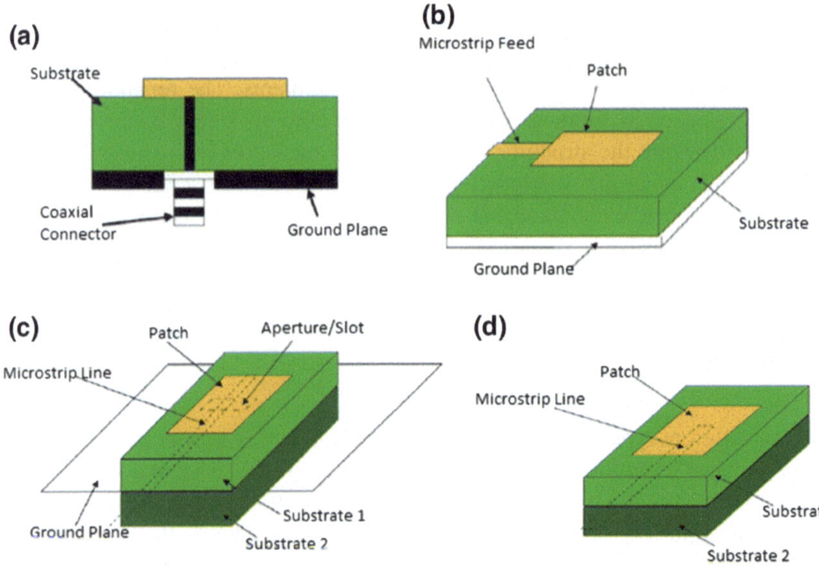

Fig. 1 Feeding methods for patch antennae: **a** coaxial feed, **b** microstrip line feed, **c** aperture coupled feed, and **d** proximity coupled feed [16, 17]

between two substrates. The optimization for the design can be achieved by changing the parameters such as feeding stub, the width and length of the patch or the dielectric of substrates. The proximity coupling feed is more difficult to fabricate but has wider bandwidth and low spurious radiation [16, 17].

Two Dimensions (2D) Transceiver Antennae

In the 2D transceiver antennae system, the transmitting antenna is placed in parallel with the receiving antenna. Each antenna might consist of several patch elements. The width (W) and the length (L) of each patch are calculated as [14, 22].

$$W = \frac{c}{2f\sqrt{\frac{\varepsilon_r+1}{2}}} \tag{1}$$

$$L = \frac{c}{2f_0\sqrt{\varepsilon_{reff}}} - 2\triangle L \tag{2}$$

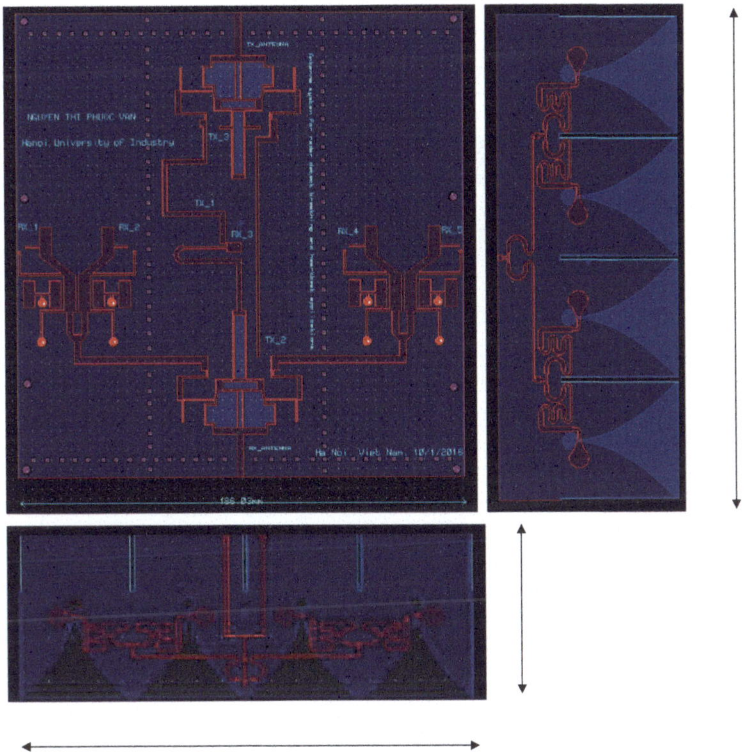

Fig. 2 Structures of the designed antennae 4× patch for RX and 3× patch for TX [23]

where c, f, and ε_r are the light velocity, operating frequency, and dielectric constant respectively. ε_{reff} is the relative and effective dielectric constants. $\triangle L$ is the patch length extension [22].

Two basic Eqs. (1) and (2) are used to estimate the size of antenna. If a 2D antenna with the gain around 10 dBi is developed, the size of the antenna can go up to 30 cm × 30 cm. Therefore, the size of transceiving antennae is larger than 30 cm × 60 cm.

To design a microstrip patch antennae for the vital-sign sensing purpose, the high gain and the narrow directivity are required because the vital signals such as respiratory rate and heart rate are very small. Moreover, in the designed antennae system, the cost-effective factor can be achieved by using low-cost material-FR4 with substrate thickness 1.2 mm and dielectric constant $\varepsilon_r = 4.4$. The modified disc sector is chosen for each patch element as presented in Fig. 2 to achieve extra band width. Figure 2 shows the 2-dimension (2D) structure of receiving and transmitting antennae. The transmitting antenna consists of 5 × 3 patch elements (see Fig. 3), so the total elements of transmitting antenna are 15. If the transmitting and receiving antennae are installed in the 2D configuration the total area of the antennae system

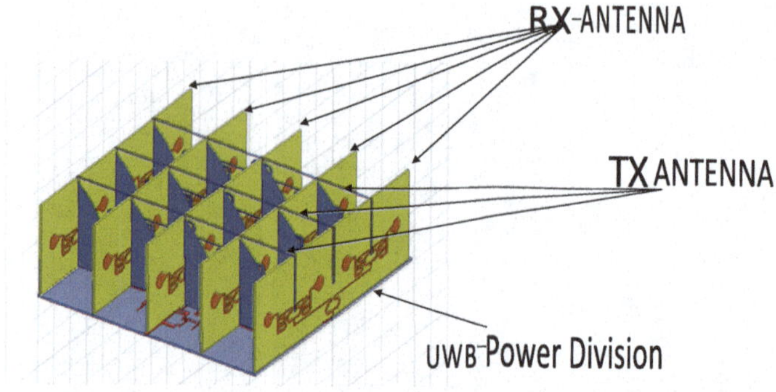

Fig. 3 3D geography of proposed antennae [23]

will be around 100 cm × 100 cm. This structure is too large for a portable vital signs detecting system. The 3D antennae structure could be a good solution for this problem.

Three Dimensions (3D) Transceiver Antennae

Fig. 3 is the receiving antennae contracted by 5 × 4 patch elements. The transmitting and receiving antennae are connected by the power splitter and power combiner. The feeding method is coaxial feed because it is easy to fabricate and to connect to the radar sensor system through the cables. The power splitter and combiner are designed based on the idea adopted from [24, 25]. With the consideration of direction, gain, compact size, and extra wide band, an orthogonal antennae system was designed and fabricated shown in Fig. 4. In comparison with 2D configuration, this novel design reduces the size about three times. The antennae development involves the use of HFSS and ADS softwares. The comparison results of the simulation and measurement are presented in Fig. 5. As it can be seen, the antennae can operate in a very wide band, from around 900 MHz to 12 GHz with $S_{11} < -10$ dBm. The measurement results of transmitter and receiver antennae on S_{11} show a better result in the frequency range from 800 MHz to 4.4 GHz and from 9.4 to 12 GHz. The difference between the design and the simulation results of S_{11} might come from the variation of substrate permittivity at different frequencies. In simulation software the permittivity is chosen as 4.4. The isolation simulation result between two antennae S_{21} is quite close to the measurement result with the value of $S_{21} < -20$ dBm.

Fig. 4 Fabricated antennae system [23]

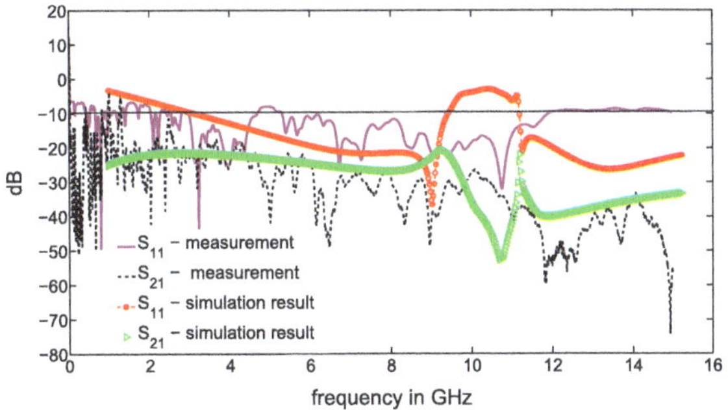

Fig. 5 Reflection at the input and isolation between two antennae [23]

Vital Signs Sensing Using Extra Wide Band 3D Antennae

System and Performance Analysis

The block system of Doppler radar sensor is shown in Fig. 6. In the continuous wave (CW) radar system a single-tone continuous wave signal ($s(t)$) as described by Eq. (3), with amplitude A_T, frequency f, wavelength λ and phase noise $\varphi(t)$ is sent to the human body [26],

$$s(t) = A_T \sin(2\pi f t + \varphi(t)) \quad (3)$$

This signal is then modulated by the chest movement $x(t)$, and the receiving signal at the RX-antenna ($r(t)$)with the amplitude (A_R) can be acquired as follows [26].

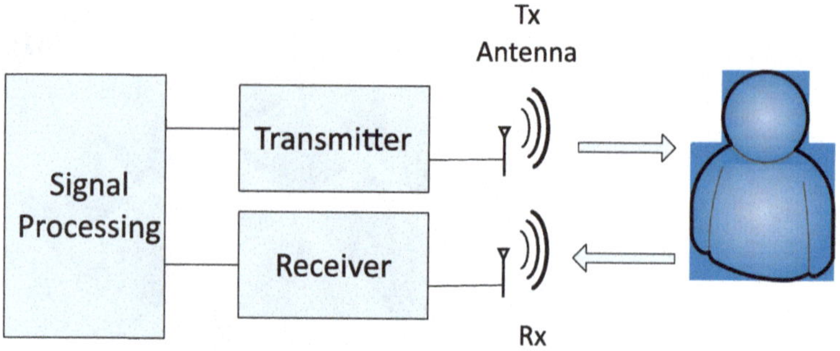

Fig. 6 Block model of the radar sensor system [23]

$$r(t) = A_R \cos\left(2\pi f t - \frac{2\pi}{\lambda}(2x(t) + 2d) + \varphi\left(t - \frac{2d}{c}\right)\right) \quad (4)$$

where d is the distance from the human location to the radar antennae, and c is the light velocity.

If the intermediate frequency (IF) is used, the radio frequency (RF) signal is down-converted to IF as in the Eq. (5) [27].

$$R_{IF}(t) = A_{IF} \cos\left(2\pi f_{IF} t + \theta_0 + \frac{4\pi x(t)}{\lambda} + \Delta\varphi(t)\right) \quad (5)$$

where A_{IF} and f_{IF} are the amplitude and frequency of the IF signal respectively. $\theta_0 = \frac{4\pi d}{\lambda} + \xi_0$ is the constant phase shift due to the distance from the target to the radar system and the hardware such as antenna, amplifier, mixer and so on. $\Delta\varphi(t) = \varphi(t) - \varphi(t - \frac{2d}{c})$ is the residual phase noise.

After the IF signal is filtered through the low pass filter (LPF) the base band output signal is given as.

$$B(t) = \cos\left(\theta_0 + \frac{4\pi x(t)}{\lambda} + \Delta\varphi(t)\right) \quad (6)$$

Obviously, when $\theta_0 = n\pi$, n is an integer number, the output $B(t)$ is calculated approximately as [28].

$$B(t) = 1 - \left(\frac{4\pi x(t)}{\lambda} + \Delta\varphi(t)\right)^2 \quad (7)$$

In this case, the receiving signal at the base band is very small, or not proportional linearly with the desired signal $(x(t))$. This issue is called the null point problem. The null point problem can be reduced by the quadrature detector. In the quadrature detector, the IF signal is divided into two parts with the phase difference of 90°. These signals are demodulated by I/Q demodulator, two base band signals $B_I(t)$ and

$B_Q(t)$ are created as follows [8].

$$B_I(t) = \cos\left(\frac{4\pi x(t)}{\lambda} + \theta_0 + \Delta\varphi(t)\right) \tag{8}$$

$$B_Q(t) = \sin\left(\frac{4\pi x(t)}{\lambda} + \theta_0 + \Delta\varphi(t)\right) \tag{9}$$

From Eqs. (8) and (9), the desired signal can be obtained by the arctangent demodulation [26].

$$\Delta\theta = \arctan\left[\frac{B_Q(t)}{B_I(t)}\right] = \frac{\sin\left(\frac{4\pi x(t)}{\lambda} + \theta_0 + \Delta\varphi(t)\right)}{\cos\left(\frac{4\pi x(t)}{\lambda} + \theta_0 + \Delta\varphi(t)\right)} \approx \frac{4\pi x(t)}{\lambda} + \theta_0 + \Delta\varphi(t) \tag{10}$$

In Eq. (10) θ_0 is a constant, if the phase noise is neglected, the output of arctangent demodulator is linearly proportional with the chest movement ($x(t)$).

$$\Delta\theta(t) \propto \frac{4\pi x(t)}{\lambda} \tag{11}$$

Signal Processing Methods

This section presents the different basic methods to extract the vital signs including Fourier Transform, Autocorrelation, Wavelet Transform, and some advanced techniques.

Fourier Transform

Fourier Transform (FT) is considered the most popular method to find the spectrum of a signal. If the output signal of the arctangent demodulator $\Delta\theta(t)$ is sampled to $\Delta\theta(n)$, n is an integer number, the Discrete Fourier Transform (DFT) is applied to find the spectral content. The spectrum of $\Delta\theta(n)$ is given as [29].

$$A_k = \sum_{n=0}^{N-1} e^{-j\frac{2\pi}{N}kn}\Delta\theta(n) \tag{12}$$

where A_k is the spectrum of the sampled signals $\Delta\theta(n)$, N is the size of DFT window. Fast Fourier Transform (FFT) is a fast algorithm to calculate DFT. In this presented research work, FFT is utilized to find the spectrum of the vital signals.

Autocorrelation

The autocorrelation function is an alternative to find the periodical patterns in a signal. The autocorrelation function of the arctangent output $\Delta\theta(n)$ is defined as [30].

$$R(k) = \frac{1}{N}\sum_{n=1}^{N-k} \Delta\theta\,(n+k)\,\Delta\theta\,(n) \tag{13}$$

where $R(k)$ is the kth lag of autocorrelation function of $\Delta\theta(n)$, N is the length of $\Delta\theta(n)$. This technique is reported to be inaccurate in finding the heartbeat when the sampling frequency is lower than 30 Hz. At the sampling rate above 80 Hz this method is quite appropriate in predicting the heartbeat [26]. To improve the performance of the autocorrelation based technique, the raw signal should go through the clutter removal and noise reduction processes. The processed signal is then employed by the random body motion detection algorithm based on autocorrelation technique. This method shows high accuracy in detecting the heart beat and respiratory rate of a stationary/non-stationary human [31].

Wavelet Transform

Wavelet Transform is a useful tool to find the instantaneous frequency of a signal in the time domain. A wavelet function ψ has zero average and is given as [32]:

$$\int_{-\infty}^{\infty} \psi(t)dt = 0 \tag{14}$$

The Wavelet transform $Wf(a, b)$ of a function $f(t)$ is given by [26, 32]:

$$Wf(a, b) = \frac{1}{\sqrt{a}}\int_{-\infty}^{\infty} f(t)\psi^*\left(\frac{t-b}{a}\right)dt \tag{15}$$

where a is the scale factor, b is the translation factor. In this application $f(t) = \Delta\theta(t)$, ψ^* is the complex conjugate of ψ. Three dimensions signal $Wf(a, b)$ can be converted into the two dimensions signal $Wf(a)$ by summating the translation coefficients as follows [26, 33].

$$Wf(a) = \int_{t_1}^{t_2} |Wf(a,t)|dt \tag{16}$$

where $Wf(a)$ is the intensity of signal frequency, t_1 and t_2 form a time span. From Eq. (16) the Time Frequency Spectrum (TFS) can be calculated as:

$$Tf(t) = \max |Wf(a,t)| \quad \text{for all } a \tag{17}$$

In Ref. [26] the Morlet is chosen as the mother Wavelet to detect the heartbeat and breathing rate. The result is remarkable and gives a higher accuracy level than the FFT method.

Advanced Techniques for Vital Signs Extraction

Besides the basic methods to extract the vital signs as mentioned above, there are many advanced signal processing methods to improve the detecting ability of a radar sensor system. For example, in 2014 Li et al. [1] presented a novel processing method in which the Curvelet transform is applied to discard the source-receiver interference noise and background clusters. After this step, the baseband signal noise is reduced through the singular value decomposition (SVD) tool. The last step is to apply the Hilber-Huang transform and FFT to get the vital signs such as respiratory rate and heartbeat.

Recently, Khan and Cho [31] proposed a statistical algorithm to detect the vital signs and reduce the breathing harmonics based on the current spectral content and previous spectra of vital signals. In their study, the Kalman filter is employed to reduce the measurement noises. Moreover, the auto-correlation method is applied to determine the random body movements and its effects on the measurement result. This algorithm shows higher accuracy in comparison with the conventional filter-based methods.

Another updated and novel method to detect the vital signs is investigated by Yo et al. [34]. In their study, the higher order cyclostationary is employed to lessen the Gaussian interference and non-periodic signals. They found that the third-order cyclostationary access results had better accuracy in terms of breathing rate and heart beat detection in comparison with the conventional FFT method.

Measurement and Discussions

Measurement Setup

In this section the performance of the proposed 3D antennae at various frequencies is tested using N5244A PNA-X Microwave Network Analyzer, 43.5 GHz. The measurement setup is shown in Fig. 7. On the left-hand side, an object (a person)

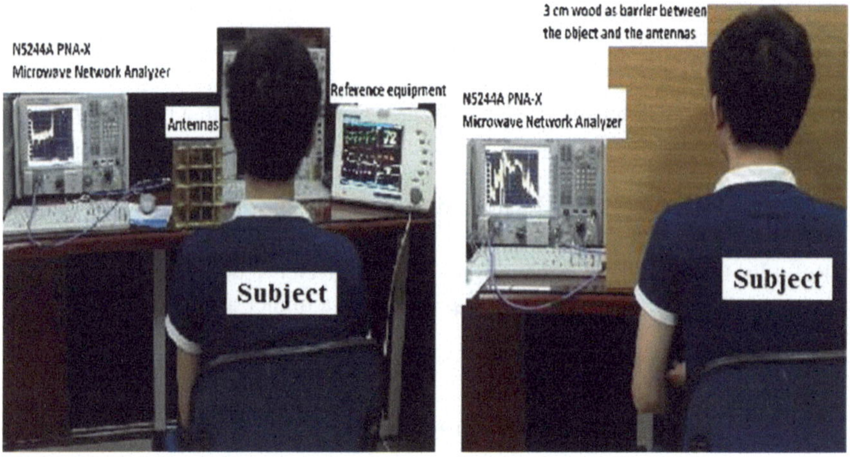

Fig. 7 The measurement setup to detect the vital signs [23]

is required to sit in front of the 3D antennae system, where the antennae are connected to two ports of the N5244A PNA-X corresponding with the transmitter and receiver. N5244A PNA-X send a continuous wave through the transmitting antenna and the changing phase of the receiving signal is determined by argtangent demodulation. The measurement was executed at wide range frequencies (0.953, 2.198, 3.362, 5.556, and 10 GHz) in the normal electronics laboratory environment. To get the stable signal, each measurement is measured in three minutes at a distance of 80 cm. The contact breathing measure equipment BK-NICO [35] was used as a reference to detect the human breathing rate. A similar measurement setup was done with a wood obstacle between the object and the antenna shown on the right-hand side in Fig. 7. The extracted data from N5244A PNA-X is processed by the common method Fast Fourier Transform-FFT get the respiratory rate [10].

Measurement Results

This part discusses the impact of antenna gain and the effect of operating frequencies on the breathing rate detection performance. Figure 8 shows that the human vital signs can be detected by the designed antennae system at different frequencies with a similar result. Average breathing rate of 17 beats per minute was measured by the commercial equipment BK-NICO [35]. Similarly, the proposed system gives the respiratory rates of 16.5, 18, 18, 15.7 (beats/min) at the frequencies of 0.953 GHz, 2.198 GHz, 3.362 GHz, and 5.556 GHz respectively. The corresponding gains are 5, 7, 5 and 4 dBi. Figure 8 also suggests that the sensitivity of the system at a lower frequency is smaller than that at a higher frequency. This outcome is demonstrated in Eq. (11). Moreover, the results shown in Fig. 8 is a good recommendation for different applications at different frequencies. For instance, high frequency benefits

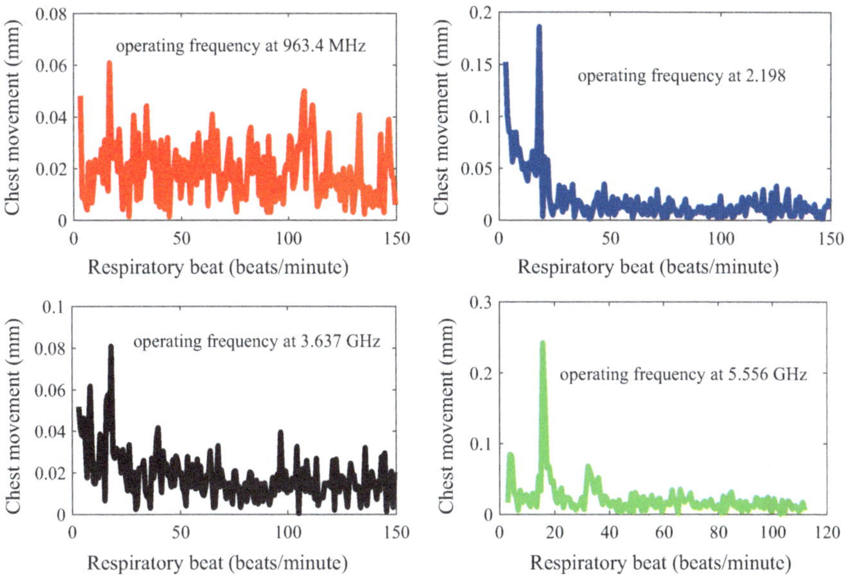

Fig. 8 Amplitude versus respiratory rate when the distance between the antennae and the object is 80 cm [23]

the medical application because of the high sensitivity of the system. On the other hand, the good penetration capacity through materials of the low frequency system handles the problem of finding survivors under the debris when there are natural hazards.

The performance of the 3D antennae system was tested when there was an obstacle (a 3 cm wooden plate) between the human and the antennae system. The object-antennae distance is 40 cm. Figure 9 illustrates the measurement results in this situation. The first point to note from this figure is that when the system operated at low frequency (963.4 MHz) it displays the poor signal to noise ratio (SNR). This result can be explained by Eq. 11. Secondly, the system behaves questionably at the highest operating frequency. Lastly, at two smaller operating frequencies 963.4 MHz and 2.198 GHz, the performance of the system is likely to be downgraded by the GSM and indoor wifi systems.

The receiving signal in time and in the frequency domain for the measurement at X band frequency are displayed in Fig. 10. The amplitude of receiving signal (0.8 mm) is stronger than the receiving signal when the system runs at smaller frequencies (in the Fig. 8). Undoubtedly, the sensitivity of the radar sensor at the higher frequency band system is enlarged.

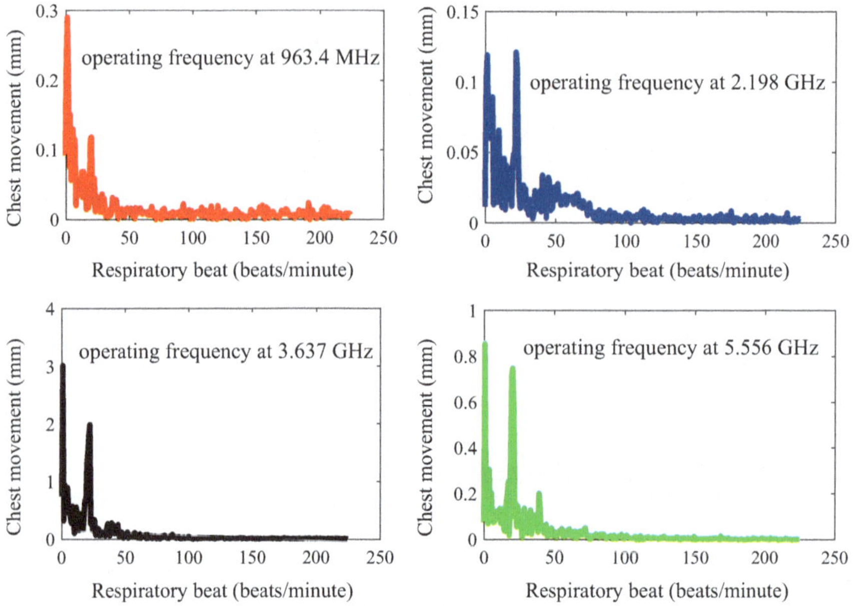

Fig. 9 Amplitude versus respiratory rate when 3 cm of wood between the antennae and the object [23]

Fig. 10 Receiving signal in time domain and in frequency domain [23]

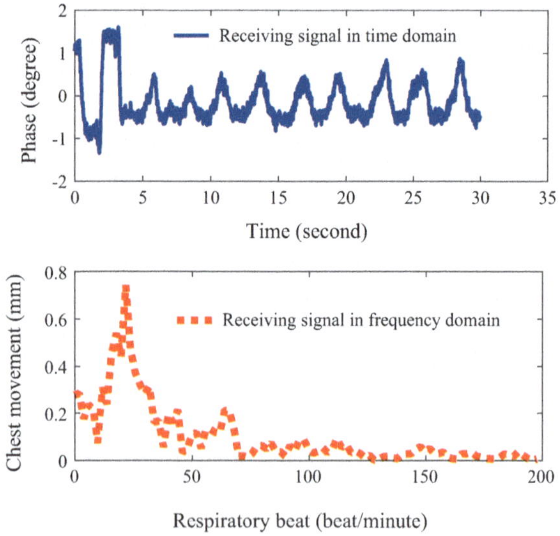

Conclusions

This chapter gives the details about the basic principles to develop the patch antennae for the vital signs detecting radar system, and the novel multiple bandwidths 3D antennae system which has high potential to be used to detect the human vital signs. The proposed antennae system can work properly at different frequencies in line-of-sign (LOS) and non-line-of-sign (NLOS) environment. This type of antennae allows the vital signs sensing radar system to operate at multiple frequencies simultaneously.

Acknowledgements This research is funded by the NZ aid program, New Zealand Ministry of Foreign Affairs and Trade, and The Faculty for the Future Program-Schlumberger Foundation.

References

1. J. Li, L. Liu, Z. Zeng, F. Liu, Advanced signal processing for vital sign extraction with applications in uwb radar detection of trapped victims in complex environments. IEEE J. Select. Top. Appl. Earth Obs. Remote Sens. **7**(3), 783–791 (2014)
2. S. Kazemi, A. Ghorbani, H. Amindavar, D.R. Morgan, Vital-sign extraction using bootstrap-based generalized warblet transform in heart and respiration monitoring radar system. IEEE Trans. Instrum. Meas. **65**(2), 255–263 (2016)
3. M. He, Y. Nian, Y. Gong, Novel signal processing method for vital sign monitoring using FMCW radar. Biomed. Signal Process. Control **33**, 335–345 (2017)
4. T.-Y.J. Kao, Y. Yan, T.-M. Shen, A.Y.-K. Chen, J. Lin, Design and analysis of a 60-GHz CMOS doppler micro-radar system-in-package for vital-sign and vibration detection. IEEE Trans. Microw. Theory Tech. **61**(4), 1649–1659 (2013)
5. L. Liu, S. Liu, Remote detection of human vital sign with stepped-frequency continuous wave radar. IEEE J. Select. Top. Appl. Earth Obs. Remote Sens. **7**(3), 775–782 (2014)
6. C. Gu, C. Li, Assessment of human respiration patterns via noncontact sensing using doppler multi-radar system. Sensors **15**(3), 6383–6398 (2015)
7. J. Lin, C. Li, Y.-C. LIU, Random body movement cancellation for non-contact vital sign detection. Sep. 8 2016, uS Patent App. 15/259,580
8. H. Bo, L. Xu, L. Hao, Y. Dou, L. Zhao, W. Yu, A single-channel non-orthogonal I/Q RF sensor for non-contact monitoring of vital signs. Appl. Comput. Electromagn. Soc. J. **31**(6) (2016)
9. T. Hall, N. Malone, J. Tsay, J. Lopez, T. Nguyen, R. Banister, D.C. Lie, Long-term vital sign measurement using a non-contact vital sign sensor inside an office cubicle setting, in *IEEE 38th Annual International Conference of the Engineering in Medicine and Biology Society (EMBC)* (IEEE, 2016), pp. 4845–4848
10. F. JalaliBidgoli, S. Moghadami, S. Ardalan, A compact portable microwave life-detection device for finding survivors. IEEE Embed. Syst. Lett. **8**(1), 10–13 (2016)
11. S.-T. Tseng, Y.-H. Kao, C.-C. Peng, J.-Y. Liu, S.-C. Chu, G.-F. Hong, C.-H. Hsieh, K.-T. Hsu, W.-T. Liu, Y.-H. Huang et al., A 65-nm CMOS low-power impulse radar system for human respiratory feature extraction and diagnosis on respiratory diseases. IEEE Trans. Microw. Theory Tech. **64**(4), 1029–1041 (2016)
12. H.-C. Kuo, C.-C. Lin, C.-H. Yu, P.-H. Lo, J.-Y. Lyu, C.-C. Chou, H.-R. Chuang, A fully integrated 60-GHz CMOS direct-conversion doppler radar RF sensor with clutter canceller for single-antenna noncontact human vital-signs detection. IEEE Trans. Microw. Theory Tech. **64**(4), 1018–1028 (2016)
13. T.-M. Shen, T.-Y.J. Kao, T.-Y. Huang, J. Tu, J. Lin, R.-B. Wu, Antenna design of 60-GHz micro-radar system-in-package for noncontact vital sign detection. IEEE Antennas Wirel. Propag. Lett. **11**, 1702–1705 (2012)

14. M.S. Rabbani, H. Ghafouri-Shiraz, Ultra-wide patch antenna array design at 60 GHz band for remote vital sign monitoring with doppler radar principle. J. Infrared Millim. Terahertz Waves 1–19
15. C.A. Balanis, *Antenna Theory: Analysis and Design* (Wiley, 2016)
16. A.B. Constantine et al., Antenna theory: analysis and design, in *Microstrip Antennas*, 3rd edn. (Wiley, 2005)
17. S. Bist, S. Saini, V. Prakash, B. Nautiyal, Study the various feeding techniques of microstrip antenna using design and simulation using CST microwave studio. Int. J. Emerg. Technol. Adv. Eng. **4**(9) (2014)
18. R. Garg, *Microstrip Antenna Design Handbook* (Artech house, 2001)
19. K. Carver, J. Mink, Microstrip antenna technology. IEEE Trans. Antennas Propag. **29**(1), 2–24 (1981)
20. K. Luk, C. Mak, Y. Chow, K. Lee, Broadband microstrip patch antenna. Electron. Lett. **34**(15), 1442–1443 (1998)
21. J.S. Sainath, K. Karthikeyan, Design and analysis of multiband hybrid coupler pentagon micro strip antenna for L-band applications, in *2015 IEEE 9th International Conference on Intelligent Systems and Control (ISCO)* (IEEE, 2015), pp. 1–5
22. P. Das, P. Kumar et al., Designing microstrip antenna with octagonal bounded elliptical slots, in *2016 10th International Conference on Intelligent Systems and Control (ISCO)* (IEEE, 2016), pp. 1–3
23. N.T.P. Van, L. Tang, N.D. Minh, F. Hasan, S. Mukhopadhyay, Extra wide band 3D patch antennae system design for remote vital sign doppler radar sensor detection, in *2017 Eleventh International Conference on Sensing Technology (ICST)* (IEEE, 2017), pp. 1–5
24. A.M. Abbosh, Design of ultra-wideband three-way arbitrary power dividers. IEEE Trans. Microw. Theory Tech. **56**(1), 194–201 (2008)
25. S.W. Wong, L. Zhu, Ultra-wideband power divider with good in-band splitting and isolation performances. IEEE Microw. Wirel. Compon. Lett. **18**(8), 518–520 (2008)
26. A. Tariq, Vital signs monitoring using doppler radar and on-body antennas. Ph.D. Dissertation, University of Birmingham (2013)
27. H. Zhao, H. Hong, L. Sun, Y. Li, C. Li, X. Zhu, *Noncontact physiological dynamics detection using low-power digital-if doppler radar* (IEEE Trans. Instrum, Meas, 2017)
28. A.D. Droitcour, O. Boric-Lubecke, V.M. Lubecke, J. Lin, G.T. Kovacs, Range correlation and I/Q performance benefits in single-chip silicon doppler radars for noncontact cardiopulmonary monitoring. IEEE Trans. Microw. Theory Tech. **52**(3), 838–848 (2004)
29. E.O. Brigham, E. Brigham, *The Fast Fourier Transform and Its Applications*, vol. 1 (Prentice Hall Englewood Cliffs, NJ, 1988)
30. M. Rahman, M. Chowdhury, S. Fattah, An efficient scheme for mental task classification utilizing reflection coefficients obtained from autocorrelation function of EEG signal. Brain Inf. 1–12 (2017)
31. F. Khan, S.H. Cho, A detailed algorithm for vital sign monitoring of a stationary/non-stationary human through IR-UWB radar. Sensors **17**(2), 290 (2017)
32. S. Mallat, *A Wavelet Tour of Signal Processing: The Sparse Way* (Academic press, 2008)
33. X. Lu, H. Liu, J. Kang, J. Cheng, Wavelet frequency spectrum and its application in analyzing an oscillating chemical system. Anal. Chimica acta **484**(2), 201–210 (2003)
34. Z. Yu, D. Zhao, Z. Zhang, Doppler radar vital signs detection method based on higher order cyclostationary. Sensors **18**(1), 47 (2017)
35. BK-NICO medical center hanoi university of technology, http://bme.hust.edu.vn, Accessed on 01 Jan 2017

Smart Sensing and Biofeedback for Vertical Jump in Sports

S. M. N. Arosha Senanayake and Abdul Ghani Naim

Abstract Vertical jump activity is a measurement of an athlete for different movements in sports required jumping. The analysis of this activity is subject to athlete's capability to Drop Vertical Jump (DVJ) landing upon achieved optimal Vertical Jump Height (VJH). This article discusses applying smart sensing mechanisms for DVJ and VJH as assistive tools to analyze DVJ and VJH and provides biofeedback for athlete during DVJ for safe landing and VJH to reach desired target. While optimization of VJH is the primary target of any sport, DVJ causes injuries on lower extremity. Hence, appropriate vibrotactile feedback during DVJ allows injury prevention and smart watches as IoTs can be used for biofeedback visualization during VJH monitoring for performance enhancement. Thus, different smart sensing mechanisms are required to analyze lower extremity kinematics jointly with the influence of muscles during DVJ and VJH. In this research, Inertia Measurement Units (IMUs), vibrotactors, Qualisys Motion Capture System and Bio-capture system consisted of muscle activity measurement sensors are integrated for DVJ and VJH measurements. The overall system design presented allows athletes to customize and to re-configure for different sport regimes requirements involving different DVJ and VJH. Integrated low powered wearable IMUs, bio-capture system, vibrotactors and smart watch are cost effective and require little infrastructure with no influence on natural human movement due to their total weight less than 500 g. This chapter will give a comprehensive system architecture of wearable embedded devices for DVJ and VJH measurements and for biofeedback during DVJ and VJH.

S. M. N. Arosha Senanayake (✉) · A. G. Naim
Faculty of Science, Institute of Applied Data Analytics, University of Brunei Darussalam, Gadong, Brunei Darussalam
e-mail: arosha.senanayake@ubd.edu.bn

S. C. Mukhopadhyay et al. (eds.), *Modern Sensing Technologies*,
Smart Sensors, Measurement and Instrumentation 29,
https://doi.org/10.1007/978-3-319-99540-3_5

Introduction and Literature Review

With the recent technological enhancements in smart sensors, actuators, IoTs, wireless communications, embedded processors and intelligent systems, sports biomechanical analysis has gained high popularity. Biomechanical analysis has become a vital tool for sports professionals, coaches and athletes in order to maintain consistency in performance and health. The ultimate goal of biomechanical analysis in sports is to improve techniques and prevent or reduce the risk of sports injuries [1]. Bartlett [2] describes the biomechanical observations for sports as the analysis of temporal, phasic features of movement, kinematic features of movement and kinetic features of movement.

While Drop Vertical Jump (DVJ) during landing places high moment demands around the knee joint and is regularly used for the assessment of dynamic knee valgus, the vertical jump height (VJH) is the jumping maneuver commonly used in current protocols evaluating athletic performance [3]. The VJH assessment in competitive athletic environment is designed to challenge the athletes' capacity to generate lower extremity power and achieve optimal VJH. High moment demands around lower extremity joints are expected as athletes exert their best effort to successfully accomplish this task. Although similar lower extremity mechanics are anecdotally seen when individuals perform both DVJ and VJH tasks, to date, the extent to which VJH can be used as a screening tool for dynamic knee valgus has not been assessed. In particular, it is conceivable that the pre-flight phase of the VJH task could also be used to assess knee injuries if knee joint valgus angles and internal adductor moments were similar and correlated with the values observed during the landing phase of DVJ. The potential to use VJH not only for athletic performance assessment, but also as an injury prevention tool, has a widespread clinical implication as discussed in [4].

In [4], authors evaluated whether individuals exhibit similar knee joint frontal plane kinematic and kinetic patterns when performing Vertical Jumps (VJs) compared with Drop Jumps (DJs). They used 11 Qualisys cameras for kinematic analysis of DJ and VJ while kinetics of DJ and VJ was measured using force plates. Figure 1 illustrates the relationship of peak knee joint adductor moment between DJ and VJ.

Identifying an approach to ensure that athletes participating in a DVJ during landing training program achieve desired kinematics is challenging. This is where the role of biofeedback systems is realized. Cowling in [5] investigated the effect of verbal instruction to improve muscle activity and reduce the risk of ACL injury during landing. Saggital plane kinematics, ground reaction forces and electromyography data were collected during landing, and verbal instructions on knee flexion and muscle activity was provided at the end of jumps. The results demonstrated a significant increase in knee flexion angle during landing at initial contact and peak contact. However, feedback received after landing tasks may limit the extent of improvement in landing kinematics, and may impose a larger effort on the subjects to concentrate and remember the feedback instruction during their performance. Munro in [6] presented a more practical approach for improving knee flexion angle at landing using audio biofeedback, with respect to a predefined training target. However, the measurement

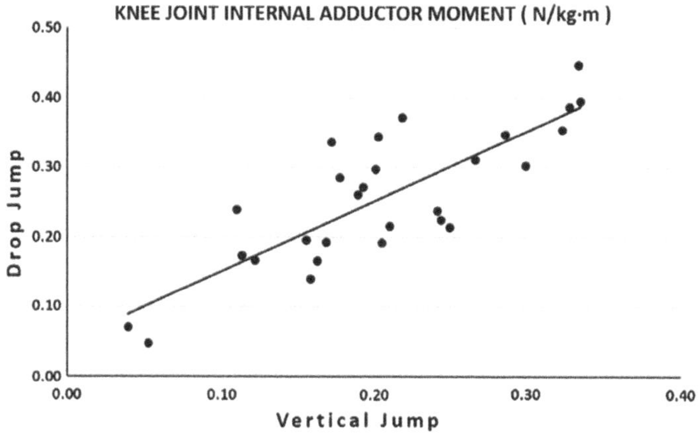

Fig. 1 Peak knee joint internal adductor moment between DJ and VJ [4]

device was not suitable to discriminate minor changes of knee flexion (<10°) and had limited use when damped by sweat and humid environmental conditions. Noyes in [7] and Onate in [8] described the impact of visual feedback using video tape information for improving knee separation during landing and increasing knee flexion and reducing peak vertical forces during landing respectively. Although landing technique has improved with augmented feedback employed in these studies, feedback has been provided after the training trials which may constrain the ability to exclusively use feedback instructions for performance enhancement. In this regard, it is clear that landing training programs may benefit from an automated, portable biofeedback system with high performance devices and a suitable feedback modality addressing the limitations encountered in existing systems.

Thus, a novel vibrotactile biofeedback system with real time functionality is introduced in this research in order to monitor and improve knee kinematics during a DVJ, providing instantaneous feedback information during training. The implementation is targeted to increase knee flexion and reduce knee abduction/adduction during landing from a jump. The real time functionality incorporated provides the system users and operators instantaneous feedback on performance and the use of vibrotactile biofeedback in place of conventional visual and auditory modalities eliminates the constraints involved in visual and auditory biofeedback systems [9].

On contrary, the goal of VJH is simply to jump and reach as high as possible. In many sports, the height to which an athlete can jump and reach is often of critical importance. In the absence of air resistance and other external forces, the upward projection of the whole body center of mass (CM) is completely determined by the vertical velocity at the instant of takeoff and the acceleration due to gravity. However, this quantity does not completely describe the overall jump and reach height that is observed. Jump and reach height may be considered to be the actual vertical height at which the athlete contacts the ball during a volleyball attack, releases a ball during a basketball lay-up, or touches the slats of a vertical jump testing device. Takeoff

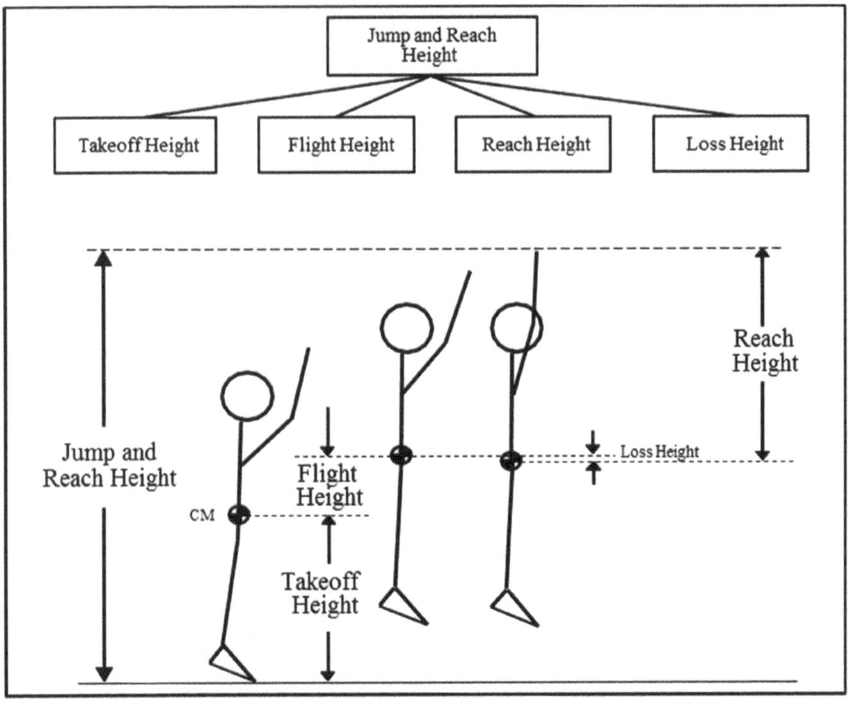

Fig. 2 Deterministic model that determines overall jump and reach height

height defines the height of the athlete's center of mass (CM) at the instant the athlete leaves the ground. Flight height refers to the actual height to which the CM is elevated during the in-flight phase of the jump. Reach height describes the vertical distance from the CM to the fingertips of the athlete at the instant the maximum height is evaluated (instant of ball contact, instant the slat is touched, etc.). Loss height refers to the difference between the absolute peak height of the CM and the height of the CM at the instant the maximum height is evaluated. This last factor can usually be attributable to a mistiming of the final reach by releasing a ball or touching the measuring device on the way down (or up). Figure 2 illustrates a deterministic model which serves to describe the mechanical or mathematical relationships that govern vertical jump and reach performances.

With vertical jumps, the muscles engage in the take-off action following the proximal-distal principle of muscle activation. In the first phase of jump when the vertical velocity of the body's center of gravity increases, the extensors of trunk and hip are the most active muscles. Thigh muscles generate the peak activation at the beginning of the hip extension [10]. In continuation of take-off action, knee extensors engage and the energy is transmitted from hip to knee. The explosive performance of vertical jumps primarily depends on the optimal co-ordination of one- and two-joint muscles. Zajac in [11] established that one-joint muscles generate the initial propul-

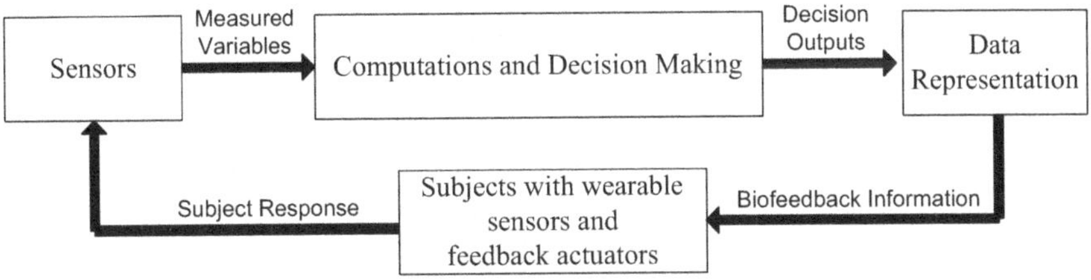

Fig. 3 Main components of a biofeedback system

sive energy for vertical jumps, whereas two-joint muscles control the inter-muscular co-ordination and the final vertical impulse. The method used for establishing the activation of muscles based on their electrical activity is electromyography (EMG) [12].

However, in spite the functionality employed in performance analysis systems, coaches and athletes receive feedback on their performance once the training sessions are completed. Revisiting one's own performance is beneficial to learn and identify improper techniques, but feedback received after the training sessions may constrain an individual's ability to accurately employ proper techniques during training. In addition, post training feedback will not have any impact on detecting improper postures during a performance which may lead to sports injuries.

Biofeedback systems in this regard play an important role in facilitating constructive real time feedback for augmenting motor performance during sports training. A biofeedback system can be effectively employed to monitor an individual's performance and provide real time response on performance statistics. Sports professionals use biomechanical analysis to understand injury prevention, evaluate clarity of technique and provide effective training protocols for performance enhancement [13]. Real time biofeedback devices compliment the trainer and the trainee during their performance by providing immediate feedback on technique and injury prone movements based on the measurement parameters through which trainees are continuously acknowledged to modify their movements.

A biofeedback system typically consists of a sensory device, a restitution device that can convert the biofeedback information to the subjects and a processing system that can perform computations, decision making and control of the input/output devices in the system. Figure 3 illustrates the main components of a biofeedback system. In this research, DVJ is analyzed using IMUs as main sensors and vibrotactors as biofeedback devices. VJH is determined using Qualisys camera system interfaced with Bio-capture system (EMGs) and smart watches as wearable sensors (IoTs) as the main device for biofeedback visualization. National female netball players are the test subjects to analyze DVJ and VJH during sports training sessions under the close monitoring of trainers and physical strength conditioning specialists.

Smart Sensing and Biofeedback for DVJ

The instrumentation system for DVJ during sports training includes two Inertia Measurement Units (IMUs) for measuring knee joint kinematics during Drop Vertical Jump (DVJ) landing, a vision module with a simple web camera for video recordings of the movements performed, custom developed system software for performing data processing and biofeedback generation and a remote vibrotactile feedback module for real time biofeedback. The aim is to obtain knee flexion and abduction/adduction during the DVJ landing phase of a jump and provide instantaneous vibrotactile feedback to increase knee flexion angle, and reduce knee abduction/adduction angle as these parameters are identified to be high risk factors causing knee injuries during landing. The accuracy of the system functionality and the impact of real time biofeedback for improving knee kinematics during landing were tested with a DVJ protocol used by netball coachers.

Each IMU is placed on the thigh and shank segments of the lower extremity to measure knee flexion and knee abduction/adduction angles during DVJ landing. A landing phase detection algorithm obtains the acquired thigh and shank angular measurements for real time landing phase detection. Feedback activation is performed using predefined training targets specified for knee flexion and knee abduction/adduction. During a landing phase, a biofeedback algorithm generates feedback commands comparing the current knee flexion and abduction/adduction measurements to the predefined training targets. If training targets have been reached, the commands are transmitted to the remote feedback module over a Radio Frequency (RF) medium for biofeedback activation. The actuators placed on the posterior and lateral sides of the shank provide feedback for knee flexion and abduction/adduction respectively based on the received feedback commands.

The implemented software tools are accessible via an Interactive Graphical User Interface (IGUI). The IGUI provides hardware device calibration settings, real time data acquisition, graphical visualization of lower extremity knee kinematics during DVJ, biofeedback generation based on individualized training targets and post training data evaluations. Figure 4 shows smart sensing and biofeedback architecture for DVJ.

The knee flexion and abduction/adduction measurements are obtained from each IMU placed on the thigh and shank segments. The sensors are aligned to provide knee flexion angle about the saggital plane using pitch (Y axis) measurements and knee abduction/adduction angle about the frontal plane using Roll (X axis) measurements. The knee flexion angle is computed as the summation of saggital plane orientation measurements of the thigh and shank segments. Abduction/adduction is obtained from the frontal plane measurements of the sensor placed on the shank segment.

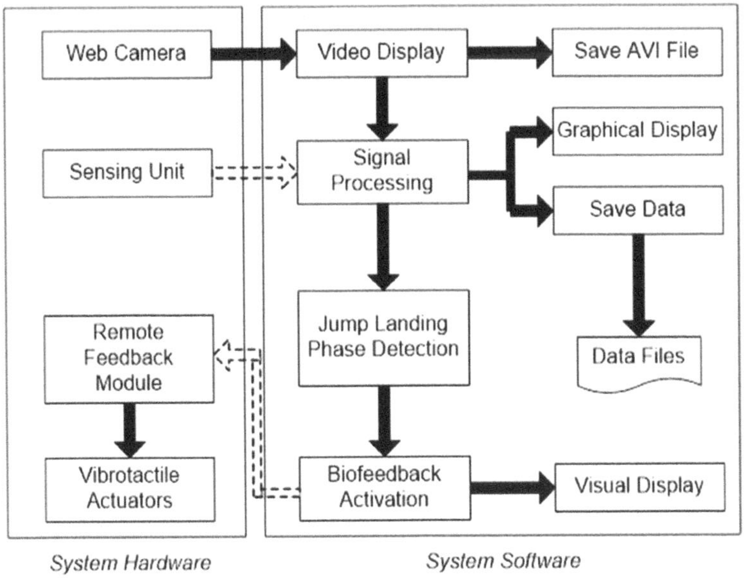

Fig. 4 System architecture of smart sensing and biofeedback for DVJ [9]

Jump Landing Monitoring and Biofeedback for DVJ

The real time landing phase detection algorithm was implemented to obtain the start and end of the landing phase of a DVJ for the activation of biofeedback. During a DVJ training session, the algorithm continuously checks for the input conditions detecting the start of the landing phase. If all conditions are reached, the landing phase is detected, and the biofeedback activation algorithm is initiated. When the algorithm detects the end of the jump, biofeedback activation is terminated. Vibrotactile biofeedback was incorporated in this system to assist subjects increase knee flexion angle whilst reducing their knee abduction/adduction angle during jump landing as these factors are identified as high risk parameters causing knee injuries during landing. Feedback is delivered via vibration stimuli to indicate if predefined targets for knee flexion and abduction/adduction have been reached during landing so that subjects can attempt to alter knee alignment during landing until feedback is received. It is intended that continuous training performed with biofeedback would result in proper kinematic adaptations during landing which will in turn reduce the risk of injuries occurring due to improper landing techniques.

For the two parameters knee flexion and abduction/adduction, biofeedback is generated using four discrete states, namely

- No feedback activation—Both knee flexion and abduction/adduction has not reached the training targets during landing

Table 1 Kinematic input conditions and biofeedback commands for jump landing

Condition	Biofeedback command
No feedback	AA
Knee flexion feedback	CA
Knee abduction/adduction feedback	AD
Knee flexion and abduction/adduction feedback	CD

- Knee Flexion feedback activation—Knee flexion angle has reached its training target but knee abduction/adduction remains outside its training target
- Knee Abduction/adduction feedback activation—Knee abduction/adduction angle has reached its training target but knee flexion remains outside its training target
- Knee Flexion and Knee Abduction/adduction feedback activation—Both parameters have reached their training targets.

The Cypress CY8C27443 microcontroller included in the remote feedback module was programmed using the PSoC programmer software provided by Cypress Microsystems [14]. To receive biofeedback commands wirelessly from the host PC, a UART module was incorporated with a baud rate of 9600 bps. Two Pulse Width Modulation (PWM) modules were included for each vibrotactile actuator to produce a 250 Hz square wave signal with the specified duty cycle. A simple ASCII protocol was developed with single byte commands to identify the vibrotactile actuator and the duty cycle. Four discrete variations of duty cycle were utilized. The variations include 0% duty cycle where vibrotactile feedback remains inactive and subsequent increments of duty cycle from 20 to 50 to 80%. Table 1 illustrates the input conditions and the feedback commands implemented for defining each feedback state. The generated commands are wirelessly transmitted to the developed remote feedback module which activates the corresponding vibrotactile actuator using PWM with an 80% vibration intensity.

An Interactive Graphical User Interface (IGUI) was developed in this system to facilitate user-friendly human-computer interaction for monitoring and improving knee kinematics during landing. The IGUI includes a variety of features essential for real time applications, and are grouped into separate frames for convenience. The frames include hardware device settings and calibration tools, jump landing monitoring and biofeedback tools and post training data evaluation tools. The hardware device settings and calibration tools are dedicated for verifying connectivity between hardware devices, reconfiguring hardware settings and aligning IMU during sensor placements. Jump landing monitoring and biofeedback tools provides the selection of individualized training targets for knee flexion and abduction/adduction (abd/add), video display, graphical representations of knee kinematics and visual feedback on the biofeedback signals delivered to subjects via a simple visual display. Post training data evaluation tools are dedicated for revisiting the kinematic and video records obtained during training sessions for thorough analysis. Using the tools, the data can be graphically evaluated and maximum knee flexion and mean knee abd/add resulted

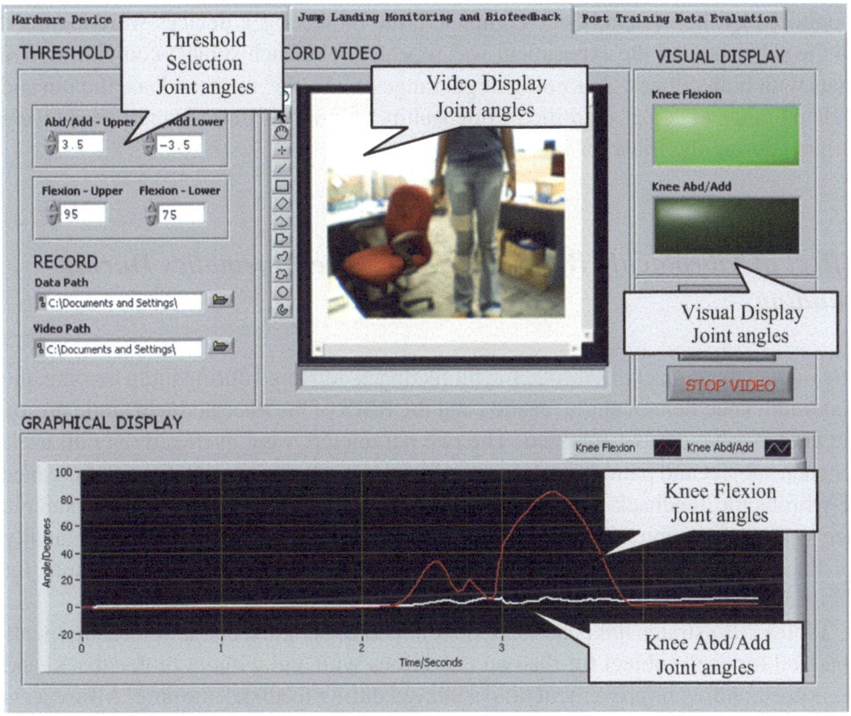

Fig. 5 Jump landing monitoring and biofeedback generation tools for DVJ

during trial session can be computed for further evaluation. Figure 5 depicts the jump landing monitoring and biofeedback tools implemented for DVJ.

The accuracy of the system implemented and the effect of instantaneous vibrotactile biofeedback was evaluated using a DVJ from a height of 30 cm to provide biofeedback for knee flexion and knee abd/add angles measured during landing. A total of 5 female subjects participated in the experiment as females are identified to be more susceptible to injuries during landing compared to their male counterparts [15]. All subjects were young, healthy individuals with an average age of 23.2 years and SD of 0.84 years.

Experimental Procedure for DVJ

Prior to the initiation of the training sessions, a brief familiarization period was carried out during which the subjects were instructed on how to perform the DVJ. Biofeedback was also introduced during the familiarization period to become accustomed to the nature of feedback. After the familiarization period, each subject performed the DVJ without the feedback system for five trials, and their knee angle kinematics was monitored by the system. Once completed, the biofeedback devices were mounted on the subjects, and the experiment was repeated with each subject completing five trials with biofeedback. A training target range of 75°–95° for peak knee flexion and abd/add range of ± 3.5° identified from healthy individuals during landing [15] were utilized during the trials for biofeedback.

Effect of Vibrotactile Biofeedback for Knee Kinematics During Landing

To evaluate the impact of vibrotactile biofeedback delivered during jump landing, the maximum knee flexion angle reached and the RMS of the knee abd/add angle during the landing phase was computed. The two parameters were averaged over all trials for each subject and paired t-tests were utilized to evaluate the statistical significance of vibrotactile feedback for improving landing knee kinematics. As a low sample size was used, Shapiro-Wilk normality tests were performed to confirm normality for knee flexion and abd/add data prior to using paired t-tests. The threshold for statistical significance was set as 0.05 [16].

Tables 2 illustrates the peak knee flexion and mean knee abd/add measurements obtained for each subject for the two conditions with and without biofeedback, and the corresponding improvement ratio achieved during feedback training. All subjects reported an increase in knee flexion angle during landing when biofeedback was in use, with subjects 3, 4 and 5 reporting an improvement above 50%. This implies that during training with biofeedback, subjects were able to interpret clearly the information provided via vibration stimuli and were capable of applying the feedback provided to improve knee flexion angle during landing [16].

For Knee abd/add during landing, subjects 2, 4 and 5 have reported an improvement with reduced knee abd/add with the inclusion of biofeedback with the highest improvement reported by subject 5 as 33.8%. Subjects 1 and 3 did not improve knee abd/add angle during landing with the use of biofeedback. This may have been due to the facts that these subjects found it difficult to concentrate on simultaneous delivery

Table 2 Maximum knee flexion and RMS knee abd/add during landing for all subjects in the DVJ

Subject no	Knee flexion			Knee Abd/Add		
	Without FB (°)	With FB (°)	Improvement ratio (%)	Without FB (°)	With FB (°)	Improvement ratio (%)
Subject 1	72	84.3	17.08	3.75	4.43	−18.13
Subject 2	70.8	94	32.77	6.54	4.53	30.73
Subject 3	52.67	82.9	57.40	4.83	5.67	−17.40
Subject 4	50.87	92.4	81.64	2.54	2.22	12.60
Subject 5	54.88	91.61	66.93	3.79	2.51	33.77

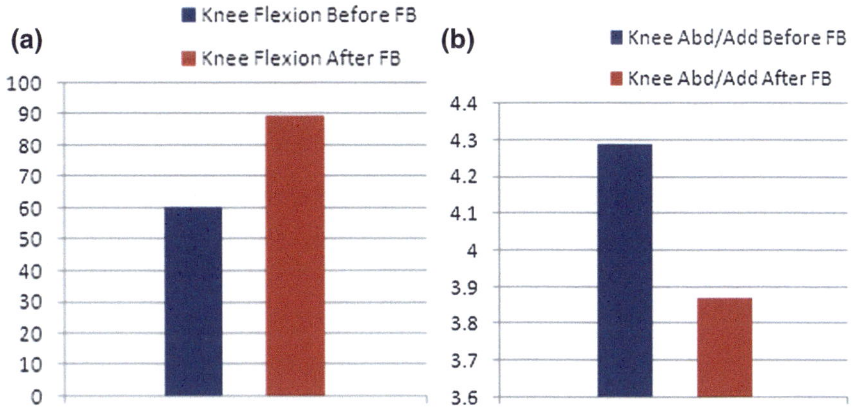

Fig. 6 **a** Mean maximum knee flexion. **b** Mean RMS knee abd/add angles computed across all participants during both conditions

of feedback signals for both knee flexion and abd/add or the inability to interpret the information provided through feedback in an effective manner to improve performance [16].

The overall mean knee flexion angle and SD for all subjects over all trials was 60.24 ± 10.3 and 89.04 ± 5.07 respectively. This increase was statistically significant at $p<0.05$ (0.00251) producing an overall improvement of 47.8%. This implies that while training with biofeedback, subjects continued to flex their knees at landing until the feedback signal was delivered indicating that the required knee flexion was reached. The mean RMS knee abd/add angle and SD for all subjects over all trials was 4.29 ± 1.50 and 3.87 ± 1.46 respectively. This reduction in knee abd/add angle was not statistically significant at $p<0.05$, however the overall reduction of knee abd/add angle by 9.8% implies that biofeedback has contributed to reducing knee abd/add angle during landing opposed to no feedback. Figure 6a, b depict the overall mean maximum knee flexion and mean RMS knee abd/add angles across all subjects computed for the two conditions with and without biofeedback.

The results clearly demonstrate the impact of real time vibrotactile biofeedback for improving knee kinematics during landing. Improved knee flexion and knee abd/add angles obtained during landing suggest that subjects were able to interpret the information provided via vibration stimuli to improve knee kinematics at landing. Increase in knee flexion angle was significant with the inclusion of biofeedback. Although not statistically significant, knee abd/add angle was reduced with the use of biofeedback, which was consistent with the idea that feedback has contributed to improving knee abd/add angle during landing. However, if a broader training target was used or if tested on a larger sample size, the impact of biofeedback for knee abd/add angle would have been more prominent [16].

The developed biofeedback prototype for monitoring and improving knee kinematics during landing is a novel approach for smart sensing and biofeedback for

DVJ. The software tools embedded allows the system operation to be customized to suit the requirements of different individuals and tasks, eliminating the applicability of the system to a single specific application.

Smart Sensing and Biofeedback Visualization for VJH

From Fig. 2, VJH is the sum of four lesser heights: takeoff height, flight height, reach height, and loss height which can be described using Eq. 1.

$$\text{VJH} = \text{Takeoff Height} + \text{Flight Height} + \text{Reach Height} - \text{Loss Height} \tag{1}$$

In standing double leg vertical jumps, takeoff height, flight height, and reach height have been found to account for about 41%, 17%, and 42% of the overall VJH respectively. Among skilled jumpers, loss height has been found to be negligible, accounting for about 0.2% of the overall VJH. It may be surprising to note that the contribution of flight height is so much smaller than that of takeoff height and reach height. Regardless, it is clear that an overwhelming percentage of the overall jump and reach height is determined *not* by the vigorous muscular effort required to propel the body upward, but simply by the physique of the jumper and the position and orientation of the body about the CM at the instant of takeoff and again when the maximum height is evaluated.

However in major sports such as netball, volleyball and basketball, VJH is the primary performance enhancement parameter in order to reach the goal with optimal VJH. Depending on the player's position in a court, VJH measurement varies based on single leg or/and double leg VJHs. Thus, the analysis of VJH is subject to single and double leg jumping to reach the desired goal.

Thus, the instrumentation system for VJH during sports training regime includes six 3D motion capture camera system (Qualisys Inc. Sweden) for joint kinematics measurements, two 8-channel BioRadio units (CleveMed Inc. USA) to measure neuromuscular activities of 8 lower extremity muscles, a vertec device (Power Systems Inc. USA) to capture the highest touch vane during VJH and Samsung smart watch for biofeedback visualization of performance enhancement as per coach's assessment criteria on VJH. The aim is to measure peak electromyography (EMG) and integrated EMG (IEMG) of the most influential 8 muscles from lower extremity using Bio-capture systems (CleveMed Inc. USA) during VJH, within the volume covered by Qualisys motion capture system in the court. Hence, 3D motion capture system facilitates to derive knee and ankle joints range of motions using joint flexion/extension, joint abduction/adduction, and joint longitudinal rotation during VJH. VJH is the measure reading on vertec device corresponding to the highest displaced vane simultaneously and synchronously capturing EMG and 3D joints' motion data during VJH.

Figure 7 shows the smart sensing and biofeedback visualization layered architecture for VJH. Data acquisition layer is responsible to acquire raw data using Qualisys

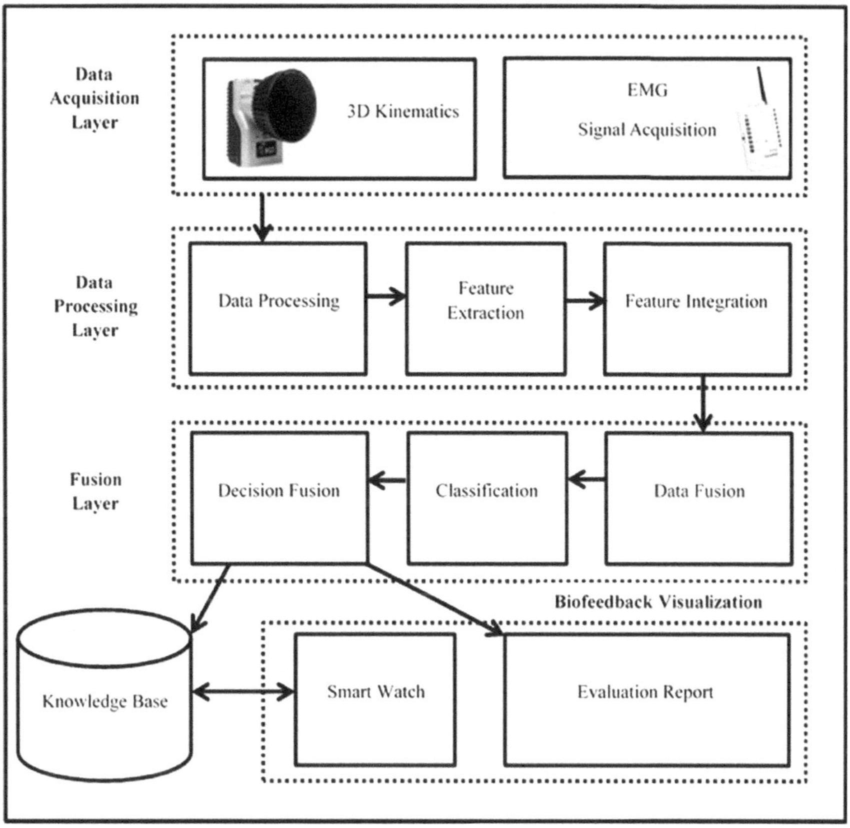

Fig. 7 Smart sensing and biofeedback visualization layered architecture for VJH

Motion Capture system set up in an indoor stadium where actual players train as per international norms under the guidance of coaches of specific games. Bioradio is worn by a test athlete under consideration in order to acquire EMG raw signals from the identified 8 muscles under the close monitoring of physical strength conditioning specialist. During a VJH, reach height is measured using vertec.

Upon finished raw data acquisition, data processing layer extracts parameters which reflect features of VJH. As 3D kinematics and EMG data are involved, feature integration is required such a way that all features are within the same envelop. Usually this is considered using matching time with frames collected using motion capture system and EMG signals encapsulated within the same time period corresponding to the frames extracted. Fusion layer is considered the most important layer of the system. The formation of integrated features in data processing layer creates huge feature set which encapsulates features not influential for VJH. Thus, different data transformation techniques are applied to integrated feature set such a way that reduced feature set will be derived during data fusion stage.

Classification of each player is subject to coach criteria on VJH performance. This criteria is usually known to the coach as norms practiced using internationally established protocols.

In order to compute and determine the classification of each player automatically and autonomously without any interference of coach, intelligent algorithms shall be applied of fused data. As coach provides the ideal criteria to categorize a player, teacher input is available. Thus, set of patters can be formed using the Eq. 2.

$$\text{Pattern}_\text{i} = \{\text{input data}_\text{i},\ \text{output}\} \tag{2}$$

where:

input data_i	ith fused data set corresponding to the integrated feature set.
Output	Current classification of each player known to the coach
Pattern_i	is formed to store in the knowledge base

Therefore, knowledge base (KB) will enrich depending on the frequency in which carries out training regime while system is set up and configured for capturing VJH raw data.

Decision fusion is carried out using the historical pattern set created and stored in KB. Therefore, with the time, KB will contain huge number of pattern set which allows the system to decide which patter set is matching with the actual pattern set formed in real time during VJH.

Smart watch and evaluation reports shall be used to assess current athletes' status for their rehabilitation or/and performance enhancement for VJH standard already established.

Experimental Procedure for VJH

With informed written consent, thirteen healthy (uninjured) female professional netball players (subjects) voluntarily participated in this study. Subjects were of mean (SD) age = 21.92 (3.59) years, Body Height (BH) = 164.67 (5.41) cm, Reach Height (RH) = 181.73 (6.33) cm, and body weight (BW) = 64.37 (7.75) Kg [17].

Aiming to touch the highest vane (Fig. 1) on a vertec device (Power Systems Inc. USA), subjects performed standing vertical jump in three trials each while simultaneously and synchronously capturing EMG and 3D joints' motion data. VJH for each trial was the recorded as the reading on vertec device corresponding to the highest displaced vane. Vertec reading was initialized to zero reading before performing the jump trials [17].

EMG activity was obtained using seventeen circular electrodes mounted on two 8-channel BioRadiounits (CleveMed Inc. USA) with a 960 Hz sampling rate which wirelessly transmitted the recorded signals onto two time-synchronized BioCapture (CleveMed Inc. USA) system units. EMG signals were recorded from eight lower body muscles including three quadriceps [rectus femoris (RF), vastus lateralis

(VL), vastus medialis (VM)], two from the hamstrings (biceps femoris (BF), semitendinosus (SM)), and three of calves [gastrocnemius medialis (GM), gastrocnemius lateralis (GL), and peroneous longus (PL)].These target muscles have been reported to be crucial sources of explosive energy during vertical jump movements [18, 19]. For normalization [20] of EMG data considering all subjects were healthy, maximum voluntary contraction (MVC) EMG data was prior collected for each subject using specific muscle-group isometric activities [17].

Anatomical joints' motion tracking was achieved through placement of 19-mm spherical reflective markers on anatomical landmarks sufficient enough for tracking motion of the shoulder, hip, knee, and ankle joints. A six 3D motion capture camera system (Qualisys Inc. Sweden) was used for joints' motion tracking at a 500 Hz capture rate while storing data on a single system unit running Qualisys Track Manager (QTM) software (Qualisys Inc. Sweden).In order to obtain calibration data, each subject's static recording of all marker coordinates was obtained before performing vertical jump trials [17].

Biofeedback Visualization for VJH Using Smart Watch

Database driven neural computing is used to classify female netball players during VJH. Having established Knowledge Base (KB) using three months intensive training regimes, real time VJH carried out by a female netball player will be classified and visualized in smart watch as the IoT worn by each athlete.

Generalized framework of smart data visualization is implemented as shown in dataflow diagram in Fig. 8. The system software developed for smart watch can be visualized as far as the classification database or/and knowledge base has been formed. Based on real time data processed and classified and updated in cloud server, smart watch will visualize the classification of athlete in real time. In this research work, netball players are classified into two; excellent (above 90%) and very good (above 80%) as per teacher input given by netball coach. Hence, decision fusion is responsible to store pattern sets of each athlete with historical classification based on the training regime undergone in KB in order to retrieve matching pattern set during real time VJH with no clue from coach. This classification was validated and tested in real time before being introduced smart watch as an IoT. Choquet integral was applied to verify and to validate the classification accuracy. Thus, smart watch is capable to visualize accurate classification of netball players VJH in real time training regime which facilitates to re-adjust the training regime on site, online and in real time.

Smart watch has been introduced as the wearable sensors for biofeedback visualization. As VJH classification is based on two types, different face in smart watch was defined for this human activity called as type 2 as shown in Fig. 9. Upon selected type 2 classification, database driven neural computing page will visualize in order to select the netball player in consideration and activity to be performed which must be matching with the KB is connected via cloud server. Amazon Web Service (AWS)

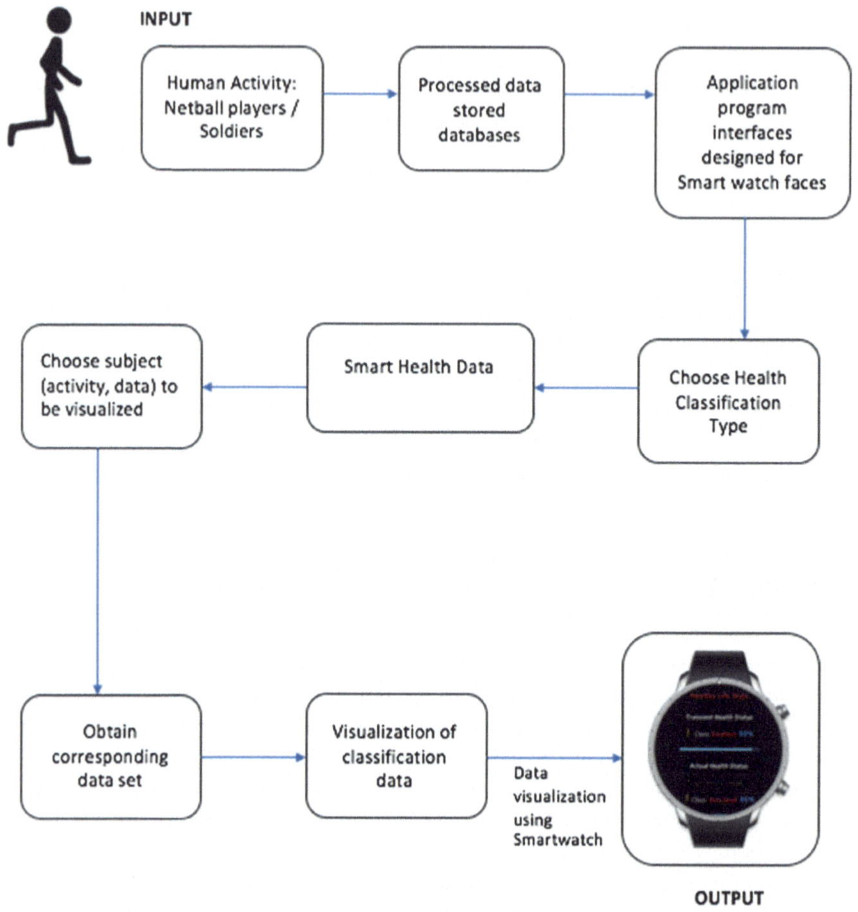

Fig. 8 Dataflow diagram of smart data visualization of human activity classification using a smart watch as an IoT

is used as the cloud platform to do necessary cloud computing. In this research KB is set up and configured only for VJH. Figure shows the result of the actual classification during VJH in real time. Actual classification is visualized in the smart watch as actual health status providing the expandability of the same application program interface for other smart health solutions. In the smart watch face as shown in Fig. 10, transient health status signifies the most recent classification stored in the watch prior to actual VJH in real time. Decision fusion runs using cloud computing in AWS in order to generate the output visualized in the smart watch. Smart watch used in this research is courtesy from Samsung Asia Pvt. Ltd, Singapore.

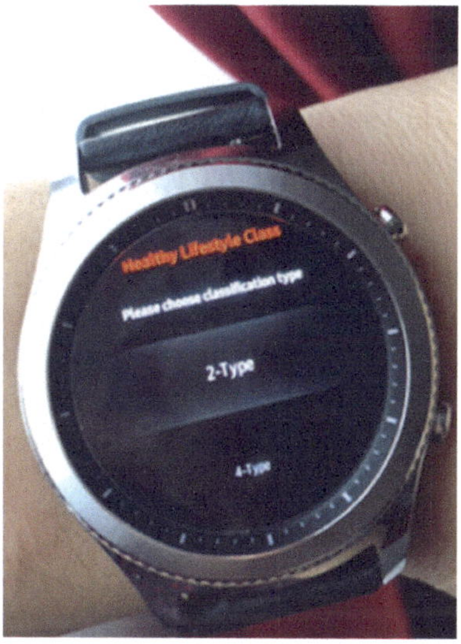

Fig. 9 Selection of classification 2-type using Samsung Smart Gear S3 watch (Courtesy from Samsung Asia Pvt. Ltd)

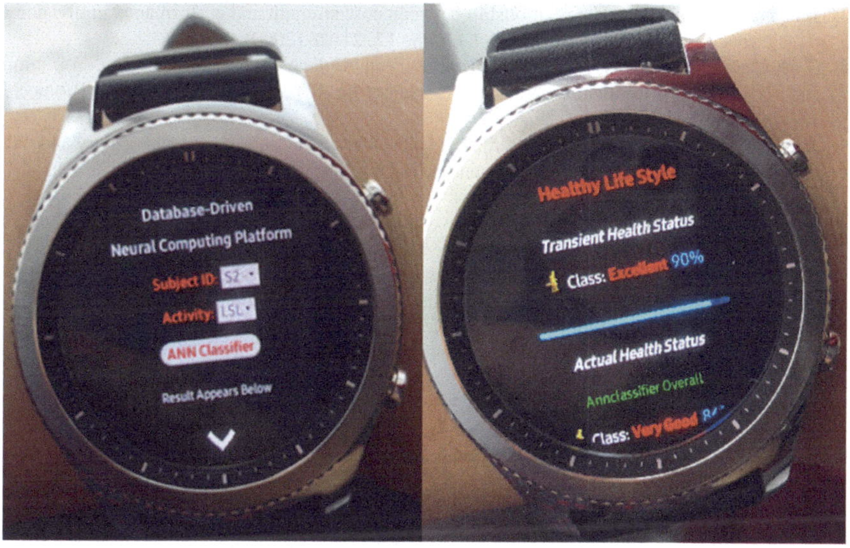

Fig. 10 Decision fusion of actual classification using Samsung Smart Gear S3 watch (Courtesy from Samsung Asia Pvt. Ltd)

Conclusions

This chapter has described in detail the implementation of smart sensing and biofeedback for vertical jump in sports. The comprehensive system has been implemented and tested for both Drop Vertical Jump (DVJ) and Vertical Jump Height (VJH) which are considered primary for injury prevention and performance enhancement of excelling sports. Vibrotactors and smart watch have been successfully integrated as biofeedback for DVJ and VJH respectively during real time training regime. Even though the prototype built was tested using national netball players, the portable system built can be easily configured and calibrated into other sports for injury prevention and performance enhancement.

References

1. P.M. McGinnis, *Introduction to Biomechanics of sport and Exercise*, 2nd edn. (Human Kinetics, 2005), p. 3
2. R. Bartlett, Aspects of biomechanical analysis of sports performance, Chapter 5, in *Sport Biomechanics: Reducing Injury and Improving Performance* (Spon Press, 1999), p. 153
3. T.W. Nesser, W.L. Lee, The relationship between core strength and performance in division I female soccer players. J. Exerc. Physiol. Online **12**, 21–28 (2009)
4. G.M. Cesar, C.L. Tomasevicz, J.M. Burnfield, Frontal plane comparison between drop jump and vertical jump: Implications for the assessment of ACL risk of injury. Sports Biomech. **15**(4), 440–449 (2016)
5. E.J. Cowling, J.R. Steele, P.J. McNair, Effect of verbal instructions on muscle activity and risk of injury to the anterior cruciate ligament during landing. Br. J. Sports Med. **37**(2), 126–130 (2003)
6. B.J. Munro et al., The intelligent knee sleeve: a wearable biofeedback device. Sens. Actuators B Chem. **131**(2), 541–547 (2008)
7. F.R. Noyes, S.D. Barber-Westin, C. Fleckenstein, C. Walsh, J. West, The drop-jump screening test, difference in lower limb control by gender and effect of neuromuscular training in female athletes. Am. J. Sports Med. **33**(2), 197–207 (2005)
8. J.A. Onate, K.M. Guskiewicz, S.W. Marshall, C. Giuliani, B. Yu, W.E. Garrett, Instruction of jump-landing technique using videotape feedback: altering lower extremity motion patterns. Am. J. Sports Med. **33**(6), 831–842 (2005)
9. A.U. Alahakone, S.M.N. Arosha Senanayake, A real time vibrotactile biofeedback system for improving lower extremity kinematic motion during sports training, in *Proceedings of the 2009 International Conference on Soft Computing and Pattern Recognition (SocPar)*, pp. 610–615, 4–7 Dec 2009. ISBN 978-0-7695-3879-2
10. C. Brockett, D. Morgan, U. Proske, Predicting hamstring strain injury in elite athletes. Med. Sci. Sports Exerc. **36**, 379–387 (2004)
11. F. Zajac, Muscle coordination of movement: a perspective. J. Biomech. **26**, 109–124 (1993)
12. M. Čoh1, M. Bračič, S. Peharec, P. Bačić, M. Bratić, M. Aleksandrović, Biodynamic characteristics of vertical and drop jumps. Acta Kinesiol. Univ. Tartu. **17**, 24–36 (2011)
13. D. Khudson, Applying biomechanics in sports and rehabilitation, Chapter 12, in *Fundamentals of Biomechanics* (Springer, 2007), p. 245
14. Cypress Perform, http://www.cypress.com/?id=1573
15. T.E. Hewett et al., Biomechanical measures of neuromuscular control and valgus loading of the knee predict anterior cruciate ligament injury risk in female athletes. Am. J. Sports Med. **33**, 492–501 (2005)

16. A.U. Alahakone, S.M.N. Arosha Senanayake, Improving lower extremity joint kinematics during jump landing using an automated vibrotactile biofeedback system. Int. J. Auton. Comput. (Inderscience Publishers) **2**(1), 39–53 (2014)
17. U. Yahya, S.M.N. Arosha Senanayake, A.G. Naim, Intelligent integrated wearable sensing mechanism for vertical jump height prediction in female netball players, in *Eleventh International Conference on Sensing Technology (ICST)*, Sydney, Australia, pp. 94–100. https://doi.org/10.1109/icsenst.2017.8304484. 978-1-5090-6526-4/17/$31.00 ©2017 Crown
18. C. Nongnapas, B. Rumpa, S. Sirod, S. Vitoon, Principal component analysis identifies major muscles recruited during elite vertical jump. Sci. Asia **39**, 257–264 (2013)
19. M.F. Bobbert, A.J. Van Soest, Effects on muscle strengthening on vertical jump height: a simulation study. Med. Sci. Sports Exerc. **26**, 1012–1020 (1994)
20. M.B.I. Raez, M.S. Hussain, F.M. Yasin, Techniques of EMG signal analysis: detection, processing, classification and applications. Biol. Proced. **8**, 11–35 (2006)

Spontaneous Facial Expression Analysis Using Optical Flow Technique

L. Sidavong, S. Lal and T. Sztynda

Abstract Investigation of emotions manifested through facial expressions has valuable applications in predictive behavioural studies. A potential application may be to impart intelligence to surveillance systems such as Closed-Circuit Television (CCTV) systems for recognition of emotional facial expressions. A facial recognition program tailored to evaluating facial behaviour for real time application can be met if patterns of emotions can be detected. An exploratory analysis of optical flow data was conducted with an aim to detect patterns and trends to differentiate between the emotional facial expressions: amusement, sadness and fear from the frontal and profile facial orientations. Analysis was in the form of emotion maps constructed from feature vectors obtained by using the Lucas-Kanade implementation of optical flow. Classification of individual emotions showed recognition of amusement was much greater in comparison to the recognition of the negative emotions, sadness and fear. Recognition was not negatively affected using reduced set of feature vectors derived from the emotion maps. Further investigation is necessary to assess the utility of emotion maps to visualise feature representations of emotional expression.

Introduction

Security is a contemporary issue in society, and this situation was exacerbated by September 11, 2001 World Trade Center destruction. As such this led to the increased use of Closed-Circuit Television (CCTV) surveillance systems to monitor the behaviour of individuals in efforts to monitor and reduce crime [1]. Despite this growth, much controversy still exists in the efficacy of such systems. There is a significant cognitive demand placed on the human operators who monitor CCTV systems [2]. It is seen that the vigilance required of human observers dramatically declines over time, resulting in oversight of crucial information that would alert them to an emerging incident [3].

L. Sidavong (✉) · S. Lal · T. Sztynda
School of Life Sciences, University of Technology Sydney, Sydney, Australia
e-mail: Lina.Sidavong@student.uts.edu.au

S. C. Mukhopadhyay et al. (eds.), *Modern Sensing Technologies* ,
Smart Sensors, Measurement and Instrumentation 29,
https://doi.org/10.1007/978-3-319-99540-3_6

In efforts to enhance security, one approach may be to impart intelligence to surveillance systems for the detection of affective states indicative of subsequent criminal behavior through facial expressions. A facial recognition program tailored to evaluating facial behaviour for forensic and surveillance applications can be met if patterns of emotions can be detected. However, research is required to link emotions to its behavioural constructs. Hence this investigation starts with the attempts to recognise the most basic emotions.

The Functions of Human Emotions

Most often, the word emotion is usually linked with behaviour. This is because emotions elicit a wide range of observable changes in an individual, frequently seen as expressed feelings and changes in the body's physical and physiological state. Emotion has been a major interest because it is believed that emotions are reflections of an individual's personality and behaviour [4]. Not only do emotions play a critical role in how people behave, they direct our engagement with the world and drive the decisions we make.

A review of research has its roots embedded in Charles Darwin's work *"The Expression of Emotions in Man and Animals"* who applied an evolutionary approach to the functions of emotion [5]. Darwin described emotions as discrete entities closely tied to specific adaptation behaviours that facilitated the survival of humans and other complex species. Furthermore, he proposed that facial expressions evolved to quickly communicate emotional states important to social survival. The hypothesis was articulated through several works by psychologist Paul Ekman who was inspired by Silvan Tomkins' work on the universal affect programs [6]. As such seven basic universal emotions have been identified to date: joy, surprise, sadness, anger, fear, contempt and disgust [7]. They have been shown to be ubiquitous in their performance and in their perception.

Facial Expression Measurement

The muscles of the face are embedded in the superficial facia, and are supplied by the 7th cranial facial nerve (CN VII). The contractions of these muscles results in temporally deformed facial features such as eyes, eyebrows, nose and lips. With typical changes of muscular activities lasting for a few seconds, between 250 ms and 5 s [8].

The measurement of human emotional facial behaviours can be acquired by using either of two methods: facial electromyography (EMG) [9, 10], or visual analysis of facial activity [11, 12]. The application for either method is dependent upon theoretical and practical considerations.

In 1978 Paul Ekman and Wallace V. Friesen published the Facial Action Coding System (FACS) which categorised facial behaviours based on the muscles that produce them [13]. The system identifies all possible visual facial movements and classifies them into Action Units (AUs), which are the unique changes produced by individual facial muscles or muscle combinations. There are 46 AUs that correspond to each independent motion of the face, 30 of which are related to contraction of specific muscles [14]. A combination of action units yields descriptive power for coding facial expressions [15].

Although FACS is an ideal system of analysis for facial behaviour, its manual execution is laborious in nature. The FACS requires approximately 100 hours of training to identify and interpret data into AUs and it takes two hours to code a one minute of video recording [16]. These drawbacks limit its suitability in situations where real-time feedback is necessary. Failing to be automated is where these systems are greatly confined.

Automatic Facial Expression Recognition

Recent advances in the field of computer vision and pattern recognition have ventured towards the development of automated systems for recognition of facial expressions. An automatic facial expression recognition (AFER) system is defined by its ability to effectively measure and classify expressions of emotion. The primary design fundamental to most systems is comprised by three components: face detection, feature extraction and expression classification [17].

Face detection initiates the recognition processes by locating the face within a scene. Pre-processing is often applied for noise removal and image enhancement to achieve a pure isolate of facial images. The feature acquisition stage deals with extracting the features best representing unique patterns residing on the face. The process usually involves transforming high-dimensional data to a lower dimensional space whilst preserving the original data. The final step is classification which uses an algorithm to build predictive models to assign the features into either respective AUs or as one of the basic emotional expressions. Common classification methods are Support Vector Machine (SVM) and K-Nearest Neighbor (KNN) procedures.

Optical Flow Feature Extraction

As emotions are displayed over time, motion based tracking has been an attractive approach to acquire temporal facial action patterns. The optical flow method is one such technique that has been applied to facial expression recognition to extract dynamic movement in image sequences [18–20]. In this context, the approach extracts motion changes in movement patterns of the skin to discriminate facial displays and is often applied using different algorithms to calculate movement from the input image window. Naghsh-Nilchi and Roshanzamir used the optical flow technique to examine image sequences available in the Cohn-Kanade database [21].

Their approach segmented the face into 6 areas before extracting motion from these components, yielding 94% recognition [21].

In a study by Uddin and colleagues, who combined optical flow techniques with Principle Component Analysis (PCA) and General Discriminant Analysis (GDA) to extract movement activity from facial expression images [22]. The feature extraction method provided a system with an average recognition rate of 99.16% for the expressions of anger, joy, sadness, surprise, fear and disgust [22].

Method of Difference Using Optical Flow

The differential method of optical flow is based on a local Taylor series approximation using the brightness constancy constraint as the common starting point for optical flow estimation [19]. It assumes that pixel intensities are translated from one frame to the next and therefore reliant on the intensity difference between the two images [22]. The explanations of optical flow are as follow:

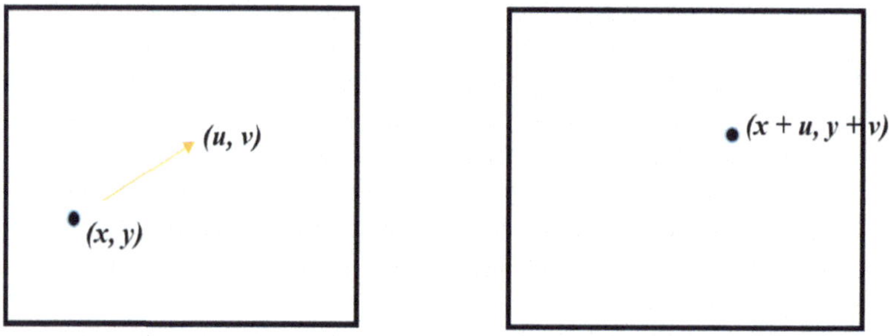

I1: *I(x ,y ,t)* I2: *I(x + u , y + v , t + 1)*

Where I (x, y, t) is the intensity of a pixel at location (x, y) in image I1 at time t and is offset by the flow (u, v) at time $t + 1$ in image I2 [19, 20]. The brightness constancy is given by:

$$I(x, y, t) = I(x + u, y + v, t + 1) \tag{1}$$

$$I(x, y, t) = I(x, y, t) + u\frac{\delta I}{\delta x} + v\frac{\delta I}{\delta y} + \frac{\delta I}{\delta t} \tag{2}$$

$$0 = u\frac{\delta I}{\delta x} + v\frac{\delta I}{\delta y} + I\frac{\delta I}{\delta t} \tag{3}$$

The partial derivatives can be substituted for the intensity derivatives:

$$Ix = \frac{\delta I}{\delta x}, \quad Iv = \frac{\delta I}{\delta y}, \quad \text{and } It = \frac{\delta I}{\delta t}$$

$$I_x u + I_y v + It = 0 \tag{4}$$

The equation can be rewritten as:

$$\nabla I \cdot \vec{v} = -It \tag{5}$$

where ∇I is the spatial intensity gradient and $\vec{v}$ is the image velocity or optical flow of pixel (x, y) at time t. Although the brightness constancy equation defines the gradient of the moving object, the boundaries of motion remain obscure. Thus, additional constraints for flow determination must be introduced [20].

The Lucas-Kanade framework assumes that motion is smooth within a local image neighbourhood [23]. The method obtains the optical flow velocities using the brightness constancy equation for a 3 × 3 neighbourhood around the pixel using the least squares criterion [24]. As such, the optimal value of flow velocities can be achieved by minimizing this error function with respect to u and v.

Opportunities

Over the last decade, automatic facial expression analysis has become an active research area that finds potential applications in areas such as:

1. Advertisement and marketing: by tracking implicit preferences of customers when being shown an advertisement of products [25].
2. Autism: systems outfitted with emotion recognition capabilities to help children with autism learn to determine the emotions of those around them to ease their integration into society [26].
3. Clinical pain management: where behavioural observation-based assessment is optimal for cognitively impaired individuals or patients who cannot communicate [27].

Another possible use could be to design a detector which can be implemented for real-time detection either as software or embedded hardware and installed in surveillance systems for security applications. There would be substantial advantages with computer aided analysis of emotional facial expressions for real-time security surveillance systems, for example CCTV. CCTV is employed as a surveillance tool to monitor the behaviour of individuals in public spaces such as malls and major thoroughfares to reduce and deter crime [1]. Whilst facilities may have already established their own practices of using CCTV monitoring of crowds, questions surround the efficacy of such operations [28].

In the case of events such as concerts, sporting matches and other organized gatherings it is reasonable to assume that CCTV monitoring requires a high level of concentration by those who monitor CCTV. However, in prolonged routine monitoring at airports, shopping malls and station concourses the observers are likely to lose concentration [29]. Automated computer aided analysis of facial expressions could aid in the observation capability to recognise signs of imminent aggression and violence, allowing operators to coordinate an immediate response to such incidents.

Challenges

Despite significant development in computer vision technology, recognising human facial expressions and emotions by a computer remains a challenge. The underlying factors that affect the appearance of facial expressions range from differences in: age, gender, ethnicity, facial hair or other occlusions, head rotations and supposition of speech-related facial deformations [30].

Most of the existing studies limit facial expression recognition to frontal facial image analysis with very few studies addressing the presence of occlusions, such as sunglasses which occlude the eye region [31, 32] or pose variations [33]. Systems amended to real world applications should allow for faces to be reliably decoded at various angles and not be limited to full features of the whole face. In reality, out of plane variation of head position is common and often accompanies change in expressions [24].

Current automatic facial expression recognition systems have favoured for their set-up the use of posed emotions such as those expressed under voluntary control. This also influences the thresholds that determine whether a feature extraction parameter belongs to one emotive state or another in the learning algorithm. The most cited and frequently used databases include the Man-Machine Interaction (MMI), Cohn-Kanade (CK and CK+), Japanese and American Female Facial Expression (JAFFE) and Face and Gesture Recognition Research Network (FG-Net) data sets [34].

Whilst expression classification works reasonably well for posed expressions, spontaneous or naturally occurring expressions are typically much more difficult to recognize. A study by Bartlett et al. who employed a trained recognizer of AUs using the FACS system on two data sets (CK and Ekman-Hager) of posed expressions and a database of spontaneous facial behaviour (RU-FACS data set), showed that recognition performance on the spontaneous database dropped by 20% [35]. A major factor contributing to the poorer performance was the general reduction of movement intensity of facial musculature. The inability to detect subtle dynamics presents a concern because, the focus of facial expression analysis is shifting towards recognising spontaneous expressions in more realistic situations [36].

The investigation focuses on spontaneous emotional facial expressions presenting results from the analysis of motion activity extracted from recorded facial responses.

Inducing Emotion

Work on eliciting emotional responses rely on the use of emotion or mood induction procedures. These include but are not limited to: the use of imagery, memory recall, pictures and sounds. Films have been known to induce emotions for longer periods of time, because they stimulate auditory and visual processes and have an ability to communicate rich contextual information that mimics real life situation [37]. Therefore, to elicit the different facial expressions, short films were used.

Results presented are from an exploratory analysis of motion activity extracted from spontaneous emotional facial expressions of the frontal and profile facial orientations [38]. Research was carried out on a data set of images extracted from video recordings of participants watching three emotion inducing short films. Each short film was intended to induce one of three emotions: amusement, sadness or fear. The Lucas-Kanade (LK) two-frame differential method of optical flow estimation was applied by image matching [39]. Image matching involves aligning two identical or similar images by the 'method of differences' to define pattern changes in facial movement most common to people when they experience an emotion. Electrodermal activity (EDA) patterns were measured along with self-reports [37] to validate the presence of emotion [40, 41].

Data Acquisition

Participants

The study was conducted on n = 75 healthy adult volunteers (Male = 22, Female = 53) aged between 18–34 years (mean = 23.6 years, SD = 4.51). All participants signed an informed consent agreement to take part in the experiment. These procedures had institutional human ethics approval from the Human Research Ethic Committee (HREC).

Protocols

Participants were seated in front of a monitor and were asked to view three short films. Each film was determined in a pilot study to elicit one main emotion: amusement, sadness or fear. The films shown to elicit either amusement, sadness or fear were a series clips from Never Say No to Panda (2010) created by Advantage Marketing and Advertising company, Last Minutes with Oden (Directed by Eliot Rausch 2009) and Lights Out (Directed by David F. Sandberg 2013) respectively. Permission for use of the short films was granted by the film directors and the Never Say No to Panda clips was acquired from the agency website. While participants were watching the short films facial movements and arousal responses were recorded.

Following stimulus presentation, subjects completed an emotion self-report questionnaire using a nine-point Likert scale. The questionnaire, adapted from the work of Gross and Levenson contained a list of sixteen emotions with an intensity grade ranging from 0 (not at all) to 8 (the most that I have felt) for each of the emotions [37]. The purpose of the questionnaire was to establish the type and extent of emotion felt during scenes from the short films, so that facial expressions corresponding to an experienced emotion could be matched to the film extracts in real time.

Facial Images

Observed arousal peaks correlating to self-report of emotional response to the film stimulus were used to establish the type and extent of emotion felt. This allowed for extraction of still facial images representative of neutral "baseline" and "peak of emotion" expressions, for each subject taken from the front, profile left and profile right sides of the face from the recorded video footage. Image pre-processing was performed manually to correctly align images and then cropped to 370×310 pixel and 370×170 pixel image sizes for the frontal and profile sides respectively (Fig. 1).

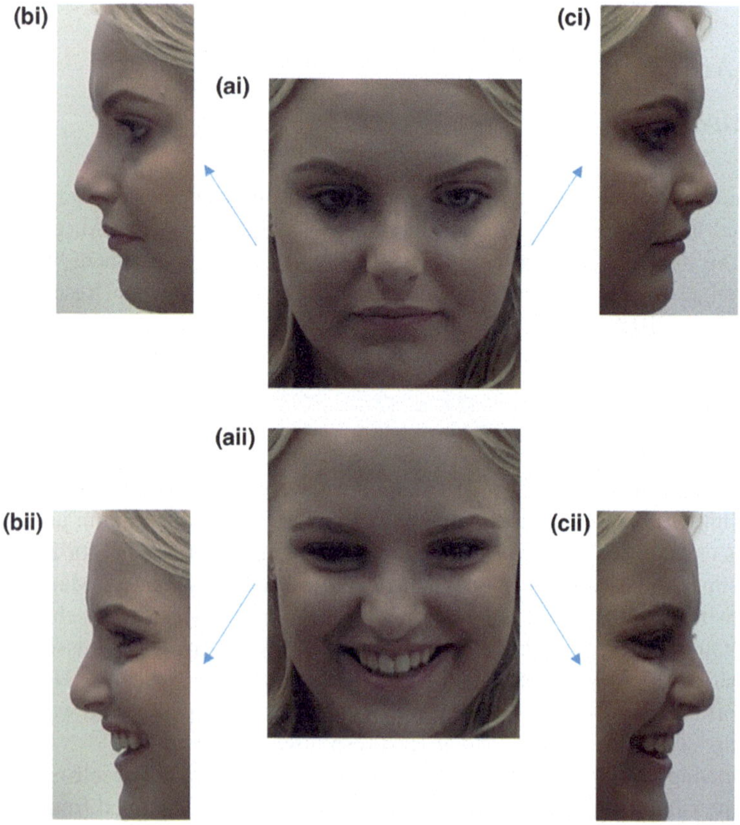

Fig. 1 Facial images. Image sets consists of facial images representative of (i) neutral and (ii) peak of emotional expression for the **a** frontal, **b** profile left and **c** profile right sides (Permission granted for use of images)

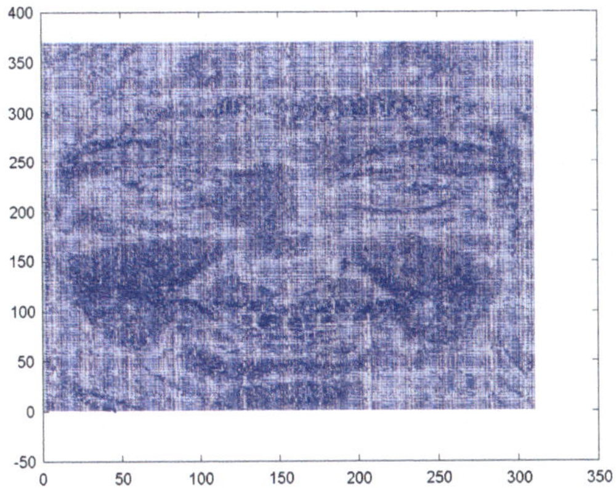

(a) Quiver plot

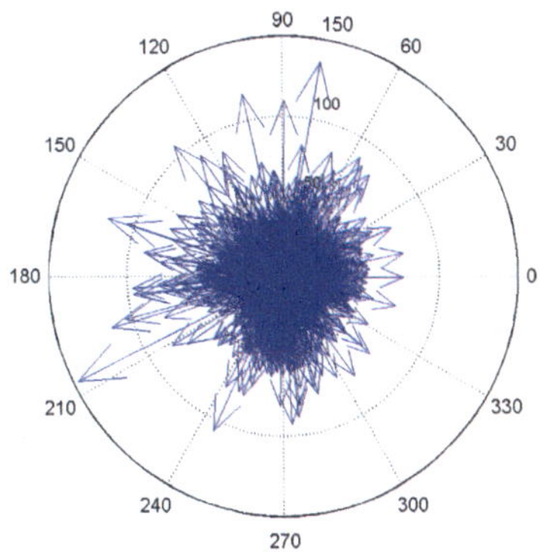

(b) Compass graph

Fig. 2 Quiver plot and compass graph. The Quiver plot **a** displays motion vectors in pixels from specific co-ordinates of origin from baseline to emotional response. Vectors from the quiver plot were used to produce a compass graph **b** that displays the total direction and magnitude of vectors. The compass diagram is separated into 12 sectors of 30° on a 360° plane

Feature Extraction

Images acquired from video footage were analysed using the optical flow technique to extract feature vectors. The LK method of optical flow was performed using the MATLAB software. When comparing two images, the algorithm first measured and quantified motion vectors which had originated as result of movement from baseline to an emotional response.

Motion vectors were mapped onto a quiver plot (Fig. 2a), which depicted the flow of facial movement. Once the quiver plot was obtained the algorithm produced a compass graph which summed all the vectors onto a circular grid, with the arrows emanating from the center of the axes (Fig. 2b). The compass graph is divided into twelve segments of 30° sectors and each sector contains 'n' number of arrows with a length (r).

Optical flow output creates twelve feature vectors by calculating the average vector magnitude for each sector. These feature vectors were used to establish emotion vector maps for specific emotions.

The emotion map is an adapted form of the compass graph starting with 0° on the right and moving counter clockwise for increasing angles and is divided into twelve even sectors. Each 30° sector of the map is colour coded for visual ease of quick segment comparisons, for commonalities and differences, between emotions (Fig. 3). These feature vectors were used as inputs into classifiers for the interpretation and categorisation of emotional expressions.

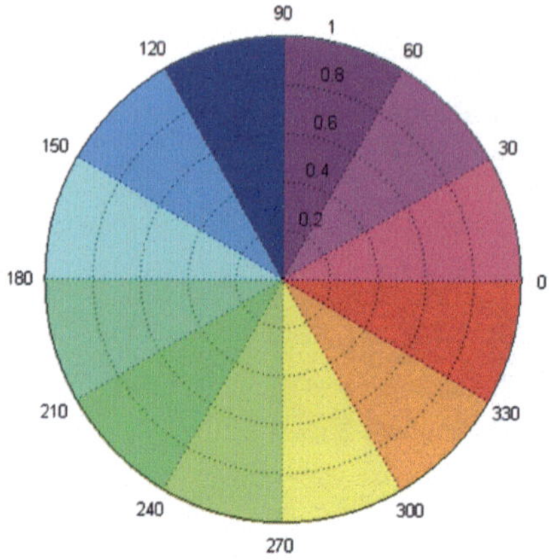

Sector	Angle	Colour	Angle	Sector	Colour
1	0° - 30°		180° - 210°	7	
2	30° - 60°		210° - 240°	8	
3	60° - 90°		240° - 270°	9	
4	90° - 120°		270° - 300°	10	
5	120° - 150°		300° - 330°	11	
6	150° - 180°		330° - 360°	12	

Fig. 3 Colour coding for emotion maps. The emotion map is divided into twelve 30° sector with each sector colour coded for visual ease of analysis. Radial distance of each sector from the centre origin is in pixel

Results

Out of the 75 participants, emotion was elicited in 65 of the participants. The distribution of emotion class in our acquired data set are seen in Table 1.

Table 1 Distribution of facial images extracted for each emotion

	Emotion Class (n = 65)		
	Amusement	Sadness	Fear
Frontal	63	42	34
Profile right	60	38	34
Profile left	61	38	33

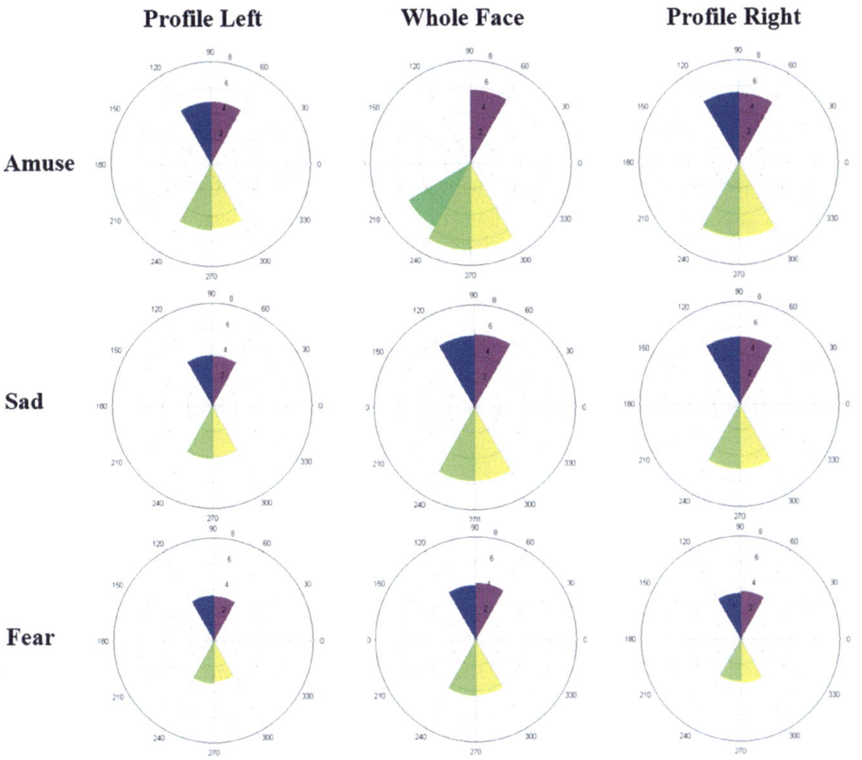

Fig. 4 Emotion maps representing higher division of movements. Emotion maps depicts average feature vector magnitude in pixels, of sectors which exhibited highest activity for the emotional expression of amusement, sad and fear from the frontal, profile left and right facial orientations

Visual analysis involved mapping feature vectors of facial expression activity onto emotion maps according to the level of activity. The feature vectors extracted for each subject were sorted into descending order of magnitude and allocated into one of three groups representing divisions of higher, moderate or lower facial activity.

Emotion maps were created showing the distribution of the sectors which exhibited the highest and lowest facial activities for the emotional expression of amusement, sadness or fear (Figs. 4 and 5). Where the radial distance (r) of the sectors from the center origin correlate to the average displacement or feature vector value (pixels).

When data were sorted to extract the highest and lowest areas of movement, pattern variations was observed in regions of lower divisions of activity. Repeated measures analysis of variance (RM-ANOVA) with a Greenhouse-Geisser correction was carried out to compare the means of common sectors in the lower areas of movement to determine whether there was a statistically significant difference in the magnitude of facial activity. Post hoc using the Bonferroni correction revealed a statistically significant difference ($p < 0.05$) in activity between the expression of amusement and sadness in sectors 1 (0°–30°) and 5 (120°–150°) for the frontal (whole face), and in sectors 7 (180°–210°) and 8 (210°–240°) for the profile right orientations. No significant differences ($p > 0.05$) were found for facial activity between the three emotions from the left profile view (Table 2).

As to whether the patterns of activity observed could be used to classify emotions, three classifiers were employed from the Classification Learner application in MATLAB: linear Support Vector Machine (SVM), Linear Discriminant Analysis (LDA), and K-Nearest Neighbor (KNN). The total accuracy of classifying all three emotions: amusement, sadness and fear using all feature vectors obtained from the optical flow module and the reduced set of feature vectors representing the lower division of movement are shown in Table 3 and Table 4, respectively.

It was shown that facial images obtained from the frontal (whole face) view yielded slightly greater accuracy than the facial images obtained from the left and right profile views. SVM and KNN classifiers showed consistency in recognition between the two sets of feature vectors. The LDA classifier boosted the accuracy of the profile views using the reduced set of feature vectors. Accuracy of classifying individual emotions using the linear SVM is shown in Table 5. It was observed that amusement had the highest recognition followed by sadness. Recognition of fear was remarkably low, particularly the profile left side which had a negative impact on the overall accuracy.

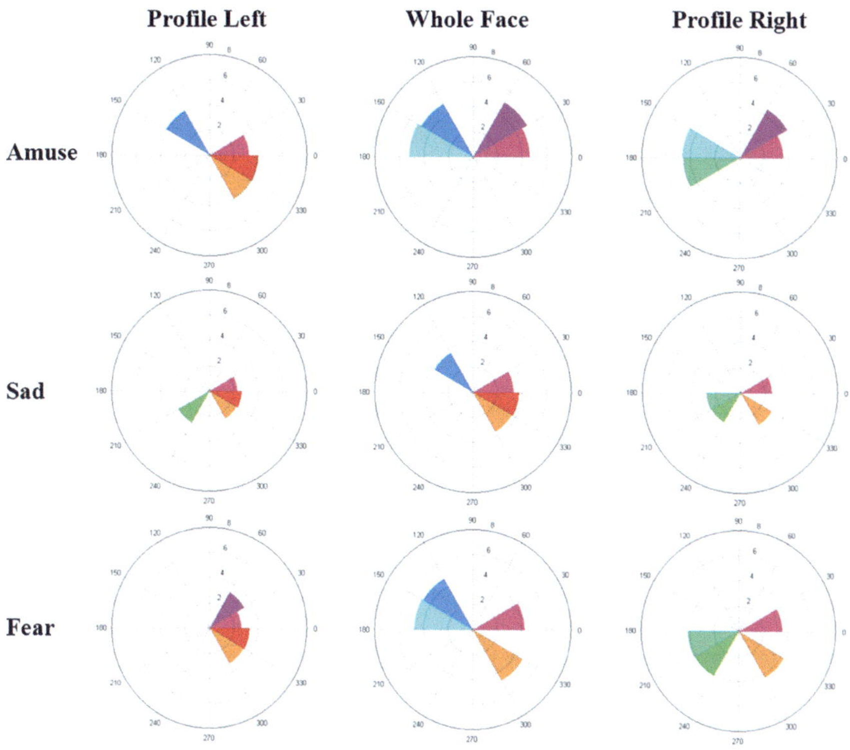

Fig. 5 Emotion maps representing lower division of movements. Emotion maps depicts average feature vector magnitude in pixels, of sectors which exhibited lowest activity for the emotional expression of amusement, sad and fear from the frontal, profile left and right facial orientations

Table 2 Significant difference results of post hoc comparison. Post hoc test using Bonferroni correction revealed statistically significant difference for mean intensity movement between amusement in sadness

	Sectors	Emotion comparison	Mean difference between emotions (pixels)	Std.error	*p* value	95% confidence interval for difference	
						Lower bound	Upper bound
Frontal	1	A − S	1.207	0.371	0.010	0.252	2.161
	5	A − S	1.248	0.404	0.014	0.216	2.279
Profile right	7	A − S	1.259	0.419	0.02	0.174	2.345
	8	A = S	1.343	0.394	0.008	0.322	2.365

Note A = Amusement; S = Sadness

Table 3 Total percentage (%) accuracy in classifying emotions using all feature vector output

	Classifiers		
	SVM	LDA	KNN
Frontal	60	59.3	57.9
Profile right	57.6	45.5	56.1
Profile left	50.8	47.7	49.2

Note SVM = Support Vector Machine; LDA = Linear Discriminant Analysis; KNN = K-Nearest Neighbor

Table 4 Total percentage (%) accuracy in classifying emotions using the lowest feature vector output

	Classifiers		
	SVM	LDA	KNN
Frontal	60	60.7	52.9
Profile right	55.3	59.1	53.8
Profile left	53.8	51.5	50.8

Note SVM = Support Vector Machine; LDA = Linear Discriminant Analysis; KNN = K-Nearest Neighbor

Table 5 Total percentage (%) accuracy in classifying individual emotions using linear support vector machine (SVM)

	Emotion class	All 12 feature vector	Lowest feature magnitudes
Frontal (whole face)	A	73	73
	S	49	52
	F	45	44
Profile right	A	65	64
	S	48	51
	F	50	55
Profile left	A	56	61
	S	48	47
	F	20	32

Note A = Amusement; S = Sadness; F = Fear

Discussions

The goal of this project was to differentiate emotions by examining the movement of facial expressions using optical flow. The study used images obtained from video recordings of subjects watching three emotion inducing short films. This investigation had two stages. The first involved using an emotion induction procedure to acquire image data sets of spontaneous emotional expressions. The second stage utilized the optical flow algorithm to evaluate and deduce compact feature representations of emotions.

Differences in Volunteer Responses

As this research intended to reflect the true nature of facial expressions, facial images chosen for the study were those demonstrating coherence between facial responses and emotions reported by the subjects.

Despite films possessing relatively high degree of ecological validity [37], one concern was the ability of the films to evoke responses congruent to intended emotions. There was considerable variability in intensity of the emotional responses observed both on the level of expression and the target emotion in the self-reports. In some cases, it was seen that subjects had shown no visible changes in facial activity despite reporting they had an emotional response to the film stimulus.

It was observed not all emotions were equally easy to elicit using films. Only in 65 out of 75 participants, amusement was induced the most intensely, followed by sadness and fear. A short film was affective in eliciting a positive emotion—amusement, successfully in 84% of subjects. Whereas, short films were less effective in inducing the negative emotions—sadness and fear, only eliciting a response in 56% and 45% of subjects respectively.

One explanation for the variation in emotional responses could be attributed to the individual's appraisal of the stimulus film. Appraisal is the idea that emotions arise from a person's evaluation of the significant events relevant for their well-being [42, 43]. Given the context of the study, how the subject evaluates the films as having personal meaning or relevance is a major factor as to the nature of the ensuing emotion. Although another possible explanation could lie in the choice of candidate films. Nonetheless, it is important to recognise the difficulties of eliciting emotions in general that even carefully selected films can elicit variable emotional responses. We contend that the appraisal of the film contents was not consistent amongst all individuals resulting in not all subjects expressing congruent emotions. The consequences of which limited the number of image sets used for analysis.

Image Analysis

Optical flow generated twelve feature vectors, each representing mean global velocity vectors of facial movement constrained within a 30° segment. Feature vectors depicting greater and lesser areas of movement were used to construct emotion maps for the expressions of amusement, sadness and fear from the whole face and profile views.

In general, greater activity was observed from emotion maps representing the frontal (whole face) rather than profile views, with amusement yielding overall greater intensity of movement. This was expected due to the decrease in visibility of the morphological changes that occur on the surface of the face when represented in profile. Visual inspection of emotion maps showed greater variations in patterns of movement in regions of lower activity for all three emotions.

Despite the observable physical facial differences between emotional expressions, visual comparisons between emotion vector maps of amusement, sadness and fear, did not vary in distribution of global flow vectors in the higher movement division. Variations in patterns of movement were more apparent in regions of lower activity between sadness and amusement. This may be attributed to the action of large, dominant facial muscles on the lesser muscles or by independent mimetic movements of muscles characteristic to each emotion.

Comparisons between the classifiers showed recognition was not negatively affected using the reduced set of feature vectors derived from the emotion maps. In fact, the SVM classifier could achieve up to 60% and using both SVM and the LDA classifiers 60.7% recognition accuracy was obtained. Classification of individual emotions showed that recognition of amusement was much greater in comparison to the recognition of fear, which may have affected the overall accuracy rate. Nonetheless, the feature vectors depicting low activity regions show potential for differentiating between emotions and warrants further investigation.

Conclusion

Automating the recognition of facial expressions has the potential for the development of a novel component in the surveillance security arsenal. A common drawback on the performance of a standard automatic facial expression recognition system is failure to detect and classify naturally occurring expressions which limits its real-world applications. We attempted to tackle this problem by implementing the optical flow technique to quantify motion changes in dynamic emotional facial expressions. Emotion vector maps were produced using the optical flow output to determine and uncover a hidden pattern for emotion.

Despite weak discrimination between the basic emotions in the data set generated from the emotion inducing procedure, the emotion map shows promise as a feature descriptor. Refined methodology and extraction designs may lead to better results and enhance the utility of the proposed maps. Measuring emotional responses to a film stimulus is a reasonable starting point for this research.

Future Directions

Research into spontaneous facial expressions is a challenging problem. Due to the variability in perception where the emotion eliciting stimulus must be re-evaluated to select films that are more effective in eliciting the desired emotion.

Current work is under way to improve the feature extraction technique by computing the optical flow of key, soft tissue, facial anthropometric landmarks. Future investigations should consider subgroup populations factoring in ethnicity, gender and age to improve model patterns for expression [44].

Acknowledgements This research is supported by an Australian Government Research Training Program Scholarship and UTS Science Faculty Research funds. Thanks to Dr. Budi Jap who proposed the feature extraction methods.

References

1. D. Wilson, A. Sutton, Open-street CCTV in Australia. Australian Institute of Criminology, Trends and Issues in Crime and Criminal Justice (2003), http://www.aic.gov.au/media_library/publications/tandi_pdf/tandi271.pdf. Accessed 5 June (2016)
2. F.M. Donald, C.H.M. Donald, Task disengagement and implications for vigilance performance in CCTV surveillance. Cogn. Technol. Work **17**, 121–130 (2015)
3. A. Isnard, Can surveillance cameras be successful in preventing crime and controlling anti-social behaviours, in *The Character, Impact and Prevention of Crime in Regional Australia Conference*, Townsville, Queensland, Australia (2001)
4. S. Freud, J. Strachey, *The Ego and the Id* (Norton, New York, 1962)
5. C. Darwin, *The Expression of the Emotions in Man and Animals* (Oxford University Press, New York, 1872)
6. S.S. Tomkins, V.E. Demos, *Exploring Affect: The Selected Writings of Silvan S Tomkins* (Cambridge University Press, 1995)
7. M.D. Matsumoto, P. Ekman, Facial expression analysis. Scholarpedia **3**, 4237–4248 (2008)
8. B. Fasel, J. Luettin, Automatic facial expression analysis: a survey. Pattern Recogn. **36**, 259–275 (2003)
9. J.T. Cacioppo, R.E. Petty, Electromyographic activity over facial muscle regions can differentiate the valence and intensity of affective reactions. J. Pers. Soc. Psychol. **50**, 260–268 (1986)
10. U. Dimberg, Facial electromyography and emotional reactions. Psychophysiology **27**, 481–494 (1990)
11. R.W. Buck, V.J. Savin, Communicating of affect through facial expressions in humans. J. Personal. Soc. Psychol. **23**, 362–371 (1972)
12. P. Ekman, G. Roper, G. C. Hager, Deliberate facial movement. Development, 886–891 (1980)
13. P. Ekman, W.V. Friesen, *Facial Action Coding System: A Technique for the Measurement of Facial Movement* (Consulting Psychological Press, Palo Alto, California, 1978)
14. P. Ekman, W.V. Friesen, J.C. Hager, *Facial Action Coding System: The Manual on CD Rom*, ed. by Human Face, Salt Lake City (2002)
15. Y.L. Tian, T. Kanade, J. Cohn, Recognising action units for facial expression analysis, in *Handbook of Face Recognition* (Springer, New York, 2003) pp. 247–275
16. M.S. Bartlett, J.R. Movellan, G.C. Littlewort, B. Braathen, M.G. Frank, T.J. Sejnowski, Towards automatic recognition of spontaneous facial actions, in *What the Face Reveals: Basic and Applied Studies of Spontaneous Expression using the Facial Action Coding System (FACS)*, ed. by P. Ekman, E. Rosenberg, 2nd edn. (Oxford University Press, NY, 2005), pp. 393–412
17. C.P. Sumanthi, T. Santhanam, M. Mahadevi, Automatic facial expression analysis: a survey. Int. J. Comput. Sci. Eng. Survey **3**, 47–59 (2012)
18. A. Sanchez, J.V. Ruiz, A.B. Moreno, A.S. Montemayor, J. Hernandez, J.J. Pantrigo, Differential optical flow applied to automatic facial expression recognition. Neurocomputing **74**, 1272–1282 (2011)

19. A. Psaltis, *Optical Flow for Dynamic Facial Expression Recognition* (Faculty of Science, Utrecht University, Master, 2013)
20. F. Zhang, Y. Gao, J.D. Bakos, Lucas-Kanade optical flow estimation on the TI C66x digital signal processor, in *IEEE High Performance Extreme Computing Conference (HPEC)* (2014), pp. 1–6
21. A.R. Naghsh-Nilchi, M. Roshanzamir, An efficient algorithm for motion detection based facial expression recognition using optical flow. Int. Sch. Sci. Res. Innov. **2**, 2725–2729 (2008)
22. M.Z. Uddin, T.S. Kim, B.C. Song, An optical flow feature-based robust facial expression recognition with HMM from video. Int. J. Innov. Comput. Inf. Control **9**, 1409–1421 (2013)
23. D. Patel, S. Upadhyay, Optical flow measurement using Lucas Kanade method. Int. J. Comput. Appl. **61**, 6–10 (2013)
24. T. Kanade, J. Cohn, Y.L. Tian, Comprehensive database for facial expression analysis, in *Proceedings of the 4th IEEE International Conference on Automatic Face and Gesture Recognition,* (2000), pp. 46–53
25. C. Grossard, O. Grynspan, S. Serret, A. Jouen, K. Bailly, D. Cohen, Serious games to teach social interactions and emotions to individuals with autism spectrum disorders (ASD). Comput. Educ. **113**, 195–211 (2017)
26. P. Werner, A. Al-Hamadi, R. Niese, S. Walter, S. Gruss, H.C. Traue, Towards pain monitoring: facial expression, head pose, a new database, an automatic system and remaining challenges, in *Proceedings of the British Machine Vision Conference* (BMVA Press, 2013), pp. 1–13. https://doi.org/10.5244/c.27.119
27. A.B. Ashraf, S. Lucey, J.F. Cohn, T. Chen, Z. Ambadar, K.M. Prkachin, P.E. Solomon, The painful face—pain expression recognition using active appearance models. Image Vis. Comput. **27**, 1788–1796 (2009)
28. T. Lawson, R. Rogerson, M. Barnacle, A comparison between the cost effectiveness of CCTV and improved street lighting as a means of crime reduction. Comput. Environ. Urban Syst. (2017). https://doi.org/10.1016/j.compenvurbsys.2017.09.008
29. J.H. Yin, S.A. Velastin, A.C. Davies, Image processing techniques for crowd density estimation using a reference image, in *ACCV* (1995), pp. 489–498
30. C.C. Chibelushi, F. Bourel, A.A. Low, Robust facial expression recognition using a state-based model of spatially-localised facial dynamics, in *Proceedings of Fifth IEEE International Conference on Automatic Face and Gesture Recognition* (2002), pp. 106–111
31. I. Kotsia, I. Buciu, I. Pitas, An analysis of facial expression recognition under partial facial image occlusion. Image Vis. Comput. **26**, 1052–1067 (2008)
32. H.K. Ekenel, R. Stiefelhagen, Why is facial occlusion a challenging problem? in *Proceedings of Third International Conference on Advances in Biometrics, ICB 2009, Alghero, Italy, 2–5 June, 2009,* ed. by M. Tistarelli, M.S. Nixon (Springer, Berlin, Heidelberg)
33. M. Dyck, M. Winbeck, S. Leigberg, R.C. Gur, K. Mathiak, Recognition profile of emotions in natural and virtual faces *PLoS ONE* 3:e3628 (2008). https://doi.org/10.1371/journal.pone.0003628
34. C. Anthica, M.K. Venkatesha, B. Suryanarayana Adiga, A survey on facial expression databases. Int. J. Eng. Sci. Technol. **2**, 5158–5174 (2010)
35. M.S. Bartlett, G. Littlewort, M. Frank, C. Lainscsek, I. Fasel, J. Movellan, Fully automatic facial action recognition in spontaneous behavior, in *7th International Conference on Automatic Face and Gesture Recognition, FGR 2006* (2006), pp. 223–230
36. T. Pfister, X. Li, G. Zhao, M. Pietikainen, Differentiating spontaneous from posed facial expressions within a generic facial expression recognition framework, in *ICCV Workshops* (2011), pp. 868–875
37. J.J. Gross, R.W. Levenson, Emotion elicitation using films, Cogn. Emot. 87–108 (1995)
38. L. Sidavong, S. Lal, T. Sztynda, Spontaneous facial expression analysis using optical flow, in *2017 Eleventh International Conference on Sensing Technology (ICST)*, Sydney, NSW, Australia (2017). https://doi.org/10.1109/icsenst.2017.8304482
39. B.D. Lucas, T. Kanade, An iterative image registration technique with an application to stereo vision, in *Proceedings of the 7th Joint Conference on Artificial Intelligence* (Vancouver British Coloumbia, Canada, 1981)

40. R.W. Levensen, The autonomic nervous system and emotion. Emot. Rev. **6**, 110–112 (2014)
41. A. Greco, A. Lanata, L. Citi, N. Vanello, G. Valenza, E.P. Scilingo, Skin admittance measurement for emotion recognition: a study over frequency sweep. Electronics **5**, 46 (2016)
42. P.C. Ellsworth, K. Scherer, Appraisal processes in emotion, in *Handbook of Affective Sciences*, ed. by R.J. Davidson, et al. (Oxford University Press, Oxford New York, 2003)
43. P.C. Ellsworth, Appraisal theory: old and new questions. Emot. Rev. **5**, 125–131 (2013)
44. B. Xia, Which Facial Expressions Can Reveal Your Gender? A Study With 3D Faces, arXiv:1805.00371 (2018)

Heart Sound: Detection and Analytical Approach Towards Diseases

Joyanta Kumar Roy, Tanmay Sinha Roy and Subhas Chandra Mukhopadhyay

Abstract Usually, physicians use an acoustic stethoscope to detect abnormalities in the heart sound and predict abnormal conditions of the human heart. As the frequency range and intensity of heart sound is very low, doctors are facing problems while detecting the cardiac sound and its abnormalities. To eradicate these severe problems, it is required to design and develop an electronic stethoscope which would assist the doctor to analyze heart sound and to detect disease of the heart. Here an acoustic stethoscope along with microphone and preamplifier module is used to increase the amplitude of the input audio signal received by the stethoscope. The soft scope of MATLAB program has also been used for analyzing the continuous set of cardiac sound and to detect its various characteristics like frequency, amplitude, etc. It is aimed to design an electronic stethoscope which would assist the doctors to analyze heart sound and identify a disease condition of the heart, but preliminarily we have achieved to detect different components of it which are lub (s1), dub (s2), s3, s4, etc. Finally, the sound signal received from the heart in the MATLAB program after filtering the noise out of it also has been plotted and analyzed in the frequency domain. As the heart sound is a complex waveform signal, harmonic distribution is used. Amplitude and phase are the two essential parameters. Thus the harmonic distribution of Amplitude and Phase are carried out. Amplitude Distribution of harmonics leads to some crucial characteristics features like RMS Value, Mean Value, Average Energy, Average Power, Mean Squared Error, Spectrogram Analysis, Periodogram Analysis, and Kalman Filtered Response. These features will readily identify and distinguish between Normal heart sound, abnormal heart sound and cardiac murmurs in Matlab programming.

J. K. Roy (✉)
System Advance Technologies Pvt. Ltd, Kolkata, India
e-mail: jkroy.cal51@gmail.com; jkroy@ieee.org

T. S. Roy
Haldia Institute of Technology, Haldia, West Bengal, India

S. C. Mukhopadhyay
School of Engineering, Macquarie University, Sydney, Australia

S. C. Mukhopadhyay et al. (eds.), *Modern Sensing Technologies*,
Smart Sensors, Measurement and Instrumentation 29,
https://doi.org/10.1007/978-3-319-99540-3_7

Introduction

Rouanet, around 140 years ago, attributed the second heart sound to the closure of the aortic and pulmonic valves. Although this explanation has generally been accepted to the present time, several theories have been explained to examine the genesis of the second heart sound. The most tenable to date suggests that closure of the aortic and pulmonic valves initiates the series of events that generate the second heart sound. The main audible components, however, originate from vibrations of the cardiac structures after valve closure. By using high-fidelity, catheter-tipped micro-manometers and echo phono-cardiography, it has been shown that the aortic and pulmonic valves close silently and that coaptation of the aortic valve cusps precedes the onset of the second sound by a few milliseconds. The second heart sound, therefore, originates from after vibrations in the cusps and in the walls, blood columns of the large vessels and their respective ventricles. The energy from these oscillations comes from the fact that sudden deceleration of retrograde blood flow of the blood in the aorta and pulmonary artery when the elastic limits of the tensed valve leaflets are met. This abrupt deceleration sets the whole cardio hemic system into a system of vibratory motion.

To understand the splitting of the second heart sound, knowledge of its relationship to the cardiac cycle is crucial. A_2 and P_2 are the actual sound coincident with the incisurac of the aorta and pulmonary artery pressure curves, respectively, and terminate left and right ventricular ejection periods. Right ventricular ejection begins before left ventricular ejection, has a slightly longer duration, and ends after left ventricular ejection, resulting in P_2 usually occurring after the sound A_2.

The differences between the aortic and pulmonary artery vascular impedance characteristics are also [1] crucial to understand the effects of respiration on the splitting of sound S_2. When the pressure curves of the pulmonary artery and right ventricle are recorded concurrently, the pulmonary arterial plot at the level of the incisura lags behind the right ventricular curve, or "hangs out" after it. The duration of the "hangout interval" is the measure of impedance in the pulmonary artery system. In the highly compliant (low-resistance, high-capacitance) pulmonary vascular bed, the hangout interval may vary from 30 to 120 ms, contributing significantly to the duration of the right ventricular ejection. In the left side of the heart, as impedance is much higher, the hangout interval between the aorta and left ventricular pressure curves is negligible which is less than or equal to 5 ms. The hangout interval, therefore, correlates closely with the impedance of the vascular bed into which blood is being injected. Its duration appears to be inversely proportional to vascular impedance.

Alterations in the impedance characteristics of the pulmonary vascular bed and the right-sided hangout interval are responsible for observed abnormalities in S_2. In a general physiologic setting, inspiration lowers the impedance in the pulmonary circuit, prolongs the hangout interval and delays pulmonic valve closure, resulting in the audible splitting of the sound A_2 and P_2. On expiration, the reverse occurs: pulmonic valve closure happens [2] earlier, and the A_2–P_2 interval is separated by less than 30 ms and may sound only once. The pulmonary circulation has a much

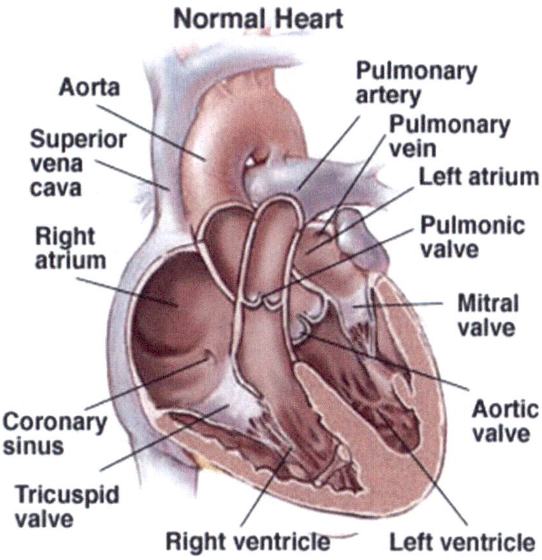

Fig. 1 Cardiovascular anatomy

lower impedance than the systemic circulation; therefore, the blood flow through the pulmonic valve takes longer time than blood flow through the aortic valve. The inspiratory split widens mainly because of delay in the pulmonic component (Fig. 1).

It was believed traditionally that an inspiratory drop in intrathoracic pressure favored the greater venous return to the right ventricle, pooling of blood in the lungs, and decreased return to the left ventricle. The increase in right ventricular blood volume prolonged right-sided ejection time and delayed P_2; the decrease in left ventricular volume reduced left-sided ejection time and caused A_2 to occur earlier than usual. The delayed occurrence of P_2 and early presence of A_2 associated with the inspiration, however, are best understood as an interplay between the [3] changes in the pulmonary vascular impedance and changes in systemic and pulmonary venous return. The net effect is that right ventricular ejection is prolonged, left ventricular ejection is shortened, and the A_2–P_2 interval widens during inspiration.

Clinical Significance

The clinical evaluation of the second heart sound has been called the “key to auscultation of the heart” [4]. It involves an assessment of splitting and a determination of the relative intensities of A_2 and P_2. Usually, the aortic closure sound (A_2) occurs before the pulmonic closure sound (P_2), and the interval between the two (splitting) widens on inspiration and narrows on expiration. With quiet respiration, A_2 will generally precede P_2 by 0.02 to 0.08 s (mean, 0.03 to 0.04 s) with inspiration. In younger subjects, inspiratory splitting averages 0.04 to 0.05 s during quiet respiration. With expiration, A_2 and P_2 may be superimposed and are rarely split as much as 0.04 s. If greater than 0.04 s on expiration splits the second sound, it is usually abnormal. Therefore, the presence of audible splitting during expiration (i.e., the ability to hear two distinct sounds during expiration) is of greater significance at the bedside in identifying underlying cardiac pathology that is the absolute inspiratory increase in the A_2–P_2 interval.

The respiratory variation of the second heart sound can be categorized as follows:

1. Regular (physiologic) splitting.
2. Persistent (audible expiratory) splitting, with normal respiratory variation.
3. Persistent splitting without respiratory modification (fixed splitting).
4. Reversed (paradoxical) splitting.

Physiologic splitting is demonstrated during inspiration in healthy individuals since the splitting interval widens primarily due to the delayed P_2. During expiration, the A_2–P_2 range is so narrow that only a single sound is usually heard.

Persistent (audible) expiratory splitting suggests an audible expiratory interval of at least 30–40 ms between the two sounds. Persistent splitting that is audible during both respiratory phases with appropriate inspirational and aspirational changes (i.e., further increase of the A_2–P_2 interval with inspiration) may occur in the recumbent position in healthy children, teenagers, and young adults. If these individuals sit, stand, or perform a Valsalva maneuver, however, the second sound often becomes single on expiration. In almost all the patients with heart disease and audible expiratory splitting in the recumbent position, expiratory [5] splitting persists when the patient is examined in the sitting or standing position. Thus, the finding of distinct expiratory splitting in both the supine and upright positions is a susceptible screening test for heart disease. Right bundle branch block (RBBB) is the most common cause of the persistence of audible expiratory splitting on standing.

Other causes of persistent expiratory splitting on standing may be due to either a delay in pulmonic valve closure or to the early closing of the aortic valve. A delay in the occurrence of P_2 is mainly attributed to the following events:

1. Delayed electrical activation of the right ventricle (e.g., left ventricular ectopic or paced beats, Wolff-Parkinson-White syndrome, and RBBB).
2. The decreased impedance [6] of the pulmonary vascular bed (e.g., atrial septal defect, partial anomalous pulmonary venous return, and idiopathic dilatation of the pulmonary artery).
3. Right ventricular pressure overload lesions (e.g., pulmonary hypertension with right heart failure, moderate to severe valvular pulmonic stenosis, and acute massive pulmonary embolus).

An early A_2 may occur in patients with decreased resistance to left ventricular outflow (e.g., mitral regurgitation or constrictive pericarditis). Moderately large ventricular septal defects may also cause extensive splitting of the second sound, but the aortic component is usually difficult to hear because of the loud holosystolic murmur.

Expiratory splitting of S_2 may occur in patients with severe congestive heart failure. The expiratory splitting usually disappears after satisfactory therapy of the heart failure. The high prevalence of expiratory splitting of S_2 in cardiomyopathy may be explained by a combination of low cardiac output, mitral regurgitation, pulmonary hypertension, right heart failure, and bundle branch block.

Fixed splitting denotes the absence of significant variation of the splitting interval with respiration, such that the separation of A_2, and P_2, remains unchanged during inspiration and expiration. An atrial septal defect, with either standard or high pulmonary vascular resistance, is the classic example of the fixed splitting of the second sound. The audible expiratory splitting in these patients is primarily a reflection of changes in the pulmonary vascular bed rather than selective volume overload of the right ventricle prolonging right ventricular systole [7]. The fixed nature of the split is due to the approximately equal inspiratory delay of the aortic and pulmonic components, indicating that the two ventricles share a common venous reservoir. Respiratory splitting of the second sound immediately returns to standard following surgical repair of an atrial septal defect, [8] although the pulmonic closure sound may remain delayed for weeks or months.

Severe right heart failure can lead to a relatively fixed split. It occurs because the right ventricle fails to respond to the increased blood volume produced by inspiration and because the lungs are so congested that impedance to forward blood flow from the right ventricle barely falls during inspiration. In anomalous pulmonary venous return without atrial septal defect, fixed splitting is not usually seen despite the simultaneous inspiratory delay in aortic and pulmonic closure.

The Valsalva maneuver [9] may be used to exaggerate the effect of respiration and obtain more explicit separation of the two components of the second sound. Patients with atrial septal defects show continuous splitting during the strain phase, and upon release, the interval between the components increases by less than 0.02 s. In healthy patients, however, splitting is exaggerated during the release phase of the Valsalva maneuver. Variation of the cardiac cycle length may also be used to evaluate splitting of S_2. During the longer cardiac cycle, patients with the atrial septal defect may show greater splitting as a result of increased atrial shunting and greater disparity between stroke volume of the two ventricles. In healthy subjects, there is no tendency to widen the splitting with longer cardiac cycles.

Pulmonary artery hypertension causes variable effects on the splitting of the second sound. Patients with ventricular septal defect who develop pulmonary hypertension may no longer have the splitting of S_2. Patients with an atrial septal defect and associated pulmonary hypertension maintain an extensive and fixed split of S_2. Splitting is shortened (less than 30 ms), but remains physiologic, in patients with patent ductus arteriosus who develop pulmonary hypertension.

Paradoxical or reversed splitting [10] is the result of a delay in the aortic closure sound. Therefore P_2 precedes A_2, and splitting is maximal on expiration and minimal or absent on inspiration. Identification of the reversed order of valve closure may be possible by judging the intensity and transmission of each component of the second sound. Often, however, the pulmonic component [11] is as loud as the aortic component because of pulmonary artery hypertension secondary to left ventricular failure. The paradoxical narrowing or disappearance of the split on inspiration is a necessary criterion for diagnosing reversed splitting by auscultation.

Paradoxical splitting [12] always indicates significant underlying cardiovascular disease and is usually due to prolongation of left ventricular activation or prolonged left ventricular emptying. The most common cause of paradoxical splitting of the second sound is left bundle branch block. Obstruction to left ventricular outflow of sufficient severity to delay aortic valve closure may also cause paradoxical splitting. In the context of aortic stenosis, such an auscultatory finding implies severe obstruction. Paradoxical splitting [13], however, occurs more commonly with hypertrophic cardiomyopathy than with aortic stenosis. Paradoxical splitting of the second sound may occur during the first few days after an acute myocardial infarction or secondary to severe left ventricular dysfunction.

A mistaken diagnosis of the abnormal splitting of the second sound must be avoided. A late systolic click of mitral valve prolapse, the opening snap (OS) of mitral stenosis, a third heart sound (S_3), or a pericardial knock may be incorrectly thought to represent the fixed splitting of the second sound. A systolic knock may vary its location in systole with specific maneuvers that change the shape of the left ventricle. The best way to differentiate between an A_2–P_2 from an A_2–OS is to have the patient stand up. The A_2–P_2 interval remains the same or narrows, whereas the A_2–OS interval widens. The third heart sound, which forms the S_2–S_3 complex, is lower in frequency range than S_2, is best [14] heard at the apex, is usually not understood at the basal auscultatory area, and occurs 0.12–0.16 s after A_2. The pericardial knock is a third heart sound that is slightly higher pitched and earlier than the usual S_3 and is also best heard at the apex.

The second heart sound can remain single throughout the respiratory cycle due to either absence of one component or to the simultaneous occurrence of the two components. Since the pulmonary vascular impedance increases with age, many normal patients over age 50 have a single S_2 or at most a narrow physiologic split on inspiration because P_2 occurs early. A unique second sound, however, is usually due to inability to auscultate a relatively soft pulmonic component. Such failure is rare in healthy infants, children, and young adults and is uncommon even in older persons under good auscultatory conditions using a rigid stethoscopic diaphragm [15].

Hyperinflation of the lungs is perhaps the most common cause of inability to hear the pulmonic closure sound. All the conditions are causing paradoxical splitting that delay A_2 may produce a single second sound when the splitting interval becomes less than 0.02 s. Inaudibility of P_2 due to an actual decrease in its intensity is relatively rare, however, and suggests tetralogy of Fallot or pulmonary atresia. The pulmonic component may be inaudible in chronic right ventricular failure, or the aortic component may be masked by the systolic murmur in patients with aortic stenosis.

The pulmonary closure [16] is completely fused with aortic closure throughout the respiratory cycle only in Eisenmenger's syndrome with a large ventricular septal defect or in cases of single ventricle, where the duration of right and left ventricular systole are virtually equal. The second sound may also be unique in a variety of congenital heart defects (e.g., truncus arteriosus, tricuspid atresia, hypoplastic left heart syndrome, transposition of the great arteries, and, occasionally, corrected transposition of the great arteries).

The loudness of each component [17] of the second heart sound is proportional to the respective pressures in the aorta and pulmonary artery at the onset of diastole. Dilatation of the aorta or pulmonary artery may also cause accentuation of the aortic and pulmonic components, respectively. The aortic component usually is of higher intensity than the pulmonic component. The aortic component, therefore, radiates widely over the chest, whereas the pulmonic component is heard mainly in the second left ICS with some radiation down the left sternal border. The higher radiation of the aortic component is probably due to the higher pressure in the aorta compared to that in the pulmonary artery. At any given level of pressure, however, the pulmonic component will be proportionately louder than the aortic component because of the close proximity of the pulmonic valve and the pulmonary artery to the chest wall. These considerations account for the relative loudness of P_2 in young patients in whom the pulmonary arteries are quite close to the chest wall. They also account for the decreased intensity of both components of the second sound in patients with emphysema in whom both arteries are displaced from the chest wall.

The pulmonic component is considered to be abnormally loud in a subject over age 20 if it is higher than the aortic component in the second left ICS or if it is audible at the cardiac apex. It may be due either to pulmonary artery hypertension or right ventricular dilatation, with part of the right ventricle assuming the position usually occupied by the left ventricle. A split second sound at the apex is, therefore, definitely abnormal. The loud P_2 commonly heard at the apex in patients with an atrial septal defect is probably due to a dilated right ventricle encroaching upon the cardiac apex.

Decreased intensity of either component of the second sound may be due to a stiff semilunar valve, reduced pressure beyond the semilunar valve [18], or deformity of the chest wall or lung. The reduced intensity of P_2 is most common in patients with chronic obstructive lung disease or valvular pulmonic stenosis. A decreased intensity [10] of A_2 is most common in patients with valvular aortic stenosis (Fig. 2).

Detection of Heart Sound

Stethoscope plays a vital role in the detection of the heart sound. Here we are going to know the history and evolution of stethoscope through ages to see the story behind detection of cardiac sounds.

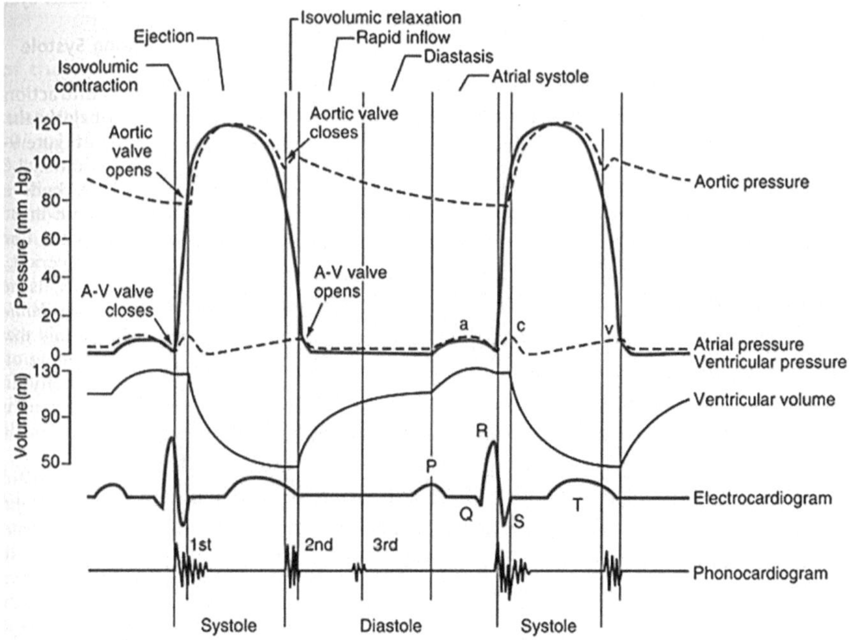

Fig. 2 Relationship of heart sound with cardiovascular system

History and Evolution of Stethoscope

The word stethoscope is derived from the two Greek words, stethos (chest) and scopos (examination). Apart from listening to the heart and chest sounds, it is also used to hear bowel sounds and blood flow noises in arteries and veins.

In the early 1800s, and before the development of the stethoscope, physicians would often perform physical examinations using techniques such as percussion and immediate auscultation. In immediate auscultation, physicians placed their ear directly on the patient to observe internal sounds.

This technique suffered from several drawbacks, the foremost being that it required physical contact between the physician and the patient and proper placement of the ear. Besides, the sounds observed by the physician were not amplified in any way, creating the possibility of missing key sounds that might indicate potential illness. Finally, the act of performing immediate auscultation could be awkward for both the physician and patient.

The invention of the first stethoscope: The stethoscope was invented in France in 1816 by René Laennec at the Necker-Enfants Malades Hospital in Paris. It comprised a wooden tube and was monaural. Laennec invented the stethoscope because he was uncomfortable placing his ear on women's chests to hear heart sounds. His device

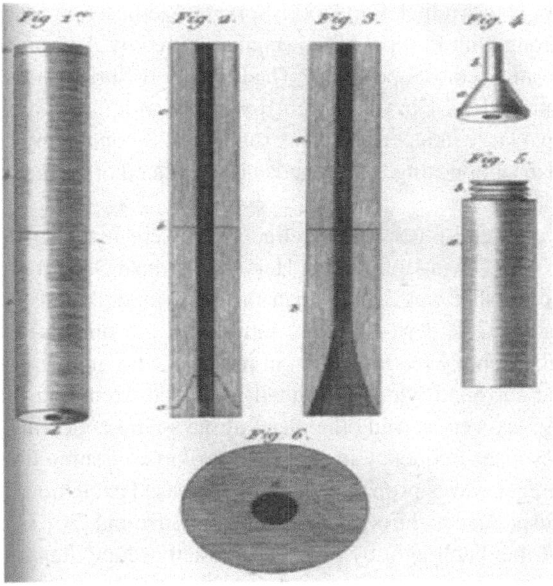

Fig. 3 Design of the first stethoscope

was similar to the common ear trumpet, a historical form of hearing aid; indeed, his invention was almost indistinguishable in structure and function from the trumpet, which was commonly called a "microphone". It was named as a stethoscope (Fig. 3).

Evolution of Stethoscope

In 1840, Golding Bird described a stethoscope he had been using with a flexible tube. The bird [19] was the first to publish a description of such a stethoscope, but he noted in his paper the prior existence of an earlier design (which he thought was of little utility) which he described as the snake ear trumpet. Bird's stethoscope had a single earpiece.

In 1851, Irish physician Arthur Leared invented a binaural stethoscope, and in 1852 George Philip Cammann perfected the design of the stethoscope instrument (that used both ears) for commercial production, which has become the standard ever since. Cammann also wrote a significant treatise on diagnosis by auscultation, which the refined binaural stethoscope made possible. By 1873, there were descriptions of a differential stethoscope that could connect to slightly different locations to create a slight stereo effect, though this did not become a standard tool in clinical practice.

Rappaport and Sprague designed a new stethoscope in the 1940s, which became the standard by which other stethoscopes are measured, consisting of two sides, one of which is used for the respiratory system, the other for the cardiovascular system. Hewlett-Packard later made the Rappaport-Sprague.

The original Rappaport-Sprague stethoscope was finally abandoned ca. 2004, along with Philips' brand (manufactured by Andromed, of Montreal, Canada) electronic stethoscope model. The Rappaport-Sprague model stethoscope was heavy and short [18–24 in (46–61 cm)] with an antiquated appearance recognizable by their two large independent latex rubber tubes connecting an exposed-leaf-spring-joined-pair of opposing "f"-shaped chrome-plated brass binaural ear tubes with a dual-head chest piece.

Several other minor refinements were made to stethoscopes, until in the early 1960s David Littmann, a Harvard Medical School professor, created a new stethoscope that was lighter than previous models and had improved acoustics. In the late 1970s, 3 M-Littmann introduced the tunable diaphragm: a very hard (G-10) glass-epoxy resin diaphragm member with an over-molded silicone flexible acoustic surround which permitted increased excursion of the diaphragm member in a "z"-axis concerning the plane of the sound collecting area. The left shift to a lower resonant frequency increases the volume of some low-frequency sounds due to the longer waves propagated by the increased excursion of the hard diaphragm member suspended in the concentric acoustic surround [20]. Conversely, restricting excursion of the diaphragm by pressing the stethoscope diaphragm surface firmly against the anatomical area overlying the physiological sounds of interest, the acoustic surround could also be used to dampen excursion of the diaphragm in response to "z"-axis pressure against a concentric fret. It raises the frequency bias by narrowing the wavelength to auscultate a higher range of physiological sounds.

In 1999, Richard Deslauriers patented the first external noise-reducing stethoscope, the DRG Puretone. It featured two parallel lumens containing two steel coils which dissipated infiltrating sound as inaudible heat energy. The steel coil "insulation" added.30 lb to each stethoscope. In 2005, DRG's diagnostics division was acquired by TRIMLINE Medical Products.

In 2015, Dr. Tarek Loubani announced an open-source 3D-printed stethoscope based on the 1960s-era *Littmann Cardiology* 3 stethoscope, which is out of patent. The 3D-printed equivalent is nearly an order of magnitude more affordable than the aforementioned non-3D-printed stethoscope and is intended to make the medical device more accessible to obtain, particularly in developing countries [21] (Fig. 4).

Golding Bird's instrument is on the left. The device on the right was current around 1855 when this image was first published [22].

Electronic Stethoscope: An electronic stethoscope (or **stethophone**) overcomes the low sound levels by electronically amplifying body sounds. However, it suffers from stethoscope contact artifacts and frequency response of the microphone, pre-amplifier, power amplifier, and speakers, which limit the overall utility. The Electronic stethoscopes require the conversion of acoustic sound waves to electrical signals. After that, the amplified signal is processed for optimal listening. Another method used in Welch-Allyn's Meditron stethoscope, made by placement of a piezoelectric crystal at the head of a metal shaft. The bottom of the metal shaft makes contact with a diaphragm. 3 M also uses a piezo-electric crystal placed within foam behind a thick rubber-like diaphragm. The Thinklabs' Rhythm 32 uses

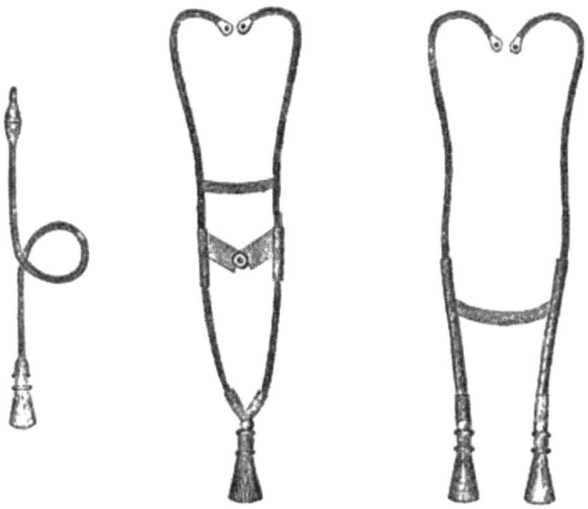

Fig. 4 Early flexible tube stethoscopes

an Electromagnetic Diaphragm with a conductive inner surface to form a capacitive sensor. This diaphragm responds to sound waves, with changes in an electric field replacing changes in air pressure. The Eko Core enables wireless transmission of heart seems to a smartphone or tablet.

Because the generated sounds are transmitted electronically, an electronic stethoscope can be viewed as a wireless device, a recording device, and can provide noise reduction, signal enhancement, and both visual and audio output. Around 2001, Stethographics introduced PC-based software which enabled a phonocardiograph, graphic representation of cardiologic and pulmonologic sounds to be generated, and interpreted according to related algorithms. All of these characteristics are helpful for telemedicine (remote diagnosis) and teaching [23].

3-D Printed Stethoscope: A 3D-printed stethoscope is an open-source medical device meant for auscultation and manufactured via means of 3D printing. The 3D stethoscope was developed by Dr. Tarek Loubani and a team of medical and technology specialists. The 3D-stethoscope was developed as part of the Glia project, and its design is open source from the outset. The stethoscope gained widespread media coverage in summer 2015.

The need for a 3D-stethoscope was borne out of a lack of stethoscopes and other vital medical equipment because of the blockade of the Gaza Strip, where Loubani, a Palestinian-Canadian, worked as an emergency physician during the 2012 conflict in Gaza. The 1960s-era *Littmann Cardiology* 3 stethoscope became the basis for the 3D-printed stethoscope developed by Loubani [23] (Figs. 5 and 6).

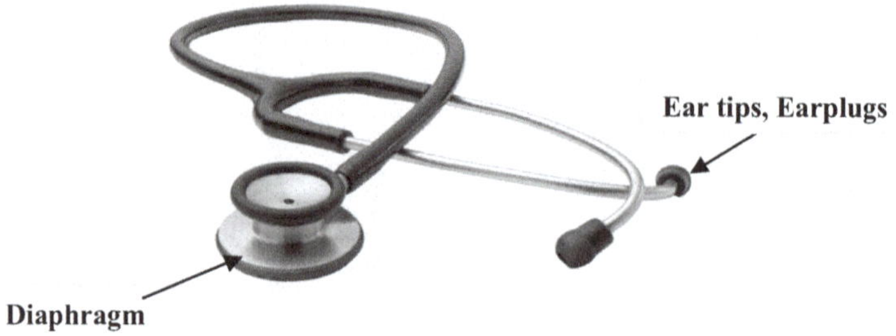

Fig. 5 Clinical acoustic stethoscope

Fig. 6 Modern acoustic stethoscope

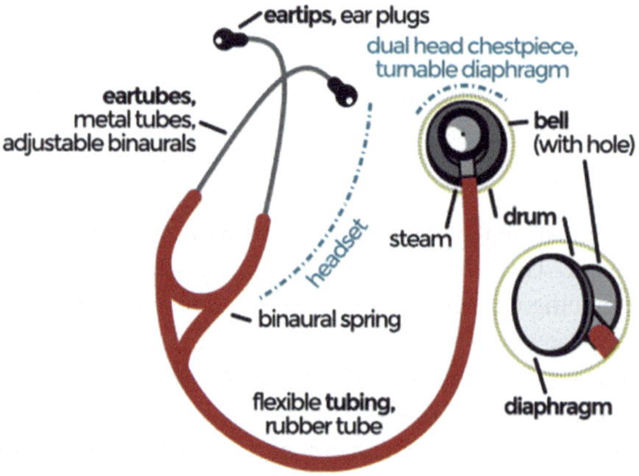

Introduction to Digital Stethoscope

A stethoscope, whether acoustic or digital, is used mainly to listen to heart and lung sounds in the body as an aid for diagnosis. Listening or auscultation has been done with acoustic [3] stethoscopes for almost two hundred years; recently, electronic digital stethoscopes have been developed over the years to withstand the high chance of dealing with cardiac related problems.

The goal of a necessary digital stethoscope is to have it retain the look and feel of an acoustic stethoscope but to improve listening performance. In addition to that, high-end digital stethoscopes offer sophisticated capabilities such as audio recording and playback. They also provide data to visually chart results by connecting to an off-instrument display such as a computer monitor. This advanced functionality increases the physician's diagnostic capability maintaining the existing acoustic stethoscope form.

Audio Signal Path

The essential components of a digital stethoscope are the sound transducer, the audio codec electronics, and the speakers. The sound transducer, which converts sound into an analog voltage, is the most critical piece in the chain. It determines the diagnostic quality of the digital stethoscope and ensures a familiar user experience to those accustomed to acoustic stethoscopes.

The analog voltage needs to be conditioned and then converted into a digital form by using a sound analog-to-digital converter (ADC) or audio codec. Some digital stethoscopes have inbuilt noise cancellation device that requires a secondary sound transducer or microphone to record the ambient noise so that it can be removed digitally. In this approach, two audio ADCs are needed (Fig. 7).

Once in the digital domain, a microcontroller unit (MCU) or digital signal processor (DSP) performs signal processing, including ambient noise reduction and filtering, to limit the bandwidth to the range for cardiac or pulmonary listening. The processed digital signal is then converted back to analog by an audio digital-to-analog converter (DAC) or audio codec [24].

A headphone or speaker amplifier conditions the audio signal before outputting to a speaker. A single speaker can be used below where the stethoscope tube bifurcates, with the amplified sound traveling through the binaural tubes to the ears. Alternatively, two speakers can be used, with one speaker at the end of each earpiece. The frequency response [25] of the speaker is similar to that of a bass speaker because of the low-frequency sound production needed. Depending upon the implementation, one or two speaker amplifiers are used.

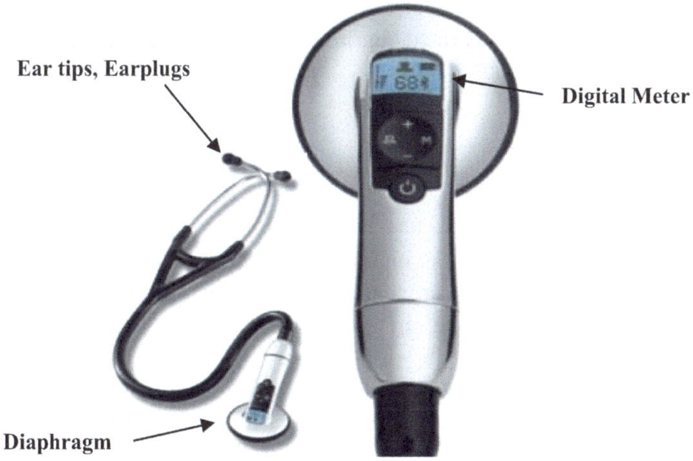

Fig. 7 A digital stethoscope

A stethoscope must be most sensitive to cardiac sound in the 20–400 Hz range and pulmonary sound in the 100–1200 Hz range. Note that the frequency ranges vary by manufacturer, and the DSP algorithms filter out sound beyond these optimal ranges.

Data Storage and Transfer

Once the captured sound is converted to an analog voltage, it can be sent out through an audio jack and played back on either a computer or through the digital stethoscope. The captured sound can also be manipulated digitally. It can be stored in the stethoscope using internal or removable nonvolatile (NV) memory like EEPROM or flash and then played back through the stethoscope's speakers, or it can be transferred to a computer for further analysis. Adding a real-time clock (RTC) facilitates tagging the recording with time and date. The sound is commonly assigned with a wired interface, such as USB, or with a wireless interface like Bluetooth or other proprietary wireless interfaces which are widely used these days.

Display and Backlighting

Some digital stethoscopes [23] have a small, simple display due to the limited space available; others have only buttons and LED indicators. Backlighting for the display is required because the ambient lighting during the procedure is often at a low level. The small display requires just one or two white light-emitting diodes (WLEDs) controlled by an LED driver or an electroluminescent (EL) panel controlled by an EL driver. Most of the user-interface buttons can be eliminated by adding a touchscreen display and controller.

Power Management

Most digital stethoscopes use either one or two AAA 1.5 V primary batteries. This design requires a step-up, or boost, switching regulator to increase the voltage to 3.0 V or 5.0 V, depending on the circuitry utilized these days for cardiac-related problems.

If a single 1.5 V battery is installed, the switching regulator will probably be on all the time, making low quiescent current a vital factor for long battery life. The longer the battery life, the more convenient the digital stethoscope is to use and the closer the experience will be to use an acoustic stethoscope [17].

When using two 1.5 V batteries in series, the switching regulator can be left on all the time or shut down when not in use. If the circuit operates from 3.6 V down to 1.8 V, then a switching regulator may not be needed. The cost will be reduced and space saved. A low-battery warning is required so that a patient's examination need not be interrupted to replace the battery.

Battery Management

Rechargeable batteries can be used; the best choice is a single-cell Li+ battery. If a rechargeable battery is used, a battery charger is required either in the digital stethoscope or a charging cradle. A fuel gauge is the best solution to determine the remaining battery life accurately. If the battery is removable, then authentication is also required for safety and aftermarket management.

The Bluetooth word mark and logos are registered trademarks owned by Bluetooth SIG, Inc., and any use of such marks by Maxim is under license.

Importance of Heart Sound Analysis

Physicians use the stethoscope for hearing heart sounds. They generally try to analyze and understand congenital heart disease by analyzing the heart sound and abnormalities in it. Naturally, the frequency and repetition of sound are significant in detecting the problems of heart disease. In case of an ordinary stethoscope, there is another problem of detecting heart sound that is the limitation of hearing music by human ears. So we need to digitalize the sound signal and analyze the sound correctly using a modern tool. Then anyone can detect precisely the wellness of heart and its various components.

Process: To develop a digital algorithm that detects first and second heart sounds, defines the systole and diastole, and characterizes the systolic murmur. Heart sounds [26] were recorded in 300 children with a cardiac murmur, using an electronic stethoscope. A Digital algorithm was developed for detection of first and second heart sounds. R-waves and T-waves in the electrocardiography were used as references for detection. The sound signal analysis was carried out using the short-time Fourier transform. The first heart sound detection rate, with reference to the R-wave, was 100% within 0.05–0.2R-R interval. The second heart sound detection rate between the end of the T-wave and the 0.6R-R interval was 97%. The systolic and diastolic phases of the cardiac cycle could be identified. Because of the overlap between heart sounds and murmur a systolic segment between the first and second heart sounds (20–70%) was selected for murmur analysis. The maximum intensity of the systolic murmur, its average frequency, and the mean spectral power was quantified. The frequency at the point with the highest sound intensity in the spectrum and its time from the first heart sound, the highest frequency, and frequency range were also determined. This method will serve as the foundation for computer-based detection of heart sounds and the characterization of cardiac murmurs [24].

Different Work Done

Renowned researchers have done previously many research works on advancement and development of stethoscope. While working on our project, we have gone through several papers and practices to gain and utilize the knowledge in the project.

1. Department of Electro-medical Engineering, Health Polytechnic Ministry of Health Surabaya, Indonesia has done a detailed research on analysis and detection of heart sound using wavelet decomposition technique. A paper in the International Journal of computer networks and communications security was published on their work. The paper was published in the year 2013, December [17].
2. In 2013 another renowned work was done by Kuldeep Singh and Preeti Abrol on Design and Development of a Digital Stethoscope for detection of Cardiac Murmur. They developed an A.V.R. based digital stethoscope to capture heart sounds and diagnose them with the help of display and analytic tools. The proposed design of the system includes an ATmega16 microcontroller, graphic LCD, memory, and MATLAB interface [25].
3. In September 2013, Mandeep Singh and Amandeep Cheema published a paper on Heart Sounds Classification using Feature Extraction of Phonocardiography Signal. They have analyzed a set of P.C.G. signal using MATLAB. They have taken the heart sound signal as the sample and analyzed in both frequency and time domain [27].
4. Anju and Sanjay Kumar have also done research work on Detection of Cardiac Murmur. They have used FPGA system to detect cardiac murmur and Artificial Neural Network to differentiate between normal and pathogenic cardiac murmur [28].
5. S. M. Debbal's work was on computerized heart sound analysis. He analyzed the sound signal using digital signal processing and Fourier transform method [11].
6. Ms. D. D. Kadam Patil and Mr. R. K. Shastri have researched and developed wireless electronic stethoscope using based on ZIGBEE transmission method. They have transmitted analyzed data from one place to another using ZIGBEE protocol [6].
7. L. B. Dahl, P. Hasvold, E. Arild and T. Hasvold have done their work on, recording of heart murmur by using a sensor based electronic stethoscope and e-mail it for remote assessment [29].
8. Bill Dillard and Wah W. Myint designed an electronic stethoscope with diagnosis capability. They have performed time-frequency analysis in every systolic and diastolic murmur [9].

Disease Categories

Electrical: Abnormal heart sounds are caused by problems with the electrical system that regulates the steady heartbeat. The heart rate may be too slow or too fast; it may stay steady or become chaotic (irregular and disorganized). Some arrhythmias are very dangerous and cause sudden cardiac death, while others may be bothersome but not life-threatening.

Circulatory: High Blood Pressure [15] and coronary artery disease (blockage in the pipes of the heart) are the main causes of blood vessel disorders. The results, such as stroke or heart attack, can be devastating. Fortunately, there are many treatment options.

Structural: Heart muscle disease (cardiomyopathy) and congenital abnormalities (problems present from birth) are two problems that can damage the heart muscle or valves.

The **third** heart sound (S_3) is a low-pitched, early diastolic sound audible during the rapid entry of blood flow from the atrium to the ventricle. S_3 can be physiologically present in patients younger than 40 years. These patients often have a thin chest wall to permit the easy transmission of S_3. In the presence of heart failure, S_3 is a bad prognostic sign. Conditions associated with pathological S_3 include the following:

- Systolic and/or diastolic ventricular dysfunction
- Ischemic heart disease
- Hyperkinetic states—Anemia, fever, pregnancy, thyrotoxicosis, AV fistula
- MR or TR
- Chronic AR with systolic dysfunction
- Systemic and pulmonary hypertension
- Acute aortic regurgitation
- Volume overload—Renal failure (Fig. 8).

The **fourth** heart sound (S_4) is a late diastolic sound that corresponds to late ventricular filling through the active atrial contraction. It is a low-intensity sound heard best with the bell of the stethoscope. Some of the conditions associated with S_4 include the following:

- Ventricular hypertrophy—LV hypertrophy (systemic hypertension, hypertrophic cardiomyopathy, AS); RV hypertrophy (pulmonary hypertension, pulmonary stenosis [PS])

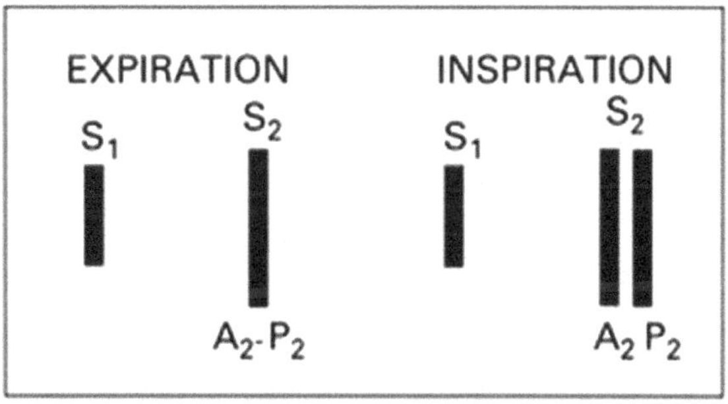

Fig. 8 Respiratory system diagram

Fig. 9 .

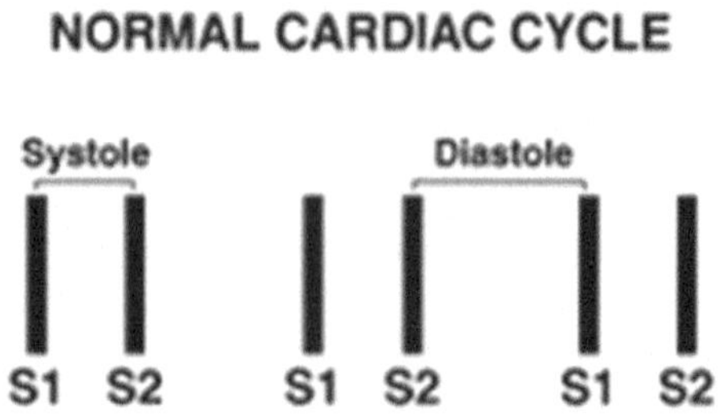

- Ischemic heart disease—Acute MI, angina
- A ventricular aneurysm
- Hyperkinetic states that cause forceful atrial contraction (Fig. 9).

S4 is a low-frequency gallop sound that results from a forceful atrial contraction during presystole that ejects blood into a ventricle which cannot expand further. S4 occurs ~90 ms before S1. The **fourth heart sound** or $\mathbf{S_4}$ is an extra heart sound that occurs during late diastole, immediately before the usual two "lub-dub" heart sounds (S_1 and S_2). It occurs just after atrial contraction and immediately before the systolic S_1 and is caused by the atria contracting forcefully to overcome an abnormally stiff or hypertrophic ventricle.

In abnormal heart, the additional sounds, called Murmurs are heard between the normal heart sounds. Murmurs are generally caused either by improper opening of the valves or by regurgitation. Heart murmurs are generated by the turbulent flow [30] of blood, which may occur inside or outside the heart. Murmurs may be physiological (benign) or pathological (abnormal). Abnormal murmurs can be caused by stenosis restricting the opening of a heart valve, resulting in turbulence as blood flows through it. Abnormal murmurs may also occur with valvular insufficiency (regurgitation), which allows backflow of blood when the incompetent valve closes with only partial effectiveness. Different murmurs are audible in different parts of the cardiac cycle, depending on the cause of the murmur.

Regurgitation through the mitral valve and Aortic valve is by far the most commonly heard murmur, producing a pansystolic/holosystolic murmur which is sometimes reasonably loud to a practiced ear, even though the volume of regurgitant blood flow may be quite small.

Stenosis of the Aortic valve and Mitral valve is typically the next most common heart murmur, a systolic ejection murmur (Fig. 10).

Fig. 10 .

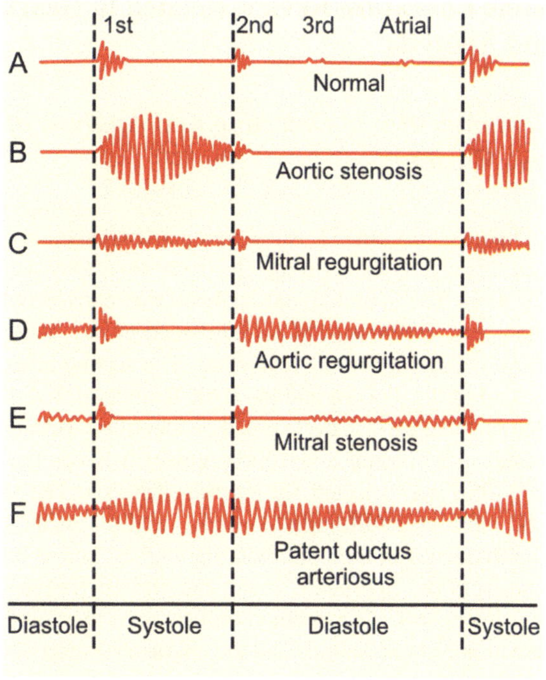

Phonocardiograms from normal and abnormal heart sounds

Multi-Signal Correlations: These are the parameters of the cardiac cycle associated with a healthy human considered in Phonocardiogram signal analysis.

- Ventricular pressure
- Aortic pressure
- Atrial pressure
- Aortic blood flow
- Venous pulse
- Electrocardiogram
- Phonocardiogram

Heart Diseases: The abnormal heart sounds in a Phonocardiogram generated due to following heart problems.

- **Heart failure**
- Coronary artery disease
- Hypertension
- Cardiomyopathy
- Valve defects
- Arrhythmia

Some Literature Review Related to Heart Sound Detection and Analysis

Computer-Based Detection and Analysis of Heart Sound and Murmur

A Digital algorithm [31] was developed for detection of first and second heart sounds. R-waves and T-waves in the electrocardiography were used as references for identification. The sound signal analysis was carried out using the short-time Fourier transform. The first heart sound detection rate, with reference to the R-wave, was 100% within 0.05–0.2R-R interval. The second heart sound detection rate between the end of the T-wave and the 0.6R-R interval was 97%. The systolic and diastolic phases of the cardiac cycle could be identified. Because of the overlap between heart sounds and murmur a systolic segment between the first and second heart sounds (20–70%) was selected for murmur analysis. The maximum intensity of the systolic murmur, its average frequency, and the mean spectral power was quantified. The frequency at the point with the highest sound intensity in the spectrum and its time from the first heart sound, the highest frequency, and frequency range were also determined. This method will serve as the foundation for computer-based [31] detection of heart sounds and the characterization of cardiac murmurs.

A Signal Processing Module for the Analysis of Heart Sounds and Heart Murmurs

A Signal Processing Module (SPM) for the computer-aided analysis of heart sounds has been developed. The module reveals essential information of cardiovascular disorders and can assist the general physician to come up with more accurate and reliable diagnosis at early stages. It can overcome the deficiency of expert doctors in rural as well as urban clinics and hospitals. The module has five main blocks: Data Acquisition and Pre-processing, Segmentation, Feature Extraction, Murmur Detection and Murmur Classification. The heart sounds are first acquired using an electronic stethoscope which has the capability of transferring these signals to the nearby workstation using wireless media. Then the signals are segmented into individual cycles as well as individual components using the spectral analysis of heart without using any reference signal like ECG. Then the features are extracted from the individual components using Spectrogram and are used as an input to a MLP (Multiple Layer Perceptron) Neural Network that is trained to detect the presence of heart murmurs. Once the murmur is discovered, they are classified into seven classes depending on their timing within the cardiac cycle using Smoothed Pseudo

Wigner-Ville distribution. The module has been tested with real heart sounds from 40 patients and has proved to be quite efficient and robust while dealing with a large variety of pathological conditions.

Heart Sounds Classification Using Feature Extraction of Phonocardiography Signal

The Phonocardiogram (PCG) signals contain beneficial information about the condition of the heart. By analyzing these signals, early detection and diagnosis of heart diseases can be done. It is also instrumental in the case of infants, where ECG recording and other techniques are challenging to implement. In this paper, [32] a classification method is proposed to classify normal and abnormal heart sound signals having murmurs without getting into the cumbersome process of segmenting fundamental heart sounds (FHS) using Electrocardiogram (ECG) gating. The proposed algorithm can be easily implemented on latest electronic stethoscopes, and therefore the unnecessary ECG can be avoided.

Heart disease is a significant health problem and a leading cause of fatality throughout the world. The treatment can be more comfortable, efficient and economical if the condition is detected early. So it would be very beneficial to detect heart diseases at an early stage. Cardiac disorders can be detected efficiently and economically using auscultation as it requires minimal equipment. Sometimes this is the only available option for diagnosis as in case of primary health care centers, where other high-end instruments for diagnosis are unavailable and also in case of infants where other techniques like ECG are challenging to implement. Conventional auscultation requires extensive training and experience and storage of records for follow-ups and future references are not possible [3]. It is the driving force for this study to move towards automatic auscultation using electronic stethoscopes. Heart sounds comprise four components among which the dominant two S1 and S2 (see Fig. 1) are commonly known as FHS and are caused by the closing of valves. After the vibration of ventricular valves causes that S3 as a result of first rapid filling, S4 occurs when atria contract during the second phase of ventricular filling. The low-pitched S1 is best heard at mitral auscultation site as it is caused by the closure of mitral and tricuspid valves. The shorter duration but louder sound S2 is generated by aortic and pulmonary valve closure and is best heard over aortic auscultation area.

The authors aims to develop a method for classification of heart sounds into normal and abnormal sounds so that state of heart could be periodically checked at home and everybody doesn't have to wait for symptoms of the disease to appear and then approach a cardiologist. The PCG signals can be an early indicator of heart problems so before worsening of the problem; a proper diagnosis can be made. The other techniques like Echocardiography could be implemented to get a better view of the problem. The proposed method is not replacing other technologies but an early indicator of the problem to prevent worst health condition. During the study, the PCG signals were analyzed for harmonic amplitude for normal, abnormal and murmur heart sounds. The result shows that the amplitude of the murmur signal is higher than normal signal between the fundamental heart sounds S1 and S2. In the experimental procedure, the segments between S1 and S2 of the cardiac cycle and consecutive [1] same segments of the sample wave form were taken for consecutive analysis. The features were extracted to differentiate between normal and murmur signals.

Heart Sound Analysis Technique

The Heart auscultation [32] is the interpretation of heart activities by the physician during cardiac diagnosis. It requires difficult skill to acquire. It is convenient to diagnosis the failure using some monitoring techniques. This paper reviews different signal processing technique for analyzing Heart Sound (HS) Vibration signals which (the interpretation by a physician of heart sounds) are used primarily to diagnose these diseases. Conventional methods for fault diagnosis are mainly based on observing the amplitude differences in time or frequency domain such as Fourier Transform (FT), Short Time Fourier Transform (STFT), and Wavelet transform. This paper includes Spectral analysis method of heart sound by using autoregressive power spectral density (AR-PSD) for discriminating normal and abnormal HS, another approach to diagnosing heart sounds such as Wavelet packet analysis and classifiers like Hidden Markov Model (HMM), Artificial Neural Network (ANN).

Signal Processing Techniques for Heart Sound Analysis in Clinical Diagnosis

It presents an overview of approaches to the analysis of heart sound signals. The paper reviews the milestones in the development of phonocardiogram (PCG) signal analysis. It describes the various stages involved in the analysis of heart sounds and discrete wavelet transform as a preferred method for bio-signal processing. Besides, the gaps that still exist between contemporary methods of signal analysis of heart sounds and their applications for clinical diagnosis is reviewed. A lot of progress has been made, but critical gaps still exist. It is observed that there is a lack of consensus in research outputs. The inter-patient adaptability of signal processing algorithm is always problematic. The process of clinical validation of analysis techniques was not sufficiently rigorous in most of the reviewed literature and as such data integrity and measurement are still in doubt, which most of the time led to an inaccurate interpretation of results. Besides, the existing diagnostic systems are too complicated and expensive. The paper concluded that the ability to correctly acquire, analyze and interpret heart sound signals for improved clinical diagnostic processes had become a priority.

Automatic Detection of Sounds and Murmurs in Patients with Lonescu-Shiley Aortic Bioprostheses

The problems faced in the automatic identification of cardiac sounds and murmurs are numerous. The phonocardiogram (PCG) is a complex sound signal produced by deterministic [32] events such as the opening and closing of the heart valves, and by random phenomena such as blood flow turbulence. In addition, background noise and the dependence of the PCG on the recording sites render automatic detection a difficult task. In the paper, we present an iterative automatic detection algorithm based on the a priori knowledge of spectral and temporal characteristics of the first and second heart sounds, the valve opening clicks, and the systolic and diastolic murmurs. The algorithm uses estimates of the PCG envelope and noise level to identify iteratively the position and duration of the significant acoustic events contained in the PCG. The results indicate that it is particularly useful in detecting the second heart sound and the aortic component of the second heart sound in patients with lonescu-Shiley aortic valve bioprostheses. It also has some potential for the detection of the first heart sound, the systolic murmur, and the diastolic murmur.

Recent Advances in Heart Sound Analysis

Cardiovascular disease (CVD) continues to be the leading cause of morbidity and mortality worldwide with an estimated 17.5 million people [33] have died from CVD-related conditions in 2012, representing 31% of all global deaths. The burden on low to middle-income countries (LMICs) is particularly worrisome, and yet high-quality diagnostics can often be difficult to obtain in much of these resources constrained regions. Although ultrasound and magnetic resonance imaging have displaced auscultation in the more vibrant economies, heart sound auscultation remains a stalwart diagnostic of the ambulatory doctor. However, with the patient to doctor ratios as high as 50,000:1 in some regions of the world, access to expert diagnosis is often impeded. A potential solution to this is to provide an automated diagnosis on the mobile phone or in the cloud [33].

In the past few decades, automated heart sound signal (i.e., phonocardiogram or PCG) analysis has been widely studied. Research shows that automated heart sound segmentation and classification techniques have the potential to screen for pathologies in a variety of clinical applications at a relatively low cost. However, comparative analyses of algorithms in the literature have been hindered by the lack of a large and open database of heart sound recordings. The 2016 PhysioNet/Computing in Cardiology (CinC) Challenge addressed this issue [33] by assembling the largest public heart sound database, aggregated from eight sources obtained by seven independent research groups around the world. The database includes 4,430 recordings taken from 1,072 subjects, totaling 233,512 heart sounds [33] collected from both healthy subjects and patients with a variety of conditions such as heart valve disease and coronary artery disease. These recordings were collected using heterogeneous equipment in both clinical and nonclinical (such as in-home visits). A total of 48 teams submitted 348 open source entries. This special issue provides an extended forum to document and discuss this challenge, and to provide follow-up and improvements.

Heart Sound Detection and Analysis in Real-Time [34]

Usually, physicians use an acoustic stethoscope to detect abnormalities in the heart sound and to detect abnormal conditions of the human heart. As the frequency range and intensity of heart sound is very low, doctors face problems while detecting the cardiac sound and its abnormalities. To eradicate these severe problems, it is required to design and develop an electronic stethoscope which would assist the doctor to analyze heart sound and to detect a diseased heart. Here an acoustic stethoscope along with microphone and preamplifier module is used to increase the amplitude of the input sound signal received by the stethoscope. The soft scope of MATLAB program has also been used to analyze the continuous set of cardiac sound and to detect its various characteristics like frequency, amplitude, etc. It is aimed to design an electronic stethoscope which would assist the doctors to analyze heart sound and detect a diseased heart, but preliminarily we have achieved to detect different components of it which are lub (s1), dub (s2), s3, s4, etc. Finally, the sound signal received from the heart in the MATLAB program after filtering the noise out of it also has been plotted (Fig. 11).

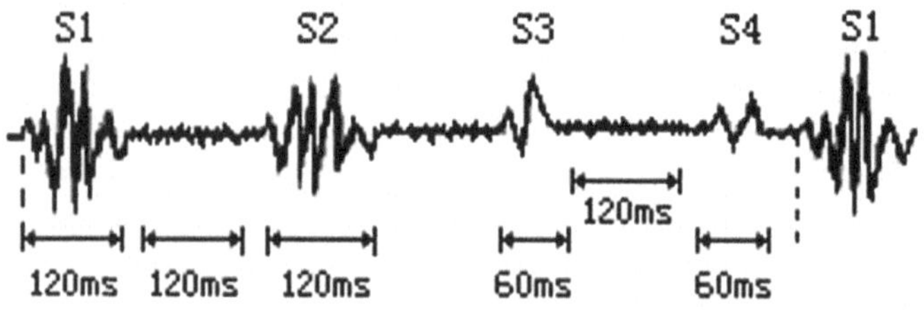

Fig. 11 Different segments of heart sound

Problem Definition and Objective

Problem Definition:

- The electronic stethoscopes available in the market are very costly with an average cost of €300–€399 [5]. This high cost makes them less than ideal for those fields where the additional functionality of the electronic stethoscopes are not required [35].
- Moreover, their battery life is also very poor with an average of 60 h of use [5]. In most of the cases, the batteries are irreplaceable, and thus the stethoscope becomes of no use once the battery life is over. In some cases, though, the batteries can be replaced which leads to additional cost after paying a heavy initial price for the electronic stethoscope.
- India has not been able to make any significant contribution to the market of the electronic stethoscope. Most of the electronic stethoscopes available in the Indian market are imported.
- Most importantly, the electronic stethoscopes available in the market are unable to diagnose heart disease by itself. It only assists the doctors in identifying heart problems quickly and efficiently. Hence these stethoscopes are not for the laymen to use. The most commonly available electronic stethoscopes in the market are 3MTHLittman Electronic Stethoscopes, E-scope, AMD-3150 and ADSCOPETH 657. All of these stethoscopes are unable to analyze heart disease by themselves and require a medical specialist for it.

Objective:

The primary aim of the research work is to develop an electronic stethoscope which will be devoid of the problems of the currently available electronic stethoscopes in the market as mentioned above. The intention is to create a stethoscope that will be able to diagnose heart disease by itself and hence can be used by the laymen. The main structure of the electronic stethoscope can be constructed by using an acoustic stethoscope along with a microphone and preamplifier module. Moreover, the aim is to evaluate the performance of the electronic stethoscope without the help of an external analog-to-digital converter (ADC). This research work also depicts the use of feature analysis to distinguish between Normal heart sound, abnormal heart sound, and Cardiac.

Methods and Materials

The entire work is a materialization of a prolonged study and thoughts. Many articles and research papers have been studied on detection of cardiac murmur, analysis of heart sound, detection of a diseased heart sound, development of electronic stethoscope, etc. In every case it has been seen through detailed and advanced work is done but the expense of developing a digitalized stethoscope is very high. It influenced us

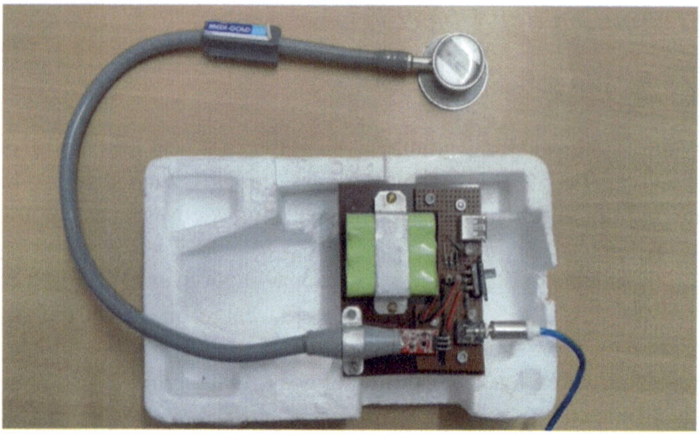

Fig. 12 Hardware circuit diagram

to develop an electronic stethoscope at low cost. It has been aimed to design an electronic stethoscope which would assist the doctor to analyze heart sound and detect a diseased heart.

Figure 12 shows the image of the hardware that has been implemented in the research work. It shows an acoustic stethoscope attached to condenser microphone of the preamplifier module with the help of a hollow rubber-like tube. The battery along with the required circuit, made using a Vero-board, is placed on a thin Bakelite board. Aluminum strip is used to hold the battery and the microphone to the stetho-tube connector to the bakelite board.

As seen from the circuit diagram, the hardware part of the circuit consists of the following components:

The acoustic sensor and the amplifier module are discussed in details in the later section. The battery used is a rechargeable battery which can be charged by connecting the USB port to a laptop or a similar device. When the battery is charged, the LED-2 glows. A resistance of 330 Ω is applied before each LED to prevent the LEDs from getting damaged. The switch used is a dual pole dual throw (DPDT) switch. When the switch is connected to point A then the power from the battery reaches the amplifier module and it starts working. At that time the LED-1 glows. The switch is connected to point B in order to charge the battery. The audio jack, when connected to the microphone plug of the laptop or CPU, transmits sound signal from the amplifier module to the laptop/computer (Figs. 13 and 14).

Testing of the stethoscope to find its frequency response

Step 1: To find out the frequency response of the developed electronic stethoscope, it needs a sound source with variable frequency. One small speaker unit of diameter same as that of the chest piece of the stethoscope and amplifier module is used to excite the speaker unit. In this testing, a smartphone app named Hz Generator is used which served as a function generator for the research work [36]. A 3.5 mm double-sided audio jack cable is used to connect the amplifier to the smartphone.

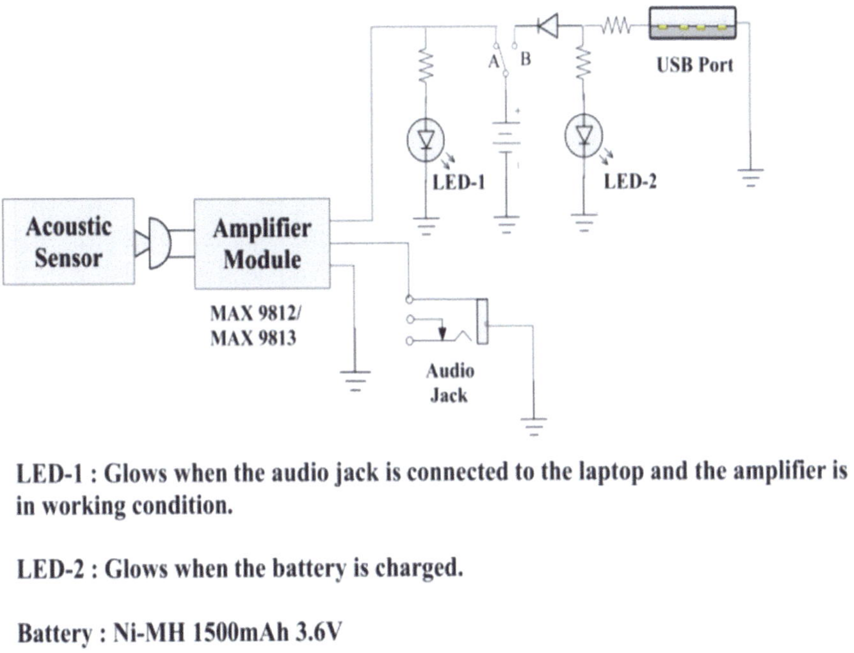

Fig. 13 Electronic circuit diagram of the implemented hardware

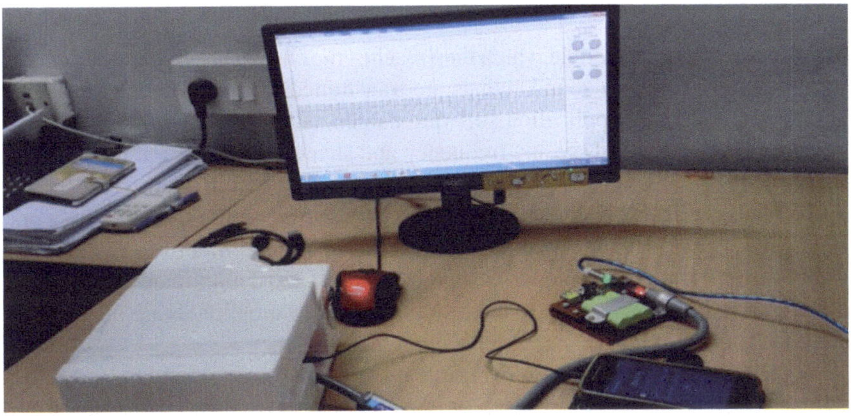

Fig. 14 Experimental set up

Step 2: The chest piece of the acoustic stethoscope is then attached to the speaker of the speaker and amplifier module with the help of cello tape.

Step 3: The setup shown in Fig. 5 is then covered with thermocol to make it soundproof. It is because thermocol is a good absorber of sound and thus prevent the ambient noise from reaching the setup. The sine wave signals of particular frequencies

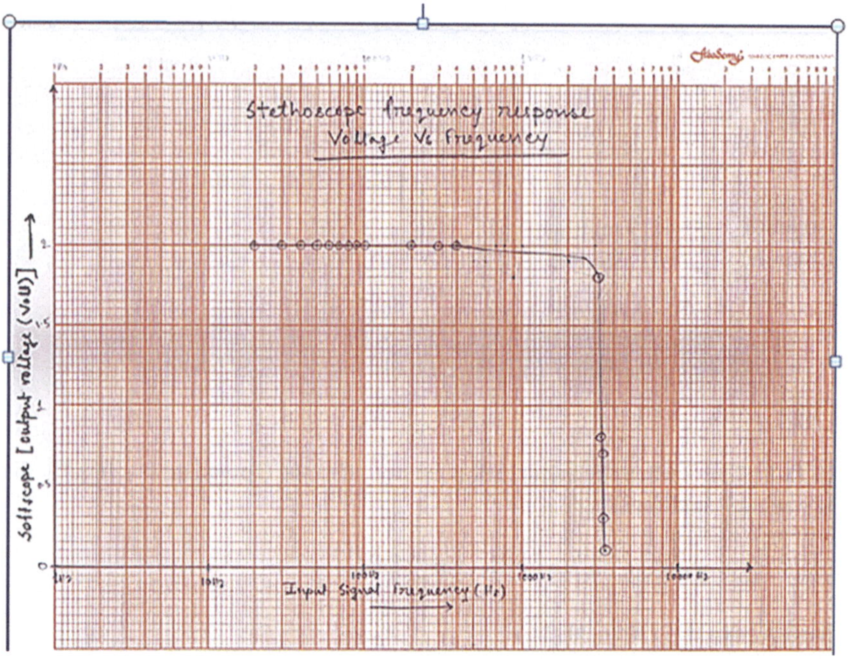

Fig. 15 Frequency response characteristics of the stethoscope

are fed to the amplifier module which then reaches the chest piece of the stethoscope. In the research work, the sound signals are transmitted from the mobile, through a cable, using cellphone application software named as Hz Generator. This software can generate sound signals of any frequency and shape as per the requirement.

Step 4: The below image shows the entire experimental setup required for finding the frequency response characteristic of the stethoscope used in the research work. The stethoscope amplifier output signal is fed to the audio input of the computer system for analysis. Then sine wave sound signals of fixed amplitude and different frequencies are sent successively, and the corresponding amplitudes are measured in the soft scope of MATLAB through the computer system.

Frequency versus amplitude data is plotted as obtained experimentally and shown in Fig. 6. It is observed from the plot, the frequency response of the developed electronic stethoscope is flat from 20 to 3100 Hz. After that, a sharp cutoff frequency has been found (Fig. 15).

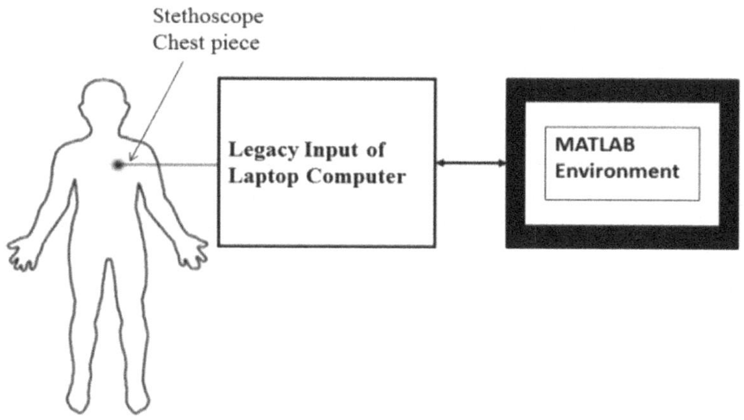

Fig. 16 System architecture of the proposed analytical system of heart sound in real-time

Software Development

Software development is the process of computer programming, documenting, testing and bug fixing involved in creating, maintaining applications and frame works involved in software released life cycle and resulting in a software product.

For this work two different approaches have been used of Software Development: (a) Matlab Scripts, (b) Simulink

It has been experimented to detect heart sound by the developed electronic stethoscope in actual human subject. Initially, it has tested individual filters by calibrated signal and then finally used to tune the measurement system.

Since the electronic stethoscope has many disadvantages like it is very costly and the sound which is heard in the earpiece of stethoscope is of very low strength and also has a lot of noise incorporated in it and hence it is very difficult to analyze.

For this reason the heart sound wave can be plotted in Matlab using Matlab scripts or Simulink and can be analyzed better by observing the spectrum. Also real time data acquisition can be done with the help of Simulink.

The filters are also designed using Matlab scripts and the Digital Signal Processing toolbox and hence it becomes cost effective and is efficient in function and gives us a noise free signal.

The Simulink Model is the second approach of the Software Development. The Simulink Model is created for a better analysis of the heart sound waveform than the Matlab Scripts. The Simulink Model can also be used for the real time data acquisition of the system. In the proposed system the developed electronic stethoscope output is fed to the microphone input of the laptop which is called legacy input of the laptop computer system. The laptop is normally have built in A/D converter with DSP (Legacy input circuit) is being exploited by MATLAB environment for heart sound data acquisition, storage and analysis (Fig. 16).

Filter Design for Removal of Noise

Filter design part is needed as it plays a significant role in eliminating noise from heart sounds. It is a pretty perfect solution, since the background noise is embedded in the low-frequency range of the spectrum, whereas the heart sound has a higher frequency. As such, a **band pass** filter can be applied to get rid of the low noise, capture most of the Heart sound, and any noisy frequencies on the higher side will get canceled as well.

Here are the steps that are implemented

1. Read in the audio file using wavread function in Matlab.
2. Plot the sound signal in time domain, for time domain analysis.
3. Take the Fast Fourier Transform (FFT) and see the frequency distribution.
4. Using (3), figure out the rough approximation of where should the cut-off frequencies be selected.
5. Design a Butterworth bandpass filter that cuts off these frequencies.

Here some standard Normal Heart Sounds have been collected from Heart-Sound Library source, and those sound files have been read in mat lab and then time domain analysis, frequency domain analysis have been performed. Finally, the filtering operation has been done of low-frequency background noise from the standard Normal heart sound using Butterworth Band Pass Filter of order-1 after selecting the proper cutoff frequencies, and then the amplitude spectrum of that signal has been plotted.

Step-1:

```
%% Read in the audio file
clearvars;
close all;
[f,fs] = wavread('PCG.wav');
```

"PCG.wav" file is the normal heart sound in wav format contains only lub dub.
f would be the signal read into MATLAB & f here is a 2D matrix.
fs is the sampling frequency of the audio signal.

Step -2:

```
%% Plot audio channel in Time Domain
N = size(f,1); % Determine total number of samples in audio file
figure;
stem(1:N, f(:,1));
title('Time Domain Analysis');
xlabel('Time');
ylabel('Amplitude');
```

The output of **Step-2** of the above written MatLab scripts is given in Fig. 17

Step -3:

FFT is used to take a look at the frequency spectrum of the signal. **FFT** can be called by specifying the input signal you want as the first parameter, followed by

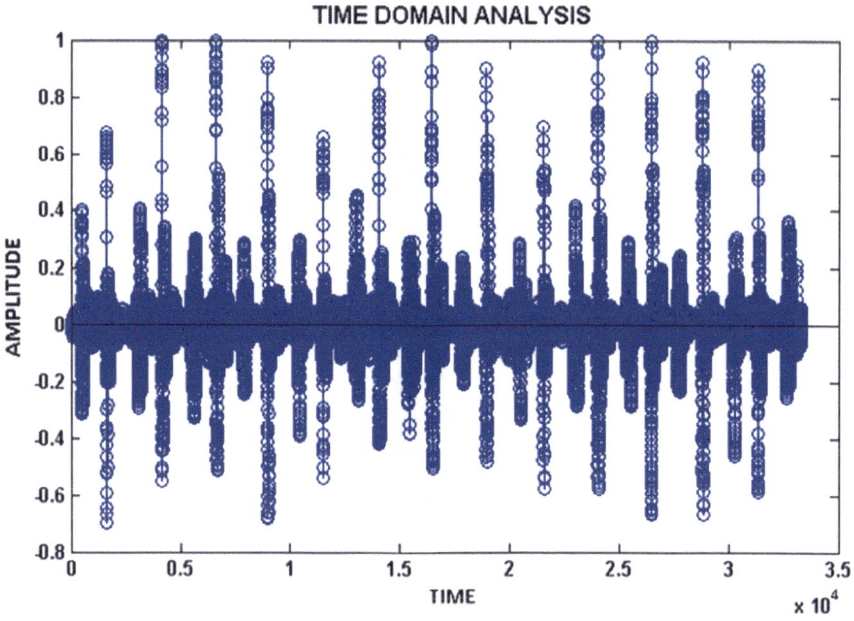

Fig. 17 Time domain analysis of normal heart sound

how many points required to evaluate at with the second parameter. It is customary that specify the number of points in the FFT to be the length of the signal. As the Frequency spectrum obtained after doing **FFT** is un-centered so take the **fftshift** to make it a centered one. When the frequency spectrum is plotted, it indicates how much contribution that frequency has with the output. That is defined with the **magnitude** of the signal. It has been done by taking the **abs** function.

%% Plot the spectrum

```
df = fs/ N;
w = (−(N/2):(N/2) − 1)*df;
y = fft(f(:,1), N)/ N; % For normalizing, but not needed for our analysis
y2 = fftshift(y);
figure;
plot(w,abs(y2));
xlabel('Frequency');
ylabel('Amplitude Distribution');
title('Amplitude Spectrum of Normal Heart Sound');
```

The output of **Step-3** of the above written MatLab scripts is given in Fig. 18.

Step -4:

The **bandpass** flag means you want to design a bandpass filter, beginFreq and end-Freq map to the normalized beginning and ending frequency you want to for the band-

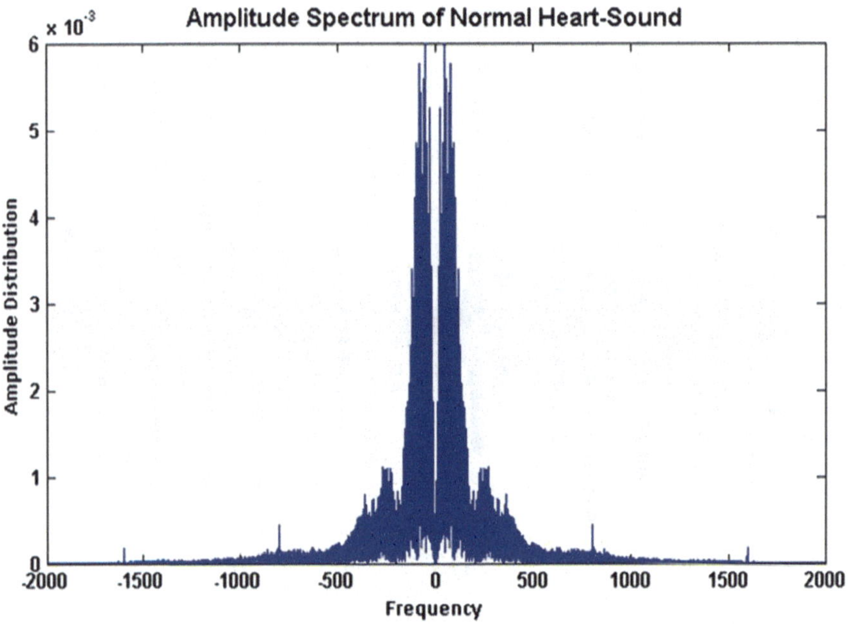

Fig. 18 Amplitude spectrum of normal heart sound

pass filter. In this work, that's beginFreq = 70/Nyquist and endFreq = 400/Nyquist. b, a is the coefficients used for a filter that will help you perform this task

```
%% Design a bandpass-filter that filters out between 800 and 1200 Hz
n = 1;
beginFreq = 70/(fs/2);
endFreq = 400/(fs/2);
[b,a] = butter(n, [beginFreq, endFreq], 'Bandpass');
```

Step -5:

"filter" is used to filter the signal using what has been obtained from Step 4. **fOut** will be the filtered signal. Finally to obtain the amplitude spectrum of the filtered signal again use fft and fftshift functions respectively. Amplitude Spectrum of the Noise Free Normal Heart Sound is shown below.

```
%% Filter the signal
fOut = filter(b, a, f);
g=fft(fOut(:,1),N)/N;
g2=fftshift(g);
figure;
plot(w,abs(g2));
xlabel('Frequency');
ylabel('Amplitude Distribution');
```

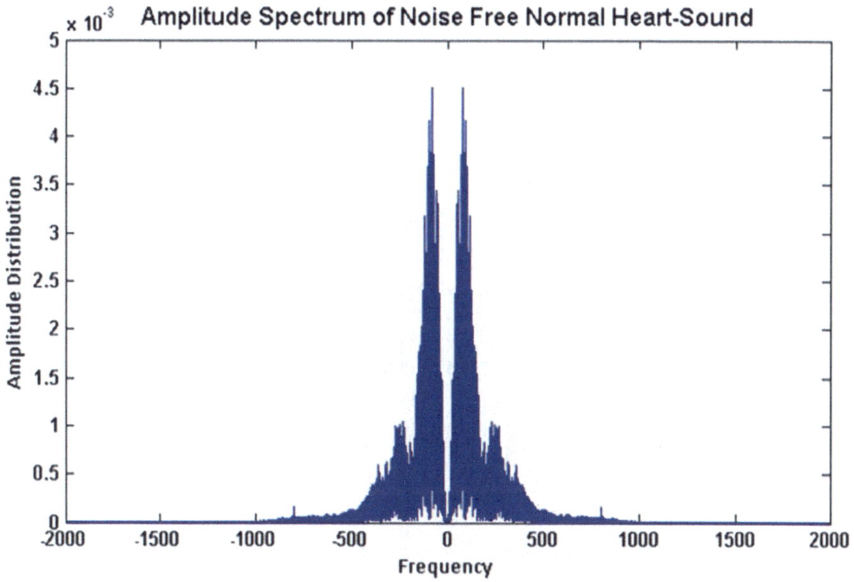

Fig. 19 Amplitude spectrum of noise free normal heart sound

title('Amplitude Spectrum of Noise Free Normal Heart Sound');

The output of **Step-5** of the above written MatLab scripts is given in Fig. 19.

Feature Analysis

After removal of noise from actual heart sound signal following features are selected for analysis

1. Mean value of Amplitude
2. Mean value of Phase
3. Kalman Filter Approach

During the analysis of Harmonic-Distribution of Amplitude and Phase, the contribution of higher order harmonics being neglected. The Harmonic Distribution can be obtained first by conversion of the time domain sound signal into frequency domain signal and then separate out the real components with the imaginary components [37].

For Simplicity, harmonics order up to 40 is considered. Following distributions depict the Harmonic-Distribution of Amplitude and Phase for Normal Heart Sound, Abnormal Heart Sound and Cardiac Murmurs. From the three Harmonic amplitude distributions as obtained from real-time analysis, it can be concluded that amplitudes of harmonics in Normal Heart Sound are higher compared to Abnormal Heart Sound

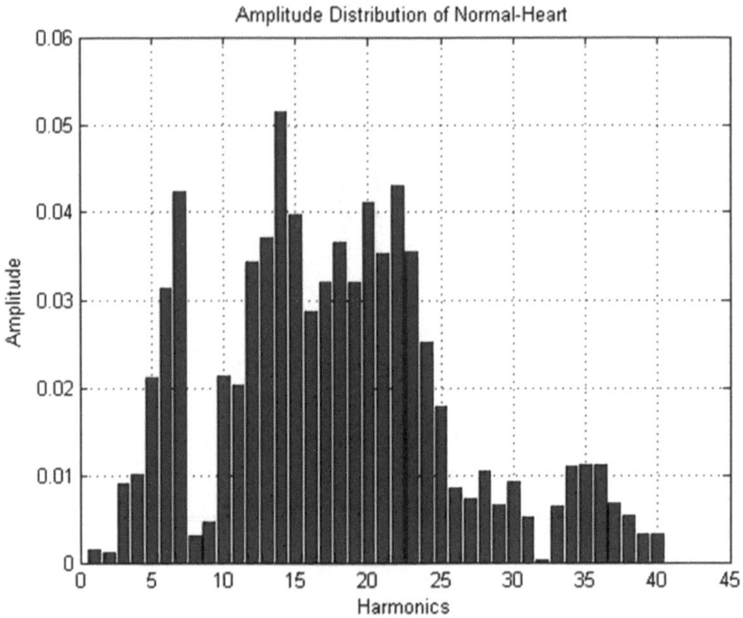
Amplitude Distribution of Normal-Heart
0.06
0.05
0.04
0.03
0.02
0.01
0
Amplitude
0
5
10
15
20
25
30
35
40
45
Harmonics

Fig. 20 .

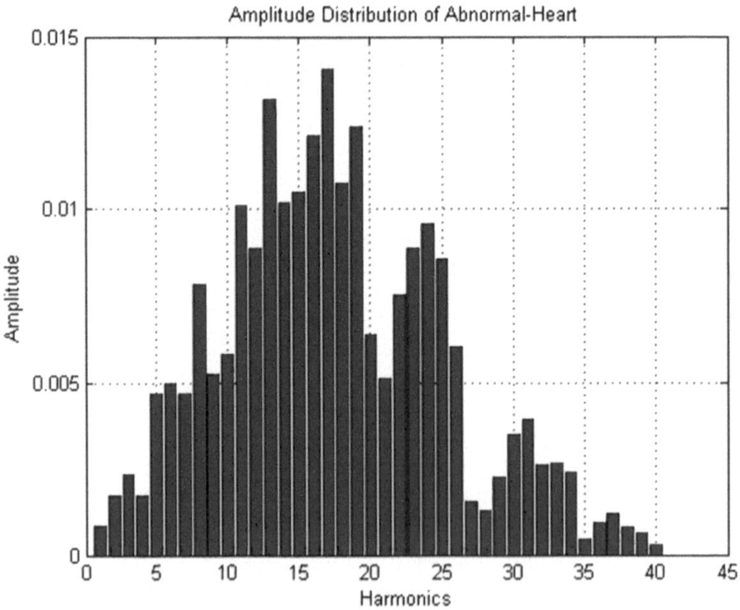
Amplitude Distribution of Abnormal-Heart
0.015
0.01
0.005
0
Amplitude
0
5
10
15
20
25
30
35
40
45
Harmonics

Fig. 21 .

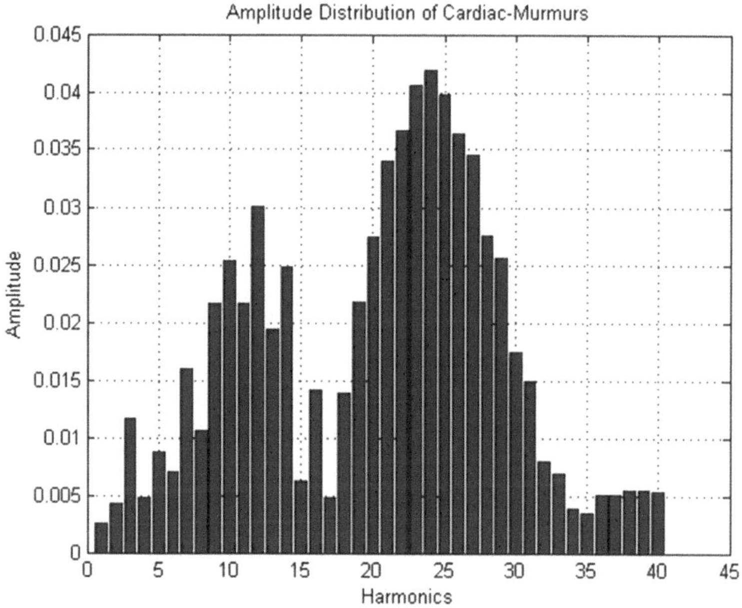

Fig. 22 .

and Cardiac Murmurs. Higher Order harmonics are neglected as their amplitudes are very small. Figures 20, 21 and 22 indicate the analysis to support the findings.

Kalman Filter Approach in Heart Sound Diagnosis

Rudolf Emil Kalman, a Professor at Standford University developed a mathematical technique called Kalman Filter widely used in digital computers of control systems, navigation, avionic and outer space vehicles to extract signal from noisy and incomplete measurements. Kalman filtering [38] is a useful tool for a variety of different applications. Kalman filtering is also sometimes called "linear quadratic estimation." Now let us think about the "filter" part. All filters share a common goal: to let something pass through while something else does not. An example that many people can relate to is a coffee filter. The coffee filter [20] will allow the liquid to pass through while leaving the solid coffee grounds behind. You can also think about a low-pass filter, which lets low frequencies pass through while attenuating high frequencies. A Kalman filter also acts as a filter, but its operation is a bit more complicated and harder to understand. A Kalman filter takes in information [39] which is known to have some error, uncertainty, or noise. The goal of the filter is to take in this imperfect information, sort out the useful parts of interest, and to reduce the uncertainty or noise [38].

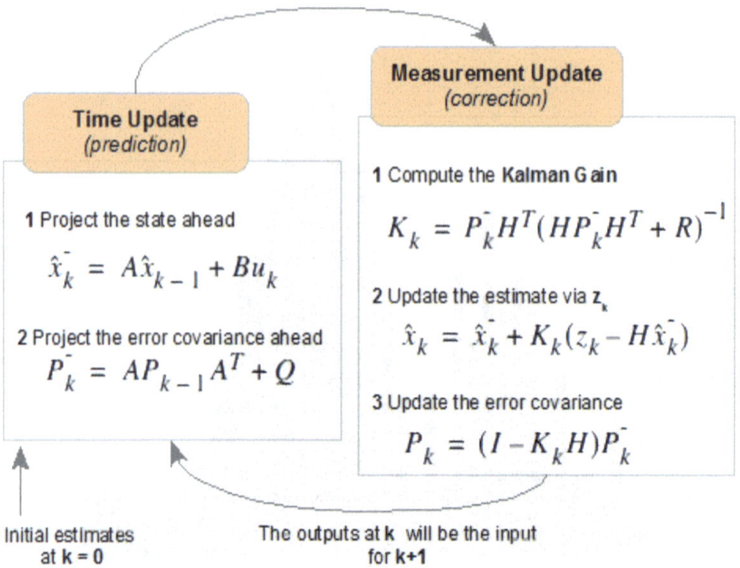

Fig. 23 Flowchart diagram of Kalman filter

The Kalman Filter is a state estimator filter which gives the result as an optimal estimator in the sense that the mean value of the sum of the estimation errors gets a minimal value [38]. The Kalman filter estimated the sum of squared errors. The filter contains a model of dynamic state process that performs a function, and as a feedback correction function method there is no requirement of the store a data and other all associate matrices like measurement system, dynamics and noise are considered to be known. Kalman filter has two steps (Fig. 23)

1. Prediction
2. Correction

In this work, Kalman filter approach has been applied to different heart sounds, and gathered information about the estimated value of the heart sound with an increase in harmonics. As the harmonics of complex heart sound gets increased, it has been found that estimated value comes closer to measured value with the decrease in gain from 1 to 0 value (Fig. 24).

Also, it has been observed that in case of Normal Heart Sound, the difference between Estimated value [40] and Measured value is more compared to Abnormal heart sound and Cardiac Murmurs (Figs. 25, 26, 27, 28 and 29).

In this chart (Fig. 30), 20 odd samples of a Normal heart sound, Abnormal Heart sound, and cardiac murmurs have been considered and compared after computation of Average of Amplitude Harmonic Distribution. It can be seen that from this comparative study of Harmonic distributions, Cardiac Murmurs can be easily distinguished and identified. It has been found that the average value of the amplitude harmonic

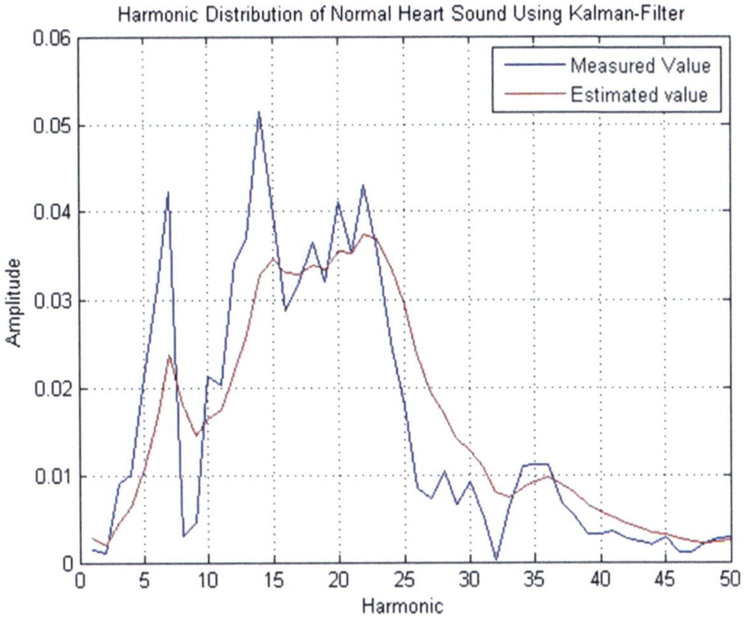

Fig. 24 Kalman filterer response of normal complex heart sound signal

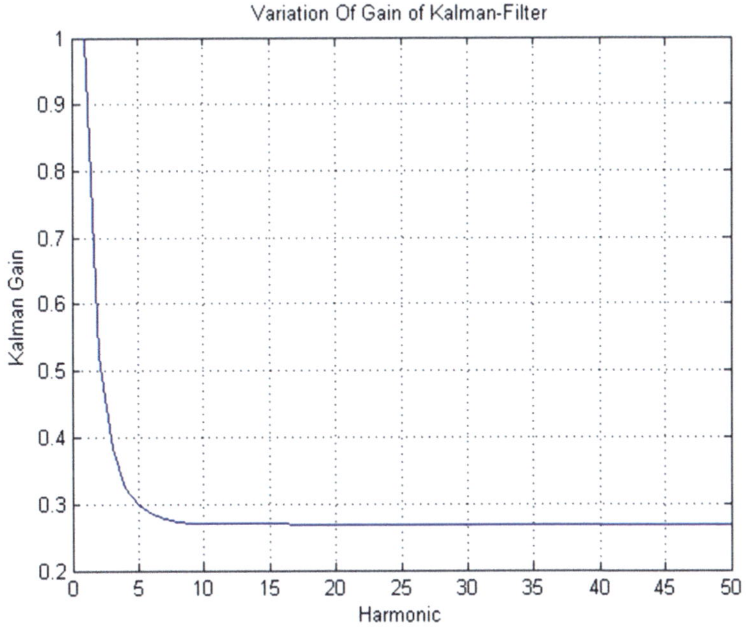

Fig. 25 Variation of Kalman-gain with harmonic distribution

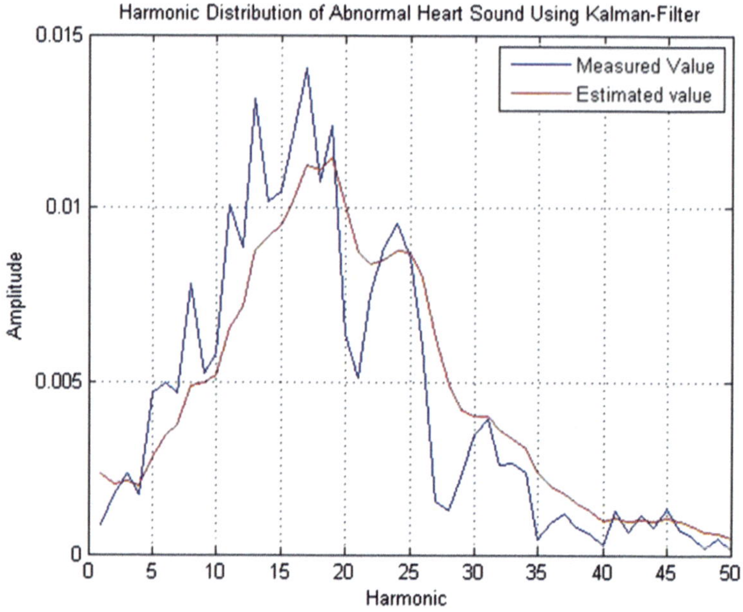

Fig. 26 Kalman filterer response of abnormal complex heart sound signal

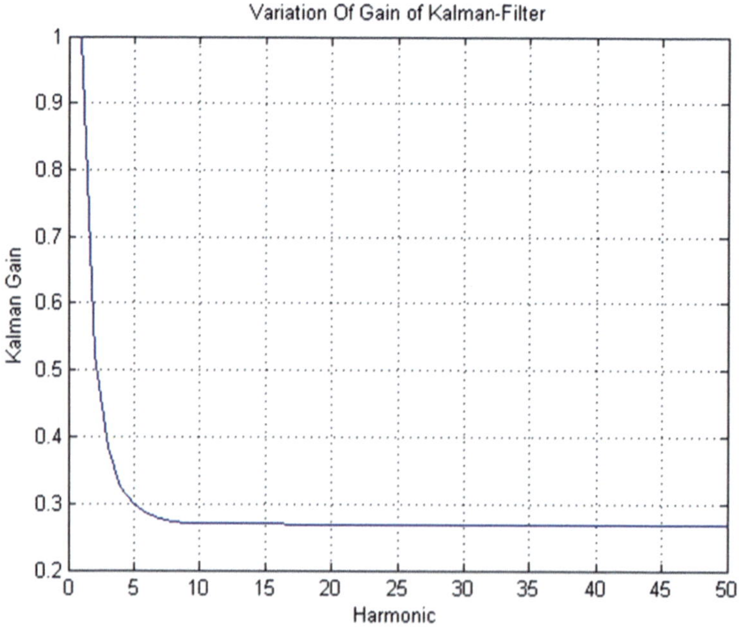

Fig. 27 Variation of Kalman-gain with harmonic distribution

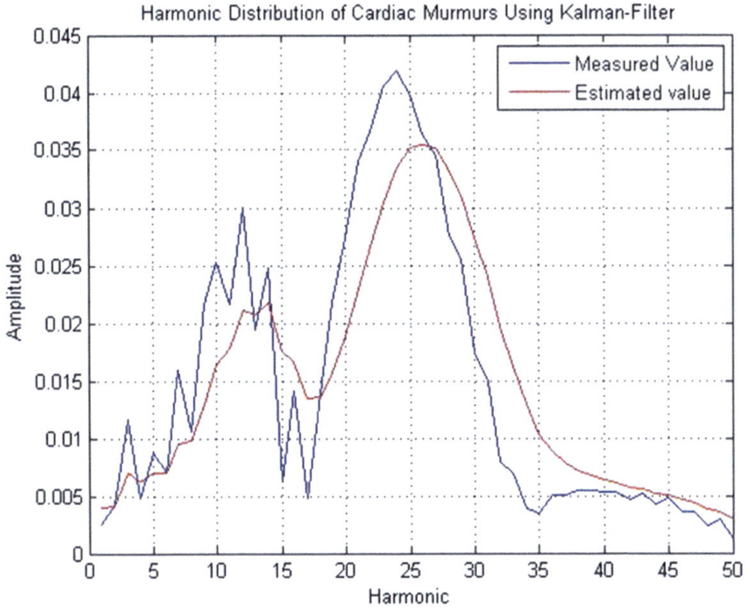

Fig. 28 Kalman filterer response of cardiac murmurs complex heart sound signal

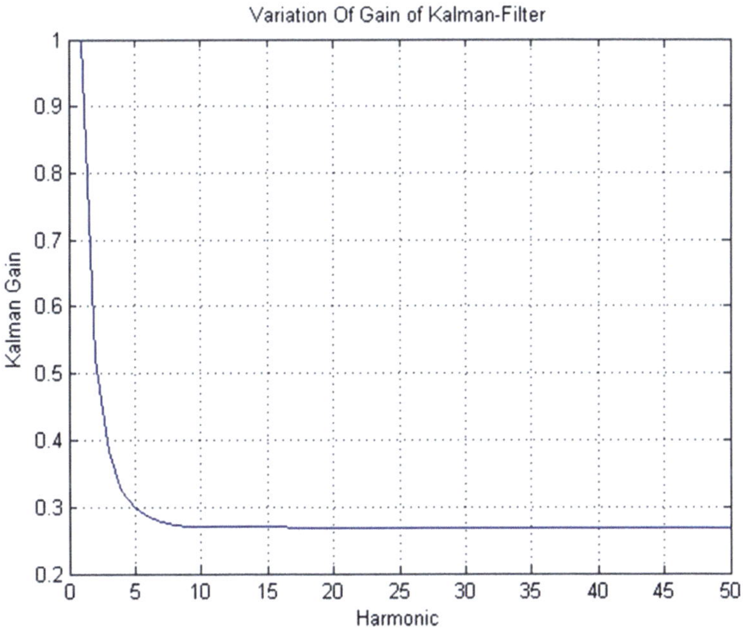

Fig. 29 Variation of Kalman-gain with harmonic distribution

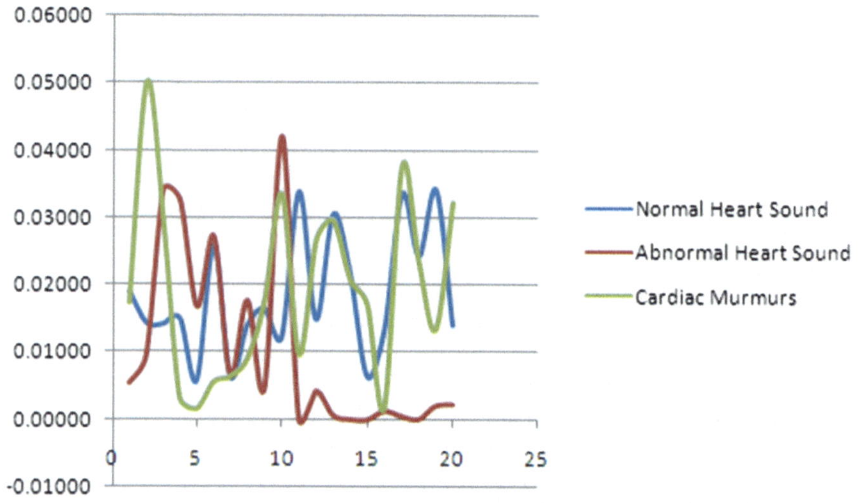

Fig. 30 Comparision of normal, abnormal and cardiac murmurs

distribution is higher in case of Cardiac murmurs compared to Normal heart sound and abnormal heart sound.

Conclusion

This electronic stethoscope can be used to detect normal heart sounds, abnormal heart sounds, and cardiac murmurs. Thus it can be used to diagnose a normal and abnormal heart sound accurately. Also with the help of filter design part, the background noise mostly gets eliminated. The instruments developed by us are working satisfactorily and detecting heart sound including murmur as well as we can use this detected signals for feature analysis [41]. The feature analysis part is in our present scope of work, but in next phase, we will carry out experimentation with our system for diseased heart sound.

This Book chapter focused on the problem phonocardiogram analysis as an aid tool to detect and diagnose any abnormalities of a heart condition. With the above-mentioned goal, the investigation can be made in a wide variety of fields ranging from cardiac auscultation, wavelet threshold de-noising, heart sound segmentation, Hilbert–Huang transform, feature extraction and classification by support vector machine. We developed a wavelet-based de-noising scheme that achieves a cleaner signal. We developed an energy-based segmentation method which is validated to perform well for both normal and typical abnormal heart sounds. We can apply Hilbert–Huang Transform to analyze the heart sounds and calculate the energy of selected IMF components that enabled us to construct the feature vector. Finally, the classification goals were accomplished using Support Vector Machine. Simulation results indicated that the proposed algorithm achieves the sought goal with a classification accuracy of 90.48% composed of 100% sensitivity and 81.8% specificity. Overall, a framework to classify normal and abnormal heart sounds is proposed using PCG. This system can be implemented in real time using PCG obtained from the new digital stethoscope technology and by the present state of the art embedded technology. Fourty one other future suggestions can be to aim to further analyze the abnormal into the different classes of heart diseases. Indeed, this goal opens up several avenues for the further research to be more accurate. First, we need to have a deep explore in Cardiac physiology, which is the foundation of all research of heart auscultation to allow for better feature recognition. Second, with the current framework, one can add more features that could present a potential to classify the different murmur signals. We need to extract the features from time domain and frequency domain and combine it with other statistical characteristics to produce a thorough analysis of the framework performance versus the different features. Third, we suggest a multi-classifier such as Multiclass SVM can be very valuable in solving such a classification problem, with the key point is the ability to create a suitable decision tree of SVM.

References

1. https://www.ukessays.com/essays/engineering/pest-detection-leafs-using-robot-processing-4110.php
2. L.-G. Durand, Y.-E. Langlois, T. Lanthier, R. Chiarella, P. Coppens, S. Carioto, S. Bertrand-Bradley, Spectral analysis and acoustic transmission of mitral and aortic valve closure sounds in dogs. Med. Biol. Eng. Comput. **28**(5), 439–445 (1990)
3. World Health Organization, Global status report on noncommunicable diseases 2014 (2014)
4. A. Leatham, Auscultation and phonocardiography: a personal view of the past 40 years. Br. Heart J. **57**, 397–403 (1987)
5. J.R. Kindig, et al., Acoustical performance of the stethoscope: a comparative analysis. Am. Heart J. **104**(2 Pt 1), 269–275 (1982); S. Lukkarinen, et al., A new phonocardiographic recording system. Comput. Cardiol. **24,** 117–120 (1997)
6. D.A. Balster et al., Digital acoustic analysis of precordial innocent versus ventricular septal defect murmurs in children. Am. J. Cardiol. **79**(11), 1552–1555 (1997)
7. R.M. Rangayyan, R.J. Lehner, Phonocardiogram signal analysis: a review. Crit. Rev. Biomed. Eng. **15**(3), 211–236 (1987)
8. https://www.gadgetsnow.com/tech-news/Govt-plans-Rs-10000cr-fund-to-create-tech-giants/articleshow/45041502.cms
9. J.R. Bulgrin et al., Comparison of short-time Fourier, wavelet and time-domain analyses of intracardiac sounds. Biomed. Sci. Instrum. **29**, 465–472 (1993)
10. P. Bentley, G. Nordehn, M. Coimbra, S. Mannor, R. Getz, Classifying heart sounds challenge. http://www.peterjbentley.com/heartchallenge/#downloads
11. M. Abella, J. Formolo, D.G. Penney, Comparison of the acoustic properties of six popular stethoscopes. J. Acoust. Soc. Am. **91**(4 Pt 1), 2224–2228 (1992)
12. Y. Wapcaplet, Diagram of the human heart (2003) (Online). http://commons.wikimedia.org/wiki/File:Diagram_of_the_human_heart_(cropped).svg
13. Anju, S. Kumar, Detection of cardiac murmur. Int. J. Comput. Sci. Mob. Comput. **3**(7), 81–87 (2014). ISSN 2320–088X

14. B. Popov, G. Sierra, L.G. Durand, J. Xu, P. Pibarot, R. Agarwal, V. Lanzo, Automated extraction of aortic and pulmonary components of the second heart sound for the estimation of pulmonary artery pressure, in *Proceedings of IEEE Engineering in Medicine and Biology Society*, vol. 2, pp. 921–924 (2004)
15. Cardiac cycle, https://en.wikipedia.org/wiki/Cardiac_cycle
16. M. Tavel, Classification of systolic murmurs: still in search of a consensus. Am. Heart J. **134**(2), 330–336 (1997)
17. C. Cortes, V. Vapnik, Support-vector networks. Mach. Learn. **20**(3), 273–297 (1995)
18. Heart murmur, National Heart & Blood Institute, https://www.nhlbi.nih.gov/health/health-topics/topics/holes/types
19. Z. Syed, D. Leeds, D. Curtis, F. Nesta, R.A. Levine, J. Guttag, A framework for the analysis of acoustical cardiac signals. IEEE Trans. Biomed. Eng. **54**(4), 651–662 (2007)
20. H. Liang, I. Hartimo, A heart sound feature extraction algorithm based on wavelet decomposition and reconstruction, in *Proceedings of 20th Annual International Conference IEEE Engineering in Medicine and Biology Society*, vol. 20; Biomed. Eng. **3**(3), 1539–1542 (1998), Towards Year 2000 Beyond (Cat. No. 98CH36286)
21. Z. Xiu-min, C. Gui-tao, A novel de-noising method for heart sound signal using improved thresholding function in wavelet domain, in *2009 International Conference on Future BioMedical Information Engineering (FBIE)*, pp. 65–68 (2009)
22. D.L. Donoho, De-noising by soft-thresholding. IEEE Trans. Inf. Theory **41**(3), 613–627 (1995)
23. A. Zaeemzadeh, Z. Nafar, S.-K. Setarehdan, Heart sound segmentation based on recurrence time statistics, in *2013 20th Iranian Conference on Biomedical Engineering (ICBME)*, no. Icbme, pp. 215–218 (2013)
24. T.-H. Hung, C.-C. Chou, W.-C. Fang, A.H.-T. Li, Y.-C. Chang, B.-K. Hwang, Y.-W. Shau, Time-frequency analysis of heart sound signals based on Hilbert-Huang transformation, in *2012 IEEE 16th International Symposium on Consumer Electronics*, pp. 1–3 (2012)
25. M. Singh, A. Cheema, Heart sounds classification using feature extraction of phonocardiography signal. Int. J. Comput. Appl. **77**(4), 975–8887 (2013)
26. H. Nygaard et al., Assessing the severity of aortic valve stenosis by spectral analysis of cardiac murmurs (spectral vibrocardiography). Part II: Clinical aspects. J. Heart Valve Dis. **2**(4), 468–475 (1993)
27. R.-E. Fan, P.-H. Chen, C.-J. Lin, Working set selection using second order information for training support vector machines. J. Mach. Learn. Res. **6**, 1889–1918 (2005)
28. D.B. Springer, L. Tarassenko, G.D. Clifford, Support vector machine hidden semi-Markov model-based heart sound segmentation time, in *Computing in Cardiology* (2014)
29. H.L. Baranek et al., Automatic detection of sounds and murmurs in patients with Ionescu-Shiley aortic bioprostheses. Med. Biol. Eng. Comput. **27**(5), 449–455 (1989)
30. Review on heart sound analysis technique. SpringerLink, https://link.springer.com/chapter/10.1007/978-81-322-1299-7_9
31. Heart sounds—Wikipedia, https://en.wikipedia.org/wiki/Heart_sounds
32. Heart sounds classification using feature extraction of …, https://pdfs.semanticscholar.org/73a7/3eb140b1f7b496f801d697174392c172c55c.pdf
33. Recent advances in heart sound analysis—Physiological …, http://iopscience.iop.org/journal/0967-3334/page/Recent-advances-in-heart-sound-analysis
34. J.K. Roy, T.S. Roy, A simple technique for heart sound detection and real time analysis, in Proceedings of ICST 2017, 2017 Eleventh International Conference on Sensing Technology (ICST) held at Maquare University Sidney, 4–6 Dec 2017, https://doi.org/10.1109/icsenst.2017.8304502
35. A. Iwata et al., Algorithm for detecting the first and the second heart sounds by spectral tracking. Med. Biol. Eng. Comput. **18**(1), 19–26 (1980)
36. The Second Heart Sound—Clinical Methods—NCBI Bookshelf, https://www.ncbi.nlm.nih.gov/books/NBK341/
37. Harvard Health Publishing, https://www.health.harvard.edu/heart-health/heart-valve-problems

38. A Kalman Filtering Tutorial For Undergraduate Students, http://aircconline.com/ijcses/V8N1/8117ijcses01.pdf
39. Heart murmur causes, https://www.nhlbi.nih.gov/health/health-topics/topics/heartmurmur/causes
40. M. El-Segaier, O. Lilja, S. Lukkarinen, L. Sörnmo, R. Sepponen, E. Pesonen, Computer-based detection and analysis of heart sound and murmur, Ann. Biomed. Eng. **33**(7), 937–942 (2005), http://www.ncbi.nlm.nih.gov/pubmed/16060534
41. S.M. Debbal, Computerized heart sound analysis. Genie—Biomedical Laboratory (GBM), Department of electronic, Faculty of science engineering, University of Abou Bekr Belkaid, Algeria
42. R.L. Donnerstein, Continuous spectral analysis of heart murmurs for evaluating stenotic cardiac lesions. Am. J. Cardiol. **64**(10), 625–630 (1989)
43. B. Maurice, E.E. Rappaport, B. Haward, M.D. Sprague, The acoustic stethoscope and the electrical amplifying stethoscope and stethograph. Am. Heart J. **21**(3), 257–318 (1940)
44. M. Nygards, L. Sornmo, Delineation of the QRS complex using the envelope of the E.C.G. Med. Biol. Eng. Comput. **21**(5), 538–547 (1983)
45. H. Nygaard et al., Assessing the severity of aortic valve stenosis by spectral analysis of cardiac murmurs (spectral vibrocardiography). Part I: Technical aspects. J. Heart Valve Dis. **2**(4), 454–467 (1993)
46. J.D. O'Toole et al., The mechanism of splitting of the secondheart sound in atrial septal defect. Circulation **56**(6), 1047–1053 (1977)
47. Gallop rhythm, https://en.wikipedia.org/wiki/Heart_sounds
48. Cardiac sound generation, http://www.wisegeek.com/what-are-different-types-of-heart-sounds.htm#
49. The circulatory system, http://lsa.colorado.edu/essence/texts/heart.html#Structure
50. Pinterest, https://in.pinterest.com/pin/407505466254590427/?autologin=true
51. Stethoscope, https://en.wikipedia.org/wiki/Stethoscope
52. D. Gede, H. Wisana, Design electronic stethoscope for cardiac auscultation analyzed using wavelet decomposition. Int. J. Comput. Netw. Commun. Secur. **1**(7), 310–315 (2013), www.ijcncs.org. ISSN 2308-9830
53. K. Singh, P. Abrol, Design and development of a digital stethoscope for cardiac murmur. Int. J. Comput. Appl. (0975–8887) **73**(22) (2013)
54. M. Singh, A. Cheema, Heart sounds classification using feature extraction of phonocardiography signal. Int. J. Comput. Appl. (0975–8887) **77**(4) (2013)
55. D.D. Kadam Patil, R.K. Shastri, Design of wireless electronic stethoscope based on ZIGBEE, Cornell University Library, https://arxiv.org/abs/1202.1680
56. L.B. Dahl, P. Hasvold, E.Arild, T. Hasvold, Heart murmurs recorded by a sensor based electronic stethoscope and e-mailed for remote assessment, Arch. Dis. Child **87,** 297–301, 19 March 2002
57. W.W. Myint, B. Dillard, An electronic stethoscope with diagnosis capability. Department of Electrical & Computer Eng., Auburn University, College of Engineering, Auburn, AL 36849 USA
58. https://datasheets.maximintegrated.com/en/ds/MAX9812-MAX9813L.pdf
59. http://telehealthtechnology.org/toolkits/electronic-stethoscopes/about-electronic-stethoscopes/technology-overview
60. J. Peter Robinson, Digital electronic stethoscopes. 3 M Healthcare Ltd, Loughborough, UK
61. History of the Stethoscope—American Diagnostic Corporation, https://www.adctoday.com/learning-center/about-stethoscopes/history-stethoscope
62. Introduction to Digital Stethoscopes and Electrical …, https://pdfserv.maximintegrated.com/en/an/AN4694.pdf
63. Automatic detection of sounds and murmurs …. SpringerLink, https://link.springer.com/article/10.1007/BF02441460

Serious Games Based on Kinect and Leap Motion Controller for Upper Limbs Physical Rehabilitation

Gabriela Postolache, Francisco Carry, Filipe Lourenço, Diogo Ferreira, Raul Oliveira, Pedro Silva Girão and Octavian Postolache

Abstract The design, implementation and tests of a system for assessment and monitoring movements, which includes the sensors from Kinect and Leap Motion Controller devices, are discussed in the present chapter. The advantages and some drawbacks of using the two devices for creating virtual environments for motor rehabilitation in which interaction of the user with virtual reality is made through natural user interfaces are described. The IoT architecture for rehabilitation environment, several serious games that our team have developed and the usability evaluation of the system are presented. Our insights, based on our research work during serious games development as well as on literature analysis, mainly focusing these on practicality of the developed serious games and on their acceptability for motor rehabilitation, are also included in the chapter. In the development of a system that includes IoT technology for movements tracking and progress evaluation during motor rehabilitation, the importance of user centered design is underscored.

Keywords Kinect · Leap Motion · Virtual reality · Usability

Introduction

Since the 1970s the console and computer games industry has been continuously increasing. The hardware, such as consoles, processors, screens, controllers and other accessories, as well as the software from games industry are expected to surpass $90 billion by 2020 in U.S. [1]. A survey conducted by Pew Research Center, have shown that in 2017, approximately one third (2.2 billion) of world population were

G. Postolache · F. Carry · P. S. Girão · O. Postolache (✉)
Instituto de Telecomunicacoes, Lisbon, Portugal
e-mail: opostolache@lx.it.pt

F. Lourenço · D. Ferreira · O. Postolache
ISCTE-Instituto Universitario de Lisboa, Lisbon, Portugal

R. Oliveira
Faculdade de Motricidade Humana, UL, Lisbon, Portugal

S. C. Mukhopadhyay et al. (eds.), *Modern Sensing Technologies*,
Smart Sensors, Measurement and Instrumentation 29,
https://doi.org/10.1007/978-3-319-99540-3_8

registered as active video gamers, and the number of video gamers is expected to further grow to 2.7 billion by 2021 [2]. In 2017, 82.5 million units of Nintendo's Wii Sports, a home video game console released in 2006, were sold worldwide [3], and the Diablo III and the life simulation video games series—the SIM, are ranked as the bestselling games of all times (Statica 2018). According to the RnR Market Research report by "Vertical (Education, Corporate, Healthcare, Retail, Media and Advertising), Application (Training, Sales, Human Resource, Marketing), Platform, End-User (Enterprise, Consumer), and Region—Forecast to 2020", the serious game market is estimated to reach $5.5 billion by 2020 [4]. Game-based learning market revenue worldwide is estimated to reach $8.1 billion in 2022 [5].

Serious games, are defined as "the games that are designed for a primary purpose other than pure entertainment" [6], or as "interactive computer applications, with or without a significant hardware component, that have a challenging goal, are fun to play and engaging, incorporate some concept of scoring and impart to the user a skill, knowledge, or attitude that can be used in the real world" [7], "games that are designed to entertain players as they educate, train or change behaviour" [8] or "games with the purpose of improving an individual's knowledge, skills, or attitude in the real world" [9]. Serious games involve different technologies as virtual reality, sensors, telecommunication technologies, human computer interfaces, dedicated server or cloud services. In the last decades increasing number of evidence have shown the capacity of serious games to increase individual motivation for performing physical exercises. In the scientific literature, interventions based on exertion games has been referred with many different names, such as exergames, serious videogames, game-based virtual reality, or active-play videogames. We used in this chapter the term exergames to define the "computer games that require players to physically move in response to game demands and/or on-screen avatar" [10]. Digital gaming systems with an interface that requires physical exertion to play the game, that include virtual environments and/or feedback on movements, have emerged as an enjoyable and motivating alternative for practicing physical exercises in the context of the motor rehabilitation [11–23]. Virtual environments are defined as "interactive, virtual image displays enhanced by special processing and by nonvisual display modalities, such as auditory and haptic, to convince users that they are immersed in a synthetic space" [24]. Exergames provide environments responsive to the actions of the user, facilitating repetitive, contextual practice and feedback consistent with user conditions, for successful motor skill acquisition [25]. A great number of studies on exergames development and their effectiveness tests were published in the last decades. Feasibility and benefits of exergames were investigated in community dwelling older adults [11–13], post-stroke patients [14] patients with heart failure [15], patients with Parkinson's disease [16], or cerebral palsy [17, 18]. The virtual reality environments from many exergames were designed having as main goals the improvement of user general physical fitness, functional mobility (i.e., muscle endurance and strength, balance performance), and for therapeutic purposes (i.e., cardiac rehabilitation, neuro-motor rehabilitation).

The use of Kinect and Leap Motion Controller for development of a system for exergaming during motor rehabilitation is discussed in the present chapter. We described the advantages and some drawbacks of using the two devices for creating virtual environments in which interaction of the user with virtual reality is made through natural user interfaces. The Internet of Things (IoT) architecture for rehabilitation environment, several serious games that our team have developed and their usability evaluation are also presented.

The chapter is organized as following: the architecture of the developed system that include Kinect, Leap Motion Controller, Internet of Things technology and serious games are presented first, followed by presentation of serious games based on Kinect that our team developed, serious games based on Leap Motion Controller, the acceptability evaluation of serious games and our insights on user centered design, focused on practicality of developed serious games as well on their acceptability and usability for motor rehabilitation.

IoT System Architecture for Motor Rehabilitation

The scheme of IoT system architecture for motor rehabilitation is presented in Fig. 1. **Microsoft Kinect** is used for tracking patient upper and lower body movements. Microsoft Kinect sensor was released in 2010 and the Software Development Kit (SDK) in 2011. In the last years numerous games based on Kinect were developed, valorising the natural user interface provided by this device. In 2015 a Kinect v2 was released that allows higher accuracy for human body detection. The SDK of Kinect V2 represent 25 skeleton joints, providing better gesture recognition by comparison with 20 joints detected by Kinect v1 [26]. The Kinect sensor include an infrared (IR) projector and RGB camera and a special microchip that generates a grid from which the location of a nearby object in 3 dimensions can be ascertained [27]. A IR laser emitter delivers infrared light beams and an infrared depth camera, 512×424 pixels, allows the extraction of depth information using the IR speckle analysis of the IR beams reflected to the sensor.

The depth information can be used to measure the distance between user body and the sensor. Inferring body position is a two-stage process: first is computed a depth map, and then is inferred the body position. The depth map is constructed by analysing a speckle pattern of IR laser light. As general principle applied in this case is the projection of a known pattern onto the scene (body) and inferred depth from the deformation of that pattern. Kinect device include the PrimeSense hardware for the depth computation. The Kinect sensor also includes a multi-array microphone that can be used for voice recognition but also to find the location of the sound source and its direction. The operational distance between the camera and target (user playing the game), where good accuracy of the detection of movements is produced, has maximum 4.5 m and minimum 0.5 m (Fig. 2). Although reliability starts to degrade at around 4.5 m, Kinect can capture images at the maximum distance of 8–10 m.

The Kinect sensor must be placed above or under the device that display the virtual reality environment. It is important to be placed on stable surface. Direct sunlight to the sensor must be avoided. Must also be avoided to be placed on an instable surface

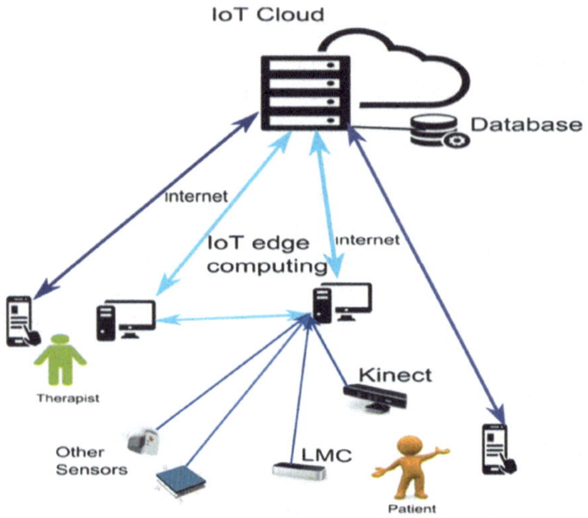

Fig. 1 IoT Rehab architecture

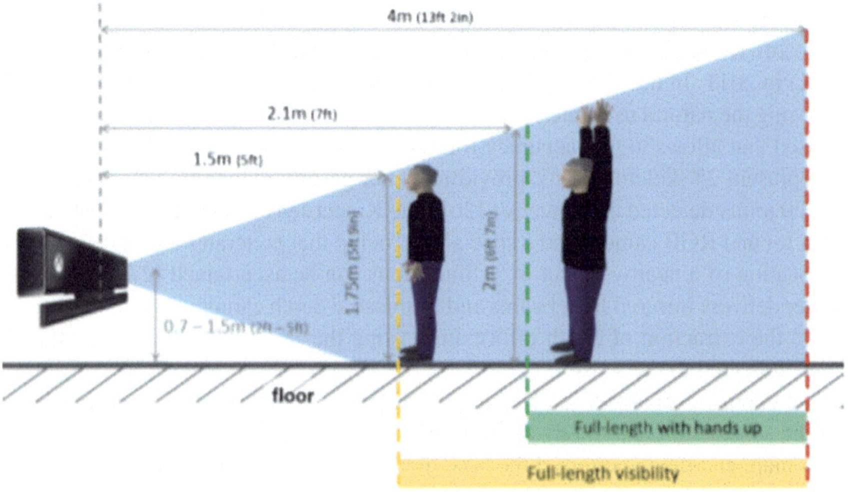

Fig. 2 Kinect V2 sensor representation of optimal position of body for different type of exercise recording

that may vibrate or make noise or in position where the risk of falling or being hit (e.g., when user perform the movements for interaction with virtual environment).

We installed the Kinect SDK 2.0 on a Windows 10 computation platform. The computation platform that is used to perform the game connected to Kinect V2, have a 64-bit (×64) processor, physical dual-core 3.1 GHz (2 logical cores per physical), a USB 3.0 controller dedicated to the Kinect for Windows v2 sensor, a Kinect Adapter

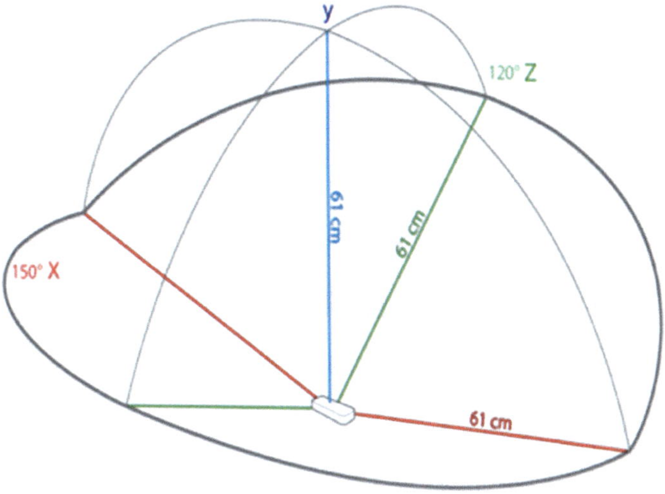

Fig. 3 Leap Motion Controller interaction area

for Windows to use with the Kinect for Xbox One, 4 GB of RAM, and a Graphics card that supports DirectX 11. In comparison with Kinect v1, the Kinect v2 SDK have improved the body, hand and joint orientation, and provide advanced facial tracking and simultaneous multi-app support.

Leap Motion Controller (LMC) is a small, rectangular device (13 mm × 13 mm × 76 mm) that weights 45 g. The Leap Motion Controller (LMC), was publicly presented for the first time in 2013. As Kinect device has low accuracy for detection of hand movements we include the LMC device in our system for providing capacity of training movements during wrist or hand rehabilitation. The LMC consists of three IR (wavelengths = 850 nm emitters) and two IR cameras [28]. It's dual platform (Macintosh/Windows), connects to a computer via USB 3.0 connection. It streams data at a variable acquisition rate, that varies between mean value of less than 40 Hz [29] to up to 120 Hz [30], under both static and dynamic conditions [29]. It has a full-functioning SDK that allow to rebuild the 3D scene and track the positions of the hands and fingers. The controller itself is accesses and programmed through Application Programming Interfaces (APIs), with support for a variety of programming languages, such as C++, Python, Unity, Unreal [31]. Positive values of the vertical y-axis increase upwards. The effective range of the controller extends from approximately 25–600 mm above the device (Fig. 3).

The activated sensor send data on position of hands and fingers during movements. The millimetre accuracy of the LMC when the object is placed below 250 mm distance from the sensors was already demonstrated [28–30]. To improve the repeatability of the system detection the participants should be informed about the optimal position of the hands above the LMC device. LMC was integrated in the developed system by using the device SDK. It provides data access through direct hand and

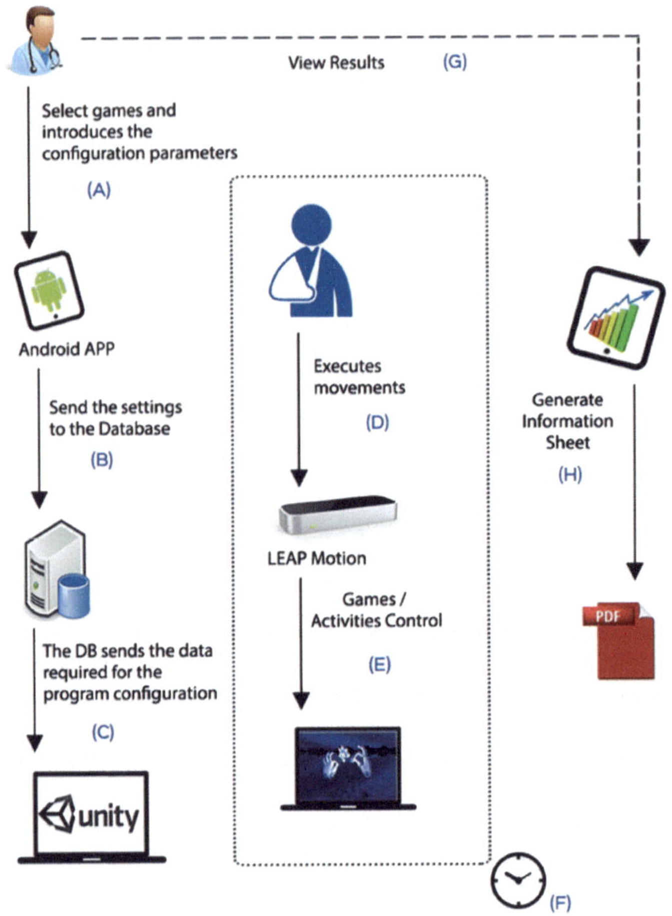

Fig. 4 Diagram with time flow of the LeaPhysio system

finger mapping detection. This feature has great importance for the development of applications that include interaction with virtual reality. In Fig. 4 is represented a time flow of the use of the LeaPhysio—Games Enhanced Physical Rehabilitation system. The letters of A-H indicate the execution steps.

After patient registration in the system, the physiotherapist can include and/or analyse data on clinical status and create an individualized training plan. The training plan setup involves parameters (*A*) as: initial and final date, duration of the game, number of repetitions per training, hand(s) that should make exercises. When a training programs is created for a patient, the mobile application sends the data to the remote database (*B*). The computer application initiates an HTTP request through the corresponding PHP script hosted on the server so that it contacts the MySQL database and sends the necessary data to the computer application according to the user input using an HTTP response (*C*). In a planned session the patient should play a serious game configured with the parameters of the training (*D*). The developed virtual environment uses LMC to replicate hands and their movements in the game scenario (*E*). Each training can have a configured time duration, and can be repeated (*F*). The physiotherapist can use the mobile application to change any parameter in the patient's training and to visualize the results obtained (*G*). The application provides to the physiotherapists or patient a report (*H*) of the exercises already performed and their respective scores.

Internet of Things (IoT) edge computing platform. A i7 512 GB, miniPC linked to a display HDMI compatible, is primary use for processing the data coming from the sensors and to run the developed serious games. The system can use the sensors from Kinect and Leap Motion Controller, but also inertial measurement unit (Shimmer) and from the force platform developed by our team. The data computed in IoT devices is delivered to the cloud server for storage and advanced data analysis. The setting of parameters for serious games can be done on mobile device applications or on a PC interface. The training settings are stored on the IoT cloud. Using the implemented software on IoT devices (edge computing), the stored settings from IoT cloud are automatically downloaded and used to setup the serious game according with training plan for an identified user.

The **framework software** includes different components such: the serious game engine based on Unity 3D; mobile applications; database. ***Unity3D*** is a multiplatform game engine developed by Unity Technologies and used for the development of video games for computers, consoles, mobile devices and websites using languages such as C++, C#, Python, JavaScript and Action Script 3. It is able to support 2D and 3D graphics. Can be associated with different computation devices such as mobile (smartphone and tablets), PC and PS4. The Unity3D is platform that is increasingly used by developers for games creation due to its powerful game engine. ***Mobile applications*** were developed for data visualization and for serious game setting for different physiotherapy plan. An integrated development environment (IDE) for Android platform were used for mobile applications development. The applications keep track of sessions, and some demographic and biometric data of patients and of physiotherapists. The data generated by games (i.e., scores but also the measured metrics associated with body movements when playing serious games) are stored in ***MySQL database***. The implemented database permits to register a new user, introducing their demographic and biometric data (e.g., name, gender, age, height, weight). The data regarding the games and exercises done by the patient (i.e., number of movements repetition, number of times played, scores) are also stores. Using the joints' position information of the user under rehabilitation, different metrics were calculated that can be used by the physiotherapists for objective evaluation of the rehabilitation process. The game customization parameters can be accessed from the game interface or from the mobile application, either by patient or by physiotherapist.

The **cloud server** is responsible for storing all the information regarding data on patients, programs of motor rehabilitation sessions, and exercise results. The LAMP (**L**inux: Operation System; **A**pache: Web Server (HTTP) software; **M**ySQL: Database component; **PHP**: Programming language) architecture was used. The main objective of the server is that all information is centralized in one place and can be accessed remotely by the system applications. By storing this data in a cloud it's possible to access the relevant data through the mobile application that can be used by the physiotherapists or the system manager. Also, good performance of the Unity application for the patients, during configurations of the games and data collection, is obtained by using a cloud server.

Serious Games Based on Kinect

Relevant information on human movements (i.e., joints movements, contribution of muscle synergies on different movements) can be captured in unobtrusive way during physical exercises, by using the 3D images provided by Kinect device. Recent literature review on Kinect applications for motor rehabilitation have shown that beside some limitations and possible errors in the motion capture, the Kinect-based rehabilitation has great potential in making therapy and alert systems financially accessible and medically beneficial for many patients [14, 23]. Our team developed several games for motor rehabilitation. The games that are presented in this chapter support understanding the benefits or drawbacks of several features related to virtual environments for potential users in rehabilitation context.

TheraSoup Game

The game *TheraSoup* was initially developed using Kinect v1. The game was adapted to Kinect v2 and new features were included. In this game a room is presented with several shelves on which are located different vegetable. In the middle of the room there is a pot where the avatar moved by movements of player should put the vegetable to make a soup. On the front wall is presented a receipt for the soup and the player should choose from the shelves the vegetable indicated in the receipt and put it in the pot (Fig. 5). The game allows customization: less shelves at low height from the floor (i.e., for people with low capacity to lift one or both arms), more shelves, higher height from the floor; less or more vegetable on the shelves; lower or higher speed of vegetable appearance on the shelves; the wall aspects; the music, and the addition or not of other sensors (IMU or force plate). The customization allows creation of a simple environment (i.e., without many things on the shelves, with less coloured objects on the walls) or a richer environment that will increased mental workload during exercising. The game support integration of inertial measurement units (Shimmer IMU) and a force platform developed by our team. Beside increasing information on posture, balance, and body segments movements, these sensors are important for the patients in motor rehabilitation that use rehabilitation equipment like crutches and walkers. The sensors attached to this equipment give relevant feedback for patient and information on progress made by patient during

Fig. 5 *TheraSoup* game virtual environment

rehabilitation process (i.e., posture and balance when using the walker or crutches during upper limbs movement training). After the patient or physiotherapists perform Login/Register (to have their data tracked and stored on the game database), by a simple menu the patients or physiotherapist can configure other features of the game than those provided by default, or can initiate the game (Fig. 6). During the game a graph representation is used for feedback on performing movements. After the game the user should indicate by using also the interface based on Kinect what was the level of pain (in a visual scale) that he/she felt in the arm(s) that were trained during the game. This information is stored together with the results on angles and velocities of the upper limbs movements. Report on progress can also be accessed using the application of this game.

Library Game

The mechanics and dynamics of the *Library* game is similar to the *TheraSoup* game. The game presents a room with paintings in the walls, a window in the front wall, a basket in the middle of the room and shelves where are located different books (Fig. 7). To give more reality to the environment the image of windows is animated (from time to time a horse pass creating an illusion of dynamic environment outside the room). The player should take the books from the shelves and put on the basket. Similar to *TheraSoup* game, this game allows customization taking into account the needs and the preferences of the players. The game application also integrates

Fig. 6 *TheraSoup* apps—initial screen

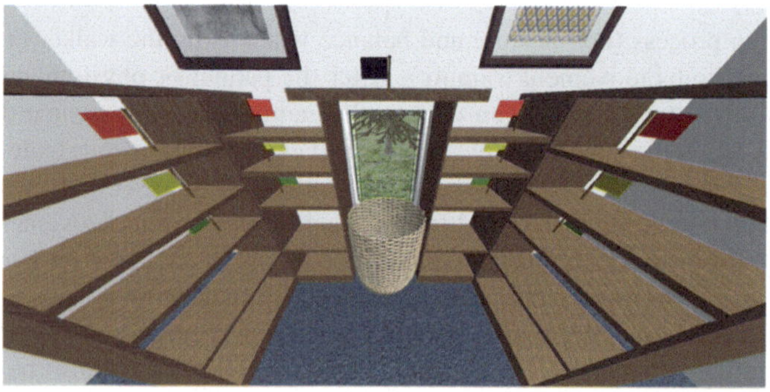

Fig. 7 Library game virtual environment

other sensors (IMU and force platform) in order to obtained more comprehensive information on posture and balance of player during reaching movements.

Step on the Tile Game

The game objective is to train the balance and movement coordination through a scenario where the user is stepping on the tiles that will be spawned on the game environment. The user should perform the lateral movements or front and back steps

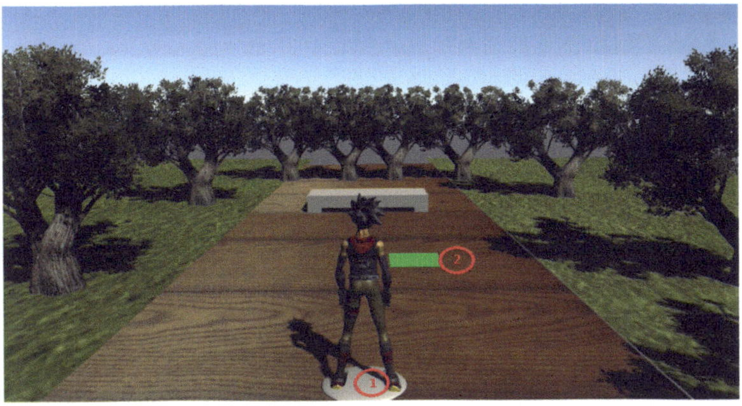

Fig. 8 Step on the Tile game virtual environment

(Fig. 8). The setting of games can be changed. The game duration, tiles spawn range (i.e., extended range will conduct to movements with higher amplitude), tiles speed (i.e., to increase velocity and coordination of the movements), and the score objective, can be configured. According with the patients' needs it is also possible to choose which leg motion will be trained. The player should step the tiles as they appear on the floor, reaching the objective score within the imposed time limit. By playing this game, the patients can train their lower limb mobility, the coordination of movements, reaction time and balance. The game may also be used for cognitive rehabilitation—the user should step only on the coloured tiles that was configured for earning points (e.g., green tales), and avoid stepping the different coloured tales (e.g., red tales) which might lead to losing points.

Before playing the game, the physiotherapist must evaluate the patients' movements and start with easy level of game. On *Start* the patient should be placed as indicated in the environment (Fig. 8) to control the avatar. The colours of the tile and the associated points that are earned or lost on stepping on tile can be configured before the game. The customization of the game allows also playing a game by side steps or front and back, by selecting "Horizontal" or "Vertical". The data on player movements and scored obtained during the game are stored in a local database and several metrics are calculated (i.e., average velocity of each movement; angle between legs when the step is performed; distance between feet when tile is stepped; number of steps for whole training period). The game has also a mobile application (Fig. 9) for configuration of the game and to see the results of each training session.

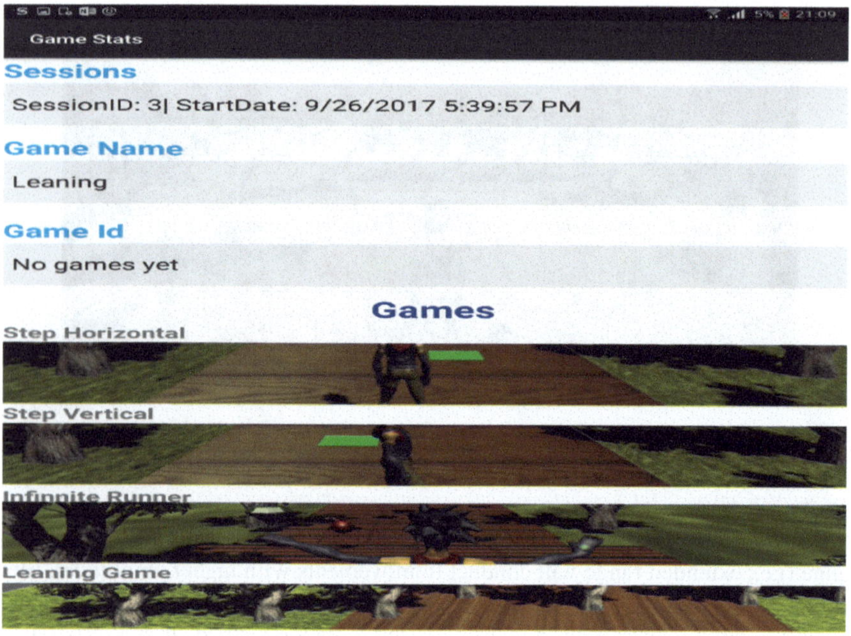

Fig. 9 Smartphone app of Step on the Tile game

Don't Hit the Wall Game

This game can be used for training upper limb movements. The player should grab a virtual ball and move it through the game map where he/she should try to reach the exit without touching the walls (Fig. 10). At beginning of game, the player has a number of lives that should be preserved during realization of movement in order to earn points. By touching the wall with the ball, the player loose one life. Different maps can be configured at the beginning of the game with different level of complexity. Mobility, coordination, endurance, muscle strength of upper limbs can be trained using this game. A smartphone apps was also developed for this game.

Infinite Runner

This game was designed also for upper limb training. The player should catch different objects by moving one arm or both arms. Along the game path, the avatar should touch different objects that appear on virtual environment, which give and take points and lives, so to win the player need to reach the established score preserving minimum one life (Fig. 11). During the game the data is collected when an object is caught with the hands—the angle of the arm according with position of

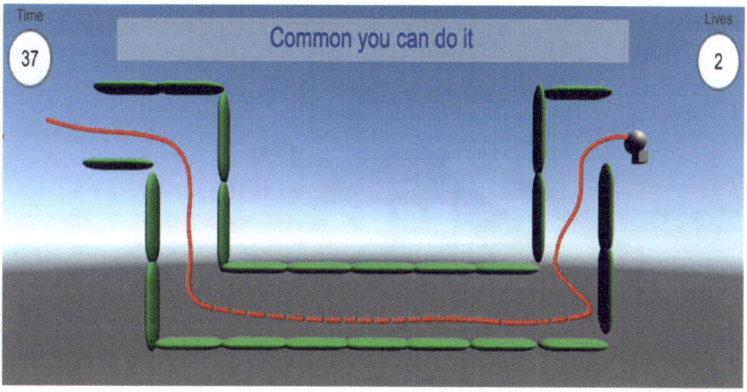

Fig. 10 Don't Hit the Wall game virtual environment

Fig. 11 Infinite Runner game virtual environment

the object is calculated and stored in the database. Game customization include the number of spawning objects, the player number of lives, the player speed and the score. The game results give information on upper limb mobility (i.e., the progress in the number of pick-up objects and the values of arm angles when objects were touched) and reaction time. The game has also an application for configuration of the game and to see the results of each training session.

Serious Games Based on Leap Motion Controller

This technology that allow unobtrusive detection of hands movements, has an important advantage for hand rehabilitation, as the patient does not need to wear data gloves or to hold any device in the hands. It has great importance mainly after hand surgery. The small dimension, the low price and millimetre accuracy of LMC

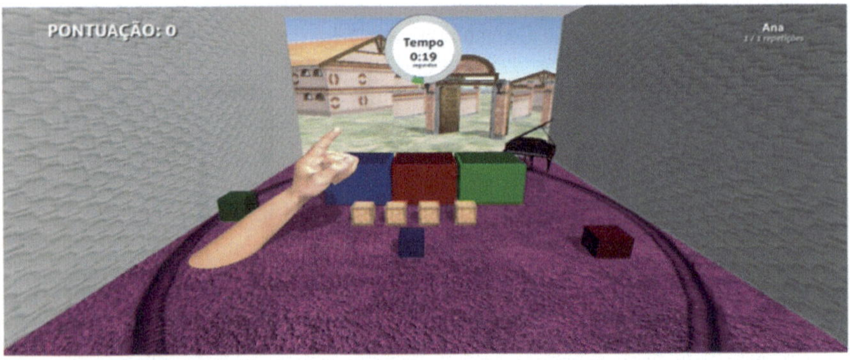

Fig. 12 CollectCube game virtual environment

are advantages that may be determinant for the development of myriad applications for motor rehabilitation environment. Study on LMC efficacy for characterization of essential tremor was recently presented [32]. Virtual environments where LMC is used for training movements of the hands of children with cerebral palsy [33] and stroke patients [34] were designed and implemented. Also, several serious games for upper limb training (VirtualRehab) based on LMC were developed by VirtualWare [35]. The goal of our work was the development of a serious game that can be used for assessment and training of fingers movements that includes an extended number of exercises relevant for motor rehabilitation.

CollectCube Game

The game goal is to pick up a cube and to introduce it according with the colour in the corresponding box to earn points. During the game, various cubes appear on the screen. For each correct action the patient earns a point and for each failure to grab a cube loose a point. Figure 12 represent a sample from the *CollectCube* game environment. This game has three levels of difficulty: (1) the scenario with only a single box with only a colour (e.g., green) and over time appear cubes with the same colour; (2) a scenario with medium difficulty has two boxes (e.g., green and red) and over time randomly appear cubes with green or red colour and player should put the cube in the box with the same colour as the cube; (3) a more complex scenario has three boxes (e.g., green, red and blue) and over time randomly appear cubes with green, red or blue colour that the player should organize in corresponding boxes. In the scenarios with higher complexity the boxes change places with each other in a pre-established period demanding always checking by the user the position of the box when it grabs a cube. This is important to increase attention of the patient during the game (to induce a flow or cognitive absorption state) that might decrease the sensation of effort and increase training time and velocity.

Fig. 13 CollectCube game with various objects that can be included in virtual environment

As many types of hand or finger impairments may be trained in motor rehabilitation session the system need to be adapted to various patient needs. The application game allows different configurations of gesture that should be used to organize the cubes. The games are adapted according to the characteristics of the patient: *Gender* (Male or Female) and *Age*. The scenario changes if the patient is male or female. The differences are obtained by changing the textures of the objects that make up the scenario, but also by exchanging certain objects for others. The game allows configuration of time for exercises and type of hands that can be used (i.e., left or right hand, robotic hands, man or woman hand). When the physiotherapist creates a training for the patient, the avatar hands are considered according with the patient gender or patient age. No visualization and interaction with virtual games of a not selected hand is produced during the games even if both hands are used. The boxes can change the colour during the game. Also, the user can choose to collect fruits, vegetables, or pets (Fig. 13). The therapist may change the setting of the game creating new scenarios for the rehabilitation sessions by changing the size of objects that should be collected, their colour and texture as well as the colour of walls that limits the virtual space of game.

The data is stored in a .CSV format in local database to facilitate future analysis. The developed software allows representation of amplitude and time interval for each different movement, standard deviation of signal amplitude, Poincaré plots, frequency of movements realized during a game by using Fast Fourier Transform (FFT) of Z and Y position signals, similarities and time delay analysis between signals from right and left hand calculated by cross-correlation and argument of the maximum. The serious game has also a smartphone app that allows configuration of the game and visualization of data from the game, and progress during various training session (Fig. 14).

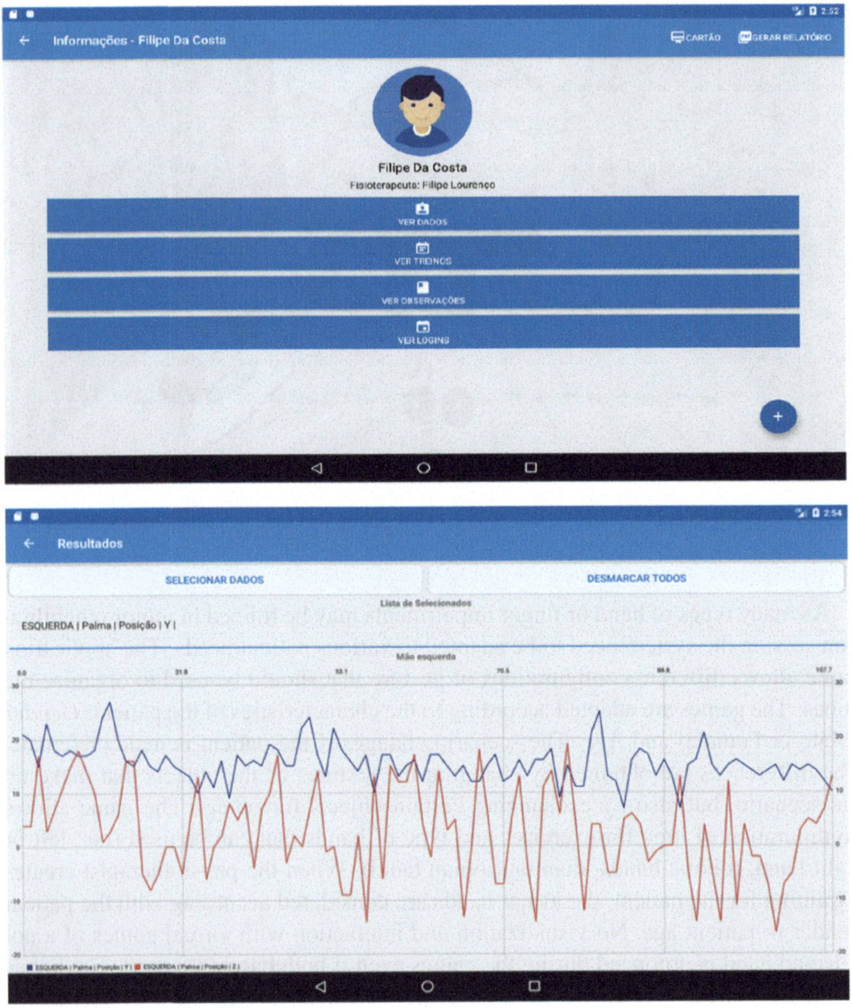

Fig. 14 Examples of screens from LeaPhysio app

The apps that run in Android OS platform was developed taking into account easier and more comfortable management of data related to patients training programs but also because smartphone or tablet support may contribute for more informed decision making on interventions and management of physiotherapy process by including many types of resources that are required in a clinical setting: (i) communication capabilities—voice calling, video conferencing, e-mail; informational resources—guidelines, medical literature; clinical software application (e.g., disease diagnosis aids). Generally, in clinics these resources are mainly provided by desktop computers, the development of the mobile apps supports therefore the need for mobility in healthcare settings.

Acceptability and Usability Tests

To test feasibility of an intervention studies several key areas of focus was proposed [36]—the acceptability, the demand, the implementation, the practicality, the adaptation, the integration, the expansion and the limited efficacy. The engineers generally test the feasibility of implementation of a system by testing the degree of execution of the objectives stablished for a development of a system, a system efficiency, speed, accuracy of measurements, factors affecting implementation, cost analysis and in the lower extent the parameters from other areas. The exergaming is a complex technique that depends on the application of knowledges of persons from different professional's fields (e.g., computer informatics, engineering, healthcare) but also on the user needs, preferences, expectation, perspectives or behaviours. Therefore, the validation of the technique based on exergaming is a complex process that should need clarifying the construct validity (what instrument measures what it is supposed to measure), the content validity (the extent to which an instrument measure all aspects of a construct). Moreover, the evidence on feasibility of exergaming should show if this technique or technology associated with exergaming might work (can it work?), if it might be efficacious under ideal or actual conditions (does it work?) compared to whatever other practices might be done instead, and if it will be effective in real-life context, settings, and cultures/populations that might adopt the intervention as practice (will it work?) [36]

Evidence have shown that exergaming can and does work for balance improvement [12–14, 16, 19–22, 37–40], upper limbs and lower limbs movements [14, 21, 22, 39, 40], muscle strain training [24, 37]. However, the studies in the area of acceptability and demand of exergaming for motor rehabilitation are scarce [41–43] and instrument used for this measurement are very diverse. Different component of acceptability and demands were measured using validated or less validated methods. A number of theories have been used to assess the acceptability of different technology. For example, self-efficacy theory is used for perceived capability for learning or performing actions at designated levels [44]. The Unified Theory of Acceptance and Use of Technology (UTAUT) and Technology Acceptance Model (TAM) are used to predict and explain the acceptance in terms of the user behavioural intention to use and their actual use of technology [45, 46]. Theory of planned behaviour (TPB) foresees and explain the human behaviour in specific context [47]. The TAM is a widely used model applied in the explanation of user motivations, attitudes, and responses to acceptance and use of technology [48]. TAM includes the attributes: perceived usefulness, ease of use, and behavioural intention to use. Usability and user experience are the most used tests for evaluation of technology acceptance. Usability is defined by the International Organization for Standardization ISO 9241-11 as "the extent to which a product can be used by specified users to achieve specified goals with effectiveness, efficiency, and satisfaction in the specified context of use". Efficiency is related to the degree to which users can complete their tasks, the effectiveness is related to the degree of accuracy with which users complete their task and the satisfaction is the extent to which the expectations are met. Commitment to usability evaluation in product design and development offers many benefits, including satisfaction and intent to continue use of a tested technology. A little different than usability concept, the user experience (UX) concept is defined by Nielsen Norma Group as all aspects of the end-user's interaction with a service or a product. Besides ISO model to measure usability, five common attributes of usability are more investigated: memorability, errors, efficiency, learnability, and satisfaction [49, 50]. The importance of assessing both acceptability and usability in a separate manner was suggested [41]. The authors say that theories of acceptability generally focus on behavioural and predictive aspect of acceptability and diagnose design problem before users have experience with a system, whereas in usability evaluation, the issues are identified after users have used the system [41]. However, there is not standardized methods for acceptability and usability and different methods are used for it measurement. For example, the interactional experience is proposed being measured as general game experience [51] whereas others suggest being measured in terms of the color, menu, and typography experience of games [52]. Some studies focus measurement of usability in terms of fun [53] and enjoyment [54]. Also, the contents of satisfaction construct and the instrument of measurement are not clearly defined. For example, the gameplay satisfaction was measured with a scale of range 1–50 [55], while another study assessed game design satisfaction on a scale from very likely to very unlikely [52]. In our work we investigated the acceptance and usability of the developed serious games by using a questionnaire that was administered after the gameplay. The usability tests were conducted using a Portuguese translation of the Questionnaire for User Interaction Satisfaction (QUIS). This questionnaire that was developed by a multidisciplinary team of researchers in the Human-Computer Interaction Lab (HCIL) at the University of Maryland at College Park, evaluates users' subjective satisfaction with specific aspects of the human-computer interface. The QUIS is highly reliable across many types of interfaces. The score of criteria from QUIS questionnaire was between 0 and 9 (i.e., for help messages on the screen the participants should evaluate if was unhelpful or helpful using a Likert scale from 0 to 9). The QUIS questionnaire was included in the questionnaire developed by the team of research project. Beside the QUIS questions the used questionnaire in our work has questions that address the emotions that the game has produced in player (by using validated Portuguese questionnaire of PANAS-VRP, Positive and Negative Affect Schedule), the arousal, and the socio-demographic data. The questionnaire for therapists and the questionnaire applied to other type of participants in usability tests (healthy volunteers or physiotherapy patients) has 23 questions, among these 13 (those related to QUIS, PANAS-VRP, and arousal scale) are equal in both questionnaires.

The tests of all serious games developed by our team, presented in this paper, were realized with 8 participants, 4 female occupational therapists (age range 29–31 years) and 4 healthy volunteers (1 female and 3 men, age range 25–48 years). In these preliminary results the participants greatly appreciated the *TheraSoup* and *Library* games, medium appreciated *CollectCube*, and the game *Step on Tile, Hit the Wall* and *Infinite Runner* receive lower scores at questions from our questionnaire. For all games, the participants positively appreciate the games configuration, the tailoring of the games to the exercises required in motor rehabilitation as well as to the user

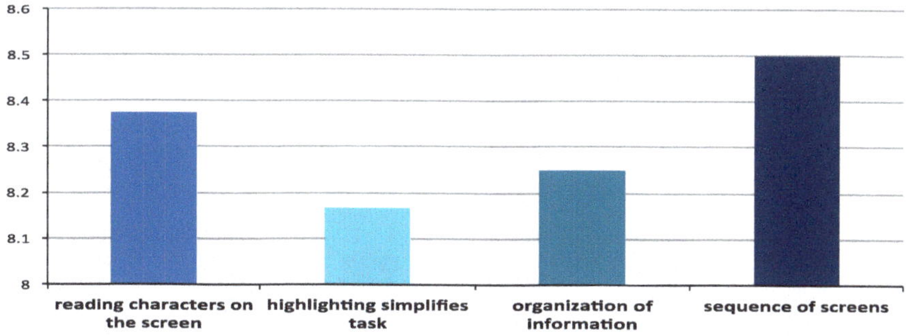

Fig. 15 Score of screen criteria for *CollectCube* exergame

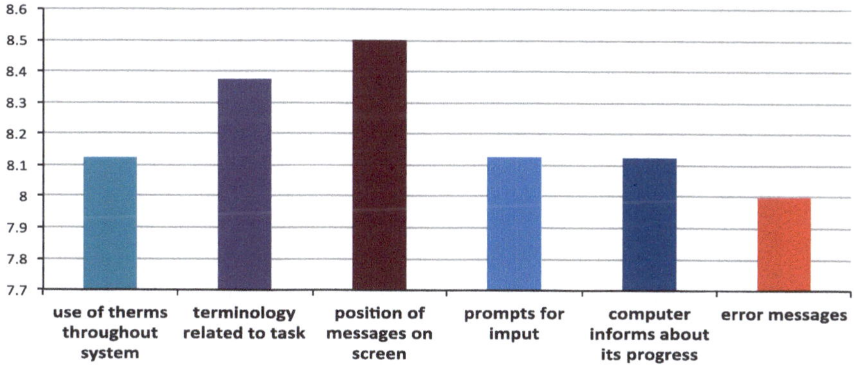

Fig. 16 Score of terminology and system information evaluation for *CollectCube* exergame

needs and preferences. The game ideas or aim of game, the scenes, colours, sound included in games as feedback, help messages, variety of games, games interaction was described as the most positive and relevant things in the developed system. Some problems were identified in *Step on the Tile, Infinite Runner* and *Hit the Wall* related the execution of the games and virtual environment design. More tests will be realized using methods that more clearly identify the cause of low acceptability, and correction of the application features will be carried out. General opinions on the serious games were good and the participants consider that this tool may increase the motivation of patients for training movements for motor rehabilitation. The results of testing acceptability and usability of *CollectCube* are presented in Figs. 15, 16, 17, 18, 19 and 20.

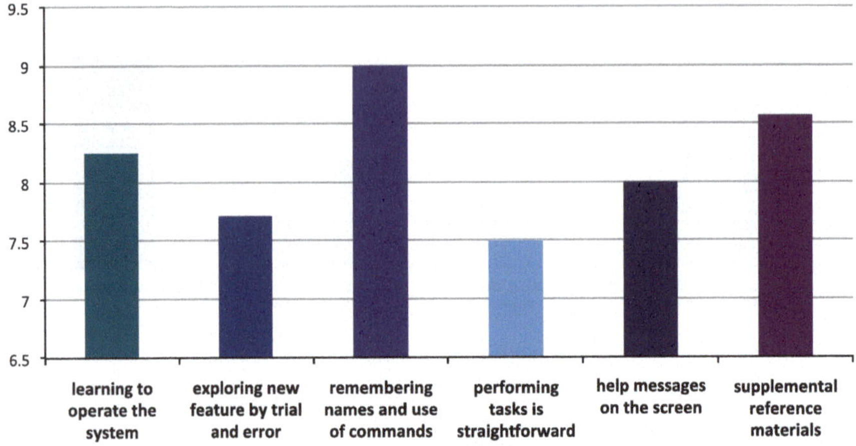

Fig. 17 Score of learning evaluation for *CollectCube* exergame

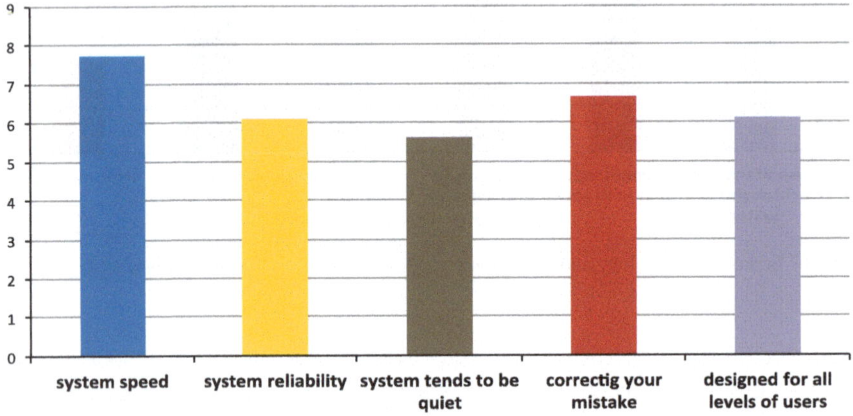

Fig. 18 Score of system capabilities evaluation for *CollectCube* exergame

User Centered Design

User-centered design (UCD) is an approach to design systems informed by scientific knowledge of how people think, act, and coordinate to accomplish their goals [56]. Presently, with researchers developing models and methods as well as heuristics for the usability or playability of games, this issue has great relevance for better understanding and development of human computer interaction [57–59]. We present here our insights, based on literature analysis and our research work during serious games development and tests. We focus on practicality of developed serious games as well on their acceptability and usability for motor rehabilitation. By practicality we mean the features of technological resources that we used in our work that are relevant

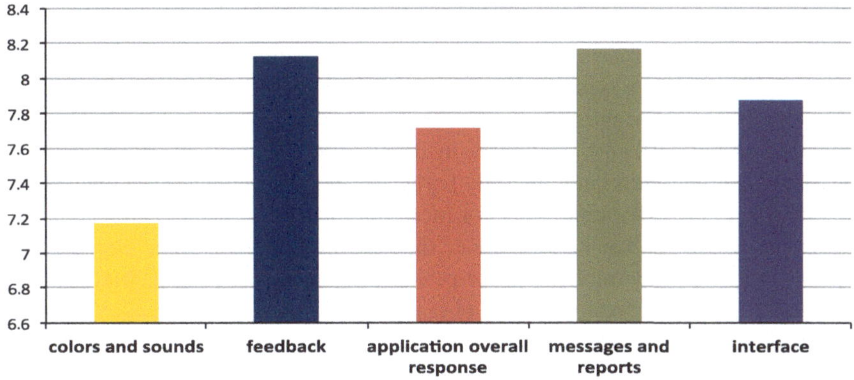

Fig. 19 Score of usability and user interface evaluation for *CollectCube* exergame

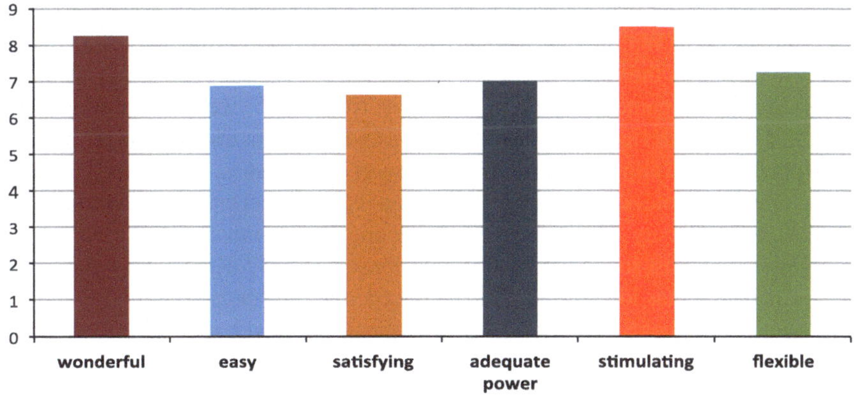

Fig. 20 Score of overall reactions to the software for *CollectCube* exergame

for motor rehabilitation compared with the most used and validated exergames. Discussion on acceptability and usability is related to perceived appropriateness, intent to continue to use, satisfaction of user related the use of serious games for motor rehabilitation.

Firstly, we enlist several **benefits** of exergames that were identified and summarized from different studies [4, 8, 12, 13, 60]: (1) can allow for task-specific, repetitive exercises to be delivered across a range of difficulty levels; (2) the gaming platform may allow remote observation of user performance, obtains objective data, provide real-time feedback and progress tracking; (3) motivate people to practice physical exercises through an attractive and interactive way and train both motor and cognitive skills when users performing dual tasks; (4) provide opportunity to interact with a variety of engaging activities, some of which may be too risky when practiced in the real world, while providing the feedback necessary for effective motor learning; (5) the players can focus their attention on the outcome of the movements in the game, not on the movements itself; (6) by engaging the player in fun and challenging virtual environments may increase overall compliance with treatment; (7) can be undertaken at home either alone or within a small group, which might make the activity more accessible and comfortable; (8) the exergame platform may provide access at any time in home-base; (9) allow setting the instructions for exercises and progress tracking; (10) the intervention may be fully replicable whether the game parameters are provided; (11) provide opportunity for more optimal use of health care resources; (12) the provided feedback could reduce the need of continuous supervision of the therapist and increased motivation for performing exercises; (13) facilitate patient self-management and independence; (14) promote well-being and functioning for elderly and people with disabilities.

The main advantages of using **Microsoft Kinect technology** are related to his features that allow natural user interfaces, and his cost. In many studies relating exergaming for motor rehabilitation, mainly commercial system (off-the-shelf exergames) were used for testing their validity and effectiveness for neuro-motor rehabilitation. The Nintendo Wii (Wii Fit with a balance board or Wii Sports with associated hand-held remotes) was the most frequently used [3, 12–14]. DanceDance Revolution (DDR) Stepmania game, Sony Playstation Eye Toy [12, 15] were also used in various studies. These devices use different sensors and inputs for tracking the movement of the player, and to allow the participant to be part of the games, that should be hand-helded (e.g., Wiimote) or are at large dimension (e.g., DanceDance Revolution). The low cost and diversity of games, made Nintendo Wii, comparing the other home PlayStation, the most sold home video game console in 2017. Kinect device has also small dimension, has low cost but more important are the capacity for natural user interface. The Kinect SDK allow ease development of novel, enjoyable, and high-quality system for motor rehabilitation. The trajectories of the user movements can be tracked by Kinect with high accuracy up to 4 m distance and a fixed Kinect sensor in a specific location can have a range of capture of roughly 8–10 m depending on space characteristics (i.e., light, obstacle to IR laser beam). The hardware advances of Kinect allow a more compact and direct human computer communication method of gesture and gesture patterns. Although, there is sufficient evidence for clinical use of Kinect for upper limbs movements tracking, mixed results have been reported for the shoulder movement caption [14]. One of three things can be the source of errors in shoulder movement detection: self-occlusion errors (which can be caused by the distance and angle between user and the sensor); specific movements such as placing a hand on one's back; or when the scene contained objects such as wheelchairs or walkers, or proportional biases which occurred in complex systems [14]. There is evidence that accuracy of movement detection by Kinect can decreased under significant user occlusion, or object interference [14, 61].

Virtual environments based on Kinect or LMC can be developed using different technologies, such as C++, C#, JAVA, Unity3D, Unreal. Many tools and information for building virtual environment can be found through internet in game community (e.g., GameMaker, GameSparks, Dev.Mag, GameArtisans, Polycount). Virtual reality as "a high-end user-computer interface that involves real-time simulation and interactions through multiple sensorial channels" [62] should induced a sense of "presence in" and "control over" the simulated environment [63, 64]. The sense of "presence in" consists of the feeling of being in an environment, even if not physically present in that environment; the sense of "control over" involves the possibility of interaction with the environment and objects giving a sense of being in that situation or environment [64, 65]. To place the user within a loop of real-time simulation, where he/she could feel being in "presence" or "control" of real situation or environment, the virtual reality systems require an output device or visual interface (e.g., computer, tablet or smartphone screen or head mounted display) and an input device for interaction (e.g., mouse, joystick, sensorized glove). Different virtual environments were designed for exergaming using Kinect. For examples, *Animal Feeder* and *Fruit Catcher* were developed by [66] that train coordination of movement in dual tasks, reaching and weight shift without movement of the feet. A game based on design of *Jewel Mine* game, to train balance and reaching movements, was tested by [67]. Commercial virtual environments based on Kinect were developed in which an avatar show the movement that a user platform should imitate (e.g., *Yugo* platform developed by BioGaming), or virtual environments with interfaces that allow different actions in environments to be induced by player movement (e.g., games developed by MiraRehab). Several frameworks for designing virtual environment for serious games were proposed. A MDA (standing for Mechanics, Dynamics, and Aesthetic) was proposed by [68]. The *mechanics* was defined as the particular components of the game and as the level of data representation and algorithms; the *dynamics* as the run-time behaviour of the mechanics acting on player inputs and each other outputs over time; and the *aesthetics* as the desirable emotional responses evoked in the player, when she interacts with the game system. The fun of the games (aesthetics) should be characterized by: sensation (game as sense-pleasure); fantasy (game as make-believe); narrative (game as drama); challenge (game as obstacle course); fellowship (game as social framework); discovery (game as uncharted territory; expression (game as self-discovery); submission (game as past-time). The dynamic of game work to create aesthetic experiences. In MDA framework the various actions, behaviours and control mechanisms afforded to the player within a game context are included in the mechanics of the game. A 6-11 framework for game analysis and design was proposed by [69]. The 6-11 framework focuses on *six emotions* (fear; anger; joy/happiness; pride; sadness; excitement) and *eleven instincts* (survival—fight or flight; self-identification; collecting; greed; protection/care/nurture; aggressiveness; revenge; competition; communication; exploration/curiosity; colour appreciation) that can contribute for engaging player in the game. A comprehensive framework for designing gamification ecosystem that enlists many methods used for development of serious games was published by [70]. Based on work of [71] they described a set of elements for game mechanics (i.e., points, levels, leader boards, badges, challenges/quests, on-boarding, engagement loops) and game dynamics (i.e., pattern recognition; collecting; surprise and unexpected delight; organizing and creating order; gifting; flirtation and romance; recognition for achievement; leading others; fame, getting attention; being the hero; gaining status; nurturing, growing). A different categorization of game elements comparing MDA framework was also presented by Werbach and Hunter [72]. They classified the game elements in *mechanics* (i.e., challenges; chance; competition; cooperation; feedback; resource acquisition; rewards; transactions; turns; win states), *dynamics* (i.e., constraints; emotions; narrative; progression; relationship) and *components* (i.e., achievements; avatars; badges; boss fights; collections; combat; content unlocking; gifting; leaderboards; levels; points; quests; social graphs; teams; virtual goods). Beside these elements in the development of virtual reality environment for serious games the components of user interface (UI) should be considered. Diegesis theory [73] allow classification of UI components in *diegetic, non-diegetic, spatial* and *meta*. The non-diegetic UI component is most often located on the screen plane above the game, is not part of the game story and the game space. Spatial UI is part of the game space but is not part of the game story (i.e., the game characters are not aware of them while the player are aware of them). Meta UI is part of the game story but is not part of the game space (i.e., components that are expressed as part of the narrative—e.g., blood spatters). Diegetic UI is part of the game story and part of the game space (i.e., provides the player with cues and information without distracting them from the narrative of the world and make experience more involving). Most of Wii game and Kinect based game used non-diegetic and spatial UI component. Considering the particularity of the patient receiving motor rehabilitation the diegetic UI should be more appropriate in order to make the experience more involving by reducing mental workload and increase focus on task demands for motor rehabilitation.

Beside the common elements for games, the serious games for motor rehabilitation require additional methods to be considered when designing these games. Three main steps were recommended by [74] for design of effective and safe therapeutic exergames: (1) a clear **identification of all the exercise requirements** (i.e., including the challenge level and progress assessment), not only in terms of goals of therapeutic intervention, but also in terms of additional constraints is necessary; (2) **collaboration** between information and communication technology experts and health professionals **for maximizing the effectiveness of exergame** implementation by inclusion of exercises in virtual environments; (3) inclusion in the virtual environments of exergames of aesthetic and dynamic elements that **maximize the entertainment of virtual environments for therapeutic exercises**.

Therefore, the answers for three important questions should be the key for better exergames designing: (1) what are the therapeutic intervention that exergame can improve in term of accessibility and quality; (2) what technology and what game mechanics and dynamics should be used to increase effectiveness of exergaming; (3) what elements of game engage more the user for exergaming.

The games should set clear therapeutic goals (i.e., forearm movement, upper extremity control, angular velocities of limbs, eye-hand coordination, endurance, speed, accuracy, range of motion, trunk movements, core posture, balance, muscle strength), set the tasks, and record and display progression so that the users feel that they are achieving their goals and managing to improve their physical function. To be effective the games should elicit specific movement characteristics in the players that are considered relevant for the function being trained [11, 74]. Also, to improve interaction between subjects and virtual environment related to a program of exercises, and to keep patient involvement in exergaming at high level, environments that facilitate ***greater transfer of learned or relearned movements*** by elderly people or people with disability to real-world situations should be implemented. Virtual ambient that are more appropriate to real living context of patient are more accepted by patients, had significant effect on perceived usability in elderly people, and produce greater transfer of movement learned during exergaming [75]. In the games that we developed we also observed that environments more appropriate of reality of living (*TheraSoup, Library, CollectCube*) was more accepted by the participants at usability tests.

Although, there is evidence that have shown that in most of studies exergaming offered similar results as conventional therapy [19, 76–78], or alternative exercising [12] and in some studies superior effect of exergaming compared with conventional therapy [11, 13, 21, 65] the information on what elements of games are more important for a category of population receiving one type of motor rehabilitation (e.g., for upper limbs movement rehabilitation in elderly patient) is scarce. The existing commercial platform for exergaming generally were not designed particularly for training movements of subjects with motor and/or neurological incapacity as for elderly people or people with motor disability. Many of the activities in the commercial games: have fast paced; the colours and moving of graphics are not age-appropriate; some games can be too colourful and with excessive visually busy game interactions for some patients (i.e., people with severe disability); music also might be unsuitable for some people; some game can have too demanding navigation through the user interface; and many games have limited game speed control or feedback on performance of exercising [11, 21, 41]. In several studies with elderly people these features have been noted as essential for compliance to exergaming [21]. For many patients, fast decision making and rapid movements required in virtual environment might be too difficult to execute. The number of mistakes was reported to increase proportionally with speed of the game [21]. In our work, special emphasis was given for the capacity to personalize the game (the player can choose an environment more for female or for male as in *CollectCube* game, can configurate the speed and colours of some components of games as in *Step on the Tile*, *Infinite Runner*, can simplified the components of virtual environments as in *TheraSoup* and *Library* games. Tailored environments to subjects needs and preferences may contribute for development of exergames that are more enjoyable and that engage more the people in performing exercises.

The **progress** can be included in the game space or in the game story or can be a report summarizing the relevant information for patient or therapist. There is a lack of studies on elements of games that may allow automatic adaptation of level of challenge and progress representation for each individual player based on their current performance. Progression should be ensured by automatically adapting the level of difficulty and the physical and mental workload, to the current performance level, to make games more challenging when the patient that perform rehabilitation exercises achieves consistently good performance, and less challenging when the difficulty of patient for exercise realization are higher. In our work this setup is made at the beginning of the game but research work is ongoing for automatically adapting the level of challenge to player performance using a diegetic user interface. Personalized

feedback has also great importance for patient in motor rehabilitation. Feedback was mainly represented in different games in the form of visual popups (e.g., points, informative messages) or audible notifications (sound effects). The feedback, is important to be adapted to the game story and patient need. Also, should be distinguishable as well as a reasonably lasting effect [66]. Users' preferences can be utilized to enable or disable type of a feedback [41]. For instance, older patient did not like showing personal data on virtual environment (e.g., body mass index (BMI) and the center of gravity showing during balance tests). Such form of personal data should be avoided mainly in multi-player settings [79]. Feedback can inform on how movements should be performed and improve engagement and adherence to the exergaming. In designing rewards and feedback, both intrinsic and extrinsic **motivation** needs to be considered [80]. Maintaining motivation and engagement have been reported to be vital to long-term exercise adherence [81]. Also, the flow or cognitive absorption when a patient "focuses" more on a game than his or her impairment has been shown to contribute on turning exercise more enjoyable and motivating [82, 83]. The provision of choice is one of the important tools of fostering intrinsic motivation and enjoyment in serious games, taking into account the self-determination theory (SDT) [84, 85]. Self-determination theory has been applied to identify which factors sustain individuals' motivation within video games [86]. SDT postulates that motivation increase if basic psychological needs for autonomy, competence, and relatedness are satisfied within a game context, making the experience more enjoyable [87]. Other theory that should be considered when designing games that should change behavior and increase motivation is Social Cognitive Theory (SCT) [88]. SCT proposes that behavior change is a function of enhanced skills and confidence (self-efficacy) in doing the new behavior [88], while modeling [88] and feedback [89] has great importance for learning skills. A comprehensive model of learning for behavior change in video games is based on SCT and the elaboration likelihood model of persuasion [90], and includes the following steps: attention, retention, production, and motivation. The role of **play in learning** was comprehensively analyzed by Vygotsky's, in his social constructivist theory [91]. Also, the concept of "flow" theory in task engagement [92], should be considered in exergames design. For example, flow theory suggests that engagement in learning is highest when perceived challenges and skills are well matched [92]. Also, was shown that learning is enhanced when participants discover and use information rather than memorize it [93]. In addition to providing individualized levels of difficulty, provisions of choice can give to the patient a sense of autonomy and control over their activity during exergaming. Therefore, all these theory of motivation and learning should be considered on designing games for motor rehabilitation in order to maximize the effectiveness of exergames for movements improvements, to increase patients' adherence and engagement to therapeutic intervention, and to facilitate greater transfer of learned and relearned movements to real-world situation. Other thing that contributes to the motivation of patient during realization of rehabilitation exercise, and which should be considered on designing and evaluation of exergames for motor rehabilitation, is often unpleasant and/or **painful** experience that can lead to decreased patient' tolerance for exercising [94]. Decreased tolerance or motivation often lead to intentional and unintentional "cheating" or, in the worst-case scenario, avoidance of rehabilitation exercises altogether [67]. For this reason, diegetic UI that creates a more immersive environment for rehabilitation may contribute to increase engagement and adherence to exergaming. Another factor that can demotivate and should be always be considered in therapeutic games, can be the absence of native language in the game. This is important mainly for older adults [79]. One more factor that should be considered is **setup support**. Inadequate setup support demotivates people to use exergames. The system for exergaming that is easy to install and setup, or that provide meaningful assistance has higher acceptability [41, 95].

Conclusions

Microsoft Kinect and Leap Motion Controller are valuable tools for development of a system for exergaming during motor rehabilitation. Integration of these two devices in serious games can increase patient engagement and adherence to therapeutic exercises by providing a natural user interface and enjoyable environment. The importance of user centered design for a system for tracking and progress evaluation of movements during motor rehabilitation was underscored. Future studies should consider in testing the feasibility of a system based on serious games for the rehabilitation, if the system might work, if it might be efficacious under ideal or actual conditions, compared to whatever other practices might be done instead, and if it will be effective in real-life context, settings, and cultures/populations that might adopt the intervention. It is important also to be considered in feasibility studies of system for exergaming for motor rehabilitation the parameters related to implementation and practicality but also parameters that measure acceptability, demand, adaptation, integration, expansion and limited efficacy.

Acknowledgements The work was supported by Fundação para a Ciência e a Tecnologia, project PTDC/DTT-DES/6776/2014 and Instituto de Telecomunicações.

References

1. Game Market, Statica (2018), https://www.statica.com/statistics/246888/value-of-the-global-video-game-market/
2. Pew Research Center—Number of Gamers. Statica (2018), https://www.statica.com/statistics/748044/number-video-gamers-world/
3. Best Selling Games, Statica (2018), https://www.statista.com/statistics/264530/all-time-best-selling-console-games-worldwide/
4. RnR, https://www.marketsandmarkets.com/PressReleases/serious-game.asp
5. Game Industry Revenue. Statica (2018), https://www.statista.com/statistics/733616/game-based-learning-industry-revenue-world/
6. C.C. Abt, *Serious Games* (University Press of America, New York, 1970)
7. B.P. Bergeron. *Developing Serious Games*, 1st edn. (Hingham Charles River Media, 2006)
8. D. Michael, S. Chen, *Serious Games that Educate, Train and Inform* (Thomson, Boston, MA, 2006)

9. D. Thompson, T. Baranowski, R. Buday. Serious video games for health. How behavioral science guided the development of a serious video game. Simul Gaming **41**(4), 587–606 (2010)
10. D.M. Harris, T. Rantalainen, M. Muthalib, L. Johnson, W.-P. Teo, Exergaming as a viable therapeutic tool to improve static and dynamic balance among older adults and people with idiopathic Parkinson's disease: a systematic review and meta-analysis. Front Aging Neurosci. **7**(167), 1–12 (2015)
11. N. Skjæret, A. Nawaz, T. Morat, D. Schoene, J.L. Helbostad, B. Vereijken, Exercise and rehabilitation delivered through exergames in older adults: an integrative review of technologies, safety and efficacy. Int. J. Med. Inform. **85**(1), 1–16 (2016)
12. L. Donath, R. Rössler, O. Faude, Effects of virtual reality training (exergaming) compared to alternative exercise training and passive control on standing balance and functional mobility in healthy community-dwelling seniors: a meta-analytical review. Sports Med. **46**(9), 1293–1309 (2016)
13. S.G.R. Neri, J.R. Cardoso, L. Cruz, R.M. Lima, R.J. de Oliveira, M.D. Iversen, R.L. Carregaro, Do virtual reality games improve mobility skills and balance measurements in community-dwelling older adults? Syst. Rev. Meta-Anal. Clin. Rehabil. **31**(10), 1292–1304 (2017)
14. D. Webster, O. Celik, Systematic review of kinect applications in elderly care and stroke rehabilitation. J. Neuroeng. Rehabil. **11**(108), 1–24 (2014)
15. V.L. Klompstra, T. Jaarsma, A. Strömberg, Exergaming in older adults: a scoping review and implementation potential for patients with heart failure. Eur. J. Cardiovasc. Nurs. **13**(5), 388–398 (2014)
16. D.M. Harris, T. Rantalainen, M. Muthalib, L. Johnson, W.P. Teo, Exergaming as a viable therapeutic tool to improve static and dynamic balance among older adults and people with idiopathic Parkinson's disease: a systematic review and meta-analysis. Front Aging Neurosci. **7**, 167 (2015)
17. P.L. Weiss, E. Tirosh, D. Fehlings, Role of virtual reality for cerebral palsy management. J. Child Neurol. **29**(8), 1119–1124 (2014)
18. B. Bonnechere, B. Jansen, L. Omelina, M. Degelaen, V. Wermenbol, M. Rooze, S. Van Sint, Jan. Can serious games be incorporated with conventional treatment of children with cerebral palsy? A review. Res. Dev. Disabil. **35**, 1899–1913 (2014)
19. Y. Laufer, G. Dar, E. Kodeak, Does a Wii-based exercise program enhance balance control of independently functioning older adults? A systematic review. Clin. Interv. Aging 1803–1813 (2014)
20. S.D. Choi, L. Guo, D. Kang, S. Xiong, Exergame technology and interactive interventions for elderly fall prevention: a systematic review. Appl. Ergon. 1–12 (2016)
21. E. Pietrzak, C. Cotea, S. Pullman, Using commercial video games for falls prevention in older adults: the way for the future? J. Geriatr. Phys. Ther. **37**, 166–177 (2014)
22. K.I. Molina, N.A. Ricci, S.A. Morais, M. Rodrigues, Virtual Reality using games for improving physical functioning in older adults: a systematic review. J NeuroEng. Rehabil. **11**(156), 1–20 (2014)
23. H.M. Hondori, M. Khademi, A review on technical and clinical impact of Microsoft Kinect on physical therapy and rehabilitation. J. Med. Eng. (Hindawi Publishing Corporation) **846514**, 1–16 (2014)
24. S.R. Ellis, What are virtual environments. IEEE Comput. Gr. Appl. **14**(1), 17–22 (1994)
25. K.J. Miller, B.S. Adair, A.J. Pearce, C.M. Said, E. Ozanne, M.M. Morris, Effectiveness and feasibility of virtual reality and gaming system use at home by older adults for enabling physical activity to improve health-related domains: a systematic review. Age Ageing **43**(2), 188–195 (2014)
26. O. Wasenmüller, D. Stricker, Comparison of kinect v1 and v2 depth images in terms of accuracy and precision, in *Asian Conference on Computer Vision—ACCV 2016 Workshops* (2016), pp. 34–45
27. MS Press, PrimeSense supplies 3-D-sensing technology to "Project Natal" for Xbox 360 (MsPress, 31 March, 2010), https://news.microsoft.com/2010/03/31/primesense-supplies-3-d-sensing-technology-to-project-natal-for-xbox-360/

28. F. Weichert, D. Bachmann, B. Rudak, D. Fisseler, Analysis of the accuracy and robustness of the Leap Motion controller. Sensors **13**, 6380–6393 (2013)
29. J. Guna, G. Jakus, M. Pogacnik, S. Tomazic, J. Sodnik, Na analysis of the precision and reliability of the Leap Motion sensor and its suitability for static and dynamic tracking. Sensors **14**, 3702–3720 (2014)
30. H. Smeragliuolo, N.J. Hill, L. Disla, D. Putrino, Validation of the Leap Motion controller using markered motion capture technology. J. Biomech. **49**, 1742–1750 (2016)
31. Leap Motion, https://www.leapmotion.com/
32. K.H. Chen, P.C. Lin, Y.J. Chen, B.S. Yang, C.H. Lin, Development of method for quantifying essential tremor using a small optical device. J. Neurosci. Method **266**, 78–83 (2016)
33. J.M. de Oliveira, R.C.G. Fernandes, C.S. Pinto, P.R. Pinheiro, S. Ribeiro, V.H.C. de Albuquerque. Novel virtual environment for alternative treatment of children with cerebral palsy. Comput. Intell. Neurosci. **8984379** (2016)
34. M. Iosa, G. Morone, A. Fusco, M. Castagnoli, F.R. Fusco, L. Pratesi, S. Paolucci. Leap Motion controlled videogame-based therapy for rehabilitation of elderly patients with subacute stroke: a feasibility pilot study. Top Stroke Rehabil. 306–316 (2015)
35. VirtualWare, virtualwaregroup.com
36. D.J. Bowen, M. Kreuter, B. Spring, L. Cofta-Woerpel, L. Linnan, D. Weiner, S. Bakken, C.P. Kaplan, L. Squiers, C. Fabrizio, M. Fernandez, How we design feasibility studies. Am. J. Prev. Med. **36**(5), 452–457 (2009)
37. M. van Diest, C.C. Lamoth, J. Stegenga, G.J. Verkerke, K. Postema, Exergaming for balance training of elderly: state of the art and future developments. J Neuroeng. Rehabil. Engl. BioMed. Central **10**, 101 (2013)
38. C.M. Bleakley, D. Charles, A. Porter-Armstrong, M.D.J. McNeill, S.M. McDonough, B. McCormack. Gaming for Health: a systematic review of the physical and cognitive effects of interactive video games in older adults. J. Appl. Gerontol. **34**, NP166–NP189 (2015)
39. D.C. Ribeiro-Papa, T. Massetti, T.B. Crocetta, L.D.C. Menezes, T.P.C. Antunes, I.M.P. Bezerra, C.B.M. Monteiro. Motor learning through virtual reality in elderly—a systematic review. Med Express **3** (2016)
40. N. Zeng, Z. Pope, J.E. Lee, Z. Gao, A systematic review of active video games on rehabilitative outcomes among older patients. J. Sport Heal. Sci. **6**, 33–43 (2017)
41. A. Nawaz, N. Skjaeret, J.L. Helbostad, Usability and acceptability of balance exergames in older adults: a scoping review. Health Inf. J. **22**(4), 911–931 (2016)
42. F.W. Simor, M.R. Brum, J.D.E. Schmidt, R. Rieder, A.C.B. de Marchi, Usability evaluation methods for gesture-based games: a systematic review. JMIR Serious Games **4**(2), e17 (2016)
43. D. Kairy, P. Lehoux, C. Vincent, M. Visitin, A systematic review of clinical outcomes, clinical process, healthcare utilization and costs associated with telerehabilitation. Disabil. Rehabil. **31**, 427–447 (2009)
44. J.E. Maddux, *Self-efficacy Theory* (Springer, New York, 1995)
45. V. Venkatesh, M.G. Morris, G.B. Davis, F.D. Davis, User acceptance of information technology: toward a unified view. MIS Q. **27**(3), 425–478 (2003)
46. F.D. Davis, Perceived usefulness, perceived ease of use, and user acceptance of information technology. MIS Q. **13**, 319–340 (1989)
47. Ajzen, The theory of planned behavior. Organ Behav. Hum. Decis. Process **50**(2), 179–211 (1991)
48. R.P. Bagozzi, The legacy of the technology acceptance model and a proposal for a paradigm shift. J. Assoc. Inf. Syst. **8**, 244–254 (2007)
49. J. Nielsen, *Usability Engineering* (Elsevier, Amsterdam, 1994)
50. B. Shneiderman, *Designing the User Interface: Strategies for Effective Human-Computer Interaction* (Pearson Education India, New Delhi, 1986)
51. K.M. Gerling, J. Schild, M. Masuch, Exergaming for elderly: analyzing player experience and performance. Mensch Comput. (2011)
52. K.-K. Kang, J.-A. Kim, D. Kim, Development of a sensory gate–ball game system for the aged people. Vis. Comput. **25**(12), 1073–1083 (2009)

53. T.C. Chan, F. Chan, Y.F. Shea, O.Y. Lin, Y.K. Luk, F.H. Chan, Interactive virtual reality Wii in geriatric day hospital: a study to assess its feasibility, acceptability and efficacy. Geriatr. Gerontol. Int. **12**(4), 714–721 (2012)
54. Y.-Y. Chao, Y.K. Scherer, Y.-W. Wu, K.T. Lucke, C.A. Montgomery, The feasibility of an intervention combining self-efficacy theory and Wii Fit exergames in assisted living residents: a pilot study. Geriatr Nurs **34**(5), 377–382 (2013)
55. A.S. Billis, E.I. Konstantinidis, A.I. Ladas, M.N. Tsolaki, C. Pappas, P.D. Bamidis. Evaluating affective usability experiences of an exergaming platform for seniors, in *Proceedings of 10th International Workshop on Biomedical Engineering*, Kos, October 2011
56. J.M. Flach, C.O. Dominguez, User-centered design: integrating the user, instrument, and goal. Ergon Des. 19–24 (1995)
57. K. Isbister, N. Shaffer, Heuristic evaluation of games, in *Game Usability: Advancing the Player Experience* (CRC Press, 2008), pp. 79–89
58. R. Bernhaupt, Evaluating user experience in games: concepts and methods, in *Human Computer Interaction Series* (2010)
59. P. Sweetser, P. Wyeth, GameFlow: a model for evaluating player enjoyment in games. Comput. Entertain. **3**, 3 (2005)
60. P. Bonato, Advances in wearable technology and applications in physical medicine and rehabilitation. J. Neuroeng. Rehabil. **2**, 2 (2005)
61. S. Obdrzalek, G. Kurillo, F. Ofli, R. Bajcsy E. Seto, H. Jimison, M. Pavel. Accuracy and robustness of kinect pose estimation in the context of coaching of elderly population, in *Proceedings of IEEE Engineering in Medicine and Biology Society (EMBC)* (2012), pp. 1188–1193
62. G.C. Burdea, Virtual rehabilitation—benefits and challenges. Methods Inf. Med. **42**(519), 1–11 (2003)
63. M. Kim, C. Jeon, J. Kim. A study on immersion and presence of a portable hand haptic system for immersive virtual reality. Sensors, **17**(5), pii: E1141 (2017)
64. B.G. Witmer, M.J. Singer, Measuring presence in virtual environments: a presence questionnaire. Presence **7**, 225–240 (1998)
65. D. Corbetta, F. Imeri, R. Gatti, Rehabilitation that incorporates virtual reality is more effective than standard rehabilitation for improving walking speed, balance and mobility after stroke: a systematic review. J. Physiother. **61**, 117–124 (2015)
66. N.A. Borghese, M. Pirovano, P.L. Lanzi, S. Wuest, E.D. de Bruin, Computational intelligence and game design for effective at-home stroke rehabilitation. Games Health J. Res. Dev. Clin. Appl. **2**(2), 81–88 (2013)
67. B. Lange, C.Y. Chang, E. Suma, B. Newman, A.S. Rizzo, M. Bolas. Development and evaluation of low cost game-based balance rehabilitation tool using the microsoft kinect sensor, in *Proceedings of IEEE Engineering in Medicine and Biology Society, EMBC* (2011), pp. 1831–1834
68. R. Hunicke, M. Le Blanc, R. Zubek, MDA: a formal approach to game design and game research, in *Proceedings of AAAI-04 Workshop on Challenges in Game AI*, July 2004, pp. 1–5
69. R. Dillon. The 6-11 framework: a new methodology for game analysis and design, in *Proceedings of GAMEON*, Singapore, Asia, March 2011, pp. 25–29
70. V. Stavljanin, M. Minovic, Gamification ecosystems: current state and perspectives, in *Open Source Solutions for Knowledge Management and Technological Ecosystems* (IGI Global, 2017)
71. G. Zichermann, C. Cunningham, *Gamification by Design. Implementing Game Mechanics in Web and Mobile Apps* (O'Reilly Media, Sebastopol, C.A., 2011)
72. K. Werbach, D. Hunter. *For the Win. How Game Thinking Can Revolutionize Your Business.* (Wharton Digital Press, 2012)
73. Dev.Mag. devmag.org.za/2012/04/19/video-game-audio-diegesis-theory-2/
74. M. Pirovano, E. Surer, R. Mainetti, P.L. Lanzi, N.A. Borghese, Exergaming and rehabilitation: a methodology for the design of effective and safe therapeutic exergames. Entertain. Comput. **14**, 55–65 (2016)
75. Y-L. Theng, A.B. Dahlan, M.L. Akmal, T.Z. Myint. An exploratory study on senior citizens' perceptions of the Nintento Wii: the case of Singapore, in *Proceedings of 3rd International Convention on Rehabilitation Engineering &Assistive Technology*, April, Singapore (2009)

76. B. Bonnechere, B. Jansen, L. Omelina, S. Van Sint, Jan. The use of commercial video games in rehabilitation: a systematic review. Int. J. Rehabil. Res. **39**(4), 277–290 (2016)
77. D. Collado-Mateo, E. Merellano-Navarro, P.R. Olivares, J. Garcia-Rubio, N. Gusi, Effect of exergames on musculoskeletal pain. a systemeatic review and meta-analysis. Scand. J. Med. Sci. Sports **28**, 760–771 (2018)
78. G. Tieri, G. Morone, S. Paolucci, M. Iosa, Virtual reality in cognitive and motor rehabilitation: facts, fiction and fallacies. Expert Rev. Med. Devices **15**(2), 107–117 (2018)
79. R. Aarhus, E. Grönvall, S.B. Larsen, S. Wollsen, Turning training into play: embodied gaming, seniors, physical training and motivation. Gerontechnology **10**(2), 110–120 (2011)
80. M.P.J. Habgood, S.E. Ainsworth, Motivating children to learn effectively: exploring the value of intrinsic integration in educational games. J. Learn. Sci. **20**(2), 169–206 (2011)
81. J. Huberty, L. Ransdell, C. Sidman, Explaining long-term exercise adherence in women who complete a structured exercise program. Res. Q. Exerc. Sport **79**, 374 (2008)
82. S.R. Wood, N. Murillo, P. Bach-y-Rita P, R.S. Leder, J.T. Marks, S.J. Page, Motivating, game-based stroke rehabilitation: a brief report. Top Stroke Rehabil. **10,** 134–140 (2003)
83. Y. Tian, Y. Bian, P. Han, P. Wang, F. Gao, Y. Chen, Physiological signal analysis for evaluating flow during playing of computer games of varying difficulty. Front Psychol. **8**, 1121 (2017)
84. R.M. Ryan, C.S. Rigby, A. Przybylski, The motivational pull of video games: a self-determination theory approach. Motiv. Emot. **30**, 347–363 (2006)
85. R.M. Ryan, E.L. Deci, Self-determination theory and the facilitation of intrinsic motivation, social development, and well-being. Am. Psychol. **55**, 68–78 (2000)
86. S. Rigby, R. Richard. Immersion and presence, in *Glued to Games: How Video Games Draw Us In and Hold Us Spellbound* (2011), pp. 81–96
87. B.D. Sylvester, M. Standage, J. Dowd, L.J. Martin, S.N. Sweet, M.R. Beauchamp, Perceived variety, psychological needs satisfaction and exercise-related well-being. Psychol. Health **29**(9), 1041–1061 (2014)
88. A. Bandura, *Social Foundations of Thought and Action: A Social Cognitive Theory* (Prentice-Hall, Englewood Cliffs, N.J., 1986)
89. M.W. Kreuter, D.W. Farrell, L.R. Olevitch, L.K. Brennan. *Tailoring Health Messages: Customizing Communication with Computer Technology* (Lawrence Erlbaum Associates, 2000)
90. D. Thompson, J. Baranowski, K. Cullen, T. Baranowski, Development of a theory-based internet program promoting maintenance of diet and physical activity change to 8-year-old African American girls. Comput. Educ. **48**, 446–459 (2007)
91. L.S. Vygotsky, *Mind in Society. The Development of Higher Psychological Processes* (1978)
92. D. Shernoff, M. Csikszentmihalyi, B. Schneider, E. Shernoff, Student engagement in high school classrooms from the perspective of flow theory. Sch. Psychol. Q. **18**, 156–176 (2003)
93. R. Garris, R. Ahlers, J.E. Driskell. Games, motivation, and learning: a research and practice model. Simul. Gaming 441–467 (2002)
94. B.H. Dobkin, Rehabilitation after stroke. N. Engl. J. Med. **352**(16), 1677–1684 (2005)
95. I-T. Chiang, Old dogs can learn new tricks: exploring effective strategies to facilitate somatosensory video games for institutionalized older veterans, in *Edutainment technologies. Educational games and virtual reality/augmented reality applications*, ed. by M. Chang, W.-Y. Hwang, M.-P. Chen, et al. (Springer, Berlin)

Microwaves and Functional Materials: A Novel Method to Continuously Detect Metal Ions in Water

Ilaria Frau, Stephen Wylie, Jeff Cullen, Olga Korostynska, Patrick Byrne and Alex Mason

Abstract Protecting water from chemical pollutants is a major societal goal. Metal ion dispersion from abandoned mines is a global concern and one of the principal causes of metal pollution in water. Toxic metals are a particular concern because they are not degraded by normal biogeochemical cycles and cause adverse environmental and human health effects even with low concentrations if there is long-term exposure. Current laboratory-based methods are not suitable for monitoring adequately water pollution in the environment. Consequently, it is necessary to develop and deploy new sensing systems to investigate water quality continuously. Microwave spectroscopy has been demonstrated as an effective method for offering continuous measurement of material properties, nevertheless, this method suffers from a lack of selectivity and sensitivity (Zarifi et al. Sens Actuators B Chem 255:1561–1568 (2018), [1]). This chapter presents a feasibility study using unique functionalised electromagnetic (EM) sensors for continuous monitoring of zinc in water. The reaction between Zn and a Bi_2O_3 based thick film that is screen-printed onto a planar interdigitated electrode (IDE) sensors starts within 30 s, and the adsorption equilibrium was attained within 10 min. The response is faster during the initial stage and slows as equilibrium is reached. Results show good linear correlations between C (capacitance), S_{11} (reflection coefficient) and Zn concentration. Also, the recovery time of sensors is evaluated to be 100–150 s demonstrating the sensors reusability and potential for continuous monitoring.

I. Frau · S. Wylie · J. Cullen · O. Korostynska
Faculty of Engineering and Technology, Liverpool John Moores University, Liverpool, UK

P. Byrne
Faculty of Science, Liverpool John Moores University, Liverpool, UK

A. Mason (✉)
Animalia AS, Norwegian Meat and Poultry Research Centre, Oslo, Norway
e-mail: alex.mason@animalia.no

S. C. Mukhopadhyay et al. (eds.), *Modern Sensing Technologies*,
Smart Sensors, Measurement and Instrumentation 29,
https://doi.org/10.1007/978-3-319-99540-3_9

Need for Research—Statement of a Problem

Water, especially freshwater, is an indispensable resource, but it is limited in quantity and in quality. Assuring a high quality of water and protecting it from chemical contamination is a major sustainable development goal [2, 3] that is becoming increasingly hard to achieve owing to factors such as climate change, overexploitation and contamination from agricultural and industrial activities.

Polluted drainage from both contemporary and historical mining and mine wastes are an issue of international concern that can result in significant water quality issues [4]. Point and diffuse sources determine the dispersion into the environment of toxic inorganic pollutants, mainly metals (zinc, lead, cadmium, copper, mercury, arsenic, etc.), in addition to problems associated with high sulphate concentrations and acidification of watercourses. Surface water, directly connected with groundwater [5], is the predominant route of dispersion of these toxic elements even at considerable distances from the source, with the consequent possibility to pollute drinking water supplies.

Metals are considered trace elements due to their presence in the environment at generally low concentrations (μg/l range to less than 100 mg/l). Their bioavailability is regulated by several physical and chemical factors, such as temperature, pH and consequent differential adsorption, phase association, thermodynamic equilibrium, complexation kinetics, and so forth [6].

Consequently, European and worldwide legislation has established environmental quality standard (EQS) for metals in surface water, which is the major route of dispersion of these pollutants. Table 1 summarises the EQS established by the UK Technical Advisory Group on the EU Water Framework Directive (UKTAG) [7] and US EPA (Environmental protection agency) [8] for toxic metals concentrations and sulphates (strongly related to them), and the typically investigated metal ion concentrations in mining catchment areas [9, 10].

Generally, the ionic metal form is considered the most toxic, although other weakly complexed species (oxides, hydroxides, and sulphates) and organic colloidal phases are also a cause for concern. The metal ionic form and its precipitation or sorption to other phases are related to the pH, the sorbate identity and concentration, the sorbent composition, and so forth [12].

Zinc (Zn) is a hazardous toxicant, which when present at high concentrations for a long-term, can cause such significant effects in humans as respiratory disorders, bronchiolar leukocytes, neuronal disorders [13]. Zinc levels in non-polluted waters range from 0.06 to 0.6 μg/l. In drinking water, it is typically around 10 μg/l but higher concentrations than 2 mg/l are found especially when the water passes through zinc pipes [14]. Globally, Zn in mine water ranges from <0.1 to >500 mg/l [10], failing the established EQS limit of 0.008–0.125 mg/l. The Water Framework Directive and the Clean Water Act, the main legal frameworks for the protection and the sustainable

Table 1 UK EQS for some metals and sulphates in surface water and global range concentrations from waters impacted by historical deep metal mining

Toxic metals	UK EQS (mg/l) for inland and other surface water [7, 11]	US EPA EQS (mg/l) for freshwater (acute and chronic) [8]	Global concentration ranges in polluted mining surface water (mg/l) [9]
Lead (Pb)	0.0072[a]	0.0015–0.065[b]	<0.0072–>12
Zinc (Zn)	0.008–0.125[b]	0.059–0.210[b]	<0.008–>500
Copper (Cu)	0.001–0.028[b]	0.017–0.034[b]	<0.001–>240
Cadmium (Cd)	0.00008–0.00025[a,b]	0.00066–0.0018[b]	<0.00008–>2.6
Iron (Fe)	1	0.3–1	<1–>1900
Sulphates (SO4)	400	250	<400–>16,000

[a]dissolved concentration (<0.45 μm)
[b]depending on the annual mean concentration of $CaCO_3$

use of freshwater resources respectively in Europe and in USA, [15] require new low-cost sensing systems for in situ monitoring, and managing water resources to minimise the environmental risk posed by mine drainage from active and abandoned mining complexes [16].

Current Methods for Metal Detection

Current methods for detecting Zn concentration (and the other toxic metals) in water are mainly laboratory-based. Examples include atomic absorption spectroscopy, inductively coupled plasma—optical emission spectrometry, inductively coupled plasma—mass spectrometry [17]. Although these methods can determine multi-metal concentrations with high selectivity and sensitivity with a low limit of detection, they have critical drawbacks. These bulky laboratory methods require sample preparation, costly chemicals and trained staff. They are unable to provide real-time and in situ measurements which limits the possibility to detect an unexpected event (i.e. the failure of a mine tailings dam spill) [18, 19] or to understand metal variations when environmental conditions change (i.e. during a flood event) [20]. Moreover, without a continuous in situ monitoring, traditional methods cannot give an answer to alterations of each metal concentration changes due to differences between high flow and low flow [21], variation in pH and temperature, presence or absence of hydrated iron oxide (HFO) on the riverbed in polluted mining areas [12, 22–24].

Water quality monitoring for emergency detection and management using advanced low-cost sensor technologies are essential for polluted water remediation. Accordingly, significant research is in progress for in situ and real-time measurements of heavy metal ions, and appreciable results have been reported using optical, electrochemical techniques and near-infrared spectroscopy [25–27].

Although these methods are sensitive and stable, they can also be cumbersome, time-consuming, expensive and laborious. Hence, the UN Agenda for Sustainable development calls for a low-cost sensing system for online and in situ monitoring [15]. Therefore, further work is needed to develop simple, reliable, and cost-effective techniques for remote detection of pollutants, to better understand the processes that regulate metal dispersion and allow for an emergency response to unexpected events.

F-EM Sensors as a Solution

A new and rapidly developing method capable of providing a continuous monitoring of metals in freshwater, namely Zn in this work, is an electromagnetic (EM) wave sensor system combined with functional materials, operating at microwave frequencies [28]. This method is highly adaptable and, depending on each scenario, can take the form of planar sensors [29–31], resonant cavities [32], flexible sensors [33], for example. Their design can be tailored to suit particular applications and is coupled with reliability and cost-efficiency.

Microwave planar printed patterns, in particular, have been increasingly used for various sensing applications due to their versatility, flat profile, low weight, simplicity, low-cost [34]. In fact, planar sensors can be easily integrated with microfluidic techniques [35], nanostructures [36], thin and thick films [37], for specific sensing purposes.

The principle of microwave spectroscopy is based on the singular interaction between incident waves at specific frequencies with the properties of the analyte presented to the sensing structure (such as conductivity, capacitance, resistance, inductance and permittivity, as well as temperature, chemical structure, molecular composition, etc.), as illustrated in Fig. 1a. The change in the S_{11} reflection coefficient can be linked to the composition and concentration of the measured solution (Fig. 1b, c). Accordingly, permittivity (ε_r) as defined in (1) relates to a material's ability to transmit an electric field and is a complex value which varies with changing frequency and accounts for both the energy stored by a material and any losses of energy which might occur.

$$\varepsilon_r = \varepsilon' - j\varepsilon'' \tag{1}$$

The response of the sensor manifests itself as a resonant frequency change or attenuation of the signal [38].

Microwave spectroscopy, however, suffers from a lack of selectivity, specificity and sensitivity. A feasible solution to this problem is to integrate EM sensors with functional materials.

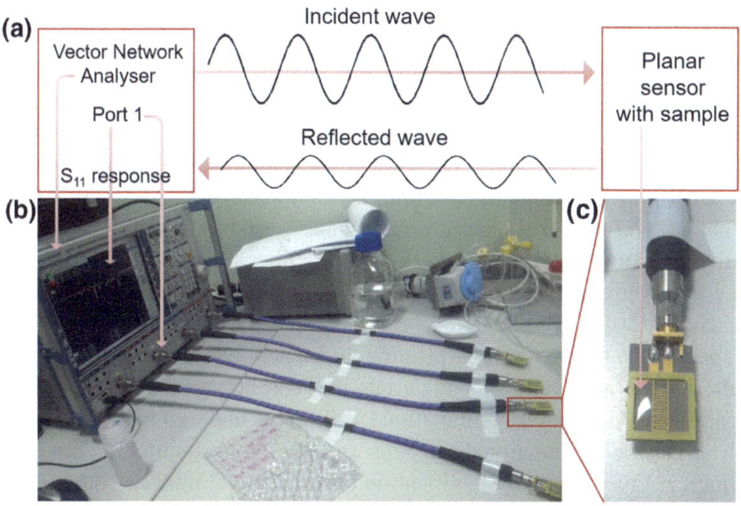

Fig. 1 (**a**) Scheme which describes the reflected signal (S_{11}) at discrete frequencies as spectrum response (**b**) with the unique interaction of the incident wave with the sample under test (**c**)

Synergy with Functional Material

Microwave spectroscopy connected with the appropriate sensing system could be able to analyse changes in properties and composition of several materials and components. Notably, the lack of selectivity is a limitation in designing a microwave sensor for a particular purpose [39]. A possible solution are planar waveguide-based chemical sensors [40]. The synergy between microwave sensing technology and chemical coatings provides interesting advantages in the field of water quality monitoring and is consequently a promising area of research and development [34, 41].

By functionalising planar EM sensors with certain sensitive materials, it is possible to obtain the desired sensitivity and/or selectivity to one or more specific analytes. This can be achieved not only with the adoption of the right sensing material but also by choosing the right thickness of the functional film and sensor geometry [42, 43]. Moreover, coatings prevent short-circuiting and oxidation when the sensor is exposed to a water sample.

The principle of using a distinct chemical coating and the interaction with selected analytes is based on the variation in the resulting electrical impedance due to the change in complex permittivity. The spectral response due to this variation will generate amplitude and/or frequency changes at specific frequency ranges [34].

Considerable progress has been made in the last decade in developing chemosensors that are able to distinguish specific metal ions, such as Zn, by using different functional materials including synthetic and biological receptors [44], zeolites, inorganic oxides [45], organic polymers, biological materials [46] and hybrid ion-exchangers [47].

Screen-Printing Technology

The goal of a cheap, sensitive and selective device to monitor metal-impacted water is achievable by the integration of chemical coatings and planar sensors using screen-printing technology [48]. These are known for their flexibility and cost-effective mass production [49]. Screen-printing is a technique where a viscous ink (0.1 to 10 Pa s) based on functional materials is set in a screen blocking layer with a stencil (a negative of the image to be printed), which defines an open area of the mesh. A squeegee is moved across the screen and the ink passes through the mesh and a thick film is created [50].

The main advantages of screen-printing as the application method for functional materials is the high diversity of materials that can be processed at low cost, and the possibility of printing a combination of different materials onto the same substrate [51]. Designing functionalised sensors with the best-suited architecture will help to increase the selective identification of multi-metal ions and other pollutants simultaneously present in water [52].

Screen-printed electrodes coupled with specific sensing materials are an attractive option for adapting diverse sensing systems to a particular purpose [50]. The screen-printing technique is used to form a thick film on top of microwave sensors and increase, by chemical or physical reaction, the sensitivity, specificity and selectivity between the EM waves and the analyte. Thick films are rugged, reproducible, inexpensive and have also been identified as useful for integration into remote monitoring systems [48, 51].

Among functional chemical compounds, bismuth (Bi) has been largely used in the last 20 years in a number of approaches for detecting metals in water. Furthermore, bismuth nanoparticles have shown in several experiments to have superior sensing characteristics compared to the Bi film electrode for detecting metals in water, mostly using electrochemical methods [53–56]. Moreover, nanomaterials as inorganic oxide compositions are considered to be advantageous, owing to their strong adsorption and rapid electron transfer kinetic [25, 45]. Nanostructured metal oxides are also interesting for their nano-morphological, functional biocompatible, non-toxic and catalytic properties. In this work, interdigitated electrodes sensors are functionalised with β-Bi_2O_3 based coatings.

The purpose of this chapter is to evaluate the feasibility of using functionalised EM planar sensors (f-EM sensors) for continuously monitoring the concentration of Zn in polluted mine water. The response time was measured for β-Bi_2O_3 based sensing film using two sensing techniques: microwave spectroscopy and capacitance measurements. Moreover, the sensors' stability and reusability were assessed by measuring their recovery time.

Experimental Setup

The time of reaction between Zn solutions and the β-Bi_2O_3 based coating, on two gold eight-pair IDE sensors on a planar PTFE (polytetrafluoroethylene) substrate was analysed using microwave and electric techniques, by measuring respectively the reflection coefficient (at 10 MHz–10 GHz) and the capacitance (at 30–200 kHz).

Fabrication of the Screen-Printed Sensors

Generally, for the development of a low-cost printable paste mixture of the desired viscosity three components are necessary: (1) a principal functional material, (2) an organic binder and (3) an organic volatile solvent. These last two components work as a matrix for developing the correct viscosity of a paste to be printed over the planar sensors.

Two sets of Au eight–pair IDE sensors were functionalised using a semi-automatic screen printer (Super Primex). A series of Ag eight-pair IDE patterns were also screen-printed in situ at LJMU on microscope slide substrates using a Silver conductor (Dupont 5064H) for a secondary cheaper development and deployment of the coatings and an initial characterization of them. Repeatability and reproducibility of the IDEs were evaluated by impedance measurements (Cp, Rp, Z).

β-Bi_2O_3 nanopowder $90 < \Phi < 210$ nm particle size, characterised by tetragonal phase and space group P 42_1 c (114), (Sigma-Aldrich 637017) was used as the principal chemical for creating the paste mixture to form the thick film. 92.5 wt% bismuth oxide nanopowders were mixed with 7.5 wt% of a polymer (Butvar B98) used as a binder, and a suitable amount of ethylene glycol butyl ether (solvent) to form the paste. The thickness of the β-Bi_2O_3 based thick film was increased by multiple screen-printing, with the set of two sensors having 4 and 6 layers printed, with suitable curing of the layers in an oven at 170° C for 1 h between each print. The schematic diagram of the screen-printing process is shown in Fig. 2.

The thickness of the final films was measured using an electronic micrometer (TESA Micromaster), a digital vernier caliper (AOS Absolute Digimatic) and a surface profiler (Taylor Hobson—Form Talysurf 120), resulting in average of 40 and 60 μm respectively for 4 and 6 layers. Morphological properties and elemental composition were analysed before and after Zn detection using a scanning electron microscope (SEM), model FEI—Quanta 200 and a X-Ray fluorescence analyser (XRF), model INCA-X-act—Oxford Instruments. Coating repeatability and reproducibility was measured through absorbance measurements of screen-printed microscope slides with β-Bi_2O_3 based paste mixtures prepared on distinct days. These measurements were performed with a UV-Vis Spectrophotometer (Jenway 7315) as shown in Fig. 3.

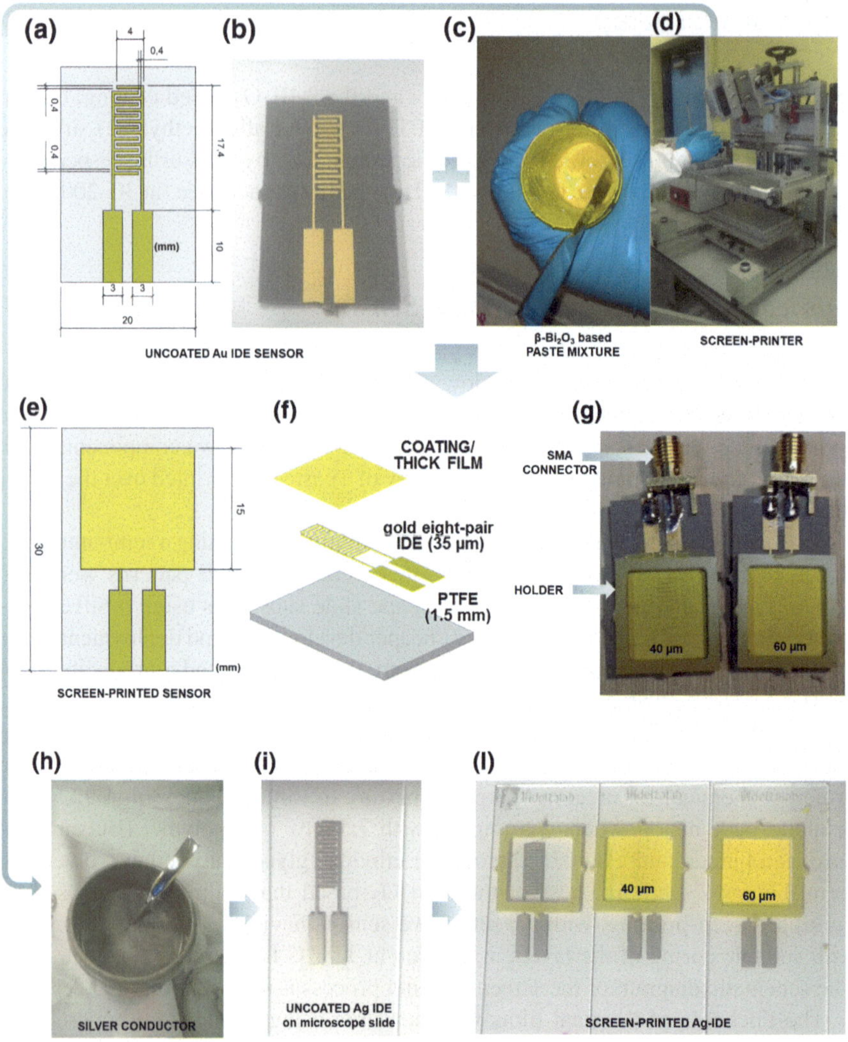

Fig. 2 Scheme which summaries the integration of functional materials with planar sensors by using screen-printing technology and consequent production of thick films. Uncoated Au IDE sensors (**a**, **b**) were screen-printed using a β-Bi_2O_3 based paste mixture (**c**) and a semi-automatic screen-printer (**d**). The finished coated sensors have 40 and 60 μm thicknesses (**e**–**g**). In-situ Ag-IDE were screen-printed on microscope slides using a silver conductor paste (**d**, **h**, **i**) and covered as described for the Au IDE sensors (**l**). The front and rear view along with the dimension of the sensor and the sensing area are shown in (**a**), (**e**) and (**f**)

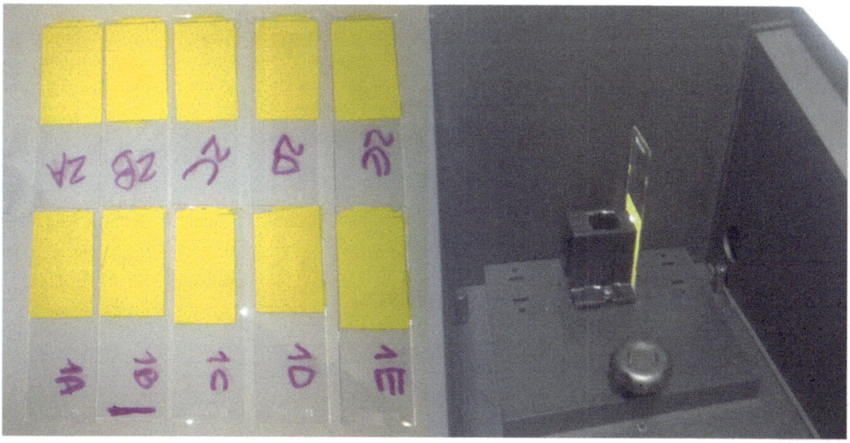

Fig. 3 Repeatability of coating method was assessed using UV-vis spectroscopy

Table 2 pH and conductivity of the zinc solutions

Zn solutions		
Zn concentration (mg/l)	pH	Conductivity
0	7.36	1.1 μS/cm
1	3.78	134.4 μS/cm
10	2.60	1.282 mS/cm
25	2.42	3.36 mS/cm
50	1.99	6.52 mS/cm
100	1.97	13.02 mS/cm

Sample Preparation

Six samples with different zinc concentrations (0, 1, 10, 25, 50, 100 mg/l) were prepared by dissolving a defined volume of Zn 1,000 ppm ICP standard solution certified (Sigma-Aldrich 18562) in deionised water. The chemical and physical parameters (pH and conductivity, Table 1) for all the samples were evaluated at constant temperature using a multi-parameter probe (PCE-PHD 1). All measurements were performed in an air-conditioned environment at a constant temperature of 18.5° C (Table 2).

Measurement Setup

Functionalised Au IDE sensors coated with the β-Bi_2O_3 based film with 40 and 60 μm thicknesses were connected to a Rohde and Schwarz ZVA 2.4 VNA (Vector Network Analyser) via a coaxial cable (Fig. 4a). In this work Molex edge mount

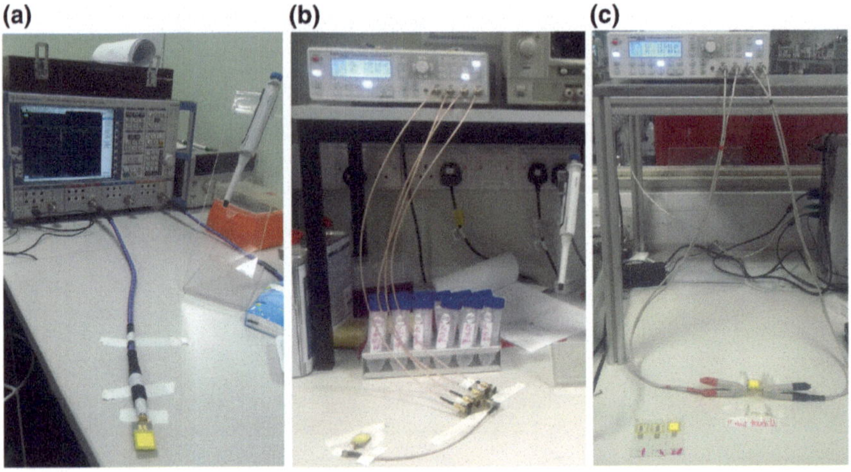

Fig. 4 Configuration of the EM measurements with **a** VNA connected via coaxial cable with a Au eight-pair IDE coated with β-Bi_2O_3 based coating. Set up of the HAMEG programmable LCR bridge connected to a coated Au IDE planar sensor (**b**) and to a coated Ag eight-pair IDE on microscope slide through a R&S HZ184 K measurable cable (**c**)

connectors were used. The reflection coefficient (S_{11}) relative to the 50 Ω impedance of the VNA was measured with a one-port configuration in the 10 MHz to 15 GHz frequency range, producing a detailed spectrum with 60,000 discrete points. For each measurement, 400 μl of Zn solution was dispensed onto the sensor using a pipette, with the solution held in place by a specific holder (Fig. 2g).

Each solution was tested (n = 3) for 30 min with both sensors. This technique permits a continuous measurement and the output was extracted every 30 s to monitor any time variation in the interaction between the coatings (40 and 60 μm) and the Zn solutions.

The sensor capacitance (C) and their interaction with Zn samples was measured using a programmable LCR bridge HAMEG 8118 configured with a bespoke coaxial probe (Fig. 4b) at the frequency range between 30 Hz and 200 kHz. Coated Ag-IDE screen-printed on microscope slide were connected with the LCR bridge through a R&S HZ184 K measurable cable (Fig. 4c). For consistency, 400 μl volume water samples were used.

Once the sensors' output in all frequency ranges was analysed using both Au and Ag electrodes, 150 Hz was selected to determine the variation in capacitance over 30 min, with a measurement recorded every 5 s. The purpose of doing this was to assess changes in electrical properties and the consequent reaction between the coated sensors and Zn solutions. All Zn samples were measured at the same temperature to eliminate possible effect of the temperature on the microwave spectra [36].

Table 3 The percentage of adsorption between β-Bi_2O_3 based film and Zn

Zn concentration (mg/l)	Concentration (mg/l) and % of adsorption					
	After 5 min		After 10 min		After 30 min	
	Conc.	%	Conc.	%	Conc.	%
0	0	0	0	0	0	0
1	0.96	4.1	0.938	6.2	0.938	6.2
10	9.6	4	9.39	6.1	9.4	6
50	48.3	3.5	47.3	5.4	47.25	5.5
100	96.7	3.3	94.9	5.1	94.8	5.2

Adsorption Experiment

The percentage of adsorption between adsorbent (β-Bi_2O_3) and adsorbate (Zn^{2+}) was estimated by measuring the difference of the initial concentration of each Zn solution and the concentration after the adsorption, by modifying the procedure described by Dada et al. [57] for simulating the capability of the coating to adsorb Zn ions. 0.02 g of the thick film (printed on a microscope slide) was put in a plastic container with 20 ml of each Zn solution (0, 1, 10, 50, 100 mg/l). The concentrations in the solution after the adsorption was measured after 5, 10 and 30 min.

The percentage of adsorption was evaluated using Eq. (2).

$$\%\ adsorption = \frac{c_i - c_e}{c_i} \times 100 \tag{2}$$

where C_i is the Zn concentration before the adsorption and C_e is the Zn concentration after the adsorption.

Results and Discussion

Results obtained with both electric and microwave measurements show how the reaction between Zn and the coating is observed to commence within 30 s of exposure and reach an adsorption equilibrium time within 600 s (10 min). At this point, the measurement of the percentage of adsorption demonstrates that approximately 4% of zinc is adsorbed on the coating about after 5 min and 6% after 10 min, reaching an adsorption equilibrium (Table 3).

Likewise, with the higher concentrations (50 and 100 mg/l), the percentage of adsorption is slightly reduced presumably due to the lower pH (<2, see Table 1). As a matter of fact, the adsorption of cations on metal-oxide minerals in the natural

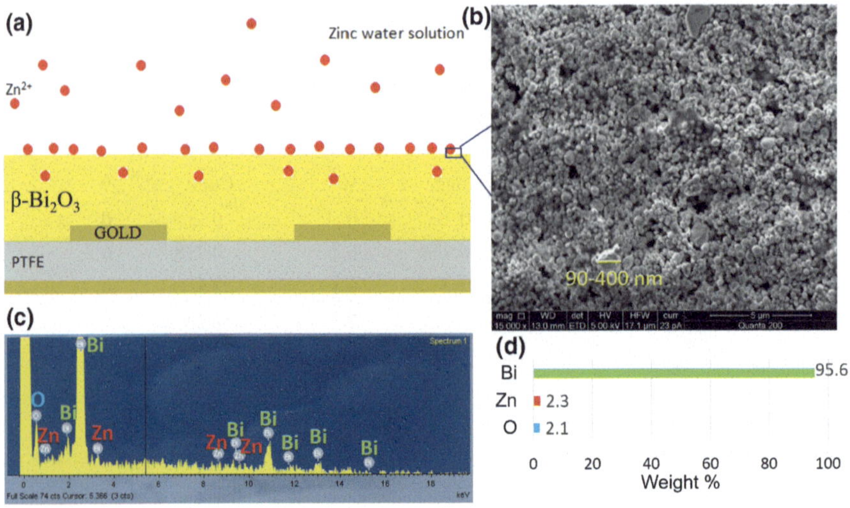

Fig. 5 **a** Partial cross-section of the functionalised sensor illustrating the sorption sensing principle; **b** Scanning electron microscope (SEM) image of the coating after the sorption reaction with the Zn ions where the particle size is highlighted as laying between 90 and 400 nm; **c** Elemental spectra of the film after the adsorption and **d** the elemental weight percentage

environment is reduced when the pH decrease [12]. Therefore, it is probable that the pH influences the Zn adsorption on the β-Bi_2O_3 based coating.

Figure 5a schematically illustrates the adsorption reaction between the Zn^{2+} ion and the β-Bi_2O_3 based film. The SEM image of the coating after measurements (Fig. 5b) and the elemental composition spectra and their weight percentage (Fig. 5c, d) obtained with the XRF confirm the presence of Zn on the coating.

The microwave spectra obtained for the Zn solutions at the concentration range 0–100 mg/l were analysed for uncoated and coated sensors with 40 and 60 μm of Bi_2O_3 based coatings. The higher linear correlation and sensitivity between the S_{11} response and Zn concentration was identified in the frequency range between 2.25 and 2.53 GHz respectively for uncoated sensors and coated sensor with 60 μm coating, as shown in Fig. 6.

Figure 7a shows the timeline responses within 30 min for the Zn solutions measured with microwave spectroscopy with the f-EM sensor with β-Bi_2O_3 based film with 60 μm thickness at the most pronounced peak (2.53 GHz). Figure 7b and c display the capacitance measurements at 150 Hz performed with the LCR bridge using respectively the same kind of sensor and a coated Ag IDE, with the same 60 μm film, printed on microscope slide in situ and connected with the LCR bridge through crocodile clips. This shows how the S_{11} and C parameters change during the first 10 min and reach stability afterwards. Capacitance and S_{11} tend to increase with Zn concentration, with most of the response occurring within the first 5 min after application of the water sample on the sensor surface; then the signal tends to stabilise by comparison and remains stable beyond 600 s. This demonstrates that the

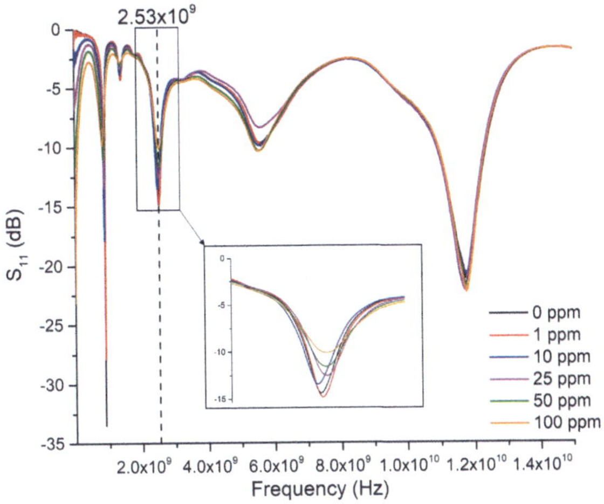

Fig. 6 Microwave spectra at the frequency range 10 MHz–15 GHz for Zn solutions 0–100 mg/l measured with a 60 μm of Bi_2O_3 based coating: the best linear correlation between Zn concentration and S_{11} response was identified at 2.53 GHz after 10 min

interaction between the functional coating and the Zn ion reach an equilibrium at this point. It is appreciable how adsorption is fast at the beginning of the experiment (first 5 min) and then progressively slows down till the equilibrium is reached.

Figure 8 shows the results obtained with f-EM sensors with 60 μm thickness in the frequency range between 1 and 3 GHz for the range of Zn solutions. A resonant shift occurs depending on the zinc concentration at two resonant frequencies, around 1.38 and 2.53 GHz. The dash-lines express the initial points, after 30 s, the dot-lines after 5 min and the solid-lines after 10 min (equilibrium). The linear correlation for the 60 μm coating for both capacitance and S_{11} response was stronger and the response more sensitive than the 40 μm coating.

Examples of linear correlations obtained for S_{11} (peak, 2.53 GHz) and capacitance at 150 Hz with 60 μm thickness coating at 30, 300 and 600 s are shown in Figs. 9 and 10. Also here is visible how the results with time change are more pronounced in the first part of measurements and minimum after 5 min. As well as the linear correlations between various Zn concentrations are similar between 300 and 600 s for both reflected power (R^2 = 0.97) and capacitance (R^2 = 0.99). Figure 10a, b compare the response obtained with the same coating thickness (60 μm) on Au IDE EM sensor and the Ag IDE printed on microscope slides. The capacitance and the R^2 are lower with the coated Ag IDE demonstrating that the Au EM sensor performs better for Zn measurements.

The sensor covered with 60 μm of β-Bi_2O_3 based coating result more sensitive than the uncoated sensor and the covered with 40 μm coating. In fact, the change in signal amplitude (expressed as dB/mg/l), at each peak (about 2.5 GHz) is higher for the 60 μm film as shown in Fig. 11 where the data are normalised. The 0 ppm value is cut off because not in line with the diverse Zn concentrations for the measured electrical property of capacitance. This is reflected also in the microwave response (S_{11}) demonstrating the ability of these two methods to identify the presence of the investigated metal.

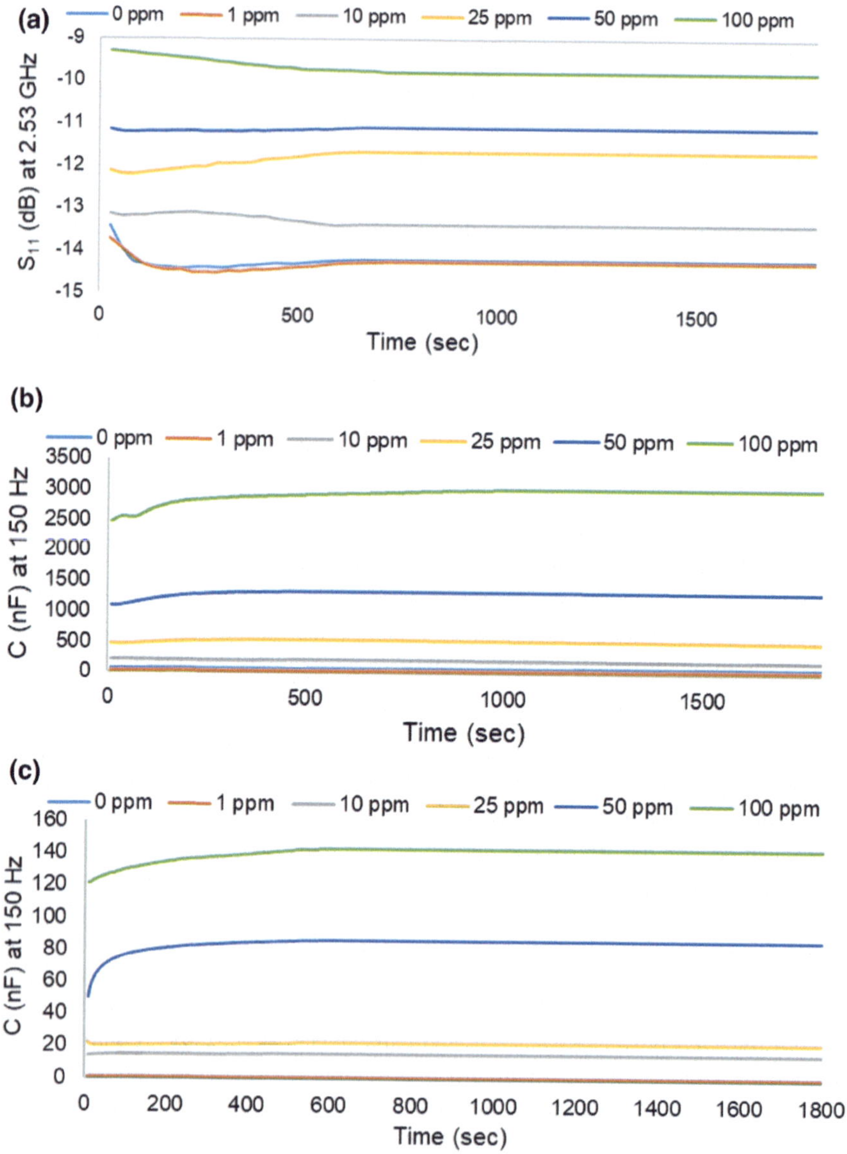

Fig. 7 Timeline response measured with a β-Bi_2O_3 based film with 60 μm thickness on the Au eight-pair EM sensor measured for 30 min with **a** microwave spectrometry at the peak 2.53 GHz and **b** with the LCR bridge at 150 Hz; **c** shows the timeline capacitance response at the same frequency, with the same thickness of the coating on an Ag IDE screen-printed on a microscope slide

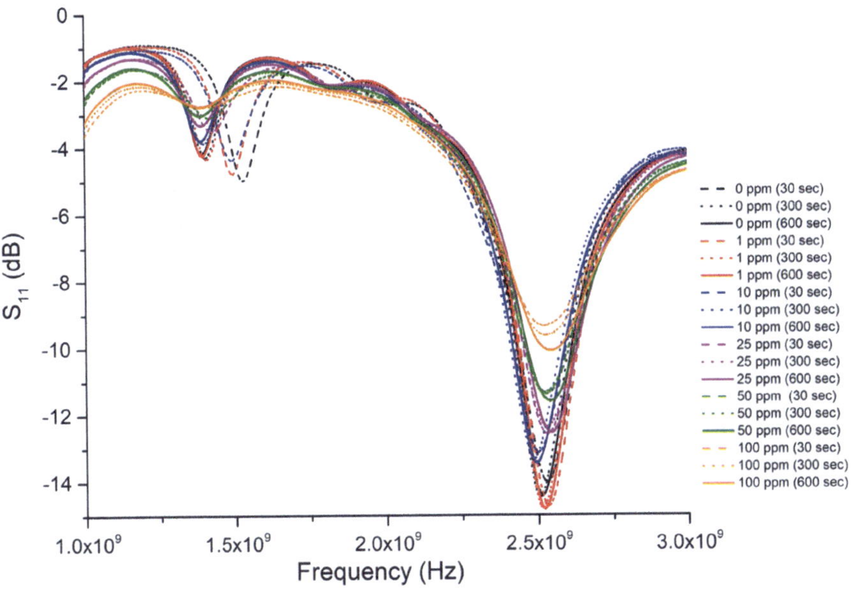

Fig. 8 Microwave output obtained with coated sensors with a β-Bi_2O_3 based thick film with 60 μm; it is been recorded the response obtained after 30 s (dash-lines), after 5 min (dot-lines) and after 10 min (solid-lines)

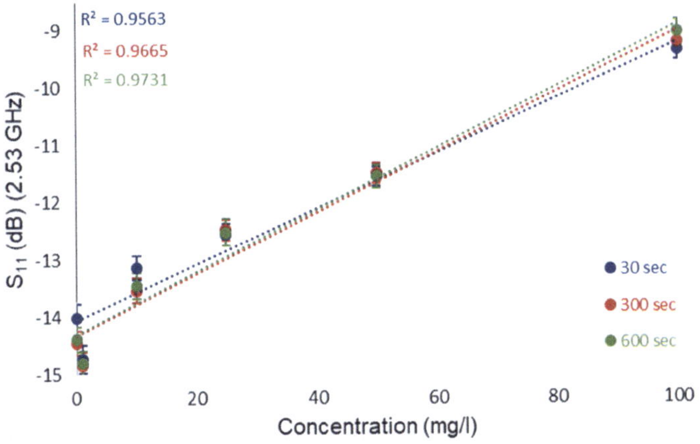

Fig. 9 Linear correlations between reflected power measured at the most pronounced peak (at 2.53 GHz) and Zn concentration at three different times: 30, 300 and 600 s measured with 60 μm thickness of the β-Bi_2O_3 based coating

In the functionalised sensor, the electromagnetic wave propagates in both: the substrate on which gold IDE are printed, and the top β-Bi_2O_3 based thick film. Therefore, the overall propagation will depend on the permittivity of each component

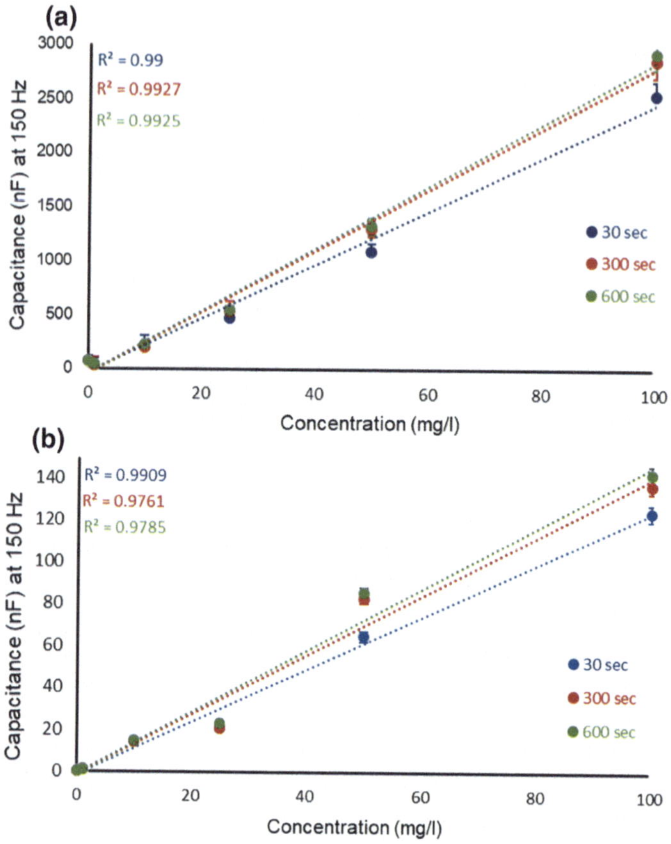

Fig. 10 Linear correlation between capacitance and Zn concentrations measured continuously, every 5 s for 30 min, and illustrate here at three specific times: 30, 300 and 600 s with 60 μm thickness of the film printed on the Au eight-pair IDE sensor (**a**) and on the eight-pair Ag IDE printed on a microscope slide (**b**)

in the system, such as a substrate permittivity and its thickness, and on the sensitive material permittivity, which will change with the adsorption of the zinc in the sensitive layer.

As a result, a continuous measurement protocol must accommodate timing. Regardless the results change with the time and Zn concentration output is related to the reaction time which links to the percentage of adsorption on the film. Consequently, it is more important to consider the relative change and not the absolute value S_{11} or C, exploring relative measurements and to measure samples for the same length of time enabling comparable results.

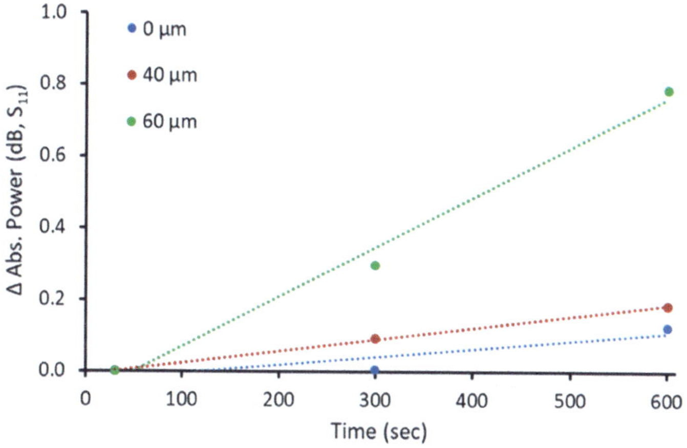

Fig. 11 Sensitivity comparison between an uncoated sensor and two coated sensors with two thicknesses (40 and 60 μm) respectively at each peak (about 2.5 GHz)

Sensor Recovery Time

It was noted that when the sensor is "dry" (unused), the reaction time is longer; when it is "wet" (previously used), it is shorter, although the final results are similar. This is possibly because the solution needs more time to saturate the voids in the film. So, when it is dry the zinc takes longer to be sorbed on the β-Bi_2O_3 tetragonal structure. Consequently, it is necessary to consider the state of the sensor when a water sample is analysed with a thick film sensor.

In this experiment, the sensor recovery time was also evaluated using the LCR bridge and the VNA as the time needed for the response of the sensor to return to the 90% of the original "air spectra". The recovery time was assessed after each measure and each f-EM sensor was rinsed in distilled water. It was recorded as being between 100–150 s, resulting in a return to the air value respectively of 31 pF and 35 pF (Fig. 12) for 40 and 60 μm thickness of films with capacitance measurements. With the microwave measurements, the results are approximately the same, returning to the air spectra after 100–150 s (Fig. 13). The reaction does not alter the coating for many hours of measurements, as also demonstrated by SEM and XRF measurements (Fig. 5b, c). It will begin to alter after about 20 h of continuous monitoring, as described in more detail in the next section.

Reversibility, Stability and Reusability of the Coating

The reversibility, stability and consequent reusability are presumably associated with weak interactions between the sensitive layer and the analyte (physical sorption), very

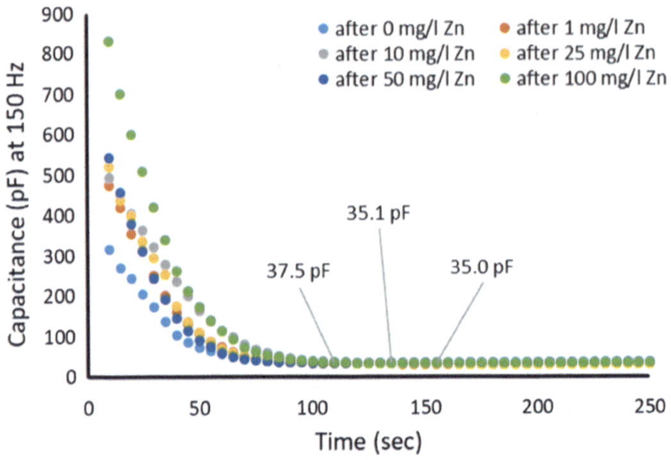

Fig. 12 Recovery time recorded every 5 s with the LCR bridge for the 60 μm thickness which returned to the air value (35 pF) after several Zn solutions measurements at different concentration

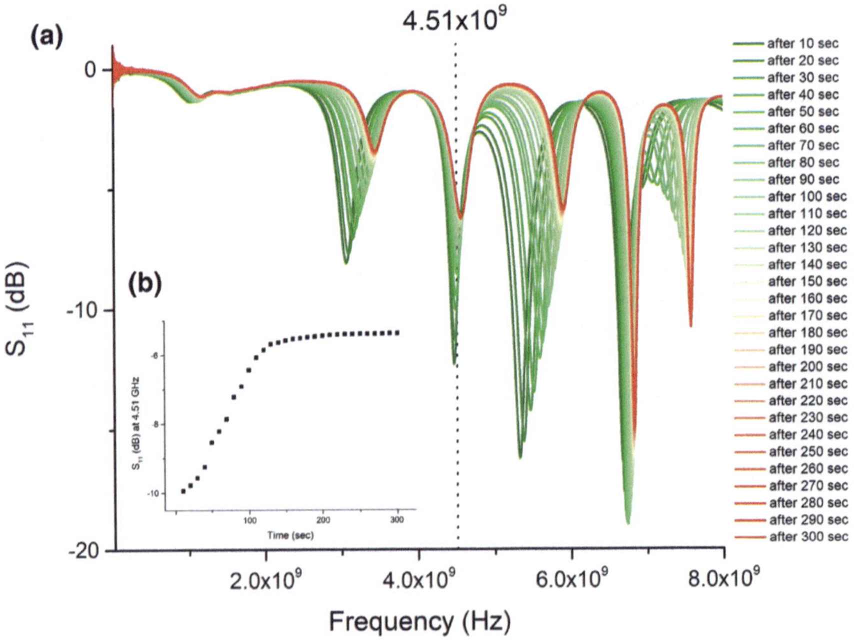

Fig. 13 Example of (**a**) a sensor recovery spectra after 10 mg/l Zn measurements captured every 10 s for 5 min using a VNA. The correlation (**b**) at 4.51 GHz between recovery time and S_{11} response is shown in the bottom left; stability is reached between 100 and 150 s

similar to van der Waals forces, which are the dominant interactions between a large range of metal oxides and cations [45]. Hence, desorption may be the ultimate result. Consequently, there is no degradation of the film for several hours of continuous monitoring and the relative standard deviation (RSD) is $<3\%$ for all measurements.

Despite this, after long usage, the film starts to be slightly modified probably because of the 1% HNO_3 of the Zn stock solution, and the consequent formation of bismuth nitrate, mostly with high concentration due to the higher amount of stock solution used for making the standards. This is deducted by a minute change observed in the air spectra and in the coating aspect.

Figure 14 shows the differences of the sensor with 60 μm coating before (a, right) and after (b, left) the sensor is deteriorated and considered disposable, after approximately 20 h of continuous use (see top images). SEM images (bottom) show the morphological surfaces of the coating for the respective sensors: clearly, with long-usage and continuous use, the sensors deteriorate. Specifically, the sensor tends to recover from repeated measurements, but after significant use, this deterioration alters the Zn characteristic response.

Considering reaction time, recovery time and sensor reusability, accomplishing continuous monitoring in situ it will be necessary to use multiple f-EM sensors, with the purpose to cover the time gaps caused by the recovery time of the sensor and obtain a set of comparable and reliable results.

Future Progression

This experimental integration between Bi_2O_3 based thick films and microwave spectroscopy demonstrates the feasibility of using this approach for increasing the performance of Zn detection using microwave spectroscopy. Future experiments will study the possibility to use this novel method for real mine polluted water and multi-metal measurements. Sensor arrays with specific functional materials for diverse toxic pollutants in water will be integrated for simultaneous and continuous monitoring of other metals in water.

Further work would focus on increasing the sensitivity, considering it has been demonstrated in the mg/l range and the EQS for Zn is freshwater is above 0.125 mg/l. However, as an early detection method, the current sensitivity may be acceptable.

Conclusions

This chapter communicates the experimental results of measuring the response and recovery times of functionalised planar type electromagnetic sensors coated with β-Bi_2O_3 for the continuous detection of Zn concentrations in water. A comprehensive set of complementary experiments using electrical and microwave detection methods demonstrate that the state of the sensor and the time of reaction are important for

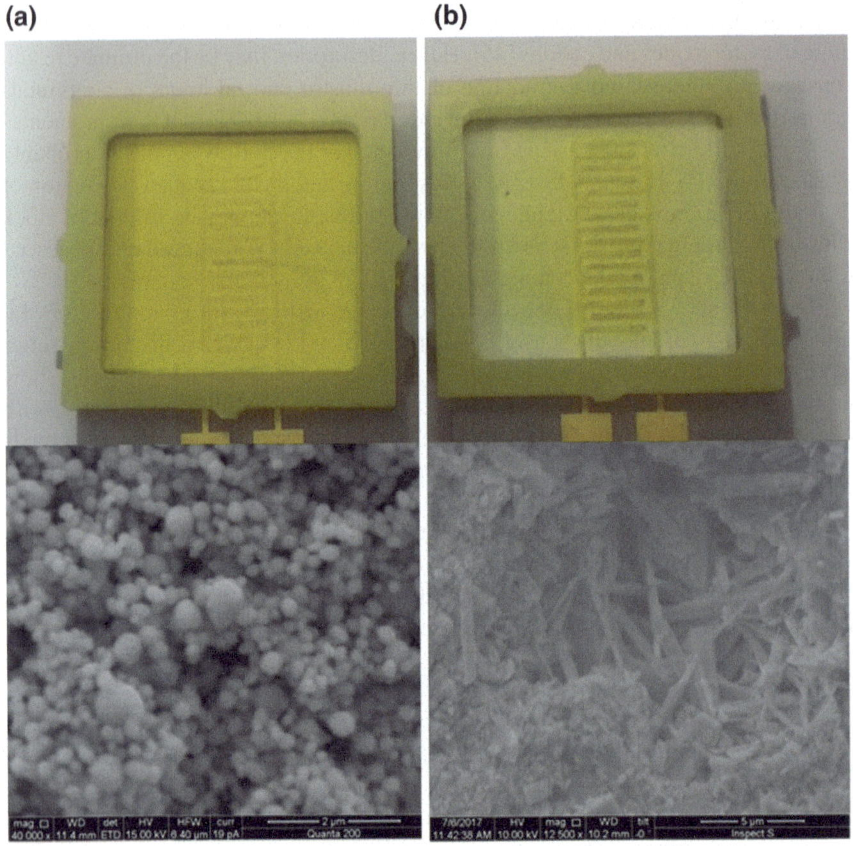

Fig. 14 Photos (top) and related SEM images (bottom) which show the differences between the sensor during the beginning of the measurement (**a**) and after 20 hours (**b**)

reaching a stability between the analysed sample, in this study zinc in water, and the β-Bi_2O_3 based film. Typically the sensor stabilises 5–10 min after applying a sample, and takes 2–3 min to recover to the original output levels after the sensor is rinsed. This would ideally be faster, but for an unattended system it is still better than sampling and for laboratory-based measurements.

The change in sensor output appears to be reversible (i.e. the sensor recovers), the sensor can be reused, and the output is relatively stable. Re-use should be further investigated since re-use is clearly finite; however, the cost of the sensor would be low in large volumes, so a semi-disposal option is feasible.

This study has confirmed the potential of a novel EM sensing approach combined with functional materials as an alternative method to current offline techniques for Zn concentration monitoring of polluted mine water, and to provide an early warning system for pollution events.

Acknowledgements The authors gratefully acknowledge the support of Liverpool John Moores University, the Faculty of Engineering and Technology Ph.D. Scholarship Programme, which allowed this research to be undertaken.

References

1. M.H. Zarifi, A. Gholidoust, M. Abdolrazzaghi, P. Shariaty, Z. Hashisho, M. Daneshmand, Sensitivity enhancement in planar microwave active-resonator using metal organic framework for CO2 detection. Sens. Actuators B Chem. **255,** 1561–1568 (2018)
2. World Health Organization (WHO) and the United Nations Children's Fund (UNICEF), *Progress on Drinking Water, Sanitation and Hygiene: 2017 Update and SDG Baselines*. Geneva (2017)
3. UNICEF and World Health Organization (WHO), *Progress on Sanitation And Drinking Water—2015 Update and MDG Assessment*. Geneva (2015)
4. W.M. Mayes, D. Johnston, H.A.B. Potter, A.P. Jarvis, A national strategy for identification, prioritisation and management of pollution from abandoned non-coal mine sites in England and Wales I. Sci. Total Environ. **407**(21), 5435–5447 (2009)
5. A. Lay-Ekuakille, I. Palamara, D. Caratelli, F. Morabito, Experimental infrared measurements for hydrocarbon pollutant determination in subterranean waters 015103 (2013)
6. P.B. Tchounwou, C.G. Yedjou, A.K. Patlolla, D.J. Sutton, Heavy metals toxicity and the environment. EXS **101**, 133–164 (2012)
7. UK Technical Advisory Group on the Water Framework Directive, UK environmental standards and conditions (phase 1—SR1-2006) (2008)
8. United States Environmental Protection Agency, *Quality Criteria for Water* (Washington, DC 20460: EPA, 1986)
9. P. Rowland, C. Neal, D. Sleep, C. Vincent, P. Scholefield, Chemical quality status of rivers for the water framework directive: a case study of toxic metals in North West England. Water **3**(2), 649 (2011)
10. P. Byrne, P.J. Wood, I. Reid, The impairment of river systems by metal mine contamination: a review including remediation options. Crit. Rev. Environ. Sci. Technol. **42**(19), pp. 2017–2077 (2012)
11. Environment Agency, H1 Annex D-basic surface water discharges, in *Annex D Surface Water Basic*, vol. 2.2 (Environment Agency, Bristol, 2011)
12. K.S. Smith, *Metal sorption on mineral surfaces: an overview with examples relating to mineral deposits* (Environmental Geochemistry of Mineral Deposits. Part A: Processes, Techniques, and Health Issues, 1999), pp. 161–182
13. L.M. Plum, L. Rink, H. Haase, The essential toxin: impact of zinc on human health. Int. J. Environmental Res. Public Health **7**(4), 1342–1365 (2010)
14. A. Léonard, G.B. Gerber, F. Léonard, Mutagenicity, carcinogenicity and teratogenicity of zinc. Mutat. Res. Rev. Genetic Toxicol. **168**(3), 343–353 (1986)
15. United Nations, *Millennium Development Goals (MDGs)* (Salem Press, 2016)
16. R. Altenburger et al., Future water quality monitoring—adapting tools to deal with mixtures of pollutants in water resource management. Sci. Total Environ. **512**, 540–551 (2015)
17. B. Bansod, T. Kumar, R. Thakur, S. Rana, I. Singh, A review on various electrochemical techniques for heavy metal ions detection with different sensing platforms, Biosens. Bioelectron. **94**,. 443–455 (2017)
18. K. Hudson-Edwards, Tackling mine wastes, (in English). Sci. Short Surv. **352**(6283), 288–290 (2016)
19. P. Byrne et al., Water quality impacts and river system recovery following the 2014 Mount Polley mine tailings dam spill, British Columbia, Canada. Appl. Geochem. **91**, 64–74 (2018)

20. G. Pattelli et al., Effects of the November 2012 flood event on the mobilization of Hg from the Mount Amiata mining district to the sediments of the Paglia river basin. Minerals **4**(2), 241 (2014)
21. S.F.L. Lynch, L.C. Batty, P. Byrne, Critical control of flooding and draining sequences on the environmental risk of Zn-contaminated riverbank sediments, J. Soils Sedim. (2017)
22. R. Cidu, F. Frau, S. Da Pelo, Drainage at abandoned mine sites: natural attenuation of contaminants in different seasons. Mine Water Environ. J. Artic. **30**(2), 113–126 (2011)
23. P. Byrne, I. Reid, P.J. Wood, Stormflow hydrochemistry of a river draining an abandoned metal mine: the Afon Twymyn, central Wales. Environ. Monit. Assess. J. Artic. **185**(3), 2817–2832 (2013)
24. A. Jones, M. Rogerson, G. Greenway, H.A.B. Potter, W.M. Mayes, Mine water geochemistry and metal flux in a major historic Pb-Zn-F orefield, the Yorkshire Pennines, UK. Environ. Sci. Pollut. Res. **20**(11), 7570–7581 (2013)
25. M.B. Gumpu, S. Sethuraman, U.M. Krishnan, J.B.B. Rayappan, A review on detection of heavy metal ions in water—An electrochemical approach, Sens. Actuators B Chem. **213**, 515–533 (2015)
26. Z. Kovacs et al., Water spectral pattern as holistic marker for water quality monitoring, Talanta **147**, 598–608 (2016)
27. J. Iqbal, D.U. Yiping, F. Howari, M. Bataineh, N. Muhammad, A. Rahim, Simultaneous enrichment and on-line detection of low-concentration copper, cobalt, and nickel ions in water by near-infrared diffuse reflectance spectroscopy combined with chemometrics. J. AOAC Int. Artic. **100**(2), 560–565 (2017)
28. A. Mason et al., Theoretical basis and application for measuring pork loin drip loss using microwave spectroscopy, Sensors **16**(2), 182 (13 pp.) (2016)
29. O. Korostynska, A. Mason, M. Ortoneda-Pedrola, A. Al-Shamma'a, Electromagnetic wave sensing of NO3 and COD concentrations for real-time environmental and industrial monitoring. Sens. Actuators B Chem. **198**, 49–54 (2014)
30. L.F. Chen, C.K. Ong, C.P. Neo, V.V. Varadan, V.K. Varadan, Planar-circuit methods, in *Microwave Electronics* (Wiley, 2005), pp. 288–322
31. M.H. Zarifi, M. Daneshmand, Liquid sensing in aquatic environment using high quality planar microwave resonator. Sens. Actuators B Chem. **225**(Supplement C), 517–521 (2016)
32. S. Cashman, O. Korostynska, A. Shaw, P. Lisboa, L. Conroy, Detecting the presence and concentration of nitrate in water using microwave spectroscopy, IEEE Sens. J. **PP**(99), 1–1 (2017)
33. A. Chahadih, P.Y. Cresson, Z. Hamouda, S. Gu, C. Mismer, T. Lasri, Microwave/microfluidic sensor fabricated on a flexible kapton substrate for complex permittivity characterization of liquids. Sens. Actuators A Phys. **229**(Supplement C), 128–135 (2015)
34. M.H. Zarifi, B. Wiltshire, N. Mahdi, P. Kar, K. Shankar, M. Daneshmand, Ultraviolet sensing using TiO2 nanotube integrated high resolution planar microwave resonator device. Nanoscale (2018)
35. M.H. Zarifi, H. Sadabadi, S.H. Hejazi, M. Daneshmand, A. Sanati-Nezhad, Noncontact and nonintrusive microwave-microfluidic flow sensor for energy and biomedical engineering. Sci. Rep. (2017)
36. M.E.E. Alahi, A. Nag, S.C. Mukhopadhyay, L. Burkitt, A temperature-compensated graphene sensor for nitrate monitoring in real-time application. Sens. Actuators A Phys. **269**(Supplement C), 79–90 (2018)
37. A. Rydosz, E. Maciak, K. Wincza, S. Gruszczynski, Microwave-based sensors with phthalocyanine films for acetone, ethanol and methanol detection. Sens. Actuators B Chem. **237**, 876–886 (2016)
38. A. Mason et al., Noninvasive *In-Situ* measurement of blood lactate using microwave sensors. IEEE Trans. Biomed. Eng. **65**(3), 698–705 (2018)
39. M.H. Zarifi, S. Farsinezhad, M. Abdolrazzaghi, M. Daneshmand, K. Shankar, Selective microwave sensors exploiting the interaction of analytes with trap states in TiO2 nanotube arrays. Nanoscale **8**(14), 7466–7473 (2016). https://doi.org/10.1039/C5NR06567D

40. A. Bogner, C. Steiner, S. Walter, J. Kita, G. Hagen, R. Moos, Planar microstrip ring resonators for microwave-based gas sensing: design aspects and initial transducers for humidity and ammonia sensing. Sensors **17**(10), 2422 (2017)
41. A. Azmi et al., Performance of coating materials on planar electromagnetic sensing array to detect water contamination. IEEE Sens. J. **17**(16), 5244–5251 (2017)
42. R. Igreja, C.J. Dias, Dielectric response of interdigital chemocapacitors: the role of the sensitive layer thickness. Sens. Actuators B Chem. **115**(1), 69–78 (2006)
43. E. Reimhult, F. Höök, Design of surface modifications for nanoscale sensor applications. Sensors (Basel, Switzerland) **15**(1), 1635–1675 (2015)
44. G. Aragay, J. Pons, A. Merkoci, Recent trends in macro-, micro-, and nanomaterial-based tools and strategies for heavy-metal detection. Chem. Rev. **111**(5), 3433–3458 (2011)
45. S. Sen Gupta, K.G. Bhattacharyya, Kinetics of adsorption of metal ions on inorganic materials: a review. Adv. Colloid Interface Sci. **162**(1–2), 39–58 (2011)
46. L. Cui, J. Wu, H. Ju, Electrochemical sensing of heavy metal ions with inorganic, organic and bio-materials. Biosens. Bioelectron. **63**, 276–286 (2015)
47. P.K. Chatterjee, A.K. SenGupta, Toxic metal sensing through novel use of hybrid inorganic and polymeric ion-exchangers. Solv. Extr. Ion Exch. **29**(3), 398–420 (2011)
48. V. Ferrari, M. Prudenziati, 8—Printed thick-film capacitive sensors, in *Printed Films*: (Woodhead Publishing, 2012), pp. 193–220
49. M. Prudenziati, M. Prudenziati (eds.) *Handbook of Sensors and Actuators*, 1st ed. (Thick Film Sensors) (Amsterdam, 1994)
50. M. Prudenziati, J. Hormadaly, 1—Technologies for printed films, in *Printed Films* (Woodhead Publishing, 2012), pp. 3–29
51. M. Kohl, G. Veltl, M. Busse, Printed sensors produced via thick-film technology for the use in monitoring applications. Procedia Technol. **15**, 107–113 (2014)
52. B. Yosypchuk, J. Barek, Analytical applications of solid and paste amalgam electrodes. Crit. Rev. Anal. Chem. **39**(3), 189–203 (2009)
53. J. Hormadaly, M. Prudenziati, 2—Materials for printed films, in *Printed Films* (Woodhead Publishing, 2012), pp. 30–62
54. G.-J. Lee, H.-M. Lee, C.-K. Rhee, Bismuth nano-powder electrode for trace analysis of heavy metals using anodic stripping voltammetry. Electrochem. Commun. **9**(10), 2514–2518 (2007)
55. I. Švancara, C. Prior, S.B. Hočevar, J. Wang, A decade with bismuth-based electrodes in electroanalysis. Electroanalysis **22**(13), 1405–1420 (2010)
56. R.O. Kadara, N. Jenkinson, C.E. Banks, Disposable bismuth oxide screen printed electrodes for the high throughput screening of heavy metals. Electroanalysis **21**(22), 2410–2414 (2009)
57. A.O. Dada, A. Olalekan, A. Olatunya, D.O. Dada, Langmuir, Freundlich, Temkin and Dubinin-Radushkevich isotherms studies of equilibrium sorption of $Zn2^{+}$ unto phosphoric acid modified rice husk. IOSR J. Appl. Chem. **3**(1), 38–45 (2012)

Reusable Surface Acoustic Wave Immunosensor for Monitoring of Mite Allergens

K. Toma, T. Arakawa and K. Mitsubayashi

Abstract Temporal information of antigen concentration is valuable especially for those samples whose concentrations fluctuate. Concentration of airborne house dust mite (HDM) allergens changes over time while being influenced by environmental conditions or human activity, and it is important to monitor this change because exposure to such allergens should be avoided for prevention of allergic diseases. In general, it is challenging for immunosensors to be used for monitoring because immunocomplex formed on the sensor surface must be once broken to regenerate the surface for a next measurement, which often causes denaturation of antibodies and thereby dysfunction of the sensor. In this chapter, we describe a reusable immunosensor for monitoring of a HDM allergen—*Dermatophagoides farinae* group 1 (*Der f* 1). The immunosensor was fabricated using a surface acoustic wave (SAW) device and pH-tolerant protein scaffold (ORLA85 protein). Capture antibodies were immobilized on the ORLA85 protein-modified SAW device surface, and *Der f* 1 was measured by detecting viscoelastic change induced by sandwich assay. In addition, differential method was employed to shorten the measurement time. It utilized a slope of the signal change as a sensor output instead of a signal shift that was used conventionally, and the measurement time was shortened by 10 min from 30 min while maintaining the sensitivity. Also a stable sensor surface to pH change during the regeneration allowed repeated immunosensing of *Der f* 1 with high reproducibility—coefficient of variation of 5.7% in ten measurements.

Introduction

Allergy is a disorder of the immune system, and it is accompanied by variety of symptoms such as rhinitis, asthma or itching [1–5]. The number of patients of allergic disease have been increasing over the years. More than 40% of children are diag-

K. Toma · T. Arakawa · K. Mitsubayashi (✉)
Institute of Biomaterials and Bioengineering,
Tokyo Medical and Dental University, Tokyo, Japan
e-mail: m.bdi@tmd.ac.jp

S. C. Mukhopadhyay et al. (eds.), *Modern Sensing Technologies*,
Smart Sensors, Measurement and Instrumentation 29,
https://doi.org/10.1007/978-3-319-99540-3_10

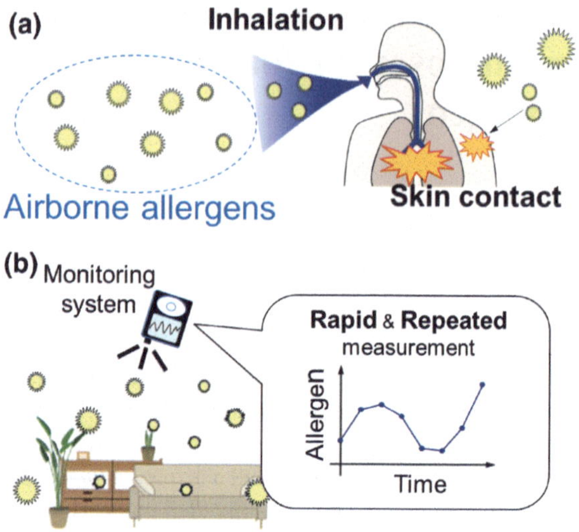

Fig. 1 **a** Main route of sensitization to allergens. **b** An airborne allergen monitoring system for semi-continuous measurement of airborne allergens

nosed with allergic rhinitis, and about 300 million people suffer from asthma in the world [3, 6]. Environmental allergens such as pollens and house dust mites (HDMs) are light and become airborne. Also the concentration of these airborne allergens fluctuates over time by being influenced by temperature, humidity or even human activities [7, 8]. However, many studies were focused on quantifying the environmental allergens in dust, but not those of the airbornes [9–12]. Apparently exposure to the airborne allergens can increase risk of inhalation and skin contact of the allergens, and thereby the sensitization. Considering this, it is important to know the concentration of the airborne allergens in an environment to reduce the risk and prevent the allergic disease. For this purpose, a monitoring system that allows rapid and repeated—semi-continuous—measurement of airborne allergens will be useful (Fig. 1).

An airborne allergen monitoring system can be developed by combining two parts: an airborne allergen sampler, and an allergen detector. In this chapter, we focus on the allergen detector, in particular an immunosensor for the monitoring. Immunosensors rely on affinity-binding between an antibody and antigen. Because of high selectivity and sensitivity, immunosensors or immunoassays such as enzyme-linked immunosorbent assays (ELISA) or lateral flow immunoassays are widely used not only in laboratories but also in practice, including measurement of HDM allergens [13–15].

One bottleneck in application of those immunosensors to a monitoring system is difficulty of repeated use. Although immunocomplex formed on a sensor surface to measure antigen needs to be dissociated for a next measurement, it also contains a risk of denaturation of proteins, which disables a function of an immunosensor. In order to overcome this challenge, we have studied and developed an immunosensor that does not lose a sensor function during a repeated measurement using a pH-tolerant protein scaffold for sensor surface modification. This chapter describes such an immunosensor for house dust mite (HDM) allergen, and an approach to shorten the measurement time for the advanced semi-continuous measurement.

Surface Acoustic Wave (SAW) Device-Based Immunosensor for Repeated Measurement of House Dust Mite Allergens

As a transducer, a surface acoustic wave (SAW) device was employed. SAW is a sound wave that travels along a surface of an elastic material, and sensitive to viscoelastic change nearby the material surface. A SAW device used in our study was composed of input and output transducers, SAW propagating path, and a reflector (Fig. 2a), and was fabricated by depositing 90 nm-thick gold on a 36Y-X quartz substrate. SAW is generated by applying alternating voltage to the input transducer on a quartz substrate, which is a piezoelectric, at a center frequency of 250 MHz. Here a shear-horizontal (SH) SAW was chosen because of less attenuation into a liquid and thus suitable to sensing in liquid phase. The generated SAW propagates along the propagation path that is used as a sensing area to the reflector, then reflected SAW was received at the output transducer and converted back to an alternating voltage. Binding of molecules to the sensing area induces a viscoelastic change in vicinity of the surface. Then this viscoelastic change alters SAW velocity, which was measured as a phase shift of the alternating voltage between the input and output transducers.

A SAW immunosensor for HDM allergen, *Dermatophagoides farinae* group 1 (*Der f* 1), was fabricated using pH-tolerant protein scaffold, ORLA85 (Orla Protein Technologies), and capture antibody (cAb) for *Der f* 1 (Fig. 2a). Figure 2b illustrates a process of SAW immunosensor fabrication: First, a mixture of ORLA85 and

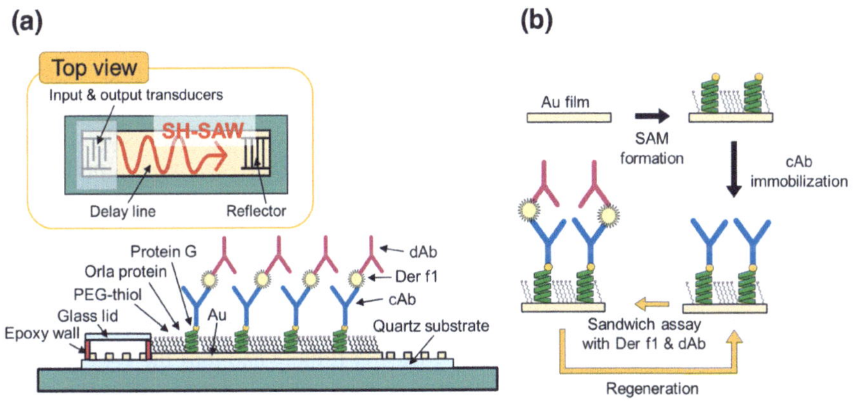

Fig. 2 **a** Schematic diagram of the SAW immunosensor. **b** A process of the surface modification to fabricate the SAW immunosensor and semi-continuous measurement of *Der f* 1. cAb, capture antibody; dAb, detection antibody. (Reprinted with permission from [16]. Copyright (2015) American Chemical Society)

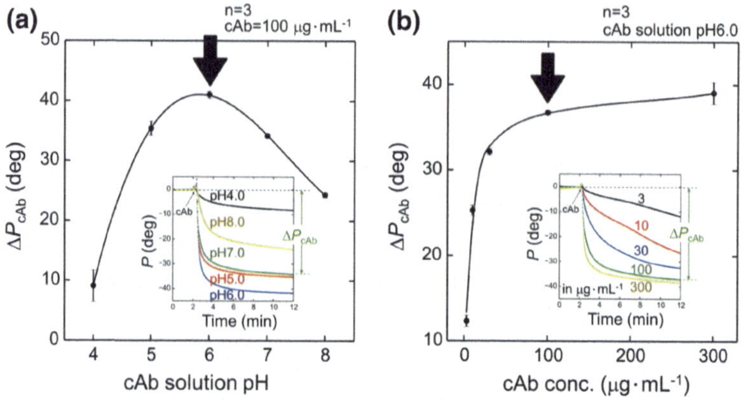

Fig. 3 Influences of **a** buffer solution pH dissolving cAb and **b** cAb concentration on the efficiency of cAb binding to protein G. The insets show the binding kinetics of cAb under various conditions. (Reprinted with permission from [16]. Copyright (2015) American Chemical Society)

polyethylene glycol-thiol (PEG-thiol) was applied to a gold surface of the sensing area in the SAW device to form a self-assembled monolayer (SAM); then, cAb was bound to protein G, a part of ORLA85, with high orientation, and was immobilized via crosslinking. Semi-continuous measurement of *Der f* 1 was conducted by the sandwich assay with *Der f* 1 and detection antibody (dAb), and surface regeneration induced by applying hydrochloride (HCl) of pH 1 in order to remove *Der f* 1 and dAb for a next measurement.

First, in binding of cAb to ORLA85, influence of pH and concentration of cAb solution on the binding efficiency was investigated. Figure 3a shows a phase shift occurred by the binding of cAb, ΔP_{cAb}, with various solution pH, which was determined as a difference between an averaged phase for the last 30 s and that before applying cAb—baseline. The cAb solution was prepared using acetate buffered saline (ABS, pH 4.0–5.0) and phosphate buffered saline with Tween 20 (PBS-T, pH 6.0–8.0). The maximum binding occurred when using pH 6.0 solution, and it was about 4-fold larger amount than that with pH 4.0 solution. This result was well agreed to a previous study concluding that a slightly acidic condition is preferable for binding between immunoglobulin G and protein G [17]. Similarly, the influence of cAb concentration was investigated. As the result shows in Fig. 3b, the phase shift ΔP_{cAb} dramatically increased at the cAb concentration less than 50 μg/mL, and showed a saturation at the concentration over 100 μg/mL. It was probably attributed to a threshold existing at around the concentration of 100 μg/mL: in the region less than the threshold, the binding rate was diffusion-limited and thus the cAb concentration dominantly influences the binding rate while surface density of cAb limited the binding rate in the region above the threshold.

After binding of cAb to ORLA85, crosslinking was taken place in order to immobilize cAb on a sensor surface. The immobilization gives such a merit that a rebinding process of cAb can be skipped in repeated measurement, and therefore

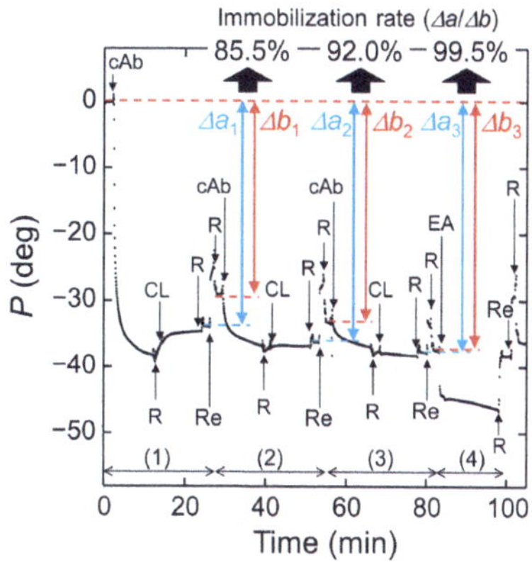

Fig. 4 Phase change during the crosslinking between cAb and protein G. CL, crosslinking with $BS(PEG)_5$; R, rinsing; Re, regeneration with HCl; EA, ethanolamine terminating the reaction. (Reprinted with permission from [16]. Copyright (2015) American Chemical Society)

the total measurement time can be shortened. Figure 4 shows a phase shift during the crosslinking of cAb. After binding of cAb, a crosslinker, 0.5 mM PEGylated bis(sulfosuccinimidyl)-suberate [$BS(PEG)_5$], was applied to the sensor surface, and reacted for 10 min. After rinsing, non-crosslinked cAb was removed by applying HCl pH 1.0 to the sensor surface. In the end of crosslinking process, unreacted active esters of $BS(PEG)_5$ was passivated by 1 M ethanolamine pH 8.4. Here an immobliza-tion rate was calculated to evaluate how much cAb was successfully cross-linked, and it was determined as a ratio between the phase shifts ΔP_{cAb} before (Δa) and after (Δb) applying HCl. The immobilization rate of the first cross-linking was of 85.5%, and it was improved up to 99.5% by repeating the cross-linking process three times. This result indicates that almost all bound cAb could be immobilized on the sensor surface.

Semi-continuous Measurement of *Der f* 1

The developed SAW immunosensor was applied to measurement of *Der f* 1. Figure 5a shows a phase shit during measurement of *Der f* 1. After a slight phase change by *Der f* 1 binding, the large phase change by dAb binding was observed, which was attributed to a molecular weight difference between two molecules. After the measurement, a regeneration by 100 mM HCl was conducted, and the phase recovered to the baseline, which indicated that bound *Der f* 1 and dAb were almost completely removed. In addition, thanks to the immobilization, the measurement time was 30 min including the regeneration, which demonstrates the suitability of the SAW immunosensor for the monitoring by considering the fact that the measurement takes almost a day by ELISA.

In order to enhance continuity of the measurement, further rapid measurement was pursued. Conventionally, the sensor output was obtained from a phase shift caused by dAb binding, ΔP_{dAb}, from a baseline (the upper graph in Fig. 5a). In a

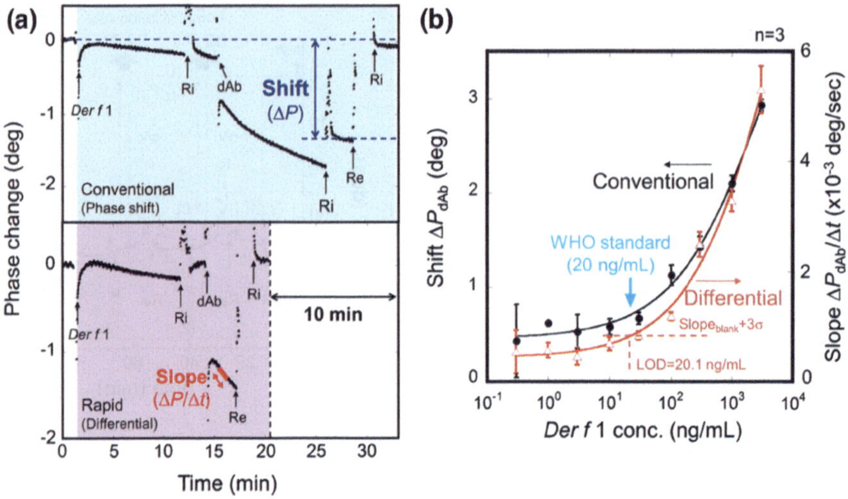

Fig. 5 **a** Phase change during the measurement of *Der f* 1 by the conventional method (upper) and the differential (lower). **b** Calibration curves of the conventional and the differential method for *Der f* 1. (Reprinted with permission from [20])

new approach—a differential method, however, the output was obtained from a slope of the phase change by dAb, $\Delta P_{\mathrm{dAb}}/\Delta t$ (the lower in Fig. 5a). This method allowed to shorten the incubation time of dAb as well as to skip rinsing process of dAb. Both rinsing of excess dAb and regeneration of the sensor surface was conducted by HCl. These resulted in shortening the measurement time by 10 min from the conventional method. The sensitivity of the sensor is also important, and we investigated if the differential method gained the rapidness at the price of the sensitivity. Sensor outputs of the two methods were compared by measuring various concentrations of *Der f* 1 (Fig. 5b). The plots were fitted by the following equation for 0.3–3000 ng/mL:

$$\Delta P_{\mathrm{dAb}}(\mathrm{deg}) = A_1 + (B_1 - A_1)/\left\{1 + ([Der\ f\ 1]/C_1)^{D_1}\right\}, \quad (1)$$

$$\Delta P_{\mathrm{dAb}}/\Delta t(\mathrm{deg/sec}) = A_2 + (B_2 - A_2)/\left\{1 + ([Der\ f\ 1]/C_2)^{D_2}\right\}, \quad (2)$$

where $A_1 = 5.8$, $B_1 = 0.46$, $C_1 = 3940$ and $D_1 = -0.58$, and $A_2 = 1.3 \times 10^{-2}$, $B_2 = 4.5 \times 10^{-4}$, $C_2 = 7040$ and $D_2 = -0.60$ are the coefficients, and [*Der f* 1] is the concentration of applied *Der f* 1 in ng/mL. The correlation coefficients of the fitting were R = 0.997 for the conventional and R = 0.995 for the differential. A limit of detection (LOD) of the differential method was determined to be 20.1 ng/mL (0.80 nM), and the dynamic range was 100–3000 ng/mL. This is comparable to the conventional method [*Der f* 1, 6.1 ng/mL (0.24 nM)] and other state of the art works: electrochemical aptasensor [*Der p* 2, 16.47 ng/mL (1.2 nM)] [18], ELISA [*Der f* 1, 0.58 ng/mL (23 pM)] [16], and surface plasmon resonance sensor [*Der f*

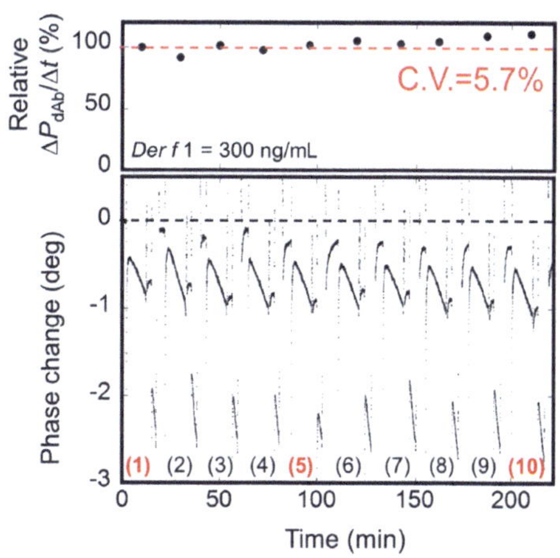

Fig. 6 Phase change over ten repeated measurements of *Der f* 1. (Figure reprinted with permission from [20])

1, 13.7 ng/mL (0.55 nM)] [19]; it was confirmed that the differential method does not degrade the sensitivity.

Now, a capability for semi-continuous measurement of *Der f* 1 was examined. 300 ng/mL *Der f* 1 was repeatedly measured ten times using the differential method, and the phase change during the measurement is shown in Fig. 6. The result shows that the sensor output through the measurement was highly reproducible, which was validated by a coefficient of variation (C.V.) of the output of 5.7%. In addition, even on the 10th measurement, the phase change and the output did not change from the 1st, which indicates that it is possible to further continue the repeated measurement by the SAW immunosensor.

Finally, the sensitivity of the SAW immunosensor was investigated. In the case of the allergen monitoring system, the sensor should be able to distinguish the target allergen from other environmental allergens. Here the selectivity was examined by measuring four different environmental allergens: mites (*Der f* 1 and *Der f* 2) and pollens (Japanese cedar pollen, *Cry j* 1, and ragweed-pollen, *Amb a* 1). Figure 7 shows the relative sensor outputs of each allergen and their mixtures to *Der f* 1. The sensor outputs were obtained only from the samples containing *Der f* 1 although the output of the mixed sample was smaller than that of only *Der f* 1, which was probably attributed to an interference from other allergens in the binding between *Der f* 1 and cAb or dAb. This result demonstrates the high selectivity of the SAW immunosensor based on the specificity of the antibodies.

Conclusions

This chapter has described a SAW immunosensor that allows repeated measurement of mite allergen, *Der f* 1, for the airborne allergen monitoring system. Based on the results demonstrating unique advantages of the SAW immunosensor including high

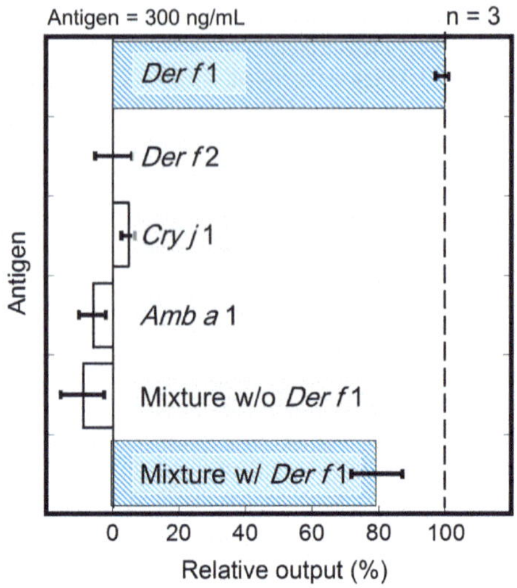

Fig. 7 Comparison of the sensor output from four different environmental allergens and their mixtures. (Reprinted with permission from [20])

sensitivity comparable to other state of the art sensors, high selectivity, short measurement time, and reusability over ten measurements, the immunosensor can provide one possible path to realize the monitoring system for prevention of environmental allergen-related diseases. In principle, the SAW immunosensor can be applied to any other target molecules by exchanging an antibody, and therefore development of a monitoring system that is able to monitor whole kinds of environmental allergens simultaneously can be expected by arraying the SAW immunosensors.

References

1. A. Togias, Rhinitis and asthma: evidence for respiratory system integration. J. Allergy Clin. Immunol. **111**, 1171–1183 (2003)
2. R. Beasley, U. Keil, E. Von Mutius, N. Pearce, Worldwide variation in prevalence of symptoms of asthma, allergic rhinoconjunctivitis, and atopic eczema: ISAAC. The international study of asthma and allergies in childhood (ISAAC) steering committee. Lancet (London, England), **351**, 1225–1232 (1998)
3. R. Pawankar, G.W. Canonica, R.F. Lockey, S.T. Holgate (eds), *WAO White Book on Allergy 2011–2012: Executive Summary* (2011)
4. S.J. Arbes, P.J. Gergen, L. Elliott, D.C. Zeldin, T. Park, Prevalences of positive skin test responses to 10 common allergens in the US population: results from the third national health and nutrition examination survey. J. Allergy Clin. Immunol. **116**, 377–383 (2005)
5. A.P. Jackson, A.P. Foster, B.J. Hart, C.R. Helps, S.E. Shaw, Prevalence of house dust mites and dermatophagoides group 1 antigens collected from bedding, skin and hair coat of dogs in South-West England. Vet. Dermatol. **16**, 32–38 (2005)
6. A.L. Wright, C.J. Holberg, F.D. Martinez, M. Halonen, W. Morgan, L.M. Taussig, Epidemiology of physician-diagnosed allergic rhinitis in childhood. Pediatrics **94**, 895–901 (1994)
7. L.D. Knibbs, C. He, C. Duchaine, L. Morawska, Vacuum cleaner emissions as a source of indoor exposure to airborne particles and bacteria. Environ. Sci. Technol. **46**, 534–542 (2012)

8. M. Sakaguchi, M. Hashimoto, N.S. Hospital, S. Inouye, R. Sasaki, M. Hashimoto, C. Kobayashi, H. Yasueda, Measurement of airborne mite allergen exposure in individual subjects. J. Allergy Clin. Immunol. **97**, 1040–1044 (1996)
9. T. Plattsmills, D. Vervloet, W. Thomas, R. Aalberse, M. Chapman, Indoor allergens and asthma: report of the third international workshop. J. Allergy Clin. Immunol. **100**, S2–S24 (1997)
10. T. Takai, T. Kato, H. Yasueda, K. Okumura, H. Ogawa, Analysis of the structure and allergenicity of recombinant pro- and mature Der p 1 and Der f 1: major conformational IgE epitopes blocked by prodomains. J. Allergy Clin. Immunol. **115**, 555–563 (2005)
11. T.J. Nuttall, J.R. Lamb, P.B. Hill, Characterisation of major and minor dermatophagoides allergens in canine atopic dermatitis. Res. Vet. Sci. **71**, 51–57 (2001)
12. E.R. Tovey, C. Almqvist, Q. Li, D. Crisafulli, G.B. Marks, Nonlinear relationship of mite allergen exposure to mite sensitization and asthma in a birth cohort. J. Allergy Clin. Immunol. **122**, 114–118 (2008)
13. U. Wahn, S. Lau, R. Bergmann, M. Kulig, J. Forster, K. Bergmann, C.-P. Bauer, I. Guggenmoos-Holzmann, Indoor allergen exposure is a risk factor for sensitization during the first three years of life. J. Allergy Clin. Immunol. **99**, 763–769 (1997)
14. A. Tsay, L. Williams, E.B. Mitchell, M.D. Chapman, A rapid test for detection of mite allergens in homes. Clin. Exp. Allergy **32**, 1596–1601 (2002)
15. M.D. Chapman, A. Tsay, L.D. Vailes, Home allergen monitoring and control–improving clinical practice and patient benefits. Allergy **56**, 604–610 (2001)
16. K. Toma, D. Miki, C. Kishikawa, N. Yoshimura, K. Miyajima, T. Arakawa, H. Yatsuda, K. Mitsubayashi, Repetitive immunoassay with a surface acoustic wave device and a highly stable protein monolayer for on-site monitoring of airborne dust mite allergens. Anal. Chem. **87**, 10470–10474 (2015)
17. B. Akerstrom, L. Bjorck, A physicochemical study of protein G, a molecule with unique immunoglobulin G-binding properties. J. Biol. Chem. **261**, 10240–10247 (1986)
18. M.C. Shen, J.C. Lai, C.Y. Hong, G.J. Wang, Electrochemical aptasensor for detecting Der P2 allergen using polycarbonate-based double-generation gold nanoparticle chip. Sens. Bio-Sens. Res. 2–7 (2016)
19. H. Huang, P. Ran, Z. Liu, Signal enhancement of surface plasmon resonance-based immunoassays for the allergen detection. Sens. Actuators B: Chem. **131**, 417–423 (2008)
20. K. Toma, M. Horibe, C. Kishikawa, N. Yoshimura, T. Arakawa, H. Yatsuda, H. Shimomura, K. Mitsubayashi, Rapid and repetitive immunoassay with a surface acoustic wave device for monitoring of dust mite allergens. Sens. Actuators B: Chem. **248**, 924–929 (2017)

Performance Enhancement of Polypyrrole Based Nano-Biosensors by Different Enzyme Deposition Techniques

D. M. G. Preethichandra, E. M. I. Mala Ekanayake, M. Onoda and K. Kaneto

Abstract This chapter presents a comparison of results from a range of experiments carried out to investigate the performance dependency of polypyrrole based nano-biosensors on fabrication and enzyme immobilization techniques. The methods compared are drop casting, co-entrapment, and electrophoretic enzyme deposition techniques. Templated polypyrrole nanotube array sensors provided high sensitivities and quick response times. The size of the template pore diameter plays a vital role in enzyme immobilizing in terms of loading capacity. Application of a polymer cross linking agent provided enhanced sensitivities. However, the response time of the crosslinked sensor was little longer than that of the physical adsorption type enzyme loading. The templated nanostructures have recorded a glucose measurement sensitivity of 62.5 mAM^{-1} cm^{-2} with a response time of 6 s. In contrast to the templated method nano-corrugated polypyrrole structures were fabricated on Indium Tin Oxide planar surface using pulse deposition method. The electrochemically developed nano-corrugated polypyrrole sensors with enzyme loaded under high electric field of 1 kV/m have displayed an extremely high sensitivity of 325 mAM^{-1} cm^{-2}. The study shows the enzyme immobilizing techniques play a great role in the conducting polymer based nano-biosensor performance.

Introduction

Enzyme based biosensors play a vital role in the fields of biomedical, food and beverage, and environmental measurement since the invention of biosensors in early 1950s by Leland C. Clark and co-researchers. In principle, enzyme biosensor converts

D. M. G. Preethichandra (✉) · E. M. I. Mala Ekanayake
School of Engineering and Technology, Central Queensland University, Rockhampton, Australia
e-mail: preethi@ieee.org; d.preethichandra@cqu.edu.au

M. Onoda
Graduate School of Engineering, University of Hyogo, Himeji, Hyogo, Japan

K. Kaneto
Department of Biomedical Engineering, Osaka Institute of Technology, Osaka, Japan

S. C. Mukhopadhyay et al. (eds.), *Modern Sensing Technologies*,
Smart Sensors, Measurement and Instrumentation 29,
https://doi.org/10.1007/978-3-319-99540-3_11

the measurand into a measurable electrical quantity by means of an electrochemical reaction where the relevant enzyme works as the catalyst. The enzyme not only work as a catalyst in the electrochemical reaction, it also works as the target selective material in the biosensor, where the response to other substances are automatically suppressed. Traditionally inorganic substrate based biosensors were developed, but with their complexity in fabrication and enzyme immobilizing process there are many research work focused on developing easy and cost effective biosensors [1, 2]. They include carbon material based sensors [3, 4], nanomaterial and inorganic composites [5, 6] and polymers [7–9]. Carbon nanotube (CNT) polymer composites have been highly investigated for biosensor development and produced convincing results due to increased conductivity and ion transfer ability [10–12].

Conducting polymer based biosensors has a wide range of application ranging from gas sensing to blood glucose sensing. They are being used in the measurement of periodic environmental gas monitoring application as the sensor can be replaced very cheaply for every single measurement resulting in high accuracy at a low cost compared to the traditional continuous monitoring sensors. They are being used in the agricultural industry as a fruit ripeness sensors to measure the volatile amines in the close proximity of fruit [13].

The fabrication process of conducting polymer based biosensors do not involve with complex fabrication facilities as compared to silicon wafer based sensors [14]. Further, the conducting polymer based biosensors are more environmental friendly as a disposable elements compared to silicon based sensors.

There is a couple of commonly used fabrication techniques to form a conducting polymer substrate including drop casting, electrochemical deposition, vapor phase polymerization, inkjet printing, and thermal deposition [14–16]. Drop-casting, inkjet printing, and electrochemical deposition are very popular due to easy and low-cost fabrication process and the film stability as well. In this chapter we discuss electro-polymerization as primary method used in our research. Conducting polymer based enzymatic nano-biosensors have to undergo a step of enzyme deposition on the substrate and this plays a crucial role in sensor performances [17].

There are different conducting polymers which have been used in the past as substrate material for glucose bio-sensing applications [18–20]. Polyaniline, polypyrrole ((PPy), and poly(3,4-ethylenedioxytheophene (PEDOT)) are some of the most investigated conducting polymers as the substrate material for glucose biosensor applications. Polypyrrole provides easy, stable and controllable substrate fabrication among them.

In this study Glucose and Hydrogen Peroxide have been chosen as target molecules, but the same method with appropriate modifications can be used for sensing other substances.

Fabricating a suitable substrate surface for enzyme entrapment and developing a matching enzyme deposition method is always a challenge. Introducing a nano-structure matching the size of enzyme is essential in this process. There are many approaches to introduce a nano-corrugated polymer surface by introducing various nanoparticles such as gold nanoparticles, silver nano-particles, carbon nanotubes, and zinc oxide crystals [21–24]. In this chapter we demonstrate how nanostructures of polymer can be developed using templating method and pulsed electrodeposition method and their influence of sensor performances. We also investigated solutions for denaturation of enzymes during the shelf lifetime.

When an enzyme is introduced to the system that reduces the activation energy for the target material. The measurements were carried out by collecting the freed electrons in the reaction by working electrode and the medea.

The same techniques of sensor fabrication and enzyme immobilization have been used in developing sensors for environmental parameter measurements and they also showed promising results.

Experimental

Material and Apparatus

Polypyrrole was used in all experiments as the substrate material. Pyrrole monomer, sodium hexafluorophosphate, Glucose Oxidase (EC1.1.3.4, 210 U/mg, from Aspergillus Niger), and Horseradish peroxidase (EC. 1.11.1.7, 225 U/mg) were purchased from Wako (Japan). Pyrrole monomer was distilled before use. D-Glucose, Gluteraldehide (wt. 50%), Na_2HPO_4 and K_2HPO_4 were purchased from Sigma Aldrich. Fresh Phosphate buffer solutions (PBS, 0.05 M, pH 6.5) were prepared for every experiment using Na_2HPO_4 and K_2HPO_4. Nano-porous Alumina disks (Anodisk with pore diameters of 100 and 200 nm) were purchased from Whatman. Industrial grade Indium Tin Oxide (ITO) substrates and other chemicals of analytical grade were used for experiments where applicable.

All electrochemical fabrication and measurement were done with Hokuto Denko HSV100 automatic polarization system. A purpose made three electrode electrochemical cell was used in templated polypyrrole nanotube array method. A two electrode parallel plate electrolytic cell was used in pulse deposited nano-corrugated polypyrrole substrate method the same cell was used in enzyme deposition under high electric field as explained in the next section. Shimadzu super scan SS-550 scanning electron microscope was used for obtaining the images of samples and JEOL quick coater (JFC 1500) was used to deposit a controlled thin layer of Platinum on the working electrode.

Fabrication Techniques

Anodiscs of two different sizes namely 100 and 200 nm in average pore diameter were used in the series of experiments. The Anodiscs were coated with a 50 nm platinum layer by placing them in JOEL quick coater where the coating thickness can be adjusted. This is a vacuum plasma deposition system and the Pt target is at

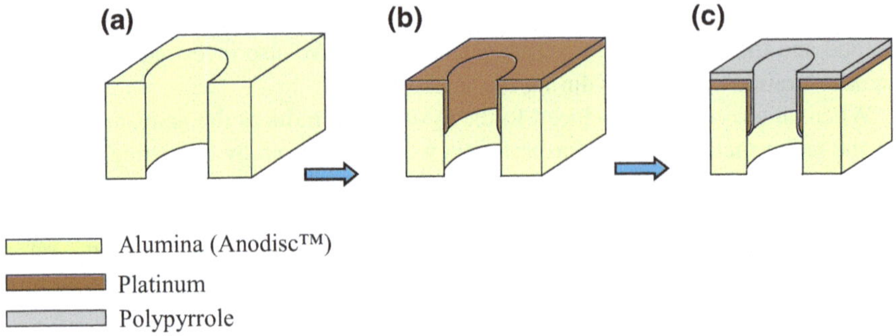

Fig. 1 Fabrication of polypyrrole nanotubes on nano-porous Anodiscs

the top of the chamber and the Anodisc was placed on the metal platform directly below it. When the system created the required level of vacuum, the high voltage is applied and the plasma is generated. Then Pt particles freed from the Pt target start travl to the bottom electrode. Since the Anodisc is sitting on that electrode with pores having their longitudinal axis vertical, some of the Pt particles enter the pore and travel inside the pore until it lands on the vertical pore-wall. Since the number of Pt particles decrease with the distance from the top surface as they travel inside, the thickness of the platinum layer deposited on the pore-wall will also be decreased. Then the Pt coated Anodisk is masked to have a 1 cm × 1 cm area exposed and the rest will be covered with a non-conducting lacquer leaving terminal connection opened from one end. Then this Pt coated Anodosc electrode is attached to the special three electrode cell as the working electrode parallel and 1 cm away from a Pt counter electrode. An Ag/AgCl electrode was used as the reference electrode.

This electrode set was then immersed in a 0.05 M pyrrole monomer solution with 0.1 M $NaPF_6$ and a polypyrrole layer was electrodeposited under constant current density of 0.3 mA/cm^2 for 90 s. Throughout the experiment the temperature was kept at 25 ± 2 °C unless otherwise stated. Figure 1 shows the schematic diagram of the cross-section of a single alumina pore during platinum deposition and polymerization process. Figure 2a shows the view from top of the polypyrrole nano-tube array on alumina Anodisc template. It can be seen that the diameter of the holes now looks smaller than the bare Anodosk porediameter of 200 nm. After 90 s electrodeposition the polypyrrole nano-tube array template on the alumina Anodisc was carefully washed in 6.5 pH buffer solution and air dried at 4 °C.

To visualize the growth structure of the polypyrrole nanostructure the Al_2O_3 template was carefully dissolved in 0.1 M NaOH. The remained polypyrrole structure was then placed on a scanning electron microscope and examined. Figure 2b shows the polypyrrole nano-tubes grown inside the Anodisc pores. The measured dimensions of the nano-tube diameter and the length gives us the increment in effective surface area has been increased by a factor ranging from 10 to 1000.

Since the templating Anodiscs cost a considerable amount, it was then investigated a new method of developing nano-corrugated polypyrrole structures using different

Fig. 2 The polypyrrole nanostructure substrate developed

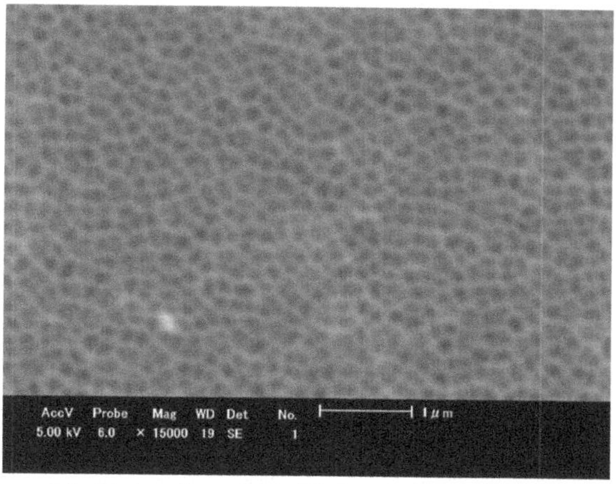

(a) Top surface view of Polypyrrole nanotubes grown inside the pores.

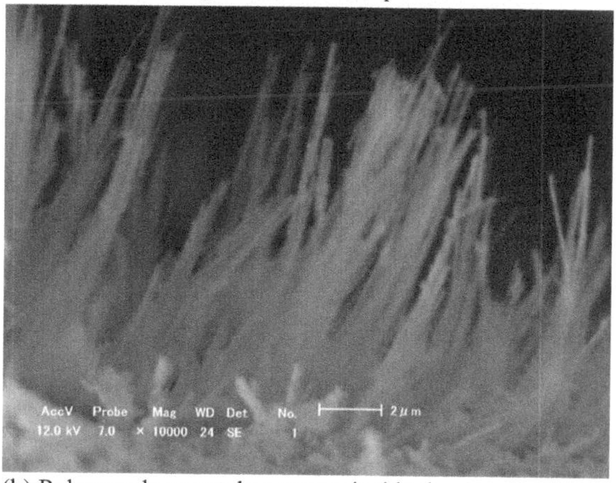

(b) Polypyrrole nanotubes grown inside the Anodisc pores

technologies. First the surface morphology of the electrodeposited polypyrrole on a ITO was investigated under scanning electron microscope. Then high electric field pulse deposited polypyrrole substrates were developed on planar ITO substrates modified with a 50 nm Pt layer. The electrolytic solution is the same as in the templated method and the fabrication was done as follows. First the Pt modified ITO plate was masked to have a 1 cm × 1 cm working electrode window. Then they were placed in the especially designed arrangement as shown in Fig. 3. The counter electrode used was a planar Pt electrode with little larger than the working electrode placed parallel to the working electrode and facing the Pt modified ITO layer. The Pt modified ITO surface faces the CE. They were 1 mm apart from each other and the terminal connections were made available to be connected to the HSV-100 polarization system. During the polymerization process the newly formed electrolytic

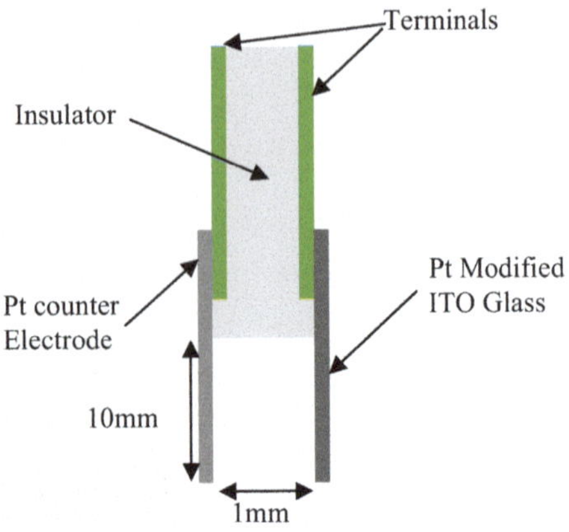

Fig. 3 Schematic diagram of the cross-section of electro-polymerization cell designed for the pulsed deposition method (diagram not to the scale)

cell was placed to have the Pt modified ITO substrate (working electrode (WE)) and the Pt electrode (counter electrode (CE)) horizontally and enough electrolyte (0.1 M NaPF6 and 0.05 M pyrrole monomer in PBS) was injected in between the two electrode. Then a train of twenty 1 Hz pulses with amplitude of 1 V and duty cycle of 50% were applied to this electrolytic cell to develop a polypyrrole layer on top of the Pt modified ITO layer. After polymerization the working electrode was washed twice in 6.5 pH PBS and cold air dried.

Enzyme Immobilization Techniques

The templated polypyrrole nanotube array substrates which were prepared as per the procedure described above and the enzyme deposition in these electrodes were done in three different methods namely drop casting, co-polymerization, and two step method.

In the first method was drop casting. A stock solution of glucose oxidase (GOx) was made to have 2 mg/ml of of GOx in 6.5 pH buffer solution. Then 5 μl of GOx solution was dropped on the 10 mm × 10 mm surface area and keep it in 4 °C to air dry. After air drying, the enzyme loaded nano-sensor was carefully washed in 6.5 pH buffer solution and kept in room temperature for measurement.

In the second method of co-entrapment, where a 1 mg/ml of GOx was dissolved in the pyrrole monomer solution and electrodeposition was done as per the normal fabrication method described earlier. This was washed in PBS solution and air dried and kept at room temperature for measurement.

The two-step method is a hybrid of the above two methods where the enzyme loading under the first step was co-entrapment and then in the second step it was drop casting. Both the 100 and 200 nm alumina Anodiscs were used in experiments

Fig. 4 Schematic representation of enzyme entrapment in three different immobilization methods

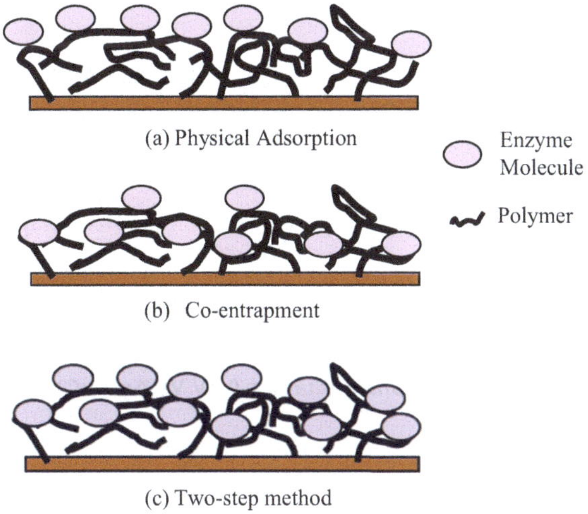

to investigate the influence of pore diameter on sensor performance. In different trials, a crosslinking agent (Glutaraldehyde) was also used.

The enzyme immobilization of pulse deposited polypyrrole substrate was done using a high electric field. Small amount of GOx (from a stock solution of 2 mg/ml) was injected in between the WE and CE and it was topped up with a very small amount of glutaraldehyde (0.1% wt.). Then a 1 V potential was applied between working electrode and the counter electrode for 20 min.

Same procedures were followed for the fabrication of nano-biosensors for hydrogen peroxide measurements with the enzyme GOx replaced by enzyme horseradish peroxidase.

Figure 4 illustrate how the enzyme molecules attached to the polymer network under three different schemes of enzyme loading. In the drop-casting method they are physically adsorbed and therefore most of the enzyme molecules will be on the surface of the polymer network. In co-entrapment method it will be within the polymer network as they get entrapped during the polymerization process. This is illustrated by Fig. 4b. In the two-step method, the enzyme molecules are within the polymer network as well as on the surface as illustrated by Fig. 4c.

Measurements

For all the samples a three electrode measurements setup as shown in Fig. 5 was used where the phosphate buffer solution was placed on a magnetic stirrer to speed-up the mixing of added glucose during the measurement steps. All measurements were started from PBS of 6.5 pH and a stable bias voltage of 0.4 V was supplied to the working electrode w.r.t. the counter electrode where an Ag/AgCl electrode was

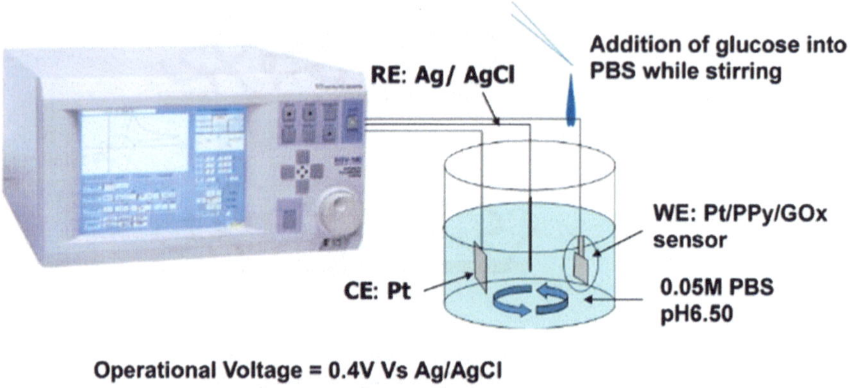

Fig. 5 Schematic of the measurement system

used as the reference electrode. Then a known amount of glucose was added and the change in current response was recorded. This was repeated in regular time intervals until the sensor response become saturated.

Similar procedure was carried out for Hydrogen Peroxide measurements where sensors used were the ones loaded with horseradish peroxidase and current response was recorded for the addition of known amount of H_2O_2 at known time intervals.

The templated nanotube array sensors were kept in the lab for seven months and measurements were taken every month from month 3 to month 7 to investigate the life time expectance of the developed sensor.

Results and Discussion

The test results for glucose measurements using templated nanotube array sensors shows relatively high sensitivities and higher linear ranges. Figure 6 shows the working electrode current variation of the templated PPy nanotube array sensor (100 nm type) in response to the variation in solved glucose concentration. It shows a sensitivity of 7.4 $mA/cm^{-2}\ M^{-1}$, a linear range of 0.5–10 mM, and a response time of less than 4 s.

The same type of sensor templated using an alumina template of 200 nm pours showed a sensitivity of 3 $mA/cm^{-2}\ M^{-1}$, a linear range of 0.5–10 mM, and a response time <3 s. Figure 7 shows the Lineweaver–Burk plot of the templated PPy nanotube array sensor (100 nm type) and the Michaelis–Menten (*Km*) value derived from that graph was 7 mM. This is little relatively lower than the standard *Km* of the GOx from Aspergillus Niger [25].

Figures 8, 9 and 10 illustrate the current response of the same type of sensor for different enzyme loading techniques. They all were subjected to similar experimental conditions other than the enzyme loading method.

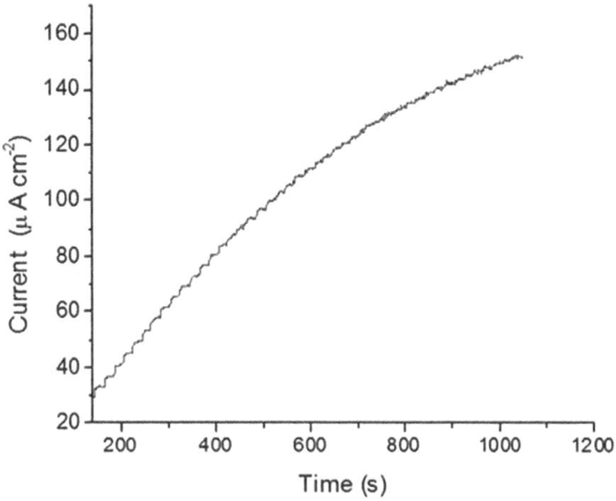

Fig. 6 Current response of the templated PPy nanotube array sensor (pore diameter = 100 nm)

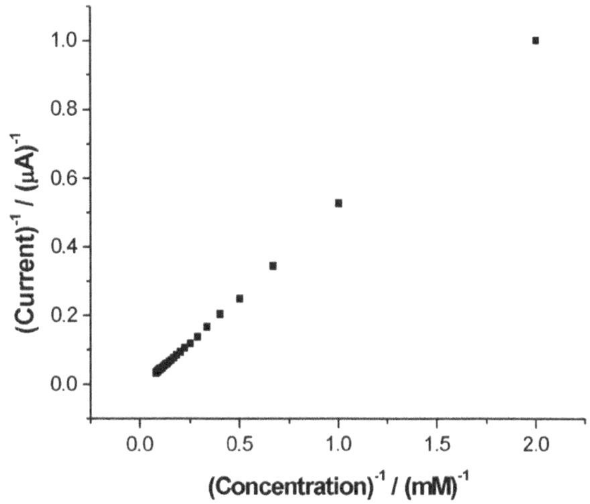

Fig. 7 Lineweaver–Burk plot from the measured current response for the PPy nanotube array sensor (pore diameter = 100 nm)

The same experiments were done with 100 and 200 nm diameter templates for all three kinds of enzyme deposition methods. Except the enzyme deposition method, all the other parameters were kept constant throughout the set of experiments. From Fig. 10 it can be clearly seen that for every addition of glucose there is a jump in response current. The gradient of the jump determines the response time and the gradient of the latter part which is almost zero indicates the long term value of the response current. The test results on 100 nm Anodisc template polypyrrole nano-tube array can be summarized after calculating all key parameters for each case as follows (Table 1).

They all displayed a very high stability at a wide range of concentrations. Figure 11 shows the current responses of one tested sensor against glucose concentrations ranging from 0.1–50 mM. This is much wider than the require range for a blood

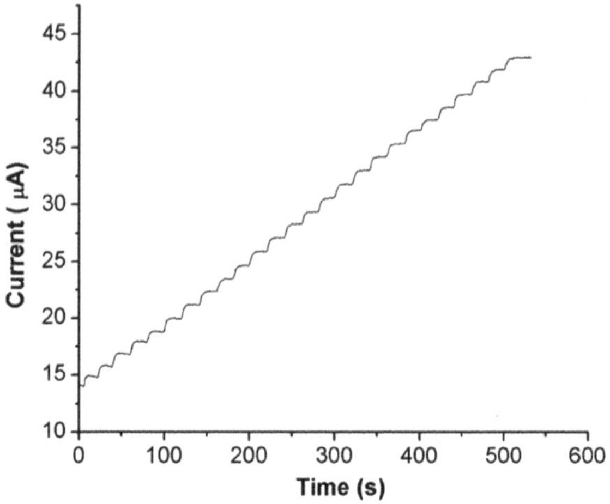

Fig. 8 Response for addition of 0.5 mM of glucose to a sensor with GOx immobilized under drop casting (physical adsorption) method

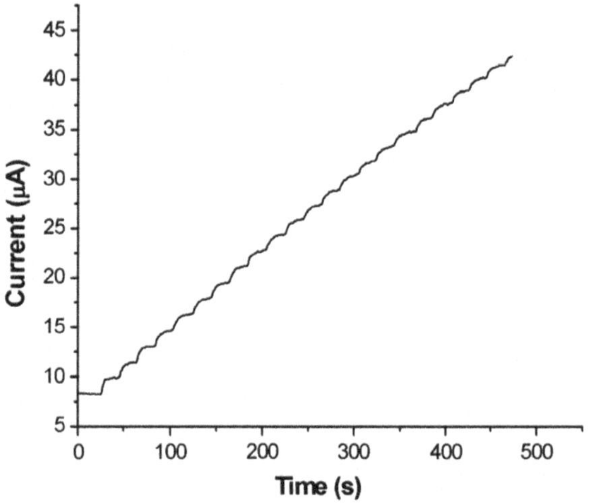

Fig. 9 Response for addition of 0.5 mM of glucose to a sensor with GOx immobilized under co-entrapment method

glucose sensor. The stability of the current response over an extensive measurement time of 100 s can clearly be seen from the curves in Fig. 11. This provides the information of the stability of glucose oxidase on to the conducting polymer substrate. The constant response current indicates that the number of enzyme sites available for the reactions to take place are the same over the 100 s period. This tells us the enzyme has properly bonded to the substrate, otherwise the current reading should decay with time as enzyme leaves the substrate.

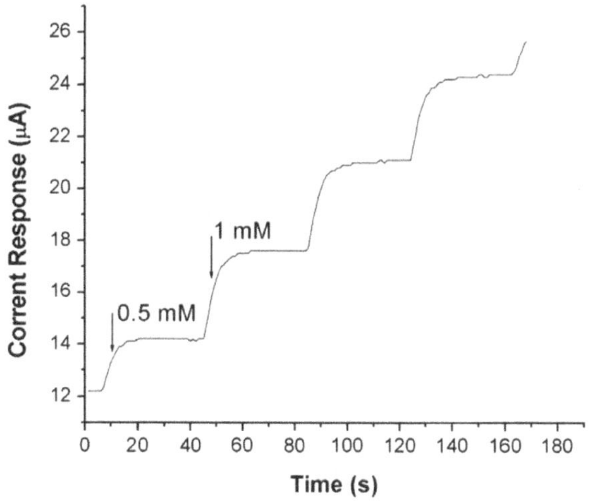

Fig. 10 Response for addition of 0.5 mM of glucose to a sensor with GOx immobilized under two-step method

Table 1 Comparison of response time and sensitivity for different enzyme immobilizing techniques (100 nm templated)

Enzyme immobilization technique used	Sensitivity ($mAM^{-1}cm^{-2}$)	Response time (s)
Physical adsorption (drop casting)	3	3
Co-entrapment	3.75	9
Two-step method	4.45	8

The shelf-lifetime against denaturizing of enzymes is a real challenge for any biosensor and we have investigated many different ways to improve that and finally found that adding a crosslinking agent is the best way to increase the shelf-lifetime. The crosslinking agent used was Glutaraldehyde and it also improved the sensitivity by a significant factor. However, there are certain factors help the denaturizing process in enzyme based nano-biosensors. They include oxidation of conducting polymer, reaction with light and decay of performance due to temperature variations in the environment.

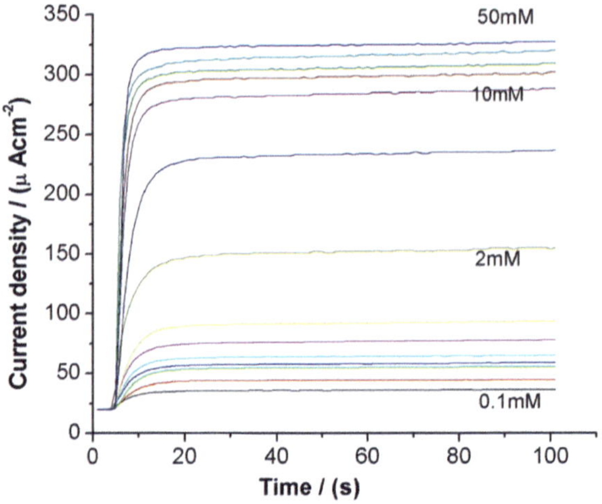

Fig. 11 Response of a templated nanotube array sensor against varying glucose concentrations

Fig. 12 Shelf-lifetime of the templated PPy nanotube array sensor

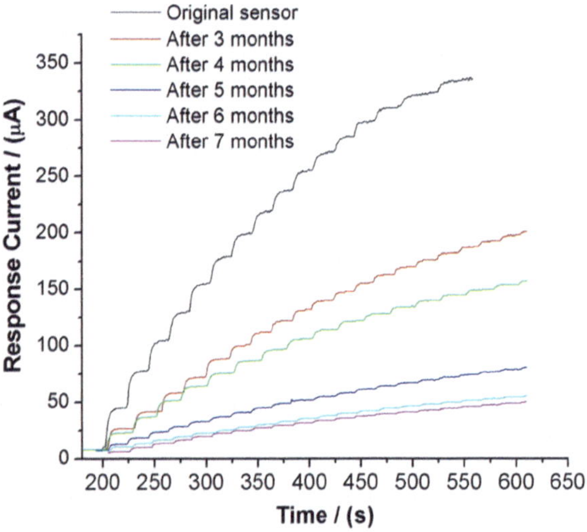

Figure 12 illustrates the shelf lifetime characteristics of the templated nanotube array type glucose biosensor. Table 2 summarizes the responses of sensors kept in the lab for different time spans with respect to the freshly made sensor. This illustrates a 33% drop in performance within the first three months and can be due to the low affinity indicated by the low *Km* value observed. However, still the after seven months the sensitivity is at a measurable level and can be used as a sensor for glucose measurements.

The following Table 3 shows the results obtained with crosslinking agent (glutaraldehyde) used during the enzyme immobilizing stage.

Table 2 Shelf lifetime of template polypyrrole nano-tube array

Lifetime (months)	0	3	4	5	6	7
Sensitivity (%)	100	66	59.5	21.4	13.8	3.5

Table 3 Comparison of sensor performances for different enzyme immobilizing techniques (200 nm templated)

Enzyme immobilization technique used	Sensitivity (mAM^{-1} cm^{-2})	Response time (s)	Linear range (mM)
Physical adsorption (drop casting)	7.4	4	0.5–13
Physical adsorption with crosslinking agent	62.5	6	0.1–5

Table 3 shows the performances of same sensor type under different enzyme immobilizing techniques. All the above discussed results are for the templated type of polypyrrole nano-tube array sensors with different enzyme immobilization techniques. They all were tested with glucose measurements. The same set of technics for fabricating nano-templated biosensors were used to make hydrogen peroxide sensors for environmental measurements [26]. The enzyme used for this set of experiments was horseradish peroxidise (HRP) and through cyclic voltammetry it was found that there are two peaks in the cycle where the sensor can operate either in anodic mode or in cathodic mode. This bi-functional behaviour provided the opportunity to make measurements of anodic and cathodic reactions at different voltages (+0.4 and −0.2 V) in ampherometric mode in response to H_2O_2 concentration. Dependent on the mode of operation the change in sensor current is also positive or negative.

Figure 13 illustrates how the response current change due to varying H_2O_2 concentrations at positive and negative bias potentials. These potentials were selected from the cyclic voltammetry test done on this sensor at 6.5 PBS solution.

Next we experimented with the ITO based pulse deposited polypyrrole nano-biosensor with GOx deposited under 1 kV/m electric field. We also used glutaraldehyde as the crosslinking agent in this experiment. Figure 14 shows the Lineweaver–Burk plot for the sensor fabricated on Pt modified ITO substrate with the Michaelis–Menten constant (*Km*) of 83 mM. Enzyme deposition of the pulse deposited polypyrrole on Pt modified ITO surface was done through a high electric field (1 kV/m) and the sensitivity obtained was 325 mA M^{-1} cm^{-2} [27]. This extremely high sensitivity value can be explained through the *Km* value of 83 which is well within the normal range of 33–110 mM [25].

Fig. 13 Response of templated polypyrrole nano-tube array sensor for H_2O_2

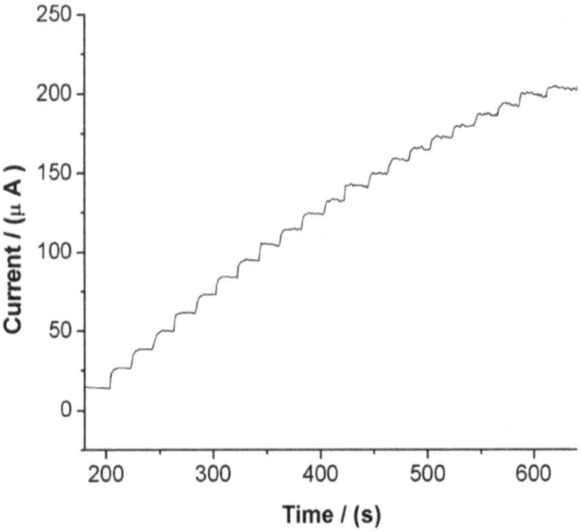

(a) Current response of the sensor for consecutive addition of 1 μl of H_2O_2 at 0.4V

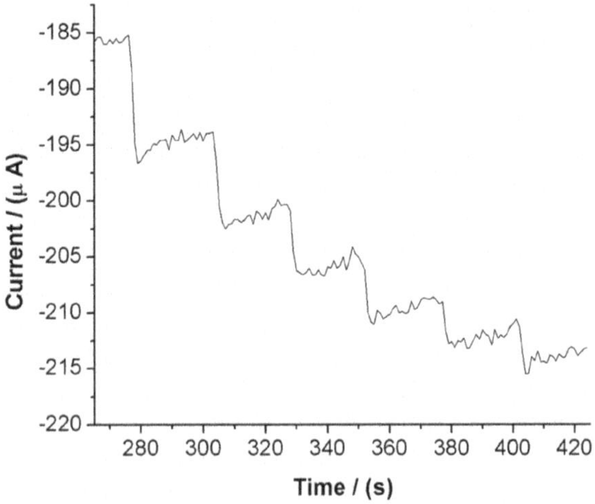

(b) Current response of the sensor for consecutive addition of 1 μl of H_2O_2 at -0.2V

Conclusions

From the results of templated polypyrrole nanotube array sensors with 100 nm and 200 nm template pore diameters it was found that the sensitivities were 7.4 mAM^{-1} cm^{-2} and 3 mAM^{-1} cm^{-2} respectively for physical adsorption of enzyme. The

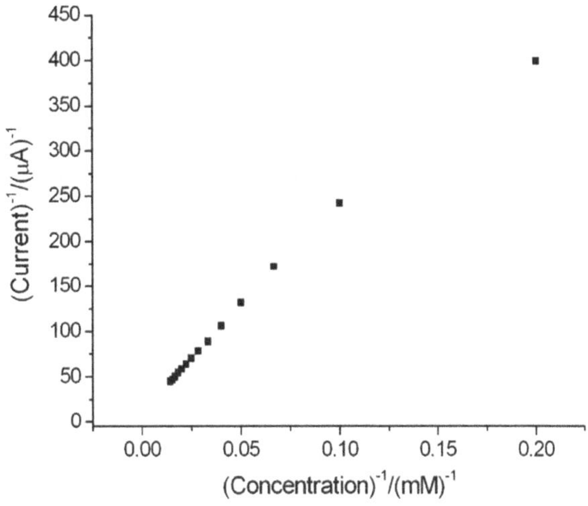

Fig. 14 Lineweaver–Burk plot from the measured current response for the pulse deposited PPy on Pt modified ITO substrate with electrophoretic enzyme deposition

response times were < 4 s and < 3 s respectively. Linear ranges were 0.5 mM–10 mM and 0.5 mM–13 mM respectively. For the 100 nm type of the sensor the sensitivity and response time were changed to 3.75 $mAM^{-1}cm^{-2}$ and 9 s for co-entrapment method and 4.45 mAM^{-1} cm^{-2} and 8 s for 2-step method respectively. The sensor substrate preparation method was kept the same for all these experiments and what changed was the enzyme immobilizing method. This indicates that the enzyme immobilizing method has a significant influence over the performance of the polypyrrole based nano-biosensors.

The pulse deposited polypyrrole nano-biosensors with the enzyme deposited under a high electric field shows the fabrication and enzyme immobilizing methods have a very high influence on the sensitivity as it showed a 325 $mAM^{-1}cm^{-2}$ for this type of a sensor.

From the whole set of the above experiments it can be concluded that the enzyme immobilizing technique plays a great role in the sensor's sensitivity and response time. In the co-entrapment method where the enzyme is mixed with the monomer increased the sensitivity. In the two step method where the polypyrrole is made in co-entrapment method and a secondary enzyme loading is done using physical adsorption the sensitivity was further improved. The crosslinking agent has increased the sensitivity by a significant factor but had to sacrifices a little in the response time. The electrophoretic enzyme deposition under a high electric field has resulted the maximum sensitivity and the summary of results clearly show the performance dependency of polypyrrole based nano-biosensor on the enzyme deposition techniques.

References

1. N. Hui et al., Nickel nanoparticles modified conducting polymer composite of reduced graphene oxide doped poly(3,4-ethylenedioxythiophene) for enhanced nonenzymatic glucose sensing. Sens. Actuators B Chem. **221**, 606–613 (2015)
2. T.A. Rajesha, D. Kumar, Recent progress in the development of nano-structured conducting polymers/nanocomposites for sensor applications. Sens. Actuators B Chem. **136**, 275–286 (2009)
3. J.D. Flower et al., Practical chemical sensors from chemically derived graphene. ACS Nano **3**(2), 301–306 (2009)
4. X.-W. Liu et al., Fabrication of metal-graphene hybrid materials by electroless deposition. Carbon **49**(2), 477–483 (2011)
5. A. Safavi, N. Maleki, E. Farjami, Fabrication of a glucose sensor based on a novel nanocomposite electrode. Biosens. Bioelectron. **24**(6), 1655–1660 (2009)
6. Y. Fan et al., Graphene–polyaniline composite film modified electrode for voltammetric determination of 4-aminophenol. Sens. Actuators B Chem. **157**(1), 669–674 (2011)
7. D.J. Macaya et al., Simple glucose sensors with micromolar sensitivity based on organic electrochemical transistors. Sens. Actuators B Chem. **123**(1), 374–378 (2007)
8. J. Heinze, B.A. Frontana-Uribe, S. Ludwigs, Electrochemistry of conducting polymers persistent models and new concepts. Chem. Rev. **110**, 4724–4771
9. C. Liu, T. Kuwahara, R. Yamazaki, M. Shimomura, Covalent immobilization of glucose oxidase on films prepared by electrochemical copolymerization of 3-methylthiophene and thiophene-3-acetic acid for amperometric sensing of glucose: effects of polymerization conditions on sensing properties. Eur. Polym. J. **43**, 3264–3276 (2007)
10. M. Braik et al., Highly sensitive amperometric enzyme biosensor for detection of superoxide based on conducting polymer/CNT modified electrodes and superoxide dismutase. Sens. Actuators B Chem. **236**, 574–582 (2016)
11. M.M. Barsan, M.E. Ghica, C.M.A. Brett, Electrochemical sensors and biosensorsbased on redox polymer/carbon nanotube modified electrodes: a review. Anal. Chim. Acta **881**, 1–23 (2015)
12. V. Pifferi et al., Synthesischaracterization and influence of poly (brilliant green) on the performance of different electrode architectures based on carbon nanotubes andpoly(3,4-ethylenedioxythiophene). Electrochim. Acta **98**, 199–207 (2013)
13. D. Restuccia et al., New EU regulation aspects and global market of active and intelligent packaging for food industry applications. Food Control **21**(11), 1425–1435 (2010)
14. M. Singh, P.K. Kathuroju, N. Jampana, Polypyrrole based ampherometric glucose biosensors. Sens. Actuators B Chem. **143**(1), 430–443 (2009)
15. E.M.I. Mala Ekanayake, D.M.G. Preethichandra, K. Kaneto, Polypyrrole nanotube array sensor for enhanced adsorption of glucose oxidase in glucose biosensors. Biosens. Bioelectron. **23**(1), 107–113 (2007)
16. R. Brooke et al., Recent advances in the synthesis of conducting polymers from the vapour phase. Prog. Mater Sci. **86**, 127–146 (2017)
17. E.M.I. Mala Ekanayake, D.M.G. Preethichandra, K. Kaneto, Effect of glucose oxidase immobilizing techniques on performances of nano scale polypyrrole glucose biosensors. Jpn. J. Appl. Phys. **47**(2), PART 2, 1321–1324 (2008)
18. V. Manzeiko et al., Gold nanoparticle and conducting polymer-polyaniline-based nanocomposites for glucose biosensor design. Sens. Actuator B Chem. **189**, 187–193 (2013)
19. J. Liu, M. Agarwal, K. Varahramyan, Glucose sensor based on organic thin film transistor using glucose oxidase and conducting polymer. Sens. Actuator B Chem. **135**, 195–199 (2008)
20. P.A. Palod, V. Singh, Improvement in glucose biosensing response of electrochemically grown polypyrrole nanotubes by incorporating crosslinked glucose oxidase. Mat. Sci. Eng. C **55**, 420–430 (2015)

21. H. Azak et al., Electrochemical glucose biosensing via new generation DTP type conducting polymers/gold nanoparticles/glucose oxidase modified electrodes. J. Electroanal. Chem. **782**, 138–153 (2016)
22. T.C. Gokoglan et.al., A novel approach for the fabrication of a flexible glucose biosensor: the combination of vertically aligned CNTs and a conjugated polymer. Food Chem. **220**, 299–305 (2017)
23. A.A. Saei et. al. Electrochemical biosensors for glucose based on metal nanoparticles. Talanta **130**, 198–206 (2014)
24. J.R. Anusha et., al., Simple fabrication of ZnO/Pt/chitosan electrode for enzymatic glucose biosensor. Sens. Actuators B Chem. **202**, 827–833 (2014)
25. https://www.sigmaaldrich.com/content/dam/sigma-aldrich/docs/Sigma/Datasheet/7/g2133dat.pdf. Accessed on 4 May 2017
26. D.M.G. Preethichandra and E.M.I. Mala Ekanayake, Nano-biosensor development for biomedical and environmental measurements, in *New Developments and Applications in Sensing Technology*, ed. by S.C. Mukhopadhyay et al., LNEE 83, pp. 279–292
27. D.M.G. Preethichandra, E.M.I. Mala Ekanayake, K. Kaneto, Development of low-cost nano-biosensors to enhance the rural healthcare services, in *IEEE 6th International Conference on Sensing Technology, ICST2012*, Kolkata, India, 18 Dec 2012, pp. 163–166

Piezoresistive Pressure Sensors for Applications in Harsh Environments—A Roadmap

Ha-Duong Ngo, Oswin Ehrmann, Martin Schneider-Ramelow and Klaus-Dieter Lang

Abstract Piezoresistive mechanical sensors play a very important role in modern industries. MEMS pressure sensor market is one of the biggest markets among all MEMS components [1]. Global pressure sensor market is growing from $6.4 billion in 2012 to $8.8 billion in 2018. The main applications are automotive, medical, consumer electronics, industry and aerospace/defense. Today, there is a growing demand for cost effective high-temperature and harsh- environment semiconductor devices, capable of operating at temperatures in the 500 °C range. Developments in energy, aircraft and space applications, automotive electronics, the oil and gas industry, the plastic and chemical industry, and the military sector are among the main drivers for research on high-temperature sensors and electronics. Existing semiconductor devices based on silicon are limited to operating temperatures below 150 °C, as thermal generation of charge carriers severely degrades device operation at higher temperatures. The development of SOI (silicon on insulator) technology helped to extend device's operating temperatures to approximately 400 °C. However, at temperatures over 400 °C, the material silicon reaches its physical limits as plastic deformation starts to occur when mechanical stress is applied. Silicon carbide is considered to be the most promising semiconductor for future high-temperature and harsh-environment applications as it features a unique combination of favorable physical, electrical, mechanical, and chemical properties. It is an extremely hard and robust material with a high thermal stability, and is chemically inert up to temperatures of several hundred degrees. Moreover, it has a higher thermal conductivity than copper, and its wide energy bandgap allows operation at high temperatures and in high radiation environments without suffering from intrinsic conduction effects. Performance and reliability of metal-semiconductor contacts, conducting paths and

H.-D. Ngo (✉) · O. Ehrmann · M. Schneider-Ramelow · K.-D. Lang
Fraunhofer Institute IZM, Gustav-Meyer-Allee 25, Berlin, Germany
e-mail: ha-duong.ngo@izm.fraunhofer.de

H.-D. Ngo
University of Applied Sciences Berlin, Wilhelminenhofstr. 75A, 12459 Berlin, Germany

S. C. Mukhopadhyay et al. (eds.), *Modern Sensing Technologies*,
Smart Sensors, Measurement and Instrumentation 29,
https://doi.org/10.1007/978-3-319-99540-3_12

the capability of etching 3D mechanical structures in SiC (such membrane or bridge) remain limiting factors for high-temperature operation of SiC electronic mechanical sensors today. In this chapter we discuss and review materials (Silicon, pSOI, SOI, 3C-SiC and 4H/6H-SiC) and technologies, which are applicable to realize MEMS piezoresistive mechanical sensor elements for applications at elevated temperatures.

Introduction

Piezoresistive mechanical sensors play a very important role in modern industries. High accuracy pressure control is a key feature in many industrial processes. MEMS pressure sensor market is one of the biggest markets among all MEMS components [1, 2]. Global pressure sensor market is growing from $6.4 billion in 2012 to $8.8 billion in 2018. The main applications are automotive, medical, consumer electronics, industry und aerospace/defense. Silicon-based MEMS sensors on the market can work only up to 150 °C [3–5]. Today, the development of microelectromechanical systems (MEMS) targets new areas of application such as high temperature, radiation, aggressive media, and other hostile environments, which extend beyond the capabilities of the well-established silicon and GaAs technology. There is a demand high temperature electronics, operating in harsh environments and at high temperatures [6]. Table 1 shows some important applications and their associated temperature ranges [6].

In this book chapter we discuss and review materials (Silicon, pSOI, SOI, 3C-SiC and 4H/6H–SiC) and technologies, which are applicable to realize MEMS piezoresistive mechanical sensor elements (not packaging issues. This topic must be discussed separately) for temperature ranges over 150 °C. The development of SOI (silicon on insulator) technology helped to extend device operating temperatures to approximately 400 °C [7–9]. However, at temperatures over 400 °C, the material silicon reaches its physical limits as plastic deformation starts to occur when mechanical stress is applied. Silicon carbide is considered to be the most promising semicon-

Table 1 High temperature applications and temperature ranges [6]

Applications	Measuring site	Temperature (°C)
Power plant	Power engine	500–600
	Turbine	800
	Waste gas cleaning	300–400
Logging	Drilling head	250–300
Plastic injection	Plasticization area or at nozzle	150–500
Petroleum	Reactor	200–1000
Medical	Instrument sterilization	150–300
Automotive	Combustion engine	150–2000

Table 2 Materials and their theoretical temperature ranges as piezoresistive sensor material [2, 3, 5, 10]

	Si	pSOI	SOI	SiCOI	Thin film SiC	Double Epi	Bulk SiC
Substrate	Si	Si	Si	Si	Si	SiC	SiC
Insulation	pn	SiO_2	SiO_2	SiO_2	pn or SiO_2	pn	pn
Piezo	p-Si	p-Si	p-Si	p-poly-SiC	c-SiC	h-SiC	c-, h-SiC
Band gap (eV)	1, 1		1, 1		2.2–2.3	2.9–3.26	2.9–3.26
Thermal conductivity (W/mK)	145	145	145	~490	490	500	500
Max T (°C)	125	250	400	450	450	600–800	600–800

ductor for future high-temperature and harsh-environment applications as it features a unique combination of favorable physical, electrical, mechanical, and chemical properties [10–12].

Table 2 gives an overview on the materials for piezoresistive mechanical sensors, their properties and their theoretical maximum working temperatures.

State-of-Art

Silicon Based Piezoresistive Sensors

Piezoresistive silicon pressure sensors utilize the piezoresistive effect to transform the pressure to be measured into an electrical signal. Single crystal germanium and silicon were the first materials widely used as piezoresistors. But silicon shows a much stronger piezoresistance effect than germanium. The applied pressure causes a pressure dependent, two-axial stress or strain field in the membrane of the pressure sensor that is measured by integrated piezoresistors (Fig. 1). The piezoresistance effect is quantified using the gauge factor, describing the relationship between the applied strain and the change in resistivity. In most applications the change in resistivity is limited to stresses and strains parallel and perpendicular to the resistive path. Figure 1 shows the longitudinal and transversal gauge factors of p- and n-type silicon materials in the (100)-planes.

To obtain a high output signal the piezoresistors are placed in areas of maximum stress or strain. For single crystalline silicon piezoresistors the sensitivity (gauge factor) depends on the conductivity type, the doping concentration, and the crystallo-

Fig. 1 Longitudinale (**a**) and transverse (**b**) gauge factors of n- and p-type single-crystal silicon in the (100)-planes

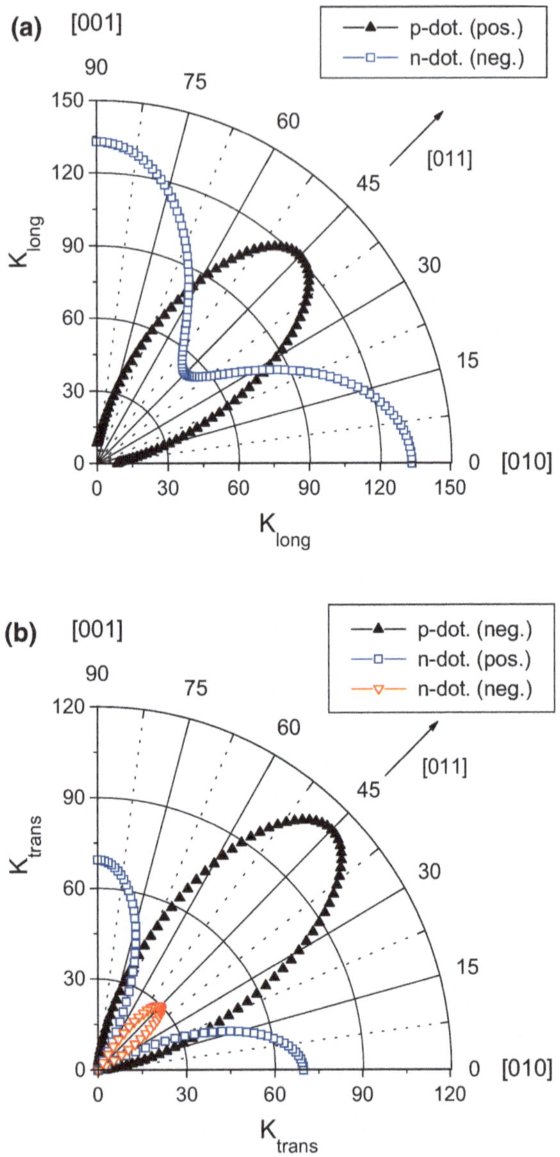

graphic orientation of the resistors on the wafer. Depending on the relative direction of the current density, the electrical field, and the mechanical stress, longitudinal, transverse, and shear effects are taken into account as well when single crystalline silicon is used.

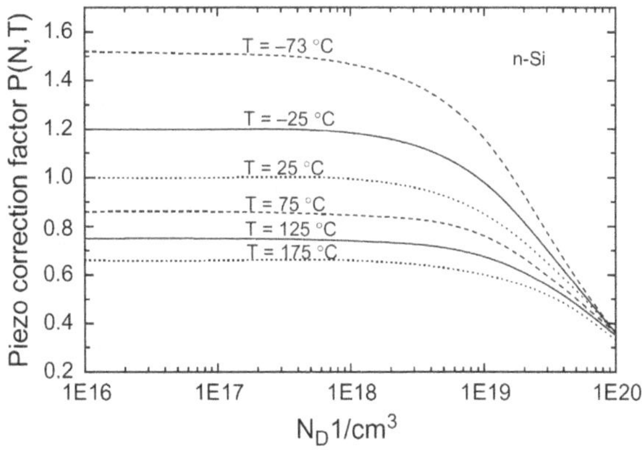

Fig. 2 Temperature dependency of n-silicon as function of doping levels and temperature [13]

The temperature dependency can be calculated by using Kanda's formulation [13]. The piezoresistance effect is decreasing with the increasing dopant concentration and temperature (Fig. 2).

The central functional element of this type of pressure sensors is a silicon membrane fabricated by either isotropic or anisotropic etching. Four piezoresistors are monolithically integrated into the membrane either by diffusion or by ion implantation. Positive and negative changes ΔR(p) in the resistance with the same absolute value of ΔR(p) are obtained by utilizing the longitudinal and transverse piezoresistive effects. The piezoresistors on the membrane are usually connected to a bridge circuit, so that exposure to pressure results in a mostly linear output signal (voltage) when the bridge is operated in the constant voltage or constant current mode. The measuring range and sensitivity of the sensors are determined by the size and thickness of the membrane. The picture bellow shows a typical silicon piezoresistive sensor (Fig. 3).

Performance of PN-Junction Based Silicon Piezoresistive Sensors at Elevated Temperatures

Restrictions concerning the maximum operating temperature are coming mainly from an insufficient electrical insulation of the piezoresistors, the thermal instability of electrical contacts, plastic deformation of used materials, a drastic increase of the intrinsic carrier concentration, electrical instabilities of passivation layers or a high output signal noise level. Figure 4 shows the insulation of standard piezoresistive silicon sensors featuring pn-junction. The substrate silicon material is typically n-type material, and the piezoresistors are made in p-type material. P-type silicon has much better piezoresistance effect than n-type material. The sensitivity is higher, and the linearity is better. The electrical isolation is determined by the leakage current through the pn-junction. It is dependent on different parameters, the whole surface of

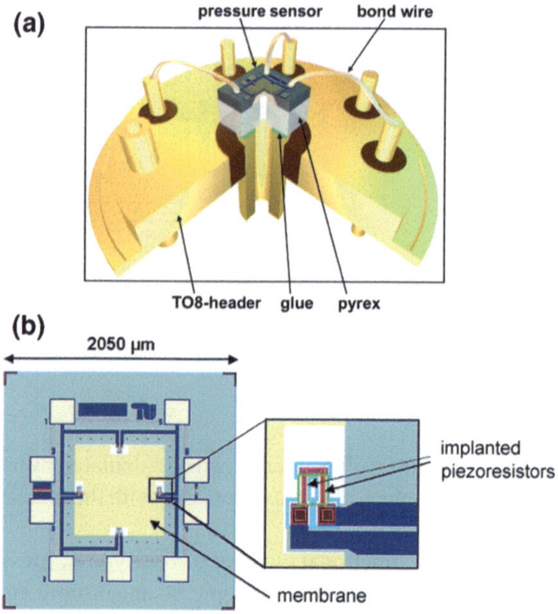

Fig. 3 Standard silicon piezoresistive sensor. **a** Sensor on TO-8 carrier wire-bonded, and **b** chip layout with 4 piezoresistors

the pn-junction, the charge concentration, the diffusion coefficient and the life time of the charges, and the temperature (Eq. 1).

The bridge circuit used for piezoresistive pressure sensors can be operated in the constant current or the constant voltage mode. The output signal is a voltage proportional to the applied pressure but changes with temperature (Fig. 5).

The leakage current through the pn-junction can be calculated by using the following formulation:

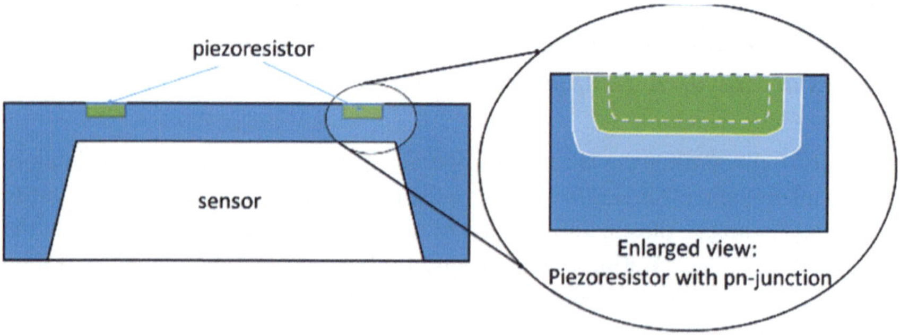

Fig. 4 Schematic view of a mechanical microsensor with integrated piezoresistors and pn-junction as electrical insulation

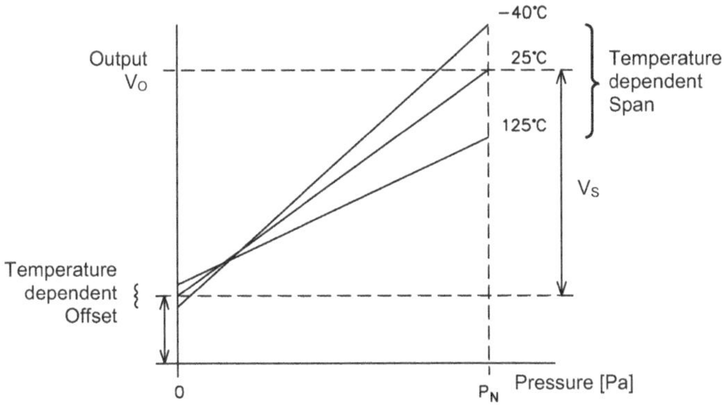

Fig. 5 Output signal of a piezoresistive pressure sensor

$$I(T) = q.A.\sqrt{\frac{\left(\frac{D_p}{\tau_p}\right)n_i^2(T)}{N_D} + \frac{qAn_i(T)W}{\tau_e}}\ [\mathrm{A}] \tag{1}$$

where:

A	pn-surface in cm^2
q	Element charge
D_p	hole's diffusion constant
τ_p	hole's life time
N_D	donator concentration
W	pn junction width
$n_i(T)$	charge concentration
$\tau_e = (\tau_n + \tau_p)/2$	effective life time of thermally generated charges

The leakage current is increasing as temperature is rising, and the silicon pn-junction lost its sufficient isolation performance at temperature around 125–150 °C [8, 9].

In order to realize high temperature sensors, three approaches can be used. The first is to use cooling concept (liquid filled sensors) to keep the sensor temperature at a low level. This method is normally used in industry with liquids, such as NaK, Mercury or oil. The pressure transmitters using cooling liquids, are called liquid-filled. This concept has advantages and disadvantages as well. The disadvantages are a difficult filling technique for the liquids, the danger of liquid leakage. It can contaminate the equipment and products. Mercury and oil will be forbidden in food and medicine products.

The second one is to isolate the piezoresistor by using a real insulator such as silicon on dioxide (SOI material systems). So the maximum thermal performance of silicon material could be exploited.

The last most sophisticated approach is to realize the electronic structures in a large band gap semiconductor material. Due to larger band gap the thermal generation of charges $n_i(T)$ is much lower.

SOI-Based Sensors

SiO_2 as Insulation for Piezoresistor

An excellent electrical insulation can also be obtained by using SOI (Silicon On Insulator) material systems. 'SOI' is a general term but means today mainly a material system with a silicon overlayer (SOL, suitable for example for the fabrication of mesa etched piezoresistors) which is electrically insulated from a silicon substrate by a buried silicon dioxide layer (BOX). SiO_2 material is one of the best insulation materials in CMOS technology [7]. This technology has many advantages compared to bulk silicon technologies:

- Low leakage current (one to two order)
- Hardness against radiation
- Latch-up free due to effective isolation
- Less parasitic components
- Low substrate leakage
- Higher operating temperature
- BOX can be used as etch stop
- High electric fields possible
- Monolithic integration.

As it can be seen in [8, 9] silicon piezoresistors with SiO_2 insulation can survive temperatures up to 300 °C (CMOS)—400 °C (Pressure Sensor). Silicon is conducting at higher temperatures. Various forms of silicon pressure sensors exist: monolithic integrated types, in SOI [7, 8] and SOS (Silicon On Sapphire) [13] technology as well as thin film types based on metal and silicon membranes with polysilicon piezoresistors.

The SOI wafers are fabricated by the SIMOX (Separation by IMplanted OXygen), the BESOI (Bond and Etchback SOI) or the Smart-Cut processes. Commercial SOI material systems with single crystalline silicon overlayers are available. Attracted by the excellent prospects for the SOI substrate market, different companies were founded (like IBIS, USA, in 1987; BCO, UK, in 1992; SOITEC, France, in 1992) and offer today SOI substrates based on different fabrication technologies with a size up to 8 inches in diameter (12 inches in the near future). Main fields of application of SOI material are integrated circuits, optoelectronics, smart power and MEMS (Micro Electro Mechanical System) devices. Important benefits of SOI material systems for electronic applications are the high temperature capability and the reduction of parasitic pn-junction capacities—an important issue for further increasing the

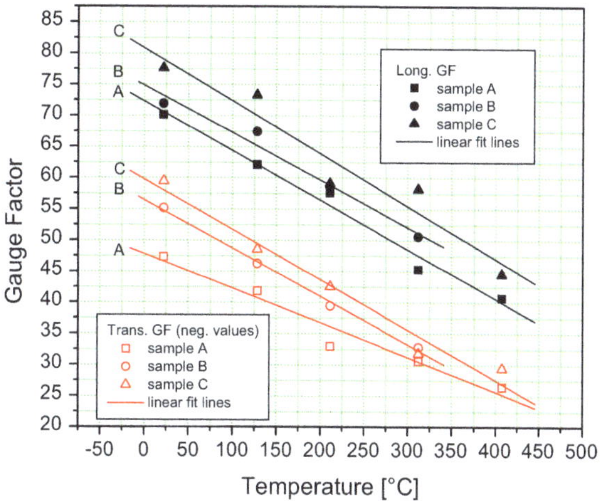

Fig. 6 The gauge factor of boron doped piezoresistors manufactured in the SOI material manufactured with Smart-Cut process [9]

switching speed of integrated circuits (ICs). Figure 6 shows the gauge factors of SOI materials manufactured with the Smart-Cut process as function of temperatures.

The polycrystalline silicon layers of PSOI pressure sensors are deposited in an LPCVD (Low Pressure Chemical Vapor Deposition) process on thermally oxidized silicon wafers (PSOI → Polysilicon-On-Insulator). Polysilicon piezoresistors result in sensitivities that are two to three times smaller than those obtained with single crystalline Si. Figure 7 shows the gauge factors of boron-doped polysilicon films as function of temperatures [14].

The technological fabrication process differs from that of conventional piezoresistive pressure sensors only in the way the piezoresistors are made. The performance of poly silicon pressure sensor could be improved by using an additional crystallization step after poly silicon deposition. New researches using laser for crystallization are on-going.

Single crystalline silicon can be heteroepitaxially deposited directly onto the insulating material sapphire, and can be structured by reactive ion etching (RIE). A great advantage is, that no interfacial layer is needed which could give rise to thermal induced stress and long term stability problems. Due to the high thermal and mechanical stability of sapphire, an excellent long term stability and temperature resistance can be obtained with sapphire based pressure sensors. The piezoresistors are formed in the SOL (Silicon Over Layer) layer by using implantation and dry etching [8]. Figure 8 shows schematically the isolation of piezoresistor with surrounded SiO_2.

As a consequence of the development of high quality material systems, pressure sensors based on SOI are commercially available today. Kulite Semiconductor Products Inc., was the first to patent a silicon-on-silicon dioxide sensor [14]. They consider a temperature of T = 480 °C as a general upper operating temperature limit of their SOI pressure sensors. But for specific applications like downhole pressure transducers (for oil well logging) based on SOI, Kulite limits the maximum operating temperature to T = 180 °C. It seems that this compromise is necessary in order to

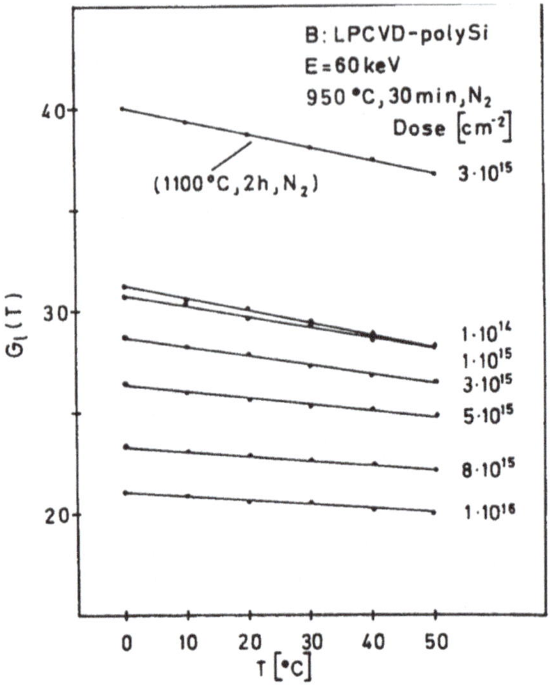

Fig. 7 The longitudinal gauge factor G_l of boron doped LPCVD polysilicon resistors [15] as function of temperatures and doping concentration

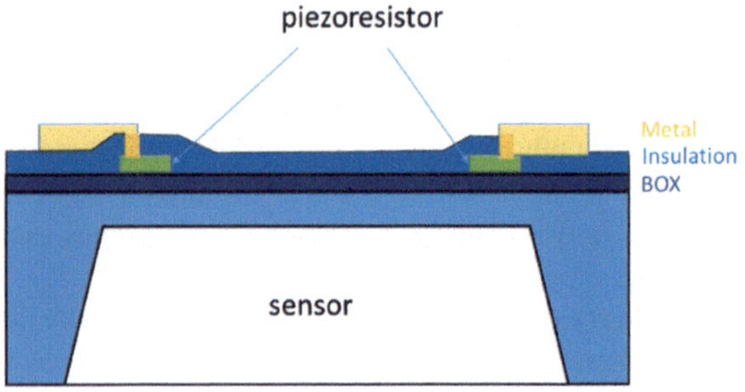

Fig. 8 Concept of piezoresistors' isolation by using surrounding silicon dioxide—BOX between SOL (Silicon Over Layer) and silicon substrate and deposited SiO_2 on top

maintain and guarantee for example the necessary long term stability during operation of the sensor.

The most advanced SOI high temperature pressure sensor concept is presented in [8]. Figure 9 shows the SOI sensor element on a ceramic carrier and a tray for temperature range up to 400 °C [8]. The sensor element is "floating" on the carrier, and not glued or bonded. So high thermal stress could be avoided. The sensor element is kept on the carrier using a push rod.

One of the crucial features for successful device performance is the availability of a suitable long-term stable high temperature metallization scheme, typically including a diffusion barrier layer, which prevents a metallurgical reaction between metallization and semiconductor substrate, a contact layer and an interconnect layer. $TiSi_2$

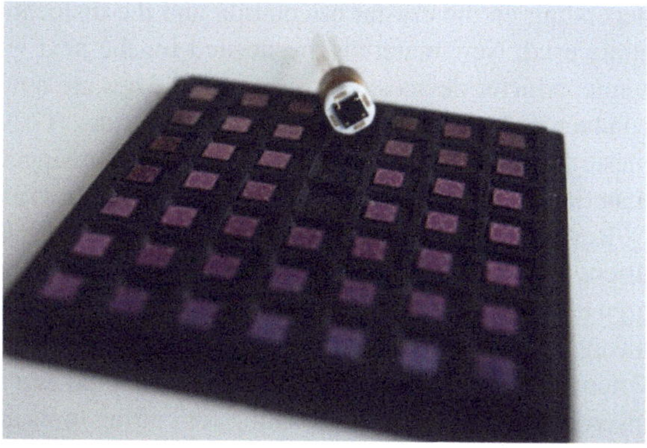

Fig. 9 SOI sensors on carrier and tray

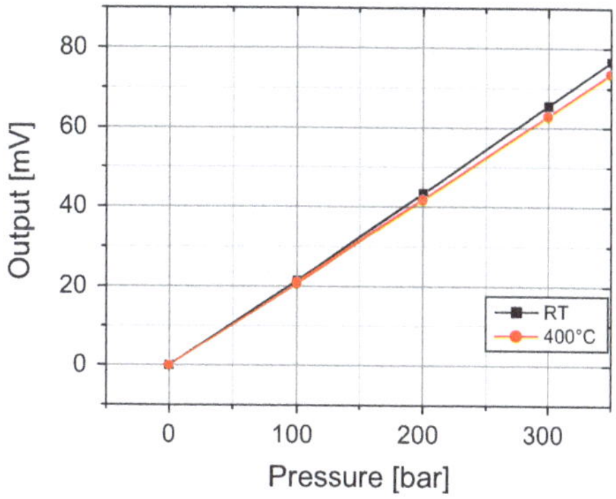

Fig. 10 Output signal of the sensor at different temperatures (RT and 400 °C) [8]

is used in microelectronics as contact material between silicon semiconductor and metallization. It allows a low and an ohmic contact to the sensor. Ternary amorphous metallic thin films have received considerable attention for high temperature diffusion barrier applications. These materials have shown ability to be chemically inert against reactions with gold and other top metals at elevated temperatures, thereby providing the low diffusivities required for a diffusion barrier.

In [8] a floating packaging concept is used. The SOI sensor element is "floating" in the carrier, and is kept by using a pushrod. This pushrod is transferring the to be measured pressure onto the silicon membrane, where the SOI piezoresistors are integrated. The output signal over pressure at 25 °C, and 400 °C, respectively, is shown in Fig. 10. The end point non-linearity at 400 °C is about 0.2% FSO.

SiC-Based Sensors

At high temperatures, the silicon reaches its physical limits. At temperatures exceeding 400 °C, plastic deformation starts to occur when high mechanical stress is applied, depending on the crystal orientation and the dislocation density of the silicon substrate used. New materials are needed for the next generation of high-temperature semiconductor devices. Promising candidates are a number of wide bandgap semiconductors including SiC, the group III-nitrides (GaN, AlN, BN) and diamond. Of these, SiC is considered to be the most promising wide bandgap semiconductor material for future application in high-temperatures and harsh environments.

Research on diamond was intensified in the 1960s, when Russian researchers discovered, that diamond cannot only be artificially produced by high pressure and high temperature (HPHT) processes (P-50-100 kbar and T-1800-2300 K) but also by means of simple and cost-effective chemical vapor deposition (CVD) processes [6]. Potential fields of application are: CVD diamond as a heat sink, as an abrasive, as an insert and/or wear-resistant coating for cutting tools and as a material for electronic and sensor devices. The piezoresistance effect of CVD diamond films was investigated [7] and a piezoresistive pressure sensor based on a heteroepitaxially deposited diamond thin film was presented [8]. But a poor and polycrystalline material quality, a missing n-type-doping technology and the lack of a convincing product which could act as a driving force for further material development led to a decreasing interest in diamond based electronic and sensor devices.

Silicon carbide is today the most matured material among the wide bandgap semiconductors. SiC replaced diamond in its role of a promising wide bandgap material and gave also rise to the research efforts spent on the group III-nitrides. Silicon Carbide is an extremely hard material (Mohs hardness of 9.1) and exhibits a high thermal stability with a sublimation point of 2830 °C (compared to melting temperature of silicon 1410 °C). At room temperature, it's chemically inert. It reacts poorly with any known material. Silicon carbide is a mechanically and chemically extremely robust material for devices operating in hostile environments. SiC devices can operate at high temperatures and higher radiation load without suffering from intrinsic conduction effects because of the wide energy band gap (Table 1). Another important advantage of SiC is the excellent thermal conductivity of 4.9 $Wcm^{-1}\ K^{-1}$ which is higher than any metal. The stacking sequence of the most common SiC polytypes are 3C (⇓-SiC), 4H, and 6H (α-SiC). A comprehensive introduction to silicon carbide crystal structure and polytypism can be found in [14].

In the beginning, Cree developed and offered substrates based on three different polytypes: bulk wafers (with epitaxial layers) consisting of the two hexagonal polytypes 4H-SiC and 6H-SiC and thin films of the cubic polytype 3C-SiC, heteroepitaxially deposited onto silicon or SOI substrates. In the middle of the 1990s, Cree stopped the fabrication of 3C-SiC, 6H-SiC and decided to concentrate on the development and production of bulk SiC wafer of the hexagonal 4H-polytypes. Currently, substrates with 4 and 6 inches in diameter are commercially available. The first 4 inch SiC wafer was presented by Cree in October 1999. Main problems related to SiC wafer production were a low growth rate and a relatively high defect and micro pipe density. Cree still holds the monopoly on large diameter, high quality, hexagonal SiC bulk wafers but other companies like SiCrystal AG, Norstel AB, Sterling Semiconductor and II–VI Incorporated catch up with the development and offer wafers (partly with epitaxial layers).

For a long time, the only SiC based commercially available products were blue (and later on also green) LEDs which are less power consuming and exhibit longer lifetimes than other light sources. But in September 2000, Infineon Technologies announced the production of SiC Schottky diodes [12]. And Cree declares that products currently under development include microwave transistors, power devices and lasers.

Initial difficulties concerning SiC wafer availability and quality are slowly being resolved. In May 2007, Cree Inc. Durham released their first zero micropipe 100 mm 4H-SiC Wafer, which can be considered a technological breakthrough although prices for such high quality SiC substrates remain high. Another option is to combine 3C single crystal SiC on Silicon substrate in order to gain advantages of both material systems—the high temperature behavior of 3C-SiC and standard process technology of Silicon.

In order to take advantage of silicon carbide's excellent physical properties for high- temperature applications, state-of-the-art fabrication and processing technologies have to be developed, e.g. SiC epitaxial growth, selective doping with both p- and n-type impurities, micromachining, passivation and isolation techniques, packaging, and stable metallization schemes. While a lot of progress has been made in most fields of silicon carbide processing technology, two major issues still need to be addressed today: the development of feasible bulk micromachining technologies, and long-term stable high-temperature metallization systems.

SiCOI Piezoresistive Microsensors

SiCOI technology (Silicon Carbide on Insulator) was introduced where epitaxial layers of single or polycrystalline SiC are deposited on SOI or oxidized silicon wafers. Typically 3C-SiC has quite low material quality, and is not very attractive for manufacturing of electronic devices. But the cubic polytype is a promising material for high temperature sensor applications. 3C-SiC films heteroepitaxially grown on silicon are not useful as a significant leakage current across the hetero pn-junction already occurs at temperatures above $T = 250$ °C. But the 3C-SiC/SOI material system seems to represent an ideal compromise:

- high quality SOI substrates are available with a size up to eight inches,
- sensor devices can be produced much cheaper compared to SOS or 6H/4H-SiC technology,
- in contrast to SOS and bulk SiC substrates, the well established silicon technologies (e.g. etching and wafer bonding) can be applied to the silicon substrates,
- a very good electrical insulation can be achieved by the buried oxide layer,
- use can be made of the outstanding properties and high temperature capabilities of silicon carbide, and
- among SiC polytypes 3C-SiC features the largest piezoresistive effect.

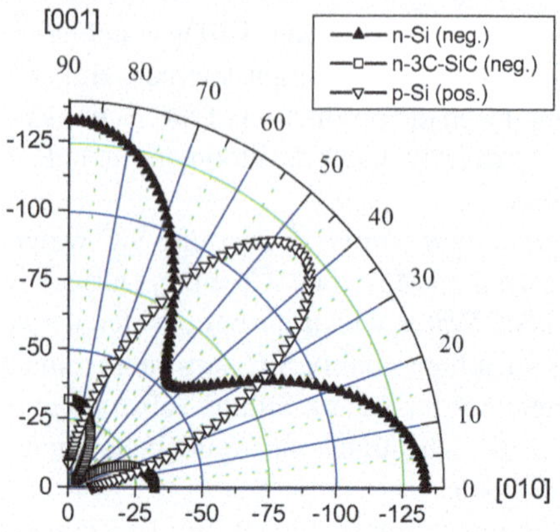

Fig. 11 Longitudinal piezocoefficients of n-type 3C-SiC and n-type silicon in the (100) plane

Similar to silicon, the conduction band of 3C-SiC possesses six minima in direction of the main crystal axes. The electron transfer theory relates the anisotropic change in resistivity to a redistribution of electrons among the multivalleys in momentum space. As the minima exhibit an ellipsoidal form, the electron mobility is highly anisotropic. But due to the symmetrical distribution of the six valleys, the net mobility is uniform. The application of a tensile strain in the <100> direction causes the multivalley minima to rise in energy in this direction, while the minima perpendicular to the applied strain will drop. This energy difference causes a redistribution of the electrons between the potential wells and this leads to an isotropic mobility and resistivity. The resistivity in direction of the applied tensile strain is decreased (longitudinal piezoresistance effect). According to this model and the way the ellipsoidal minima are affected, the transverse piezoresistance effect (resistivity measured perpendicular to the applied strain) is smaller and has a different sign. Figure 11 compares the longitudinal gauge factors of 3C-SiC and silicon in (100) plane.

The transverse piezocoefficients of 3C-SiC and silicon in the (100) plane are shown in Fig. 12. The maximum value in <100> direction has a positive sign and is smaller compared to the absolute value of the longitudinal piezocoefficient. A second, smaller extreme value with a negative sign is observed in <100> direction. The ratio between the transverse piezocoefficients and the transverse gauge factors is comparable to those of the longitudinal values.

A variety of SiCOI-based sensors have been successfully demonstrated in literature [15], but such devices still suffer from the physical limitations of the silicon substrate material and can only be seen as intermediate step toward high- temperature stable devices for the 500 °C range (Fig. 13).

A main problem related to the 3C-SiC growth on UNIBOND SOI wafers is the formation of cavities in the SOL (and the BOX). One reason is, that silicon at its melting point of T = 1410 °C exhibits a poor wetting behavior on silicon dioxide films (the wetting angle is 87° for Si on SiO_2, close to complete instability). The

Fig. 12 Calculated transverse piezocoefficients of n-type 3C-SiC and n-type silicon in the (100) plane

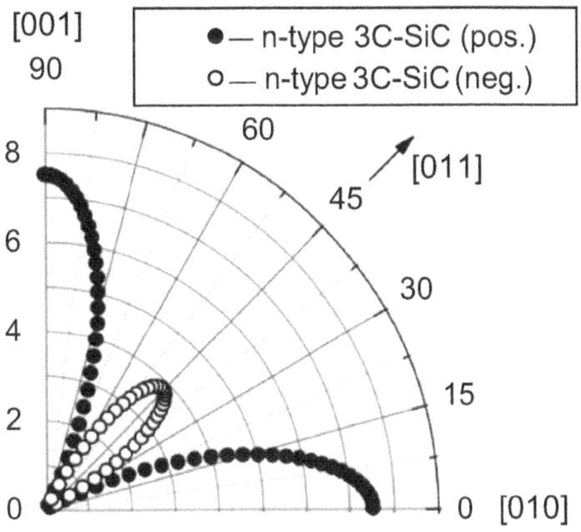

situation is improved for silicon on silicon nitride films (the wetting angle is 25° for Si on Si_3N_4). SICOI sensors for high and low pressures have been developed [15]. Figure 14 shows output signal of a high pressure sensor at 400 °C.

New on-going strategy is transferring 3C-SiC single crystal layer onto insulator substrate by using wafer-bonding techniques. The goal of this approach is to avoid the issues of deposition of SiC directly on Silicon surface at high temperatures.

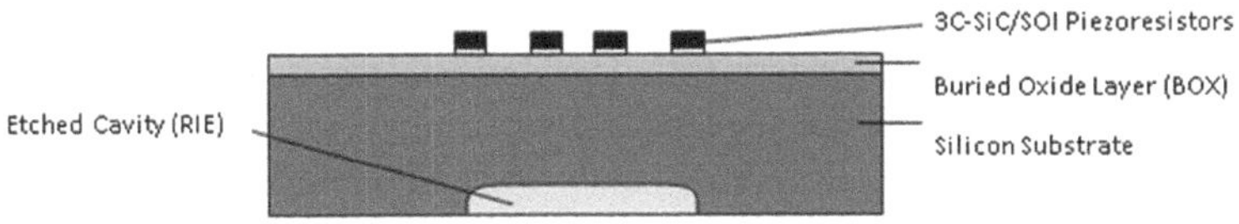

Fig. 13 Cross section of a SICOI pressure sensor cell. A SOI wafer is structured by isotropic Reactive Ion Etching (RIE). The 3C-SiC layer is deposited onto the SOL of the SOI wafer. The piezoresistors are structured by RIE [15]

Fig. 14 Output signal of a SICOI-high pressure sensor at 400 °C. The sensor shows a sensitivity about 20 μV/V bar [15]

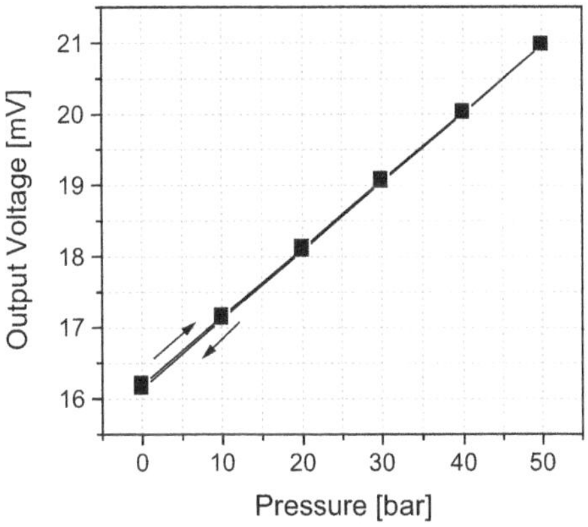

Table 3 Piezoresistive gauge factor of 4H and 6H silicon carbides in literatures [16–19]

Polytype	Type	Dopant	Carrier concentration (cm^{-3})	GF at room temperature	GF at high temperature	Orientation	Stress type
Single 4H	n	Nitrogen	1.5×10^{19}	20.8	–	(0001) plane	Uniaxial
Single 6H	n	Nitrogen	3.8×10^{18}	−29.4	−17 (250 °C)	(0001) plane	Uniaxial
Single 6H	n	Nitrogen	2×10^{19}	−22	−11 (250 °C)	(0001) plane	Uniaxial
Single 6H	n	Aluminium	2×10^{19}	27	12 (250 °C)	(0001) plane	Uniaxial

Double Epi and Bulk SiC Piezoresistive Microsensors

Among the silicon carbide polytypes, 3C-SiC features the largest piezoresistive effect. However, because single-crystalline 3C-SiC wafers are not available, 4H and 6H-SiC are most suitable for bulk piezoresistive sensors towards 500 °C range and higher. SiC technology is still in research stage.

The piezoresistive effect in n-type and p-type silicon carbide materials was investigated [16–18]. The n-type SiC can be formed using nitrogen, while aluminum and boron are employed to make p-type SiC. The gauge factors can be seen in Table 3. At room temperature, the absolute gauge factor of single crystalline SiC is approximately 20–30 and decreases to approximately 10 at high temperatures.

In case of 6H-SiC, the gauge factors of low-doped and medium-doped materials are comparable to crystalline 3C-SiC. The temperature variation between room temperature and 300 °C causes a reduction in gauge factors by nearly 40% [19]. At temperatures below 100 °C, 6H-SiC has a negative temperature coefficient, while the TCR becomes positive at temperatures above 100 °C (Fig. 15).

4H-SiC was investigated in [18]. The transverse gauge factor of 4H-SiC (doping 1.5×10^{19} cm^{-3}) was reported to be 20.8. Piezoresitive effects in SiC at high temperatures still need to be investigated further in future works [19].

Some SiC bulk piezoresistive sensor systems have been developed. Figure 16 shows the two sensor elements published in [16, 17]. For the sensor in [16] a double epitaxy 6H-SiC wafer (n-p-n) was used. The piezoresistors were realized in the n-type top layer by using RIE etching. A pn-junction was formed between the n-type SiC layer and a p-type SiC layer underneath, and served as isolation for the piezoresistors. This concept has also an advantage, the second p-type layer can seal the micro pipes in SiC wafer. Micro pipes are common defects in SiC wafers. The sensor showed a very high TCE value (reduction about 56% from RT to 350 °C). A burn-in at 380 °C for 10 h lead to an irreversible reduction of sensitivity about 6%. Also an irreversible change in sensor offset was shown. For the second sensor [17,

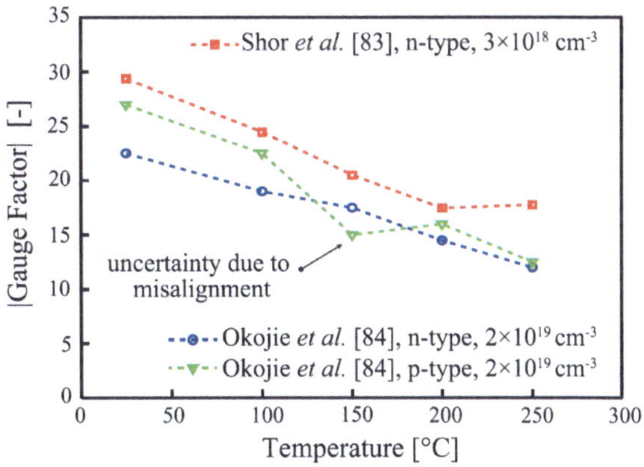

Fig. 15 Temperature dependency of gauge factors in 6H-SiC materials [19]

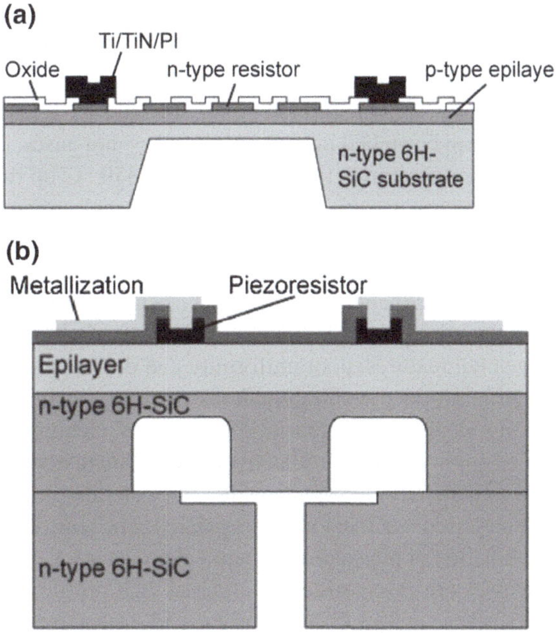

Fig. 16 Realized SiC piezoresistive pressure sensor elements in [16, 17]

18] a similar SiC material was used. The sensor war tested at 400 °C and showed a sensitivity reduction about 43%. Figure 17 shows the sensor signal for RT and 400 °C.

One of crucial problems to be solved before the true potential of SiC-based sensors and electronics can be tapped, is a long-term high-temperature stable metallization system. New metallization systems [6] features both electrical and thermos dynamical long-term stability for temperatures around 500 °C have been developed.

The other serious problem to be solved is a feasible bulk micromachining technology. Due to its hardness, silicon carbide is one of the most widely used lapping and

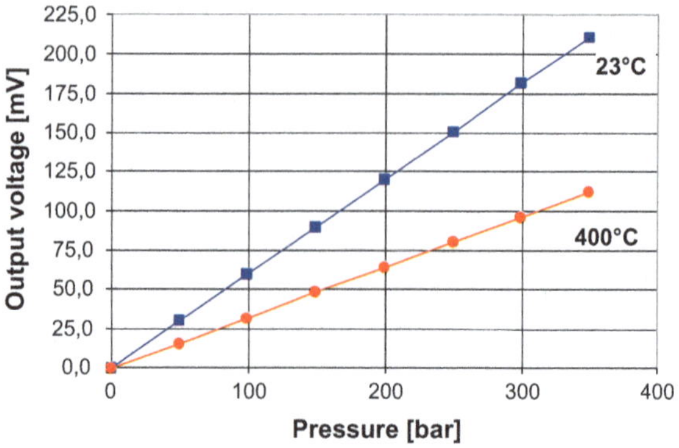

Fig. 17 Output signals of the sensor in [17, 18] at RT and 400 °C

polishing abrasives for metals, metallic components and semiconductors wafer [20, 21]. This very property makes it extremely difficult to etch silicon carbide in typical acid or base solutions such as KOH or HNA systems for silicon material. In its single crystal form, silicon carbide is not attacked by acids at room temperature. Silicon carbide can be etched in molten salt fluxes, hot gases, electrochemical processes and at high temperatures, typically between 350 °C up to 900 °C [22–30]. These high temperature, corrosive mixtures do have some important limitations, for examples the need of Pt beakers and holders, and the inability to etch masked samples. An other disadvantage is the isotropic etch behavior. Dissolution of silicon carbide in acids can be enhanced by photoelectrochemical etching method [31]. This method has some disadvantages, including rough surface morphologies, inability to pattern very small features, poor uniformity and difficulty to perform on wafer level.

Different deep etching methods, such dry etching, electrochemical etching and drilling were investigated in [17, 18]. Dry etching is a very effective method and can be applied to create selectively three dimensional features in silicon. It could be used to etch carbide substrates too. Dry etching of SiC is mainly achieved through plasma-based reactive ion etching (RIE). The removal of SiC in RIE occurs through a combination of physical and chemical processes. Controlling of these two competing processes are essential for creating an etch profile with the desired side-wall angle and surface quality [32]. Many critical issues are still to be solved regarding dry etching of SiC, such as chemistry, mask, reactor type and compatibility to CMOS etc.

Some fluorine based chemistries mixed with O_2 have been used to etch SiC. The presence of oxygen in fluorinated plasma provides more reactive species for removing SiC. Best results have been achieved with SF_6 and O_2 [23]. This process typically uses metal masks to achieve high selectivity. However, metal masks are known to create micro masking problem. This results in grass-like structures and other defects on the etch surfaces Fig. 18.

Up to now there is no deep etching technology, which is suitable to manufacture high quality SiC micromechanical structures such membranes, spring and center-

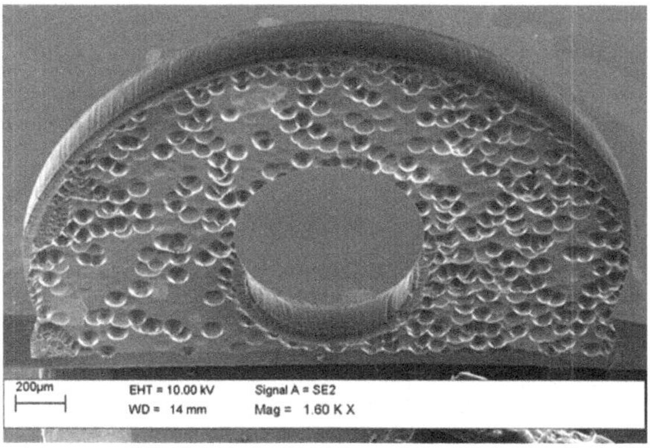

Fig. 18 SiC membrane etched with SF_6/O_2 and metal masks [23]

boss structures. In future, more effort has to be done here to develop standard bulk SiC MEMS technology.

Conclusion

In this book chapter we discuss the state-of-the-art of piezoresistive pressure sensors. We review materials and technologies, which is suitable for manufacturing this kind of sensors. We show the actual silicon technology for standard applications in industries. Silicon technology is still the most sophisticated technology for piezoresistive sensors. It has many advantages in compare to other technologies. We also show the limitation of using silicon as piezoresistors at elevated temperatures. The maximum working temperature of silicon-based piezoresistive sensors is about 125–150 °C (depending on silicon material and processes used in manufacturing). For higher temperatures silicon piezoresistive sensors have to be cooled. Cooling liquids such as oil and Mercury have high contamination potential in critical applications in medicine and food industry, and will be forbidden in future sensor systems.

For temperature range over 150 °C are material systems such as SOI, pSOI, SOS and SiC suitable. SOI and SOS materials are high quality single crystalline materials, and can be used to realized piezoresistive sensors up to 400 °C. The material pSOI is a low-cost alternative for temperature up to 250 °C. The material can be realized by standard processes such as LPCVD.

Among the materials mentioned above, the SiC system seems to be the most promising semiconductor for high temperature electronic and piezoresistive sensors in future. Due to its large bandgap (2.2–3.2 eV, compared to 1.1 eV for Si) SiC enables MEMS with significantly higher operating temperatures. Furthermore, SiC exhibits high chemical stability and thermal conductivity. Piezoresitive effects in SiC materials at high temperatures are still not fully understood and need to be investigated further. In compare to silicon technology, there are no standard bulk SiC technology, metallization schemes and packaging solution for high temperatures over 400 °C available. Therefore, the development or improvement of SiC material systems, new processing technologies and new advanced packaging concepts is necessary in order to achieve an accurate sensor performance at high temperatures.

References

1. Yole report (2016), http://www.yole.fr/
2. K.E. Petersen, Micromechanical membrane switches on silicon. IBM J. Res. Develop. **23**(4), 376–385 (1979)
3. M. Willander, H.L. Hartnagel, *High Temperature Electronics* (Chapman & Hall, 1997). ISBN: 0-412- 62510-5
4. D.A. Maxwell, R.H. Beeson, D.F. Allison, The minimization of parasitics in integrated circuits by dielectric insulation. IEEE Trans. Electron. Devices **12**(1), 20–25 (1965)
5. J.C. Doll, *Piezoresistors Design and Applications* (Springer, 2013). ISBN 978-1-4614-8516-2
6. H.-D. Ngo, A WSi–WSiN–Pt Metallization Scheme for Silicon", in MDPI micromachines. Micromachines (Special Issues Silicon Carbide-Based High Temperature Microsystems) **7**, 193 (2016). https://doi.org/10.3390/mi7100193
7. O. Kononchuk, B.-Y. Nguyen, *Silicon-On-Insulator (SOI) Technology* (Woodhead Publishing, 5th June 2014). ISBN: 9780857095268
8. H.-D. Ngo, Liquid-free, piezoresistive, SOI-based pressure sensor for high temperature measurements up to 400 °C, in *IEEE Sensors Conference*, Taipei, Taiwan (2012)
9. H. Kappert, "High Temperature SOI CMOS Technology and Circuit Realization for Applications up to 300 °C", paper IEEE, 978-1-4799-8391-9/15/
10. G.L. Harries, *Properties of Silicon Carbide* (INSPECT, 1995). ISBN: 0 852 96 870 1
11. R. Cheung, *Silicon Carbide Micro Electromechanical Systems for Harsh Environments* (ICP, 2006). ISBN: 1 86094 624 0
12. M. Shur, *SiC Materials and Devices* (World Scientific Publishing, 2006). ISBN: 981 256 835 2
13. Y. Kanda, A graphical representation of the piezoresistance coefficients in silicon. Electron Device **29**(1), 64–70 (1982)
14. V.M. Stuchebnikov, Silicon-on-Sapphire structures as a material for piezoresistive mechanical transducers. Commun. Technol. Electron. **50**(6) (2005)
15. E. Obermeier et al., Polysilicon as a material for microsensor applications. Sens. Actuators A **30**, 149–155 (1992)
16. J.S. Shor, L. Bemis, A.D. Kurtz, Characterization of monolithic n-type 6H-SiC piezoresistive sensing elements. IEEE Trans. Electron Devices **41**(5), 661–665 (1994)
17. R.S. Okojie, A.A. Ned, A.D. Kurtz, W.N. Carr, Characterization of highly doped n- and p-type 6H-SiC piezoresistors. IEEE Trans. Electron Devices **45**(4), 785–790 (1998)
18. T. Akiyama, D. Briand, N.F. de Rooij, Design-dependent gauge factors of highly doped n-type 4H-SiC piezoresistors. J. Micromech. Microeng. **22**(8), 085034 (2012)
19. H.-P. Phan et al., The piezoresistive effect of SiC for MEMS sensors at high temperatures: a review. IEEE J. MEMS **34**(6) (2015)
20. J.A. Powell, P. Pirouz, W.J. Choyke, *Growth and Characterization of Silicon Carbide Polytypes for Electronic Applications, Semiconductor Interfaces, Microstructures, and Devices: Properties and Applications* ed. by Z.C. Feng (Institute of Physics Publishing, Bristol, United Kingdom, 1993), p. 257
21. S. Zappe, Pressure sensors based on 3C-SiC on Si-on-insulator for high temperature applications, in *Microelectronics, Microsystems and Nanotechnology (MMN 2000)* (World Sciencetific Publishing Co. Pte. Ltd, 2000). ISBN: 981-02-4769-9

22. R.S. Okojie, α(6H)-SiC pressure sensors at 350 °C, in 1996, International Conferences on Electron Devices Meeting, 8–11 Dec 1996, pp. 525–528
23. M.G. Wieczorek, SiC based pressure sensor for high-temperature environments, in *IEEE Sensors*, 28–31 Oct 2007, pp. 748–751
24. G. Wieczorek, H.-D. Ngo, Micromachining of bulk SiC using dry etching and ultrasonic drilling, in *Eurosensors XX*
25. A. Barker, Advances in back-side via etching of SiC for GaN device applications, in *CS MANTECH Conference*, 13–16 May 2013, New Orleans, Loisiana, USA
26. R. Cheung, *Silicon Carbide Micro Electromechanical Systems for Harsh Environments* (Imperial College Press, 2006). ISBN: 1-86094-624-0
27. C.-M. Zetterling, *Process Technology for Silicon Carbide Devices* (INSPEC Verlag, 2002). ISBN: 0 85296 998 8
28. D. Buckley, J. Vac. Sci. Technol. **A3**, 762 (1985)
29. T.L. Chu, J. Electrochem. Soc. **112**, 955 (1965)
30. W.K. Liebmann, J. Electrochem. Soc., **12**, 885 (1964)
31. E.D. Wolley, J. Appl. Phys. **37**, 1588 (1966)
32. M.B.J. Wijesundara, *Silicon Carbide Microsystems for Harsh Environments* (Springer, 2015). ISBN: 978-1-4419-7120-3

Noncontact Temperature Sensing of Heated Cylindrical Rod by Laser-Ultrasonic Method

I. Ihara and A. Kosugi

Abstract A new noncontact method to measure both surface and internal temperatures of a heated cylindrical rod is presented. In the method a laser ultrasonic technique that provides noncontact measurements of ultrasonic waves in such heated rod is effectively employed. To quantitatively determine both the surface and internal temperatures near the rod end, an ultrasonic thermometry that is a technique for measuring temperature by ultrasound has been developed by considering the direction and path of the ultrasonic waves propagating in the rod. The thermometry is basically a combined method consisting of ultrasonic wave velocity measurements based on pitch-catch configurations and a one-dimensional finite difference calculation for unsteady heat conduction. The advantage of the method is that there is no need to know the thermal boundary condition at the rod end which is a heating surface where the thermal state is often unstable and normally unknown. To demonstrate the feasibility of the proposed method, experiments with an aluminum rod whose end surface is heated by a gas burner are carried out. A laser ultrasonic system consisting of a pulsed laser generator (Nd:YAG, 1064 nm, 180 mJ) and a laser Doppler vibrometer (He–Ne, 633 nm, <1 mW) is used for making noncontact measurements of shear and longitudinal waves propagating near the rod end during the gas burner heating. It has been shown that the ultrasonically determined temperature distributions almost agree with those measured using thermocouples. This chapter presents the basic principle of the ultrasonic thermometry and its experimental verification.

Introduction

Measuring temperature is one of the most fundamental and important issues in the various fields of materials science, engineering and industries. This is basically because material properties relating to mechanical, electrical, and chemical behaviors are

I. Ihara (✉) · A. Kosugi
Department of Mechanical Engineering, Nagaoka University of Technology, Nagaoka, Niigata 940-2188, Japan
e-mail: ihara@mech.nagaokaut.ac.jp

S. C. Mukhopadhyay et al. (eds.), *Modern Sensing Technologies*,
Smart Sensors, Measurement and Instrumentation 29,
https://doi.org/10.1007/978-3-319-99540-3_13

closely related to temperature and often show a strong temperature dependence. Thus, quantitative information on temperature of heated materials is indispensable to ensure the quality and reliability of the materials. In particular, for materials processing and manufacturing, it is strongly required to monitor the temperature distribution and its transient variation of the material being processed at high temperatures because the temperature state in the material crucially influences the quality and productivity of final products. Therefore, an in situ or in-process monitoring technique of temperature distributions of heated materials during their processing could be quite beneficial not only for conducting basic researches in materials science but also for realizing an effective process control in materials manufacturing.

It is known that ultrasound, because of its high sensitivity to temperature, has the potential as an effective means for measuring temperatures. Actually, ultrasonic thermometry, a method to estimate temperature by ultrasound, is expected to be a promising tool for providing advanced temperature measurements, and several works on ultrasonic temperature measurements have been made extensively [1–4]. Such ultrasonic thermometry provides not only noninvasive temperature measurements of materials but also internal temperature estimations with faster response time and higher spatial resolution. In addition, using laser ultrasonic techniques for the thermometry, it is even possible to make a noncontact temperature monitoring of higher temperature and/or moving objects.

In our previous works, effective ultrasonic thermometry for measuring internal temperature distributions of heated materials were developed [5–10]. The method can successfully measure internal temperatures of heated materials based on ultrasonic pulse-echo measurements of longitudinal or shear waves in the materials. When the ultrasonic thermometry is applied to a heated material with a single sided heating, an ultrasonic sensor should be placed at a material surface opposite and parallel to the heating surface so that efficient ultrasonic pulse-echo measurements could be performed appropriately. However, such restriction on the sensor configuration causes difficulty in applying the thermometry to some practical cases where the sensor installation location is spatially restricted. It should be noted that it is not always possible to have such material surface appropriate for making ultrasonic pulse-echo measurements. Such problematic situations often occur in manufacturing and/or heat treatment processes of materials in industries. One of the typical examples for such practical situations can be found in a rotating tool used in a friction stir welding (FSW) shown in Fig. 1. The left side of Fig. 1 shows a schematic of FSW. This is a kind of advanced welding technique known as a solid state joining process to join two workpieces, where friction heat between the rotating tool and the workpiece plays an important role in joining the workpieces. Although it is quite important for making a successful joining process to know the temperature distribution around the superheated region as well as the temperature of the friction surface, there is little information about such temperatures because it is very difficult to estimate those temperatures not only theoretically but also experimentally. Actually, there exist no suitable techniques available to measure such temperatures because of the difficulty in installing any temperature probes in the vicinity of the superheated region. Unfortunately, no conventional techniques such as thermocouples or infrared radiation

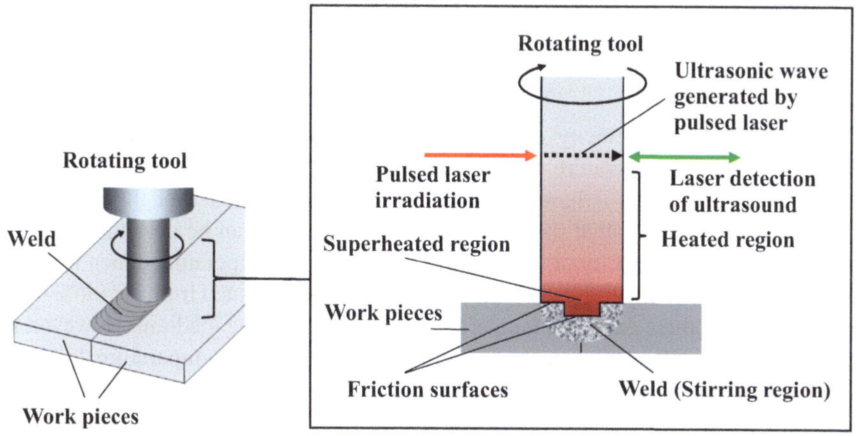

Fig. 1 Schematic of friction stir welding (left) and the cross-sectional view (right) showing the temperature distribution of a superheated rejoin around the bottom end of the rotating tool, with an ultrasonic wave propagating through the tool

methods are available for measuring temperatures near heating region of the rotating tool. Although infrared radiation method enables noncontact measurements of the surface temperatures of the tool, it does not allow to obtain internal temperatures. In addition, accurate measurements using infrared radiation are often hindered by the different emissivity and reflection of infrared radiation from other heat sources. It is noted that similar problems in such temperature measurements are also found in not only FSW process but also other machining processes such as tuning and milling. To overcome such common problem, the ultrasonic thermometry with a laser ultrasonic technique had been applied to a rotating cylinder [11]. In the application, a longitudinal ultrasonic wave generated by pulsed laser irradiation, shown in the right side of Fig. 1, was successfully used for estimating the internal temperature distribution in the diameter direction of the rotating cylinder. However, the temperatures near the superheated region still remain unknown because of difficulty in measuring those temperatures very near to the friction surface. In particular, it is extremely difficult to measure the temperature at the friction surface which is the bottom end of the cylinder. In this work, an attempt to use an oblique incident wave that may be able to reach to the bottom surface of the cylinder has been made to obtain the temperature at the bottom surface. The ultrasonic thermometry using a laser ultrasonic technique is applied to a heated cylindrical rod to measure the bottom surface temperature and internal temperature distribution near the rod end. An effective ultrasonic thermometry based on the use of oblique incident waves is proposed and its feasibility is demonstrated experimentally.

Application of Ultrasonic Thermometry to Heated Cylindrical Rod

It is known that the velocity of ultrasonic wave propagating through a medium changes with the temperature of the medium. The principle of temperature measurement by ultrasound is based on the temperature dependence of the ultrasonic wave velocity. It means that the temperature of a medium can be determined from the ultrasonic wave velocity of the medium as long as the temperature dependence is known. In fact, the temperature of a medium is simply determined from the ultrasonic wave velocity if there exists no temperature distribution in the medium. However, it is quite difficult to determine the temperature if there exists any temperature distribution in a medium. In such case, quantitative determination of temperature distribution from ultrasonic velocity is like trying to solve a kind of complicated inverse problem. To determine the temperature distribution in a heated medium quantitatively, combined estimation methods consisting of ultrasonic wave velocity measurements and a one-dimensional finite difference calculation of heat conduction had been developed [5–11]. Using those methods, temperature distributions were successfully determined for one-dimensional and axisymmetric heat conduction states.

In this work, we have modified such method to make temperature profiling near the bottom end of a cylindrical rod whose end is heated, as shown in Fig. 2 where the cylindrical rod imitates a FSW tool whose bottom surface is heated. In such situation, two ultrasonic waves, an oblique incident ultrasonic wave onto the bottom surface and a direct ultrasonic wave propagating in the diameter direction of the rod, are generated by pulsed laser irradiation on the side surface of the rod. Simultaneously, the two waves, a reflected wave from the bottom surface and the direct wave thorough the rod, are detected by a laser displacement meter. It should be noted that such ultrasonic generation and detection are performed in a noncontact configuration.

Generation of Oblique Incident Waves

The important point to note is that the oblique incident wave is generated at the side surface of the cylindrical rod and reflected at the bottom surface of the rod so that the temperature at the bottom surface and temperature distribution in the vicinity of the surface are determined by the method described in the next section. The reason why such oblique incident wave can be generated is understood from the directivity patterns of laser generated ultrasound [12–14]. Figure 3 shows schematics of typical directivity patterns of the amplitude of ultrasonic waves generated in a metal by a pulsed laser irradiation in ablation regime or thermos-elastic regime. It can be seen that the maximum amplitude of the longitudinal wave occurs around 0° (normal direction to the material surface) and the maximum of the shear wave occurs around 45° (inclined from the normal) for ablation regime as shown in Fig. 3a. On the other hand, the direction in which the maximum amplitude of longitudinal

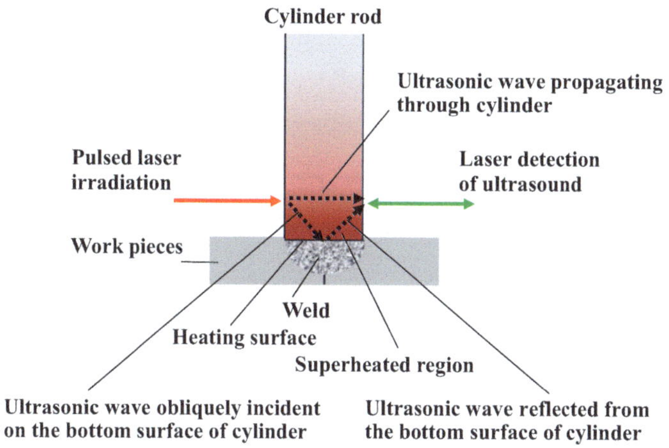

Fig. 2 Schematic model for the temperature measurement of a cylindrical rod which imitates a FSW tool, showing laser generation and detection of ultrasonic waves for the cylindrical rod whose bottom surface is heated, with an oblique incident ultrasonic wave onto the bottom surface and a reflected wave from the surface

wave or shear wave occurs is inclined from the normal for thermo-elastic regime as shown in Fig. 3b. It is noted in the experiment of this work that ablation regime is normally dominant over thermos-elastic regime in generating ultrasonic waves by laser irradiation. Therefore, it is highly expected that longitudinal wave components mainly propagate in a direction of the normal of the material surface, whereas shear wave components propagate in a direction of approximately 45° with respect to the normal of the material surface, as shown in Fig. 2. Using an appropriate displacement measurement technique such as a laser interferometer or laser Doppler vibrometer, it is possible to detect those two waves in a noncontact configuration.

Ultrasonic Thermometry with Oblique Incident Wave

We consider that the end surface of a cylindrical rod is uniformly heated as shown in Fig. 4a. In a certain region L, assuming a one-dimensional unsteady heat conduction with a constant thermal diffusivity, heat conduction equation in the region can be defined by

$$\frac{\partial T}{\partial t} = \alpha \frac{\partial^2 T}{\partial x^2} \tag{1}$$

where α is thermal diffusivity coefficient, and x axis is taken as the direction of heat transfer. Figure 4b shows a schematic of staggered grid model for a one-dimensional finite difference calculation analysis for determining temperature distribution in the

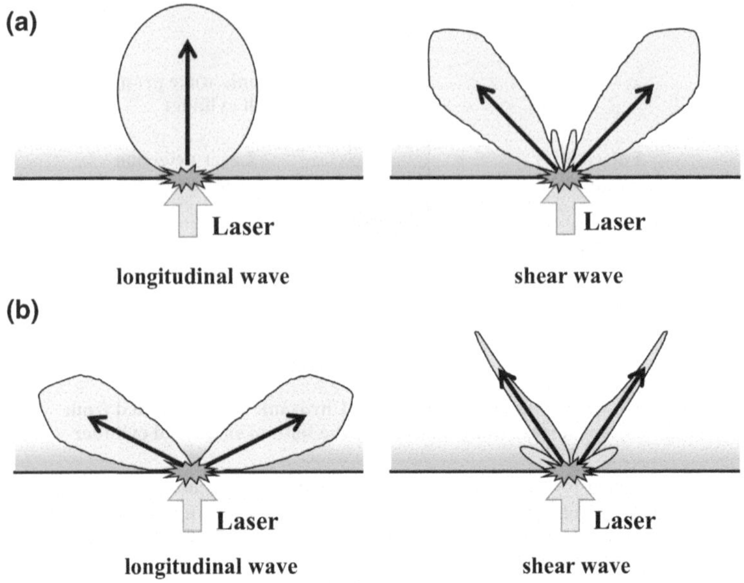

Fig. 3 Schematics of sound fields (directivity patterns) of longitudinal and shear waves generated by laser irradiation onto metal surface, **a** by ablation regime and **b** by thermo-elastic regime

region L. In the model, when all temperatures, $T_1^n \sim T_{N-1}^n$, at a time step n are known as the initial condition, the following internal temperatures, $T_1^{n+1} \sim T_{N-1}^{n+1}$, at a time step $n+1$ can be determined from Eq. (1) using a finite difference analysis. However, temperatures, T_0^{n+1} and T_N^{n+1}, are still unknown at this moment. To determine those temperatures, we consider to utilize two ultrasonic waves, BW1 and BW2, shown in Fig. 4a. To generate those two waves, a pulsed laser is irradiated to the point A at the side surface of the rod. The generated two waves, BW1 and BW2, propagate through the rod as shown in Fig. 4a and are then detected at the point C using an appropriate detection apparatus. It is noted that the wave BW1 is obliquely incident to the rod end which is a heating surface and refracted back to the point C, whereas the wave BW2 propagates to the point C directly. Using the BW1 and BW2, temperatures, T_0^{n+1} and T_N^{n+1}, are determined, respectively. Actually, temperature at $i = N$, T_N^{n+1} can easily be determined from the velocity of BW2 propagating between A and C when the temperature dependence of the ultrasonic wave velocity of the material is known, because the temperature in the wave propagation region from A to C is uniform. Although it is assumed here that no temperature distribution exists in the region, this assumption is probably valid for the single side heating of a metal rod. On the other hand, a significant temperature gradient should exist in the propagation region of the BW1 due to the single side heating at the rod end. The transit time of ultrasonic wave from A to C through B is then given by

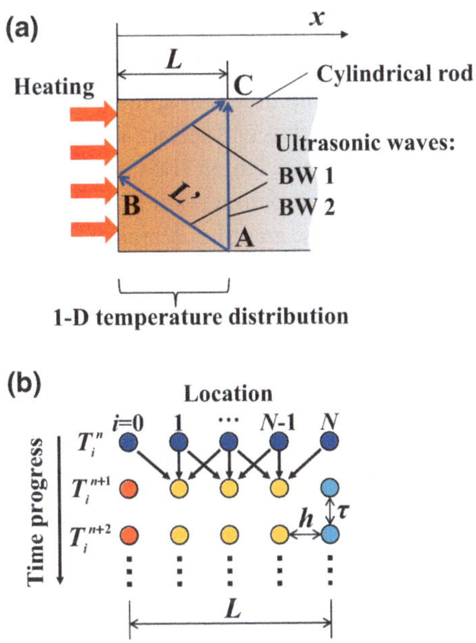

Fig. 4 Analysis model used for temperature distribution determination. **a** Schematic of the rod end portion with ultrasonic waves. **b** Schematic of the staggered grid model for the finite difference calculation analysis for one-dimensional heat conduction

$$t = 2\int_0^{L'} \frac{1}{v(T(x))} dx \tag{2}$$

where L' is the propagation distance of the oblique incident wave BW1 and v is the wave velocity which is a function of the temperature distribution $T(x)$ in the region L'. This also means that the wave velocity is a function of the location x in the region L'. It is noted that the temperature distribution in the region A to B is the same as that in the region C to B. According to the temperature analysis procedure developed in the previous works [5–8], the temperature at the heated surface, T_0^{n+1}, can be determined from the combined method consisting of the Eq. (2) and the finite difference calculation for one-dimensional unsteady heat conduction based on the staggered grid model as shown in Fig. 4b, where the temperature at $i = N$, T_N^{n+1} determined above is used as the boundary condition at each time steep. It should be noted in the determination that staggered grid points of the wave velocity for estimating the propagation time of the wave are taken on the line A to B which corresponds to the propagation region of the oblique incidence wave BW1. Thus, the temperature distribution in the region L including the temperature at the heating surface can be obtained, as long as such two waves, BW1 and BW2 are continuously being measured.

Experimental Setup

Figure 5 shows the experimental setup used. An aluminum cylindrical rod of 30 mm in diameter is used for a specimen and its end surface is heated by a gas burner. Using a pulsed laser generator (Nd:YAG, wavelength 1064 nm, energy 200 mJ/pulse, pulse width 3 ns), pulsed laser is irradiated onto a point at 13 mm from the rod end, and then generated two ultrasonic waves, an oblique incident shear wave and a normal incident longitudinal wave, are detected by a laser Doppler vibrometer (He–Ne, wavelength 633 nm, power < 1 mW) at the opposite side. This configuration of the laser generation and detection of ultrasonic waves makes the oblique incident shear wave at 41°. It is noted that it is experimentally verified that the incident angle of 41° is most effective to generate appropriate ultrasonic waves with higher signal-to-noise ratio, by the preliminary experiments with different incident angles. The center frequency of the ultrasonic waves is about 2 MHz. The waves are continuously acquired every 0.2 s with a PC based real-time acquisition system with sampling rate of 100 MHz. The transit time of each wave is precisely determined by taking the cross correlation of the detected signals of the waves, and then used for the analysis to determine temperatures near the rod end. Five thermocouples of 0.5 mm diameter are inserted into the rod at 1, 3, 5, 10, 15 mm from the rod end to obtain reference values of temperature gradient for comparing with the ultrasonically determined values, as shown in Fig. 5b. External views of a pulsed laser generator and laser Doppler vibrometer used are shown in Fig. 6 with their specifications.

Experimental Results

Figure 7 shows the measured waveforms of ultrasonic waves generated by the pulsed laser generator and detected by the laser Doppler vibrometer, before the heating starts. A direct longitudinal wave BW2 propagating in the diameter direction of the rod and an oblique incident shear wave BW1 are clearly seen in Fig. 7. It is obvious that the BW1 is the shear wave reflected at the bottom end which is a heating surface. Although some other waves are also found in the measured results, the two waves, BW1 and BW2 are important here because they can be used for temperature determination. The BW2 is used for determining the average temperature in the cross section at the laser irradiated point of the rod while the BW1 is used for determining the internal temperature distribution near the rod end as well as the surface temperature.

Temperature dependences of the shear and longitudinal wave velocities of the aluminum rod used were precisely determined by the preliminary experiments in advance, and then used for the temperature determinations by the present method in this work. It has experimentally been found that the both velocities change linearly with temperature in the temperature range up to 130°C. Therefore, the following linear relations, $v_{\mathrm{LW}} = -1.042T + 6402$ (m/s) for longitudinal wave (BW2) and v_{SW}

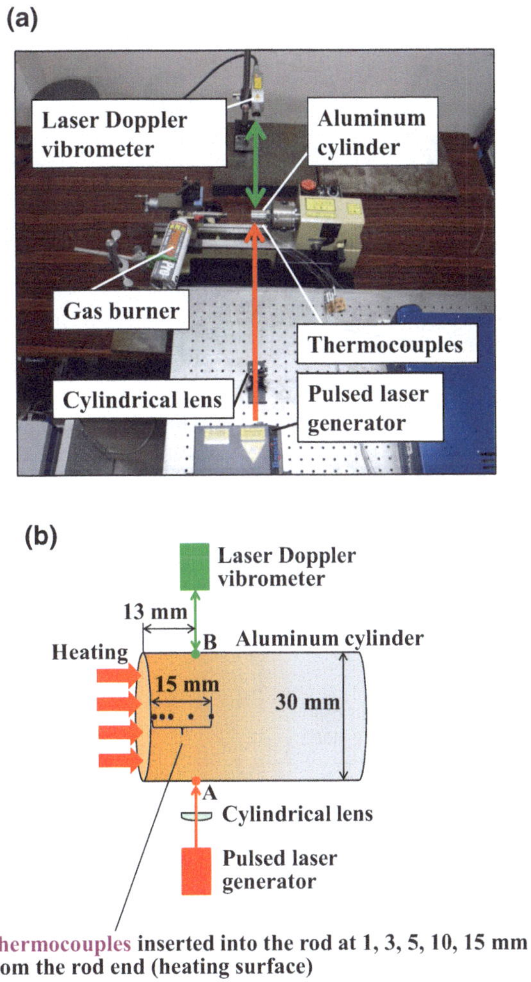

Fig. 5 Experimental setup used, **a** External view of the specimen (aluminum cylinder) and a laser ultrasonic measurement system. **b** Schematic of the detail of the measurement region of the aluminum cylinder

$= -0.989T + 3247$ (m/s) for shear wave (BW1) are employed for the temperature determination, where v and T are the velocity and temperature (°C), respectively. The average temperature in the cross section of the rod mentioned above is simply calculated from the linear relation of the BW2 if the velocity of the BW2 is precisely known by the measurement.

The transit time of each wave is precisely determined by taking the cross correlation of the detected waves. The variations in the transit time of the BW1 and BW2 with the elapsed time during heating are shown in Fig. 8a, b, respectively. In each figure, the variation in the temperature measured using the thermocouple at a certain portion is also depicted for the reference. As we expected, the transit time of the ultrasonic wave increases as the temperature increases. It is noted that the increasing

(a)

Laser	Nd:YAG Q-switch pulse ClassIV
Wave length	1064 mm
Repetition rate	30 Hz
Energy	180 mJ
Pulse width	3-5 ns

(b)

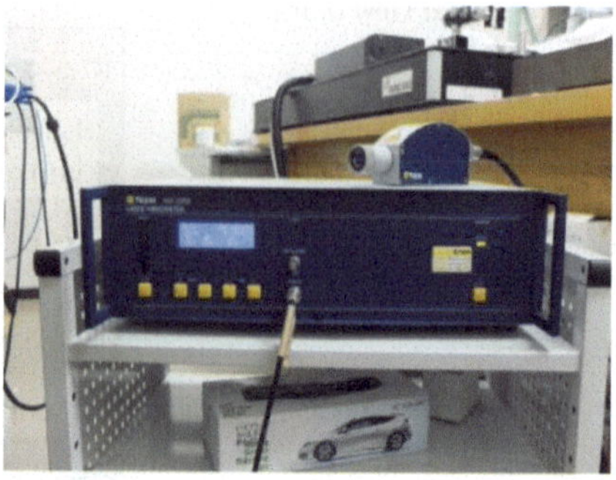

Laser	He-Ne CW Class II
Wave length	633 mm
Energy	<1 mW
Spot size	25 μm
frequency range	≦2.5 MHz

Fig. 6 External views and specifications of **a** pulsed laser generator and **b** laser Doppler vibrometer used

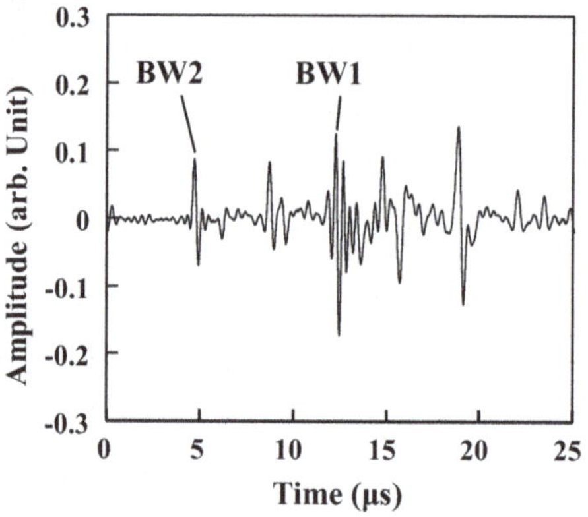

Fig. 7 Measured ultrasonic waves before heating starts. BW1 and BW2 are the oblique incident shear wave and the direct longitudinal wave propagating through the rod, respectively

rate of the transit time of BW1 is higher than that of BW2 because the material temperature in the propagation region of BW1 and its increasing rate are higher than those in the propagation region of BW2. Such transit time data at each elapsed time are then used for the analysis to determine both the internal temperature distribution and the rod end temperature, and then the temperature variations with the elapsed time are obtained.

Figure 9 shows the estimated temperature distributions at 0, 1, 5, 10, 20 and 30 s after the heating starts. The spatial resolution and time resolution of the present

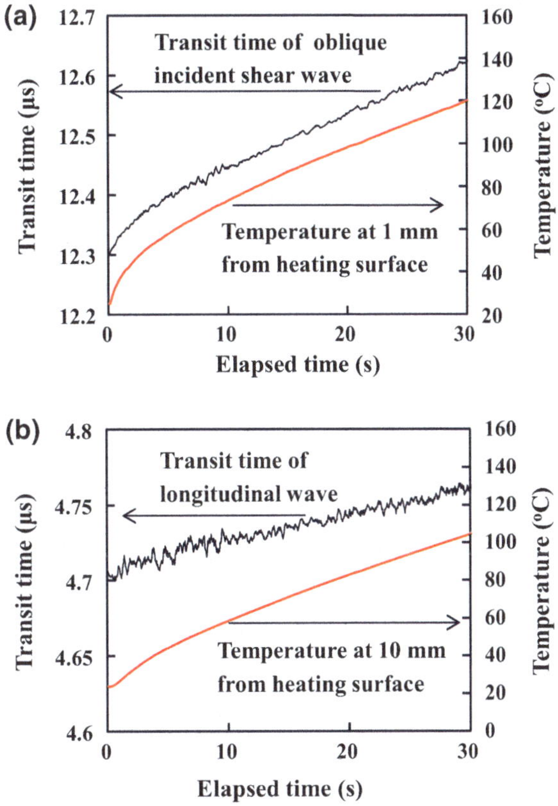

Fig. 8 Variations in the temperature and the transit time of **a** oblique incident shear wave BW1 and **b** longitudinal wave BW2, with the elapsed time during heating

ultrasonic estimation are 2.6 mm and 0.033 s, respectively. Temperature distributions estimated by the ultrasonic method almost agree with those measured using the thermocouples. It can be seen in the early stage of the heating that there is a small discrepancy between the ultrasound and thermocouple. In other words, the temperature rise estimated using the thermocouples is slightly delayed from that by the ultrasound. This is because the ultrasonic method has a higher time response than the thermocouple in measuring temperature.

Figure 10 shows the variations in the estimated temperatures with the elapsed time, at different three locations where ultrasonically estimated results are compared with those estimated using thermocouples. We can see that both results by the ultrasound and thermocouple almost agree with each other and shows the same tendency. Large deviation in the ultrasonically estimated values at the heating surface is observed. This deviation is basically caused by a considerable amount of fluctuation of the estimated transit time due to relatively low signal-to-noise ratio of the ultrasonic waves, BW1 and BW2. It is expected to improve the signal-to-noise ratio in measuring those ultrasonic waves in future. It should be noted that the shear wave BW1 does not always propagate in a straight line because there is a possibility that

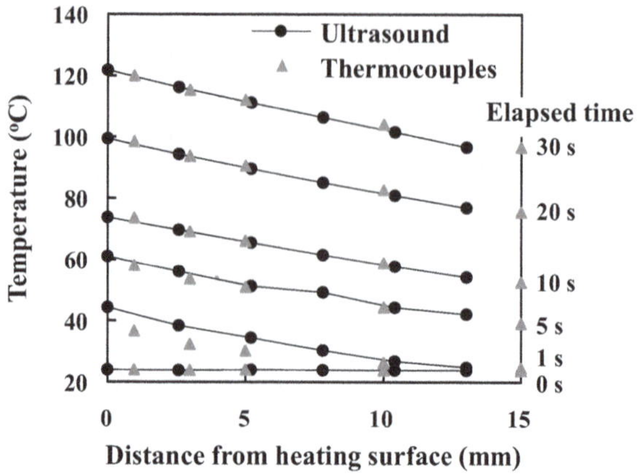

Fig. 9 Temperature distributions of the aluminum cylinder and their variations with the elapsed time after the heating starts, measured by the ultrasonic method and thermocouples

the propagation path of such oblique incident wave is slightly curved owing to the temperature gradient exiting in its propagation region. Although it seems that such influence of the distortion of the propagation path on the temperature estimation is not clearly found in the Figs. 9 and 10, it is necessary to study such influence on temperature measurements quantitatively.

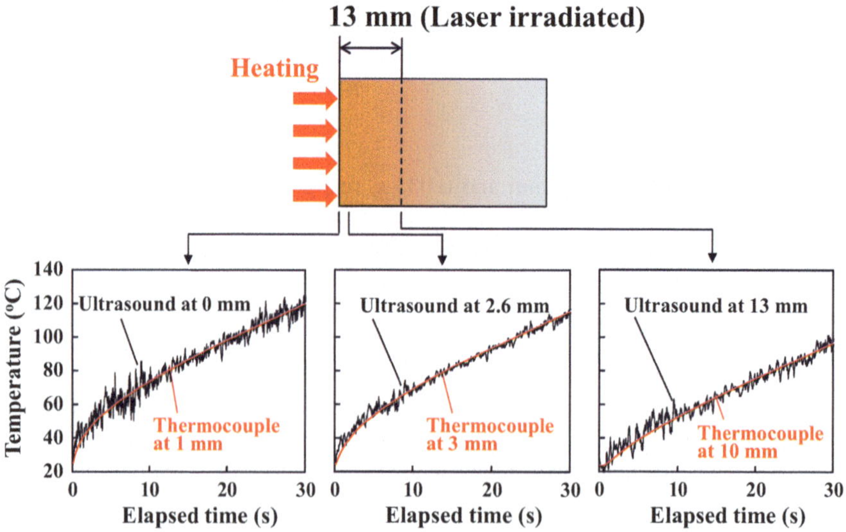

Fig. 10 Variations in the estimated temperature at each location with elapsed time

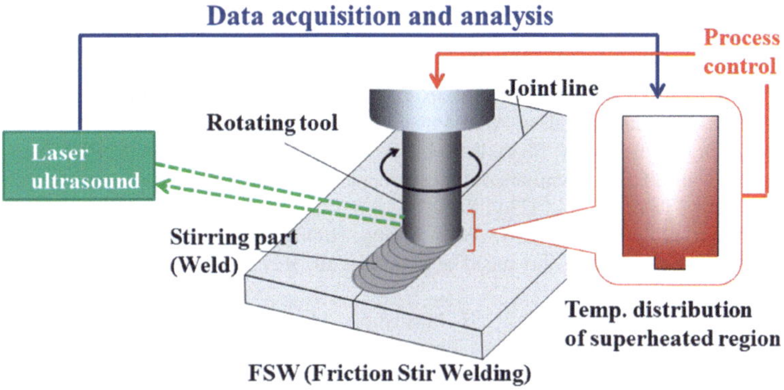

Fig. 11 A schematic showing an advanced process control using a laser-ultrasonic technique for a friction stir welding (FSW) process based on online non-contact temperature measurements inside the rotating tool

Conclusions

This chapter has presented a laser ultrasonic thermometry using an oblique incident ultrasonic wave for monitoring temperature distribution of a heated cylindrical rod. The method is applied to a heated aluminium cylindrical rod to measure both the surface and internal temperatures near the rod end. Temperatures estimated by the present method almost agree with those measured by thermocouples. Thus, the feasibility of the method has been demonstrated. In the future, based on the result, an attempt of applying the laser ultrasonic thermometry to a rotating cylindrical rod to monitor the temperatures of the heating portion of the rod will be made, which is similar to the former works [10, 11] but more practical. One of our final goals is to establish an intelligent process monitoring of materials using the ultrasonic thermometry. Figure 11 shows a schematic of an advanced process control using a laser-ultrasonic technique for a friction stir welding (FSW) process based on online non-contact temperature measurements. Non-contact ultrasonic thermometry is highly expected to be a promising tool for realizing such advanced process monitoring.

Acknowledgements Support from JSPS KAKENHI (Japan Society for the Promotion of Science, Grant-in-Aid for Scientific Research (B), Grant Number 16H04507) is greatly appreciated.

References

1. F.L. Degertekin, J. Pei, B.T. Khuri-Yakub, K.C. Saraswat, In-situ acoustic temperature tomography of semiconductor wafers. Appl. Phys. Lett. **64**, 1040–1338 (1994)
2. K. Balasubramainiam, V.V. Shah, R.D. Costley, G. Boudreaux, J.P. Singh, High temperature ultrasonic sensor for the simultaneous measurement of viscosity and temperature of melts. Rev. Sci. Instrum. **70–12**, 4618–4623 (1999)
3. K.N. Huang, C.F. Huang, Y.C. Li, M.S. Young, High precision fast ultrasonic thermometer based on measurement of the speed of sound in air. Rev. Sci. Instrum. **73–11**, 4022–4027 (2002)
4. K. Mizutani, S. Kawabe, I. Saito, H. Masuyama, Measurement of temperature distribution using acoustic reflector. Jpn. J. Appl. Phys. **45–5B**, 4516–4520 (2006)
5. I. Ihara, M. Takahashi, Non-invasive monitoring of temperature distribution inside materials with ultrasound inversion method. Int. J. Intell. Syst. Technol. Appl. **7–1**, 80–91 (2009)
6. M. Takahashi, I. Ihara, Quantitative evaluation of one-dimensional temperature distribution on material surface using surface acoustic wave. Jpn. J. App. Phys. **48,** GB04-1-5 (2009)
7. I. Ihara, T. Tomomatsu, In-Situ measurement of internal temperature distribution of sintered materials using ultrasonic technique. IOP Conf. Ser. Mater. Sci. Eng. **18**, 022008 (2011)
8. H. Yamada, A. Kosugi, I. Ihara, Non contact monitoring of surface temperature distribution by laser ultrasound scanning, Jpn. J. App. Phys. **50**, 07HC06-1-5 (2011)
9. A. Kosugi, I. Ihara, I. Matsuya, Accuracy evaluation of surface temperature profiling by a laser ultrasonic method. Jpn. J. Appl. Phys. **51**, 07GB01-1-8 (2012)
10. A. Kosugi, Y. Ono, I. Matsuya, I. Ihara, Application of laser ultrasound to noncontact temperature profiling of a heated hollow cylinder. J. Phys: Conf. Ser. **520**, 012015 (2014)
11. I. Ihara, Y. Ono, A. Kosugi, I. Matsuya, Noncontact temperature profiling of rotating cylinder by laser-ultrasound, in *Sensing Technology: Current Status and Future Trends III, Smart Sensors, Measurement and Instrumentation*, ed. by A. Mason et al., vol. 11 (Springer International Publishing, 2015), pp. 327–339
12. C.B. Scruby et al., Laser generation of ultrasound in metals, in *Research Techniques in Nondestructive Testing*, ed. by R.S. Sharpe, vol. 5 (Academic, New York, 1982), pp. 281–327
13. D.A. Hutchings, Ultrasonic generation by pulsed lasers, in *Physical Acoustics*, ed. by W.P. Mason, R.N. Thurston, vol. 28 (Academic, New York, 1988), p. 21
14. B. Yuhai et al., Experimental studies of directivity patterns od laser generated ultrasound in neutral glasses. Ultrasonics **33–6**, 429–436 (1995)

Development of a PVDF Multi-resonance Vibration Sensor with a Wide Range of Frequency Resolution

Mousa Hadipour, Murat Tahtali and Andrew J. Lambert

Abstract An electromechanical concept in the dynamics of piezoelectric polyvinylidene fluoride-(PVDF) films is employed to develop a multi-resonance vibration sensor. This real-time sensor has the capability of adjusting its resonant frequency to the ambient vibration frequency. Using only one sensor, we are able to measure the vibration at a wide range of frequencies and overcome many sources of error associated with traditional sensors such as hysteresis, variation in tension and geometry with time, and influence of temperature on electromechanical properties. The equation of motion of piezoelectric films shows that the bias voltage changes the resonant frequencies of the PVDF films significantly. A prototype is built to investigate the effect of the bias voltage on a ribbon PVDF sensor experimentally. Results confirm that the bias voltage also improves the amplitude of vibration, which is proportional to the generated voltage. The proposed sensor is simulated in an FE-software package to study its mechanical performance and electrical output. The results from finite element simulation show excellent agreement with the experimental data.

Introduction

Among the smart materials, the piezoelectric materials are known to provide an opportunity to represent the interaction between the electrical and the mechanical states. They are widely employed in sensors, actuators, and transducers [1–3]. In this article, the emphasis is given to piezoelectric sensors and especially vibration sensors. Piezoelectric sensors are the most favourable type of sensors to measure physical quantities in acoustics, mechanics, and electronics. They are applied to a number of areas such as structural health monitoring and mechanical vibration measurement of rotary machinery. In mechanics, they are also used as spectrum sensors to monitor mechanical vibration as a predictive measurement to prevent damage and unexpected

M. Hadipour (✉) · M. Tahtali · A. J. Lambert
School of Engineering and Information Technology, University of New South Wales (UNSW), Canberra, Australia
e-mail: m.hadipour@student.unsw.edu.au

S. C. Mukhopadhyay et al. (eds.), *Modern Sensing Technologies* ,
Smart Sensors, Measurement and Instrumentation 29,
https://doi.org/10.1007/978-3-319-99540-3_14

shutdowns in structures and equipment. The accuracy of the sensor measurement is the most important factor in vibration analysis since more accurate measurements would result in more reliable condition interpretation of systems. Nowadays, there are several types of sensors to measure the mechanical and structural vibration. An easy electronic interface and having moderately high resolution of the frequency are important features of sensor networks.

Piezoelectric films are one of the most suitable parts used in sensors and actuators. Thanks to the lightweight, lower stiffness and lower acoustic impedance of polymer piezoelectric materials rather than piezo-ceramic, poly (vinylidene fluoride) (PVDF) films have been adopted in many fields of science [4–8]. In sensing technology, PVDF sensors are used to measure the vibration spectrum of rotary machinery and structures which are under dynamic loads, such as aircraft and bridges [9]. An array of spectrum sensors, with different resonant frequencies in a sensor network, is attached into the vibration source to collect the vibration signature and avoid drifting toward the resonant frequency of the systems [10]. When the vibration frequency is matched to the resonant frequency of the designed sensor for that range of frequencies, the output voltage is strengthened. Therefore, designing several sensors with different resonant frequencies on a system to be able to measure a wide range of input vibrations is necessary. These type of sensors are associated with several sources of error and need to be calibrated frequently. Hysteresis, variation in tension and geometric with time, and influence of temperature on electromechanical properties of sensors are some of the sources of error [10]. To overcome the resonant frequency variation, we introduce a novel multi-resonance sensor whose resonant frequency could be changed in real-time to mitigate the sources of error in PVDF sensors. The bias voltage is the key concept in this type of sensor. Using the bias voltage we are able to manipulate the resonance region of a sensor and adjust it with the input vibration. It would expand the frequency resolution and enhance the output signal of the PVDF sensors.

The structure of this chapter is as follows: in section "Theory" the effects of bias voltage on the vibration characteristics of the piezoelectric films are investigated in theory. In section "Experiments" a prototype of PVDF ribbon sensor is created to show the effect of bias voltage on the mechanical properties and electrical behaviour of the sensor. The finite element model of a PVDF vibration sensor is developed in section "FE Simulation" to simulate the dynamics of the film and its output voltage. In the last two sections, results are discussed, and the conclusion is drawn.

Theory

Equation of Motion

The linear equation of motion of a piezoelectric thin film can be extracted using Newton's second law as follows [11]:

$$A_1\frac{\partial^4 w}{\partial x^4}+2A_2\frac{\partial^4 w}{\partial x^2 y^2}+A_3\frac{\partial^4 w}{\partial y^4}+A_4\frac{\partial^2 w}{\partial x^2}+A_5\frac{\partial^2 w}{\partial y^2}=\rho h\frac{\partial^2 w}{\partial t^2} \tag{1}$$

where

$$A_1=-\frac{h^3}{12}\left(c_{11}^{\epsilon}+\frac{e_{31}^2}{\xi_{33}}\right)=-D_{px}$$

$$A_2=-\frac{h^3}{12}\left(c_{12}^{\epsilon}+2c_{66}^{\epsilon}+\frac{e_{31}e_{32}}{\xi_{33}}\right)=-D_{pxy}$$

$$A_3=-\frac{h^3}{12}\left(c_{22}^{\epsilon}+\frac{e_{32}^2}{\xi_{33}}\right)=-D_{py}$$

$$A_4=e_{31}V_{rel}+P_x h$$

$$A_5=e_{32}V_{rel}+P_y h$$

w and h are the transverse deformation and the thickness of the piezoelectric film. ρ is the piezoelectric material density. D_p represents the flexural rigidity of the piezoelectric film that is a function of piezoelectric constants, e_{ij}, impermittivity constant in thickness, ξ_{33}, and stiffness constants in a constant electric field, c_{pq}^{ϵ}.

The partial differential Eq. 1 expresses the dynamic behaviour of the piezoelectric film as an integration of the dynamic of continuous plates and membranes. If the effect of stiffness is much more than the in-plane tension, such as in piezoelectric ceramics or thick PVDF films, the piezoelectric film behaves similarly to a plate; otherwise, the in-plane tension and electrical properties play an important role in determining the vibration characteristics of the film, i.e. resonant frequencies and mode shapes, as well as the stiffness. Usually, thin PVDF films behave the same as the latter because of their light weight and low stiffness.

Resonant Frequency

At resonance, the vibration strength increases dramatically, and designers try to avoid it. In this research, we are manipulating the resonant frequency of the sensor to match it to the ambient vibration frequency. As the partial differential Eq. (1) shows, the stiffness, in-plane tension, and bias voltage V_{rel} form the coefficient matrix of the eigenvalue equation to extract resonant frequencies. Since the stiffness coefficients, i.e. A_1, A_2 and A_3, are associated with the material and manufacturing process, it is not possible to change them easily after manufacturing. On the other hand, the coefficient $eV_{rel} + Ph$ expresses that the application of a bias DC to the piezoelectric film works similar to the in-plane tension.

Figure 1 represents the schematic of a thin PVDF ribbon stretched along its length and fixed at the ends. Figure 2a shows that the application of the constant electric field, ε, in the same direction of piezoelectric polarization would result in the expansion of strip in direction x while the clamped ends restrict it and an in-plane compression force is induced. Consequently, the resonance, which directly depends on the in-plane tension, happens at a lower frequency. On the other hand, opposite electric field and polarization directions result in contraction along the strip and then an in-plane tension is induced due to the fixed ends, as illustrated in Fig. 2b. In this configuration, the resonance happens at higher frequencies.

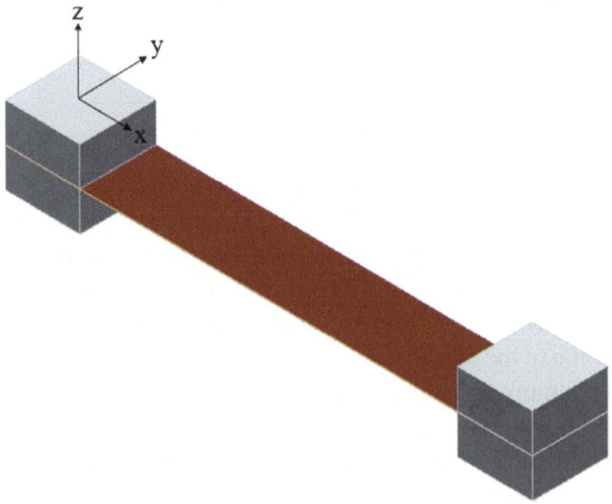

Fig. 1 Schematic of a stretched PVDF film clamped at the ends

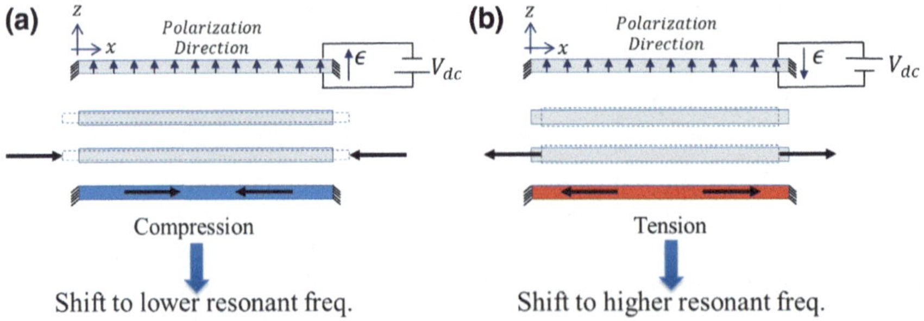

Fig. 2 PVDF film deformation under the application of bias voltage **a** the electric field is in the same direction of polarization and induces in-plane compression **b** the electric field is opposite to the direction of polarization and induces in-plane tension

Therefore, we can easily change the resonant frequency of a PVDF ribbon via the application of bias voltage. In next sections, the effect of bias DC on the vibration attributes and the output voltage of the PVDF film sensor is investigated experimentally and numerically

Experiments

Experimental Setup

A stretched 25 mm × 9 mm × 28 μm clamped PVDF film is cut as a ribbon shape and clamped on a mounting table attached to the shaker, as shown in Fig. 3a. The PVDF film surfaces are metallized with 70 nm Ni-Cu layers. To clamp the film, four rivets seated in a groove prevent the leakage of electrically induced tension produced along the PVDF film. Figure 3b depicts the setup and vibration measurement procedure. The output of the DC supply is amplified by 20 times and applied to both sides of the PVDF film. The shaker exerts a sinusoidal displacement to the boundaries of the clamped sensor. It simulates the input ambient vibration to the film. A Polytech scanning vibrometer, PSV-400, scans the sensor surface and collects the velocity data. The input data to the laser vibrometer is processed, and displacement and acceleration spectra are generated as well. The laser vibrometer has the capability of surface scanning to extract motion of the ribbon.

Vibration Measurements

The electrical output of the piezoelectric film is directly related to the vibration characteristics of the film. Thus, the vibration analysis of the film helps to understand the sensor response to the input excitation. To do so, an external vibration including a range of frequencies is exerted to the film and its response is measured using the vibrometer. The input force to the film is swept at 16-s intervals from 200 to 800 Hz with an amplitude of 20 V_{pp} that is applied to the shaker. The vibrometer measures the spectrum of the oscillation at the central point of the film, and the fundamental natural frequency of the film at 495.44 Hz is detected. The surface deflection is also scanned at the resonance to extract the fundamental mode shape of vibration. Figure 4a, b illustrates the first mode of the sensor at peak and valley positions, respectively. The scanned surface of the sensor, as shown in Fig. 4c, demonstrates that the deflection in regions close to the free edges is greater than in other areas, and a saddle shape is created on the surface. It effects on the output voltage. The reason will be discussed in the next subsection.

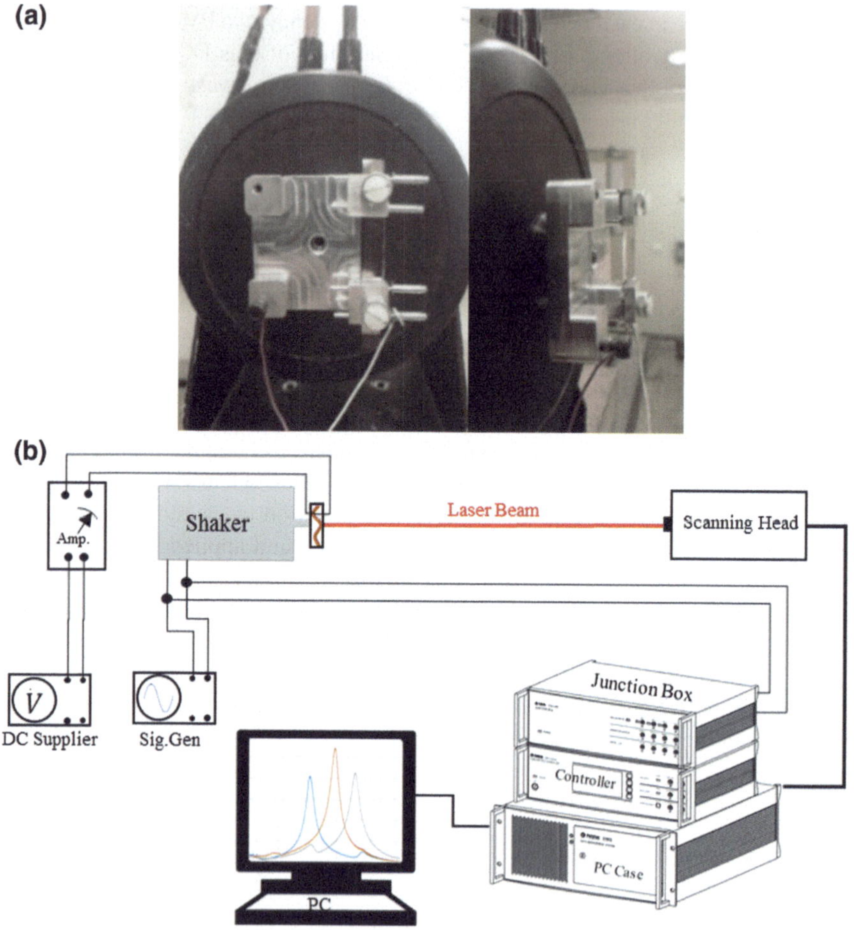

Fig. 3 Experimental setup **a** prototype, **b** vibration measurement procedure

Figure 5 illustrates the Fast Fourier Transform (FFT) spectra of displacement at the centre point of the sensor surface. It shows that by applying ±200 V DC to the film the fundamental resonant frequency shifts down or up depending on the electric field direction. As discussed in Section "Theory", application of constant electric fields in the opposite direction of the polarization orientation induces in-plane tension in the film and results in higher resonant frequency. Here, once 200 V DC is applied to the film, the resonant frequency is increased by roughly 68 Hz and reached to 563.81 Hz. On the other hand, the same bias voltage but in the same direction with polarization has reduced the resonant frequency to 407.13 Hz. The stress field variations under the bias voltage are studied numerically in Section "FE Simulation".

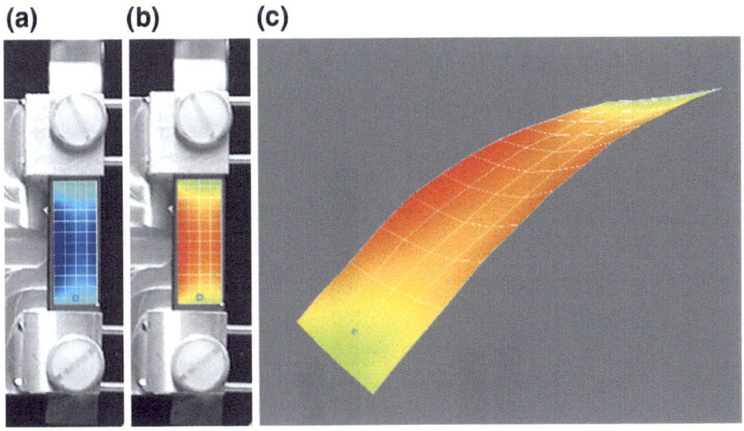

Fig. 4 The back and forth deflections and the scanned surface of the sensor measured by vibrometer

Fig. 5 Displacement spectra under different bias DCs measured by vibrometer

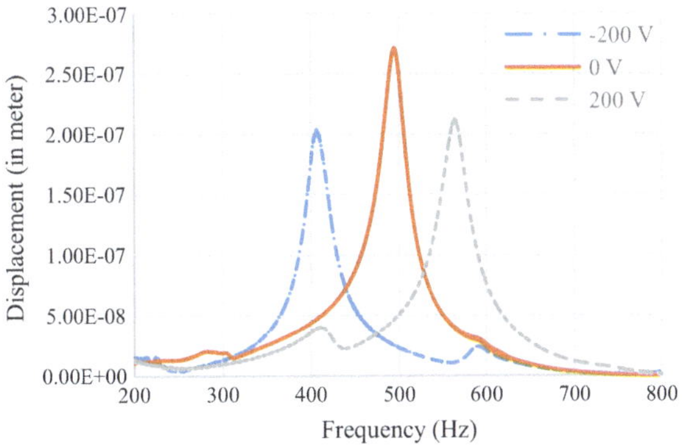

The displacement spectra depict that there is a small peak around 600 Hz. It remains constant when the electric field is changed so that it is an induced excitation to the film because of the resonant frequency of the mounting setup. Moreover, because of the uneven distribution of the tension in the film and the inertia in oscillation, other small peaks at frequencies apart from the resonant frequency are seen in each spectrum. These peaks influence on the displacement amplitude at resonant frequencies.

The bar chart is shown in Fig. 6a compares the deflection of the film at resonant frequencies under different applied bias DC voltages. The amplitude at the lower frequency, i.e. 407.13 Hz, is almost 5 microns higher than the unpowered film. The reason of this is that the displacement amplitude of the shaker head at lower frequencies is greater than the higher frequencies. The mounting table amplitude at 407.13, 495.44, and 563.81 Hz is measured as 1.47, 0.94, and 0.68 μm, respectively. Therefore, the normalized values of displacement to the shaker amplitude are more reasonable to reveal the dynamic behavior of the film. The normalized amplitudes are

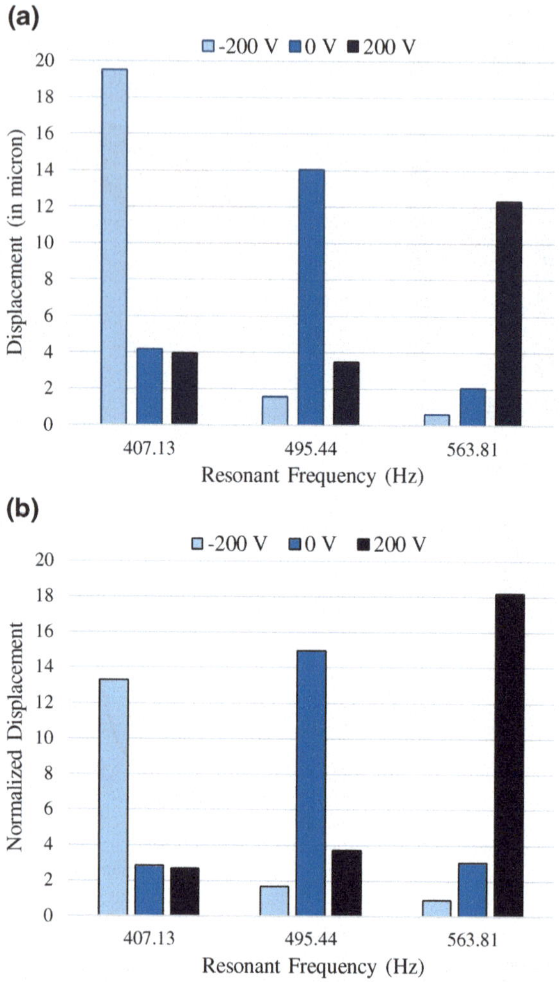

Fig. 6 Centre point displacement of the sensor at different resonant frequencies when −200, 0 and 200 V DC are applied, measured by vibrometer, **a** the centre point amplitude in micron, **b** the normalized amplitude to the input base motion

presented in Fig. 6b. It shows that the higher electric field in the opposite direction of the polarization results in increasing the resonant frequency with greater amplitude.

Sensor Output

As a proof of concept, the designed setup as a vibration sensor is integrated with the electronics. The electrical output of the sensor is filtered via a DC filter circuit, and the generated voltage is measured using a dSPACE A/D converter. Frequency is swept with time to find the resonance region of the film. At resonance, the output signal of sensor forms a superposition of harmonics instead of a pure sinusoidal signal, as shown in Fig. 7. The reason for that is shown in the FFT plots of Fig. 8 measured at three different points on the PVDF film surface. They reveal that harmonic frequencies are excited at the resonance as well as $1\times$ frequency. This would happen because of imperfections in the clamped boundaries and the orthotropic behaviour of PVDF materials that result in uneven distribution of tension along the film. Moreover, the loss of tension along the width of the film because of the negative value of piezoelectric constant e_{32} adds the harmonic vibrations to the fundamental mode. The concave profile of the fundamental mode shape along the width of the film scanned using the vibrometer, Fig. 4c, depicts the reason for uneven distribution of tension. Due to this tension distribution, the fundamental resonant frequency is dominant at resonance while there are the components of its harmonics as well. The tension decreases close to edges and causes the influence of 2x frequency to be increased, as shown in Fig. 8. Thus, we see the combination of harmonies in the output voltage that causes lower amplitude than a pure sinusoidal wave, as shown in Fig. 7. The possible solution is modifying the surface profile of the PVDF film so that it is narrower in the middle to distribute the stress field evenly along the length. It is notable that the electrical output is proportional to the film surface. To avoid the harmonics excitation, we need to weigh up the surface area and electrical output.

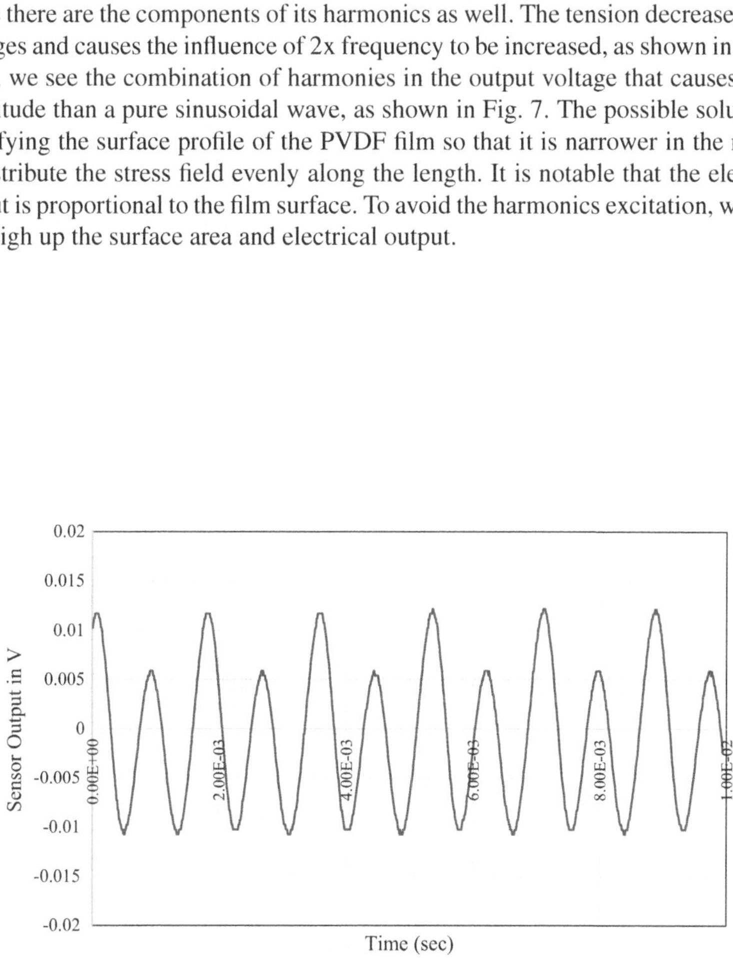

Fig. 7 Electrical output of the vibration sensor measured at 495 Hz

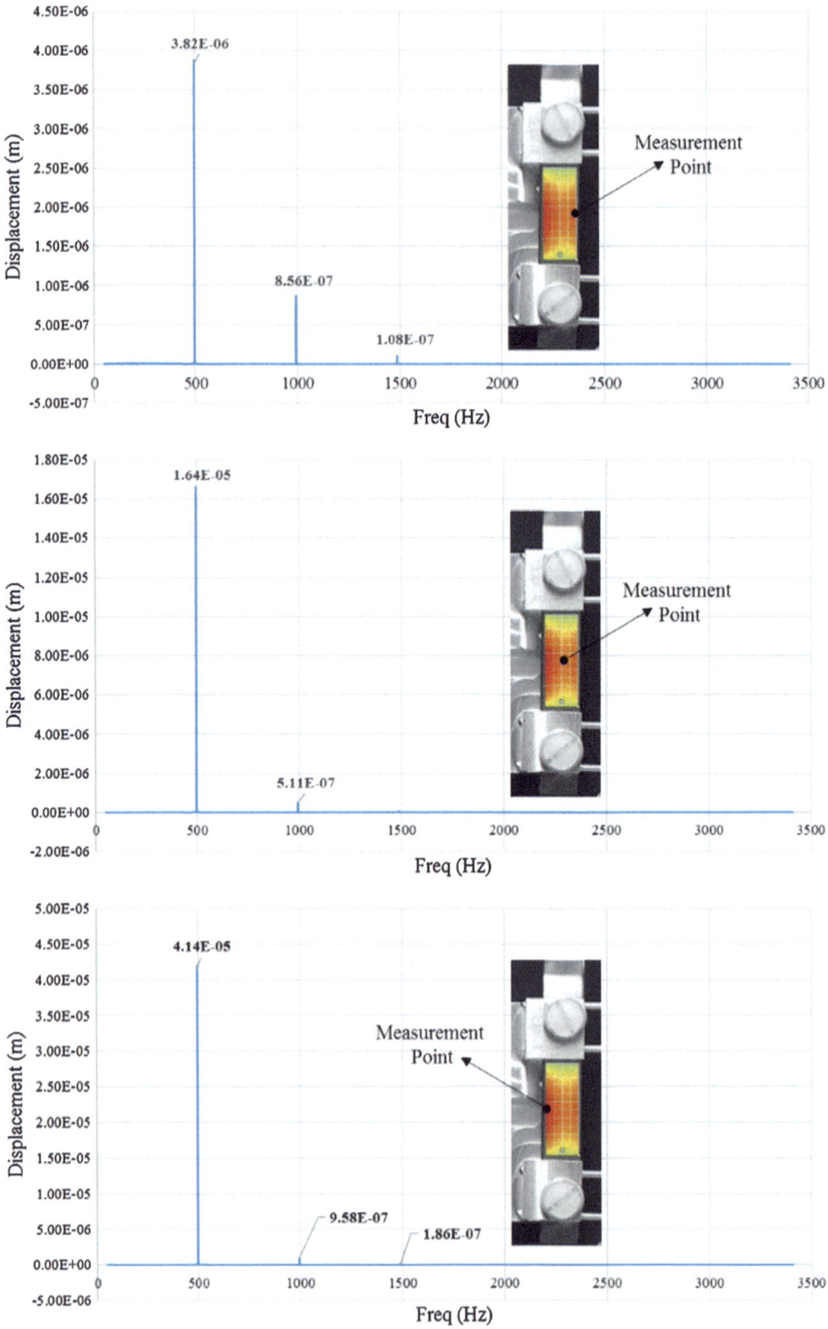

Fig. 8 FFT of vibrating PVDF film at 495.44 Hz measured at three points. It is shown that the second and third harmonics of the fundamental frequency are also excited

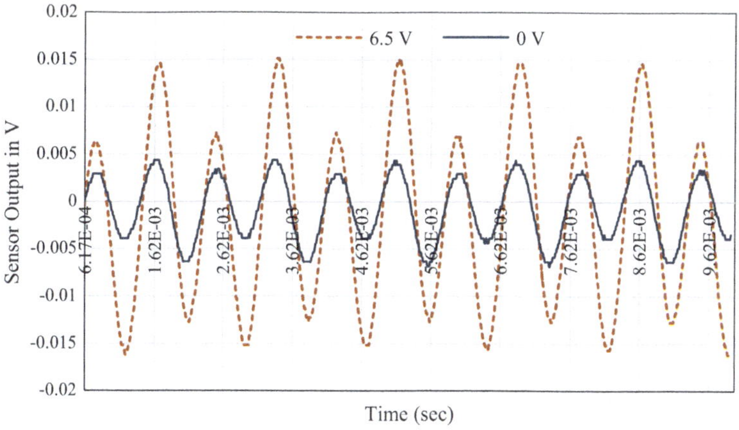

Fig. 9 Electrical outputs of the vibration sensor measured at the biased resonant frequency, i.e. 502 Hz which is the resonant frequency of the film under 6.5 V DC. The solid line signal shows the output of the sensor in the absence of bias DC voltage

In addition to the stress distribution, non-linearity of the PVDF film influence on the output signal. The equivalent electric circuit of a piezoelectric PVDF film is a voltage source in series with a capacitor. Since the dielectric has ferroelectric properties, it exhibits a highly non-linear behaviour and induces harmonics to the output signal.

Measured electrical outputs show that when a positive 6.5 V DC is applied to the sensor, the resonance happens at a higher frequency than the primary resonant frequency. In our case, it increases by 7 Hz and reaches 502 Hz. The output signals of two different states, i.e. 0 and 6.5 V DC, are compared in Fig. 9. This illustrates that the positive bias DC increases the amplitude of the output signal. According to Fig. 9, the PVDF sensor without the bias DC measures a signal with the amplitude 0.01 V peak to peak at 502 Hz, while applying the bias DC to the sensor increases the amplitude by a factor of 3, i.e. 0.03 V peak to peak. Therefore, the bias DC strengthens the output of the sensor and helps to measure the input vibration at frequencies far from the resonance.

FE Simulation

Modelling

In this section, the finite element analysis is conducted to expand knowledge on the dynamics of the proposed sensor and to suggest some modifications to improve the sensor output.

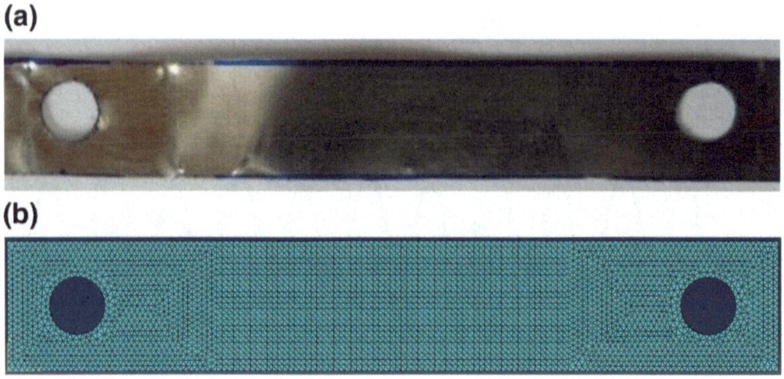

Fig. 10 PVDF film modelling. **a** real film **b** simulated film meshed using tetrahedral element

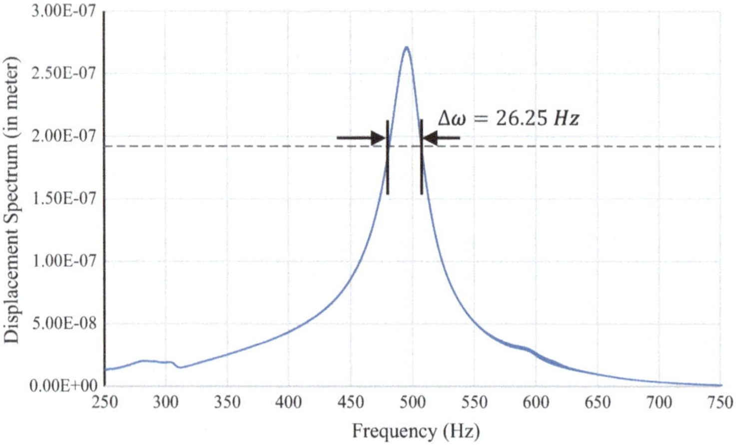

Fig. 11 Half-power bandwidth of the FRF in absence of bias voltage

The stretched 25 mm × 9 mm × 28 μm clamped PVDF ribbon, under 1MPa in-plane tension, is simulated in a FE-simulation software package, ABAQUS®. Figure 10 illustrates the real sensor and the simulated one.

Real sensor possesses a kind of energy dissipation because of damping that reduces the amplitude of vibration and consequently decreases the electrical output of sensors. Thus, it is of importance to calculate the damping of the sensor and apply it to the finite element model. Half-Power Point (HPP) method is employed to approximate the damping ratio from the frequency response of the sensor. Figure 11 shows the bandwidth of the frequency response of the PVDF ribbon measured by vibrometer. The dashed line crosses 0.192 μm that is $1/\sqrt{2}$ times the deflection at the resonant frequency. The damping ratio of the associated mode can be approximated as follows:

$$\zeta = \frac{\Delta\omega}{2\omega_n} \tag{2}$$

Table 1 Damping ratio of the film under different electric field situations

Bias voltage (V)	ω_n(Hz)	$\Delta\omega$(Hz)	ζ
−200	407.13	25.25	0.031
0	495.44	26.25	0.026
200	563.81	29.38	0.026

Table 2 Properties of the simulated mono-oriented PVDF

Property	Value
Young's modulus (E)[a]	3 GPa
Poisson's ratio (ν)[a]	0.34
Density (ρ)[a]	$1780\,kg/m^3$
e_{31}[b]	$0.042\,C/m^2$
e_{32}[b]	$-0.006\,C/m^2$
e_{33}[b]	$-0.033\,C/m^2$
ξ_{33}[a]	106e−12 F/m

[a]Ref. [13]
[b]Ref. [14]

where ω_n represents the resonant frequency of the mode.

Table 1 gives the damping ratios of the PVDF ribbon under the application of different bias voltages.

To obtain the Rayleigh damping coefficients, Eq. 3, which expresses the relation between the fraction of critical damping, Alpha, and Beta, is introduced as [12]:

$$\zeta_n = \frac{1}{2\omega_n}\alpha_M + \frac{\omega_n}{2}\beta_K \tag{3}$$

Therefore, having the damping ratio at two resonant frequencies, we are able two calculate Rayleigh coefficients. Assuming $\zeta = 0.026$ at different resonant frequencies of the PVDF film, α_M and β_K are obtained as 2.75 and 0.0022, respectively.

The PVDF film is modelled as an isotropic material with the electromechanical properties that are summarized in Table 2.

Static Analysis

The static analysis of the model gives us the effect of bias voltage on the stress field inside the PVDF film. The initial stretch determines the fundamental frequency of the film in absence of bias voltage. In Section "Resonant Frequency", the experimental results showed that the application of the bias voltage changes the in-plane tension and consequently results in resonant frequency variations. The static analysis of the

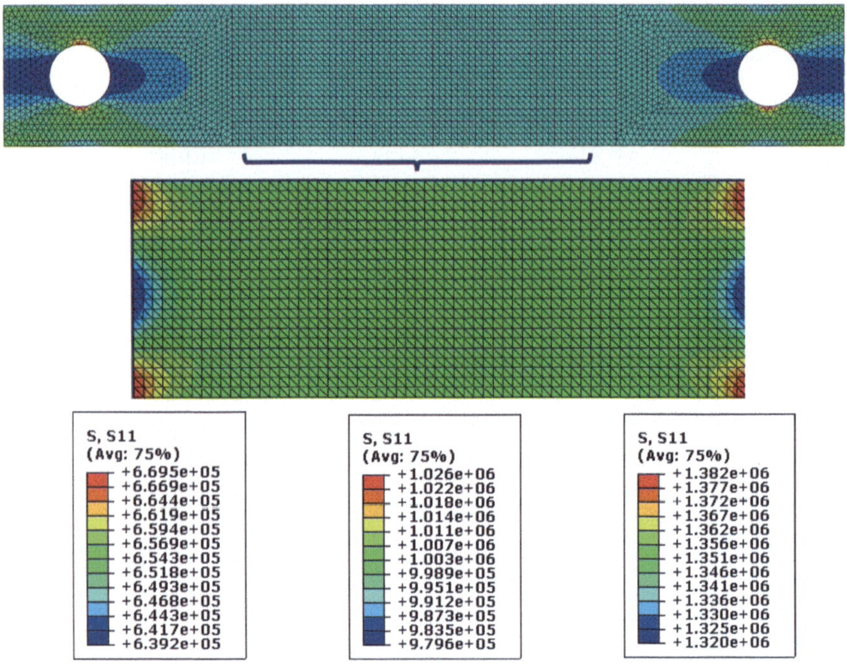

(a) -200 V (b) 0 V DC (c) 200 V

Fig. 12 Finite element stress field in direction 1 (σ_{11}). The effective portion of the whole PVDF ribbon that generates the electric signal induced by the ambient vibration has been focused. **a** −200 V DC decreases the dominant stress field (the green area) to average 0.65 MPa **b** in absence of bias voltage, the dominant stress is equal to the stretch load, i.e. 1 MPa **c** 200 V DC increases the dominant stress field to 1.35 MPa

film in the finite element simulation proves this fact. Stress fields along the stretched direction under different bias DC voltages are illustrated in Fig. 12. The effective area of the sensor is cut from the fixed boundary condition to show the stress contour clearly. The stress contours look same in all conditions, but the stress values are different. It is evident that applying ±200 V DC along the film's thickness releases (Fig. 12a) or increases (Fig. 12c) the in-plane tension by roughly 400 kPa, depending on the direction of the electric field. Therefore, the bias DC would alter the in-plane tension, so that it eases the manipulation of the eigenvalues of the PVDF sensor.

Modal Analysis

This subsection deals with the dynamic simulation of the PVDF ribbon. Modal and steady-state dynamic analysis of the ribbon are performed and the frequency responses at a broadband signal are compared with the experimental data. Firstly,

Table 3 Comparison of resonant frequencies

Bias voltage (V)	Experimental Freq. (Hz)	FE Freq. (Hz)	Error (%)
−200	407.13	406.35	0.19
0	495.44	495.38	0.01
200	563.81	570.31	−1.16

to validate the model, resonant frequencies of the film under bias DC voltages are compared with the experimental results. Performing the modal analysis results in 406.35, 495.38 and 570.31 Hz resonant frequencies associated with the desired mode shape for sensor under the application of −200, 0, and 200 V DC, respectively. They are compared with the experimental results in Table 3 and show their excellent agreement.

Extracting other resonant frequencies of the sensor and their corresponding mode shapes are of interest in designing stage. The mode shape with the highest mean value of maximum displacement produces the strongest electrical output. The interference of modes with the mode shape that releases the maximum electric charge will reduce the output voltage of the sensor. The first four mode shapes of the sensor in different bias voltages are extracted and shown in Fig. 13.

It is interesting that the bias voltage changes the sequence of mode shapes of the sensor as well. The negative voltage creates a smaller resonant frequency than the desired mode for the sensors at 235.90 Hz, as shown in Fig. 13a. This frequency cannot be seen in FFT spectrum since the centre point is located on the nodal line. On the other hand, the second mode shape of the positively biased sensor happens at 1027.2 Hz that is far enough away from the fundamental mode, as shown in Fig. 13c.

Since the second mode shape of the negatively biased sensor and the fundamental mode shapes of the unpowered and positively biased sensor have the maximum mean value of displacement, they give maximum electrical output.

Steady-State Response

Steady-state dynamics gives us the spectrum response of the simulated sensor to the applied external load. To simulate the ambient vibration, a displacement load is applied to the fixed ends of the film. The frequency of applied displacement is swept from 250 to 750 Hz, and the response of the sensor is calculated. Figure 14a, c show that the frequency responses of the model under bias voltage and in absence of the bias voltage are well correlated with the experimental results. Figure 14b, c respectively show the vibration energy dissipation around 400 Hz and 600 Hz in the experimental measurements. As a result, the amplitude of displacement at resonance has been decreased in the experimental data.

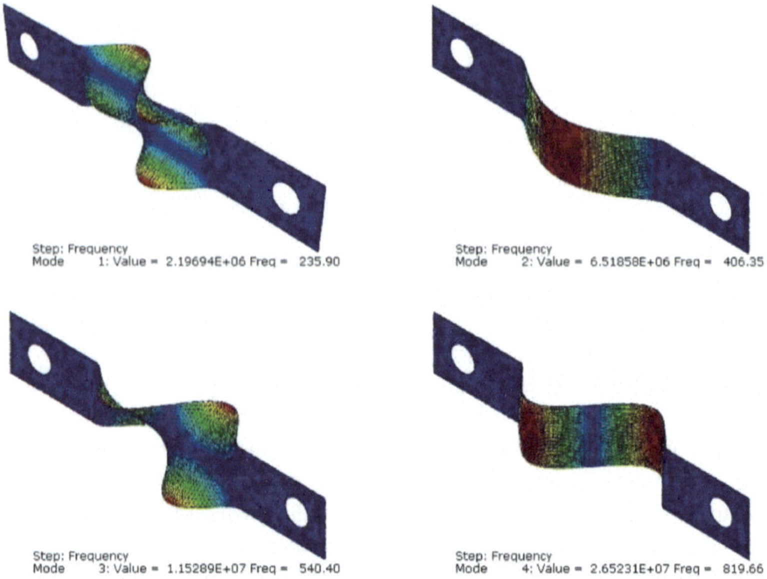

a) The vibration mode shapes of the PVDF sensor under -200 V DC

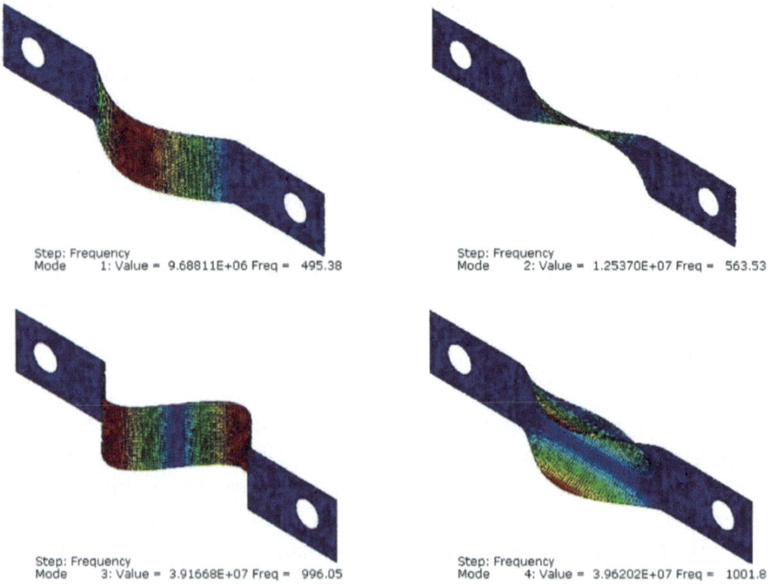

b) The vibration mode shapes of the PVDF sensor in absence of DC voltage

Fig. 13 Mode shapes of the sensor under different bias voltages

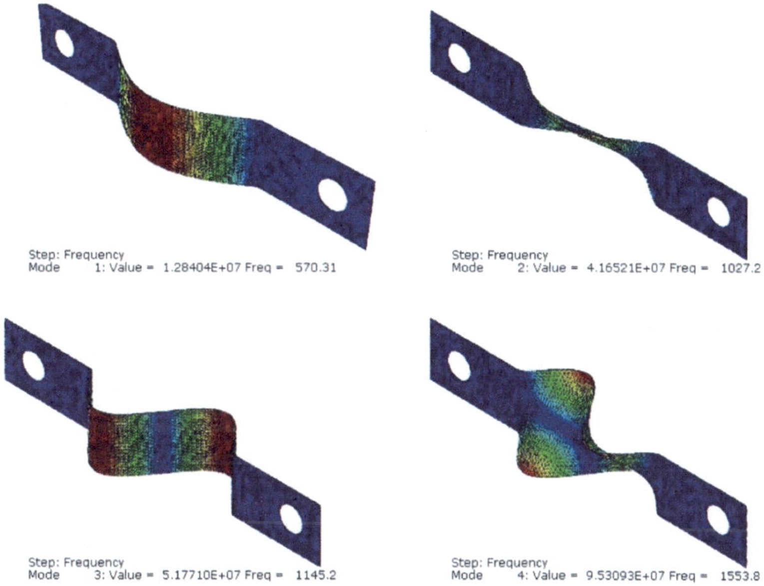

c) The vibration mode shapes of the PVDF sensor under 200 V DC

Fig. 13 (continued)

Figure 14d compares the displacement spectra obtained from the finite element simulation. As it was expected from the experimental measurements, the bias voltage changes the amplitudes of displacement in a way that the resonant amplitude increases at higher frequencies.

Output Voltage

Vibrating piezoelectric films are modelled as a capacitor in parallel with an alternating supply, and the deflection of the film alters the net charge on its surfaces. Here, the electric potential output of the sensor is extracted to study the electrical behaviour of the PVDF film in existence of constant electric field through its thickness. To do so, the frequency of an applied dynamic load to the film is swept from 250 to 800 Hz and the generated electric potential is monitored. Figure 15 represents the FE electric potential spectra. The highest value of the electric potential amplitude is roughly 0.26 mV that happens around the resonant frequencies. Comparison of Fig. 14d and Fig. 15 shows that there is a direct relation between the output voltage of the sensor and the film's deflection at the selected mode. In other words, greater deflection results in greater electric charge on the PVDF surface.

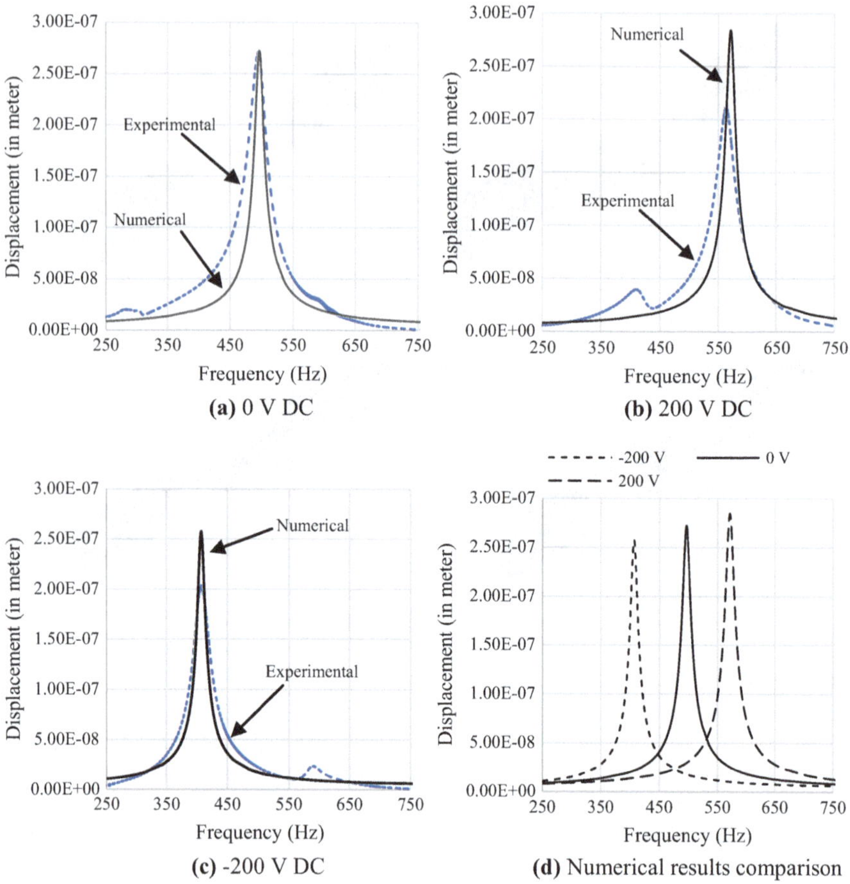

Fig. 14 FRF comparison of experimental and numerical results

Results and Discussion

The equation of motion showed that applying a constant electric potential along the thickness of piezoelectric films would alter their resonant frequencies. Depending on the direction of the electric field, the film would be stretched or contracted. If the piezoelectric film is firmly constrained at the ends, the resonant frequency would be shifted up or down.

A stretched PVDF ribbon prototype as a vibration sensor was designed to monitor its deflection and its electrical output under different input vibrations. Experimental results showed that a ±200 V DC changes the resonant frequency of the film from 407.13 to 563.81 Hz. The sensor was integrated with an electric circuit to measure its output voltage. A low DC voltage, 6.5 V DC, was applied to the sensor under constant vibration input and using an A/D converter the electrical output of the sensor

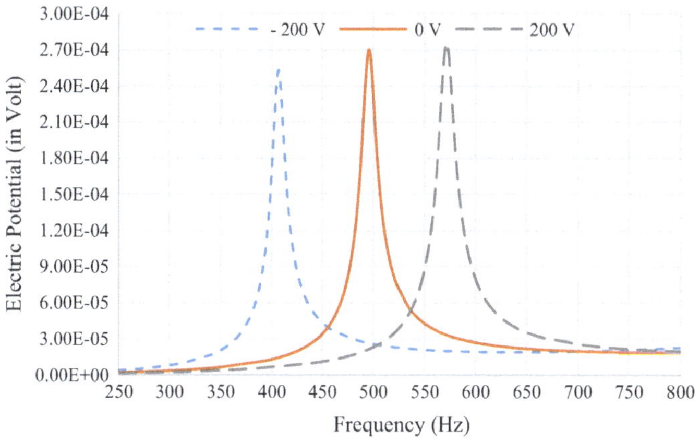

Fig. 15 The FE electric potential spectra under different bias DCs

was monitored. It was shown that the 6.5 V bias DC strengthened the output signal of the sensor by a factor of 3 at the secondary resonance.

To support the theory and experimental results, we simulated the PVDF film in a finite element software and studied the mechanical and electrical behaviour of the film numerically. Finite element results illustrated that applying a DC voltage alters the stress field inside the film and causes variations in vibration characteristics. It was shown that the bias voltage changes the order of mode shapes. Comparison of FE and experimental results show their excellent agreement.

Therefore, shifting the resonance region towards the frequency range of the ambient vibration exerts the highest deflection on the PVDF ribbon, and consequently, the highest sensor output is generated. It would increase the frequency resolution of the sensor to detect and measure the input vibration. In systems with high threshold voltages or when the input vibration is not strong enough to be detected using a stretched PVDF film, varying the value of the bias DC by sweeping would enhance the sensor's output signal. Moreover, using the introduced multi-resonance vibration sensor it is possible to change the resonant frequency of the sensor in real-time to mitigate the nonlinearity and the sources of error in PVDF sensors that alter the resonant frequency. Hysteresis, variation in tension and geometry with time, and influence of temperature on electromechanical properties of sensors are known as sources for this type of phenomena that can be addressed using the proposed sensor.

Finally, the following points are recommended to design and improve the sensor performance:

- The geometry and initial tension of the sensor have to be designed in a way that the fundamental resonant frequency in absence of the bias voltage is close enough to the ambient vibration frequency.

- The flow of stress in the film should be smooth to avoid harmonics in the output signal. To do that, there has to be a trade-off between the sensor shape and the effective surface area to generate the maximum possible output voltage.
- The resonant frequencies of the mounting parts should be far away from the frequency resolution of the sensor.
- The modal analysis of the sensor under bias voltage is needed to be carried out so as to avoid the mode shapes superposition. It is notable that the second resonance is far enough away from the fundamental one under positive bias voltage while the negative bias voltage creates another resonant frequency lower than the desired resonance for the sensor application.

Conclusion

In this chapter, the effect of a bias DC voltage on the output of a PVDF vibration sensor was studied. The theory behind this idea was presented and a prototype sensor was built to measure its vibration characteristics and electrical outputs. The PVDF piezoelectric sensor was also simulated in a finite element package to approximate its mechanical dynamics and electrical behaviour under various constant electric fields. All results demonstrated that we can manipulate the resonant frequency in a wide range and enhance the output signal of the sensor. Moreover, the manipulation of the resonant frequency would make it possible to mitigate the intrinsic negative nonlinear effects that change the resonance regions of the sensor. The proposed sensor could be easily deployed in monitoring the structural vibration of machinery and civil structures without affecting their responses due to its negligible weight and tunable resonant frequency.

References

1. P. Muralt, Ferroelectric thin films for micro-sensors and actuators: a review. J. Micromech. Microeng. **10**, 136 (2000)
2. S.R. Khaled, D. Sameoto, S. Evoy, A review of piezoelectric polymers as functional materials for electromechanical transducers. Smart Mater. Struct. **23**, 033001 (2014)
3. R.A. Steven, A.S. Henry, A review of power harvesting using piezoelectric materials (2003–2006). Smart Mater. Struct. **16**, R1 (2007)
4. M. Hadipour, M. Tahtali, A. Lambert, Dynamics of vibrating piezo-membrane mirror, in *Imaging and Applied Optics 2016*, Heidelberg (2016) p. AOT1C.3
5. M. Hadipour, M. Tahtali, and A. J. Lambert, Vibrating membrane mirror concept for adaptive optics, in *SPIE Astronomical Telescopes+Instrumentation Edinburgh*, United Kingdom (2016)
6. Y. Bormashenko, R. Pogreb, O. Stanevsky, E. Bormashenko, Vibrational spectrum of PVDF and its interpretation, Polym. Test. **23**, 791–796 (2004)
7. F. Yao, C.H. Erol, K.G. Muralidhar, M.S. Geoff, Design, fabrication and testing of piezoelectric polymer PVDF microactuators. Smart Mater. Struct. **15**, S141 (2006)
8. P.N. Karanth, V. Desai, S.M. Kulkarni, Modeling of single and multilayer polyvinylidene fluoride film for micro pump actuation. Microsyst. Technol. **16**, 641–646 (2010)

9. Y. Xin, H. Sun, H. Tian, C. Guo, X. Li, S. Wang et al., The use of polyvinylidene fluoride (PVDF) films as sensors for vibration measurement: a brief review. Ferroelectrics **502**, 28–42 (2016)
10. R. Mahidhar, K. Prakash, R.N. Aswathi, M.V.N. Prasad, S. Sanjiv, Vibration spectrum analyzer using stretched membranes of polymer piezoelectrics for sensor networks. Meas. Sci. Technol. **24**, 055108 (2013)
11. M. Hadipour, M. Tahtali, A.J. Lambert, H.R. Pota, Study on a hybrid model for the vibration control of flexible piezoelectric thin films (under review)
12. A. Preumont, *Vibration Control of Active Structures*, 3rd edn. (Springer, Berlin Heidelberg, 2011)
13. Piezo Film Sensors-Technical Manual (Online). http://www.te.com
14. H. Schewe, Piezoelectricity of uniaxially oriented polyvinylidene fluoride, in *1982 Ultrasonics Symposium* (1982), pp. 519–524

Unique and Unclonable Capacitive Sensors for Nature-Inspired Secure Sensing

C. B. Karuthedath and N. Schwesinger

Abstract With growing popularity of Internet of things (IoT) more and more sensors are getting connected to public network. Identification of sensors is a permanently growing concern. The conventional way of adding a cryptophytic module along with a sensor is vulnerable to direct attacks; unauthorized sensor replacement, malicious signal insertion, etc. are possible. A new sensing architecture inspired from nature's sensing systems can prevent such direct attacks. The idea is to develop sensors with unique and unclonable characteristics and enroll the unique sensor characteristic into the verifier, for instance a cryptography module. As the verifier is tuned to the specific characteristic of a particular sensor, direct attacks are difficult. In order to develop such highly secure sensing modules, unique and unclonable sensors that are similar to sensors found in nature are required. Unique and unclonable sensors respond to specific mechanical parameter and each sensor output is different from others. Random structural variations present in the sensor structures are sources of the uniqueness. As random structural variation is difficult to measure, model or duplicate, such sensors are unclonable. Such sensors can be identified from their output. Design, modeling, simulation, fabrication, testing and application of such unique and unclonable sensors are discussed in this chapter.

Security of Sensors

Information from the sensors is of critical importance for many applications. For instance, armed forces deploy MEMS sensors to track the use of their weapons [1]. Additionally, security agencies, including the UN (United Nations), need sensors to monitor various international research facilities, whilst in cars, ABS (Anti-lock Braking Systems) control units require information from different sensors, to decide on the breaking torque factor [2]. In such situations, as the data collected from the sensors are used to make a decision on further crucial actions, the trustworthiness of

C. B. Karuthedath · N. Schwesinger (✉)
Technical University of Munich, Munich, Germany
e-mail: schwesinger@tum.de

S. C. Mukhopadhyay et al. (eds.), *Modern Sensing Technologies*, Smart Sensors, Measurement and Instrumentation 29,
https://doi.org/10.1007/978-3-319-99540-3_15

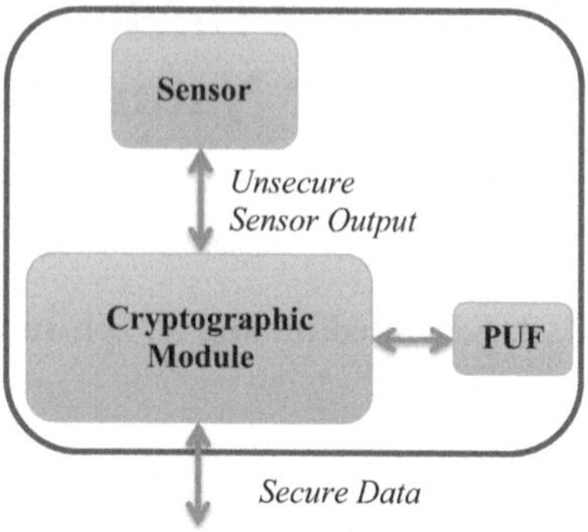

Fig. 1 Block representation of a conventional secure sensing module

the sensor and its data is a major concern. The verifier or the corresponding authorities must guarantee that the information is from a specific sensor and not spoofed.

The common way to provide security for a sensing system is to use an additional cryptographic module along with the sensor, as shown in Fig. 1. The cryptographic module encrypts the sensor output and sends the data to the verifier. The verifier authenticates and decrypts the sensor data. Classical cryptographic techniques developed for computers, cannot be implemented in sensor systems, due to their resource limitations. Fortunately, it has been shown that PUF (Physical Unclonable Function)-based security architectures are well-suited to low-cost, lightweight applications, including sensor systems [3].

The Physical Unclonable Function (PUF) is a simple, low-cost and superior alternative to memory-based identifiers. PUFs are function based on physical characteristics which are unique, unpredictable and impossible to duplicate. Similar to biometrics, a PUF uses random physical variations inside the device as a 'signature' for identification. These variations can either be intrinsic within the devices or intentionally introduced during the fabrication processes [4]. A PUF exhibits challenge-response behaviour, such that when it is excited with an external stimulus, called a 'challenge' (*CH*) and it generates an output, called a 'response' (*RS*). Due to random physical variations, the *RS* of each PUF is unique for a specific *CH*. The unique challenge-response pair (*CRP*) of a PUF can be used as an identifier. PUF based cryptographic modules (in Fig. 1) utilize the CRPs to generate encryption/decryption keys. Since PUFs base on structural uniqueness, they can offer a high level of security without memory and processing overhead. PUFs can be implemented in various ways. Delay-based PUFs [5], memory-based PUFs [6] and transistor PUFs [7] utilize uncertainties in the silicon fabrication processes for generating unique CRPs. The optical PUF discussed in [8] consists of a transparent layer filled with randomly distributed light-scattering particles. The refraction pattern from this layer is unique and unclonable. The coating PUF reported in [9] uses a layer with random dielectric patterns to generate a unique and unclonable capacitance output.

In the security architecture shown in Fig. 1, the cryptographic module is separate. Therefore, the system is vulnerable to direct sensor attacks, such as sensor tampering and sensor substitution. For instance, in a remote video surveillance sys-

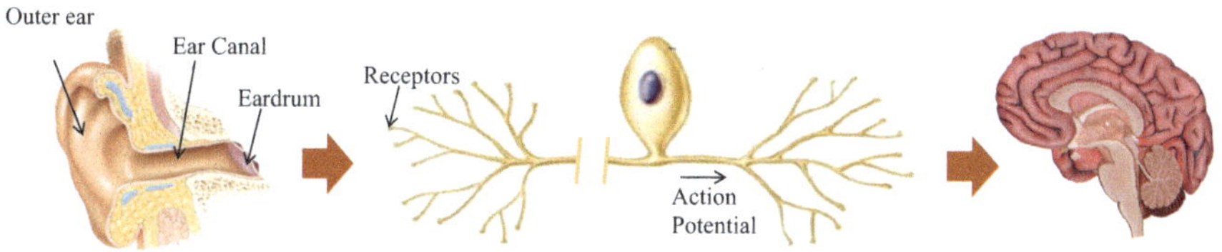

Fig. 2 Sensing in nature

tem, an attacker can disconnect a Closed Circuit Television camera and insert any video signal into the cryptographic module. In [10], the authors discussed methods employed to insert bogus information into the sensors, using intentional electromagnetic interference. Such attacks can inhibit pacing or induce defibrillator shocks in Cardiac Implantable Electrical Devices. Measurements from magnetic rotational sensors used by Antilock Braking Systems can be easily corrupted using a simple external magnetic field, which in turn may cause life-threatening situations [2]. The security model shown in Fig. 1 does not have the ability to detect or prevent this kind of direct sensor attack. In network connected sensors, the attacker can use it as a weak entry point and gain some form of control over the system by manipulating the measurand.

Sensing Systems in Nature

In nature, sensing systems are capable of understanding whether the information is coming from the intended sensor or not. A typical natural sensing system is shown in Fig. 2. It consists of an auxiliary structural element, such as an ear, a cornea, a hair cell structure, etc., designed to accept the external stimulus, which, when collected, is passed to the sensory receptors. These sensory receptors convert the stimulus into electro-chemical signals. Neurons are designed to accept these electrochemical signals and carry them as an 'action potential' to the brain, which in turn distinguishes the type of sensory signal and the intensity of the stimulus from the identity of the neuron and frequency of the signal. After processing, the brain compares the received signal with patterns stored in the memory, to create the sensor perception.

Unlike conventional sensors, sensors in nature have unique characteristics. Even though the overall characteristics of sensors are similar, the exact characteristics of each sensor are different from others. For instance, consider auditory sensing systems in humans. The overall structure and functionalities of auditory sensor structures are similar. However, the exact shape and structure of the outer ear is unique for every person [11]. Ear canal morphologies are also different. The ways in which the outer ear and the ear canal modify sound waves are different for every person, resulting in unique eardrum vibrations. Hence, every person perceives the same sound signal in a slightly different manner. Similarly, the shape and texture of the tongue are also

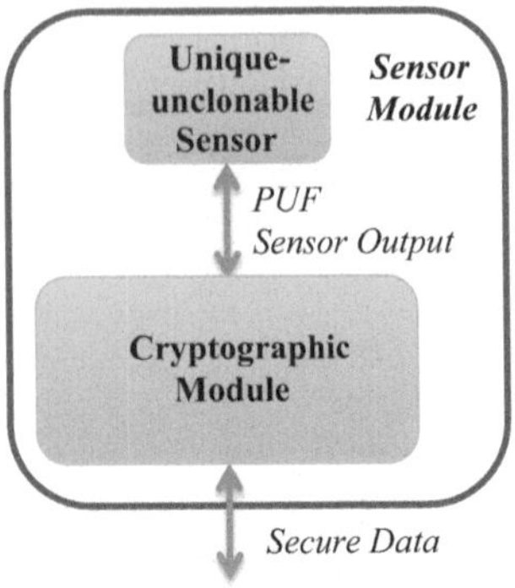

Fig. 3 Block representation of nature inspired secure sensing module

unique [12]. Even within a sensor array found in nature, characteristics of each sensor elements differ. One typical example is hair cell sensors in lateral lines. Almost all fishes and aquatic creatures have hair cell sensors in a line called 'lateral line'. These hair cell sensors detect pressure, tilt and acceleration, which in turn enable the fishes to navigate and find prey in noisy waters. The characteristics of each hair cell element in the lateral line are also unique. One advantage of combining sensors with different characteristics is to obtain high sensitivity over a wide bandwidth. The uniqueness properties of natural sensors results from uncontrolled evolvement processes, which introduces random minor variations in the sensor structures. This makes the sensor characteristics difficult to predict, model and duplicate; hence, sensors in nature are not only unique, but also unclonable.

In natural sensing systems, the brain is tuned to the unique properties of the sensor. Hypothetically, if a sensor is replaced, it badly affects the sensing ability. For instance, in [13] it is showed that the replacement of outer ear affects the hearing and sound localization ability of human. However, through training, brain gets tuned to the new ear characteristics, and the abilities get regained.

Nature Inspired Sensor Security

In nature, it is impossible to replace the sensors without the 'authorization' of the brain. The sensing system need to go through a training process in order to enroll the unique characteristics of the new sensor into the brain. Sensing architecture similar to nature can be employed to prevent direct sensor attacks. Such a secure sensing is shown in Fig. 3. It consists of a nature-like unique and unclonable sensor. The overall properties of the sensors are similar however, the exact characteristics of each sensor is different from others. The unique characteristics of each sensor are enrolled with a cryptographic module. Here, from the sensor output, the cryptographic module determines whether the information coming from the authorized sensor. Hence, direct attacks are difficult.

This concept is contradictory to conventional Silicon based sensing systems that rely on highly precise sensors. Precision of sensors is important because if the characteristics of the sensors are the same, the mapping function that maps the sensor's output to the measured quantity will also be the same. Therefore, similar sensing systems can be mass-produced without the need of calibration steps. Furthermore, it gives the flexibility of replacing sensors and associated circuitry without any calibration. It is possible to build highly sensitive sensing systems using sensors with a high uniqueness, too. Here, each sensor must be separately calibrated/enrolled. For any identification process, an enrollment of the unique identifier is unavoidable. For example, in fingerprint identification, fingerprint of all concerned parties need to be enrolled with the system. In case of unique and unclonable sensors, the calibration function is the unique identifier. Calibration is same as enrollment and hence it is not an additional overhead.

Unique and Unclonable Capacitive Sensor (UUCS)

A nature-like unique and unclonable sensor is the basis of the secure sensing architecture discussed above. While conventional sensor technology is trying to produce precise sensors, unique and unclonable sensors are not available on market. Even though engineers copied nature's sensors, both functionally and structurally, to best of our knowledge unique and unclonable sensors have not been developed. Sensors that are somewhat similar to unique and unclonable sensors are PUF sensors. It is a relatively new concept. All reported PUF sensors fall into two different categories. The first category exploits random variations in several existing sensors, in order to derive PUF characteristics. In this scheme (shown in Fig. 4a), sensor identification and/or key generation are carried out in a separate *CH-RS* cycle, before or along with the measurement. The sensor output ($f(m)$) is proportional to the measurand (m). Here, the function $f(m)$ is the same for each sensor (not device-specific), only the *CH-RS* is unique. For instance, variations of impulse response and inherent offset values of a commercially available accelerometer sensor are used as PUF [14]. The verifier/cryptographic module measures the offset value at 0 g and/or impulse response and derives the cryptographic key. Similarly, the possibility of using frequency modes, capacitances and quadrature signals of a gyroscope for generating cryptographic keys is explored in [15]. The second category uses conventional PUFs as sensors. Here, the vulnerability of PUFs to parameters such as temperature, voltage and pressure, which is unwanted, is exploited to measure these parameters. The output of the PUF sensor ($f(m, CH)$) depends on the CH and the measurand m. Such PUF sensor implementations can be employed for the secure transfer of information over an untrusted channel, even without cryptographic modules [16]. In [17, 18], voltage sensors exploiting the unwanted effects of voltage on a ring oscillator (RO)-PUF and a glitch-PUF are reported. Similarly, a PUF temperature sensor is prototyped, using an exclusive OR gate for the secure transfer of information over an untrusted channel [16]. The pressure sensitivity of an electrostatically actuated

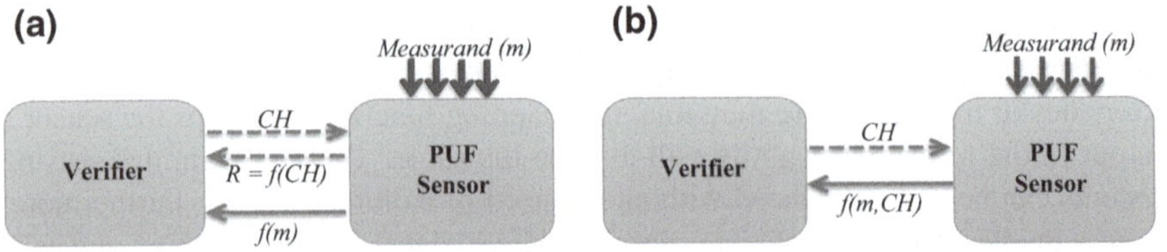

Fig. 4 Block diagram showing two categories of PUF sensors: **a** an existing sensor is used as a PUF and **b** a conventional PUF is used as a sensor

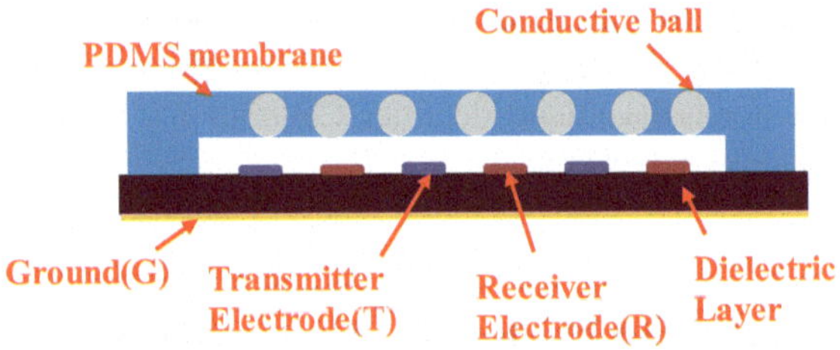

Fig. 5 First generation unique and unclonable capacitive sensor

micro-electro-mechanical relay based NOT gate of RO PUFs is used for secure pressure measurements [19]. Another approach involves developing a dedicated, unique, and unclonable sensor and the integration with a conventional PUF to form a PUF sensor [20]. In this case, sensor design and fabrication process can be optimized to gain maximum variability. Output of these PUF sensors are unique and unclonable. However unlike sensors in nature, PUF sensors always require an additional challenge (CH). In unique and unclonable sensors measurand is the challenge.

Nature like unique and unclonable capacitive sensors are first reported in previous works [21, 22]. The sensor has similar functions like hair cell sensors. It responds to pressure, tilt or acceleration and provides a unique and unclonable capacitance output. The sensors structure is shown in the Fig. 5. It consists of a PDMS (Poly-Di-Methyl-Siloxane) membrane filled with randomly distributed conductive balls. The membrane deflects for tilt, pressure or acceleration. A capacitive electrode structure converts the deflection of the membrane into a change of the capacitance. The random distribution of conductive balls modifies the deflection characteristic of the membrane and hence the electrical field pattern from the electrode structure of each sensor in a unique way, which results in a unique output capacitance for each sensor. In this design, only conductive balls actively participate in sensing, covering 30% of the active membrane volume. The PDMS constitutes the remaining 70% of the volume. As the dielectric constant (ϵ_r) of PDMS is low ($\epsilon_r = 2.75$), 70% of the active membrane volume has only negligible effects. Hence, the surface area required to achieve a high sensitivity is large for many space-limited application. The degree of uniqueness is also low.

The design of a miniaturized capacitive sensor with unique and unclonable characteristic is presented in this section. The sensor uses a Conductive PDMS (CPDMS) membrane with conductive balls on top. Here, the whole region of the membrane

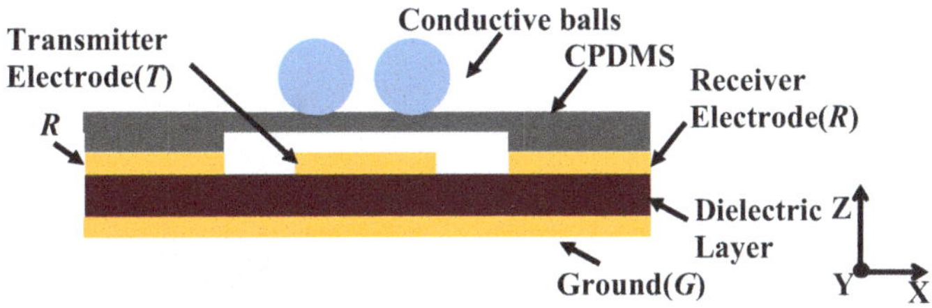

Fig. 6 Unique and unclonable capacitive sensor design

actively contributes to the sensing process. Due to that, the sensitivity increases along with a reduction of the overall size.

Sensor Structure

The detailed 2D structure of the miniaturized capacitive sensor is shown in Fig. 6. A CPDMS membrane with conductive balls on top represents the active sensing element. The balls are randomly distributed on the membrane. The sensor has three electrodes: A transmitter (T), a receiver (R) and ground (G). The membrane is attached on top of the electrode structure and electrically connected to the receiver R. As the membrane is conductive, it has the same potential as the receiver R. The CPDMS membrane above the transmitter *T* is the active area. The transmitter *T* and the CPDMS membrane form a parallel plate capacitor. External forces such as pressure, tilt or acceleration deflect the CPDMS membrane. Owing to random ball distribution, mass distribution is different on each sensor, and therefore the deflection of the CPDMS membrane for the same force is different for each sensor. This deflection of the CPDMS membrane varies in its distance from the transmitter T, and results finally in unique capacitance output.

The electrical equivalent model of the proposed sensor is shown in Fig. 7. C_{TP} represents the capacitance between terminal *T* and the CPDMS membrane, when the CPDMS membrane is in the non-deflected position. The deflection of the CPDMS membrane changes the value of C_{TP}. The movement of the CPDMS membrane towards *T* decreases the distance between them, and C_{TP} increases. Similarly, the movement of the CPDMS membrane away from *T* decreases C_{TP}. Therefore, the capacitance between *T* and the CPDMS membrane can be represented as $C_{TP} \pm \Delta C_{TP}$. ΔC_{TP} is the change of the capacitance due to tilt. The other capacitors C_{TR}, C_{ST} and C_{RG} represent capacitances between the terminals *T* and *R*, *T* and *G*, as well as *R* and *G*, respectively. These capacitors are constant. The membrane is fabricated using CPDMS, which is not a pure conductor and has a certain electrical resistance. The resistor R_P represents the resistance of the CPDMS membrane. Here, sensor capacitance is measured with help of the receiver current i_R. The capacitors C_{TG} and C_{ST} are connected across V_S. Hence, the receiver current i_R is independent

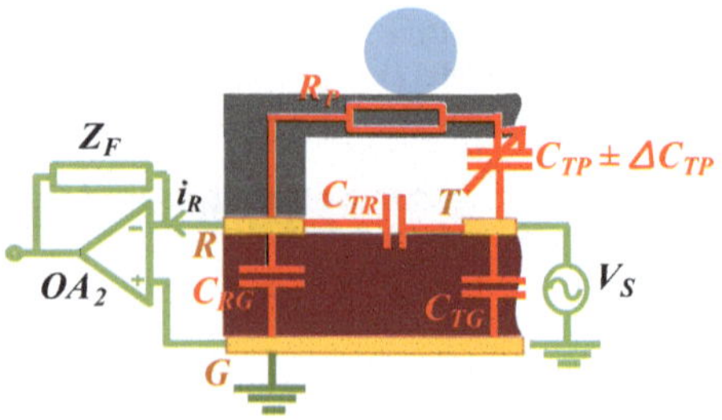

Fig. 7 Electrical equivalent diagram of unique and unclonable capacitive sensor

of these capacitors. To measure i_R, R is connected to the inverting terminal of the current-to-voltage converter (built around OA_2). As the inverting terminal of the converter has the floating ground potential, G is grounded and no current flows through C_{RG}. Hence, in such an electrode set-up, the sensor capacitance is (R_P neglected):

$$C_S = C_{TR} + C_{TP} \pm \Delta C_{TP} \tag{1}$$

where C_{TR} and C_{TP} are fixed components and ΔC_{TP} is the variable component.

Analytical Model

It is not easy to develop the analytical model of the variable capacitance $C_{TP} \pm \Delta C_{TP}$ if the balls are randomly distributed. Therefore, in order to develop the model, it is assumed that the mass is equally distributed on the membrane. Tilt along the X-axis induces forces $F_g \cos\theta$ and $F_g \sin\theta$ along the Z-axis and the Y-axis respectively. Hereby, F_g is the force due to gravity and θ is the tilt angle. These forces try to deflect the membrane in the Z- and Y-direction. As the deflection in the Y- (and also X-) direction is restricted, the CPDMS membrane moves parallel to T. The change of capacitance due to such a membrane deflection is small and negligible. The sensor is only sensitive to deflections perpendicular to the electrode structure (Z-axis). Membrane deflections along the Z-axis change the distance between T and the CPDMS membrane, and also the capacitance. The capacitance between T and the CPDMS membrane can be calculated as:

$$C_{TP} + \Delta C_{TP} = \varepsilon_0 \iint \frac{dS}{d_0 - w(S)} \tag{2}$$

where ε_0 the permittivity of is free space, d_0 is the distance between T and the CPDMS membrane and $w(S)$ is deflection of the CPDMS membrane surface S along the Z-axis.

For a circular structural element, the membrane deflection at a distance r from the center can be expressed as

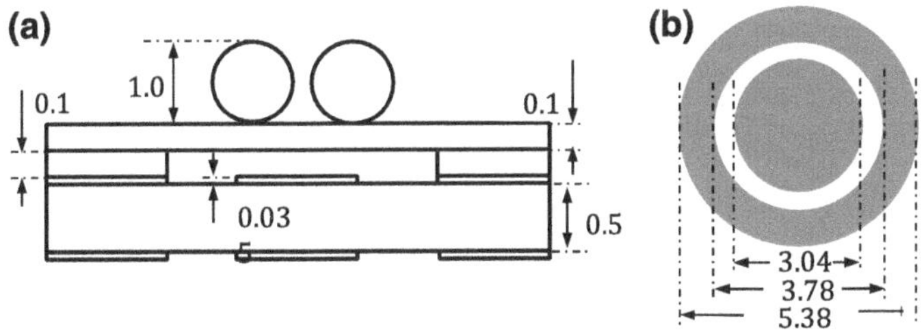

Fig. 8 Sensor dimensions: **a** sensor structure, **b** electrode

$$w(r) = w_0\left(1 - \frac{r^2}{R_m^2}\right)^2 \tag{3}$$

where R_m is the radius of the circular membrane [23]. w_0—the deflection at the center ($r = 0$) of the circular membrane is given as

$$w_0 = \frac{3R_m^4(1-\mu^2)}{16E_Y t^3 A} F_g \cos\theta \tag{4}$$

μ is the Poisson's ratio, E_Y is the Young's modulus, t is the thickness and A is the area of the membrane.

$$C_{TP} + \Delta C_{TP} = \varepsilon_0 \int \frac{rdr}{d_0 - w_0\left(1 - \frac{r^2}{R_m^2}\right)^2}$$

$$C_{TP} + \Delta C_{TP} = \left\{ \begin{array}{l} \frac{\pi \varepsilon_0 R_m^2 arctanh\sqrt{\frac{w_0}{d_0}}}{\sqrt{w_0 d_0}},\ if\ w_0 \geq 0, \\ \frac{\pi \varepsilon_0 R_m^2 arctan\sqrt{\frac{w_0}{d_0}}}{\sqrt{w_0 d_0}},\ otherwise \end{array} \right\} \tag{5}$$

By substituting (3) in (2) one will get an expression for the capacity values C_{TP}.

Simulation Studies

The sensor design was verified using the finite element analysis software package COMSOL Multiphysics. Sensor dimensions are shown in Fig. 8. Selected materials, domain and boundary conditions are illustrated in Table 1. As the material CPDMS is not available in the COMSOL library, PDMS was selected and the electrical (conductivity) and mechanical (Young's modulus) properties were correspondingly modified.

Table 1 FEA simulation details

Description	Material	Domain	Boundary
Active element	CPDMS	Linear elastic dielectric, gravity	
Distance holder	CPDMS	Linear elastic dielectric gravity	Fixed constraint
Electrodes	Copper		Terminal
Balls	Steel	Linear elastic dielectric gravity	Floating potential
Dielectric layer	FR4	Linear elastic dielectric	Fixed constraint

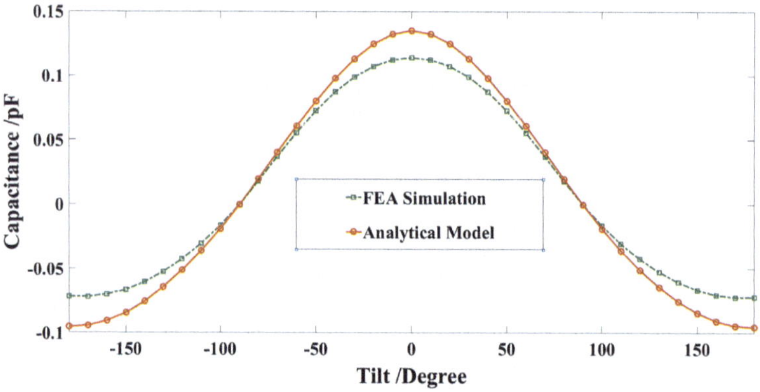

Fig. 9 Variable capacitance (ΔC_{TP}) recorded from FEA model and analytical model for different values of tilt

Initially, to validate the analytical model, three-dimensional structures of the sensor were simulated. In this case, T was set to 1 V, R to 0 V, G was grounded, and the mass on the membrane is evenly distributed. A tilt from $-180°$ to $+180°$ was simulated with a variable gravity load applied to the membrane. At $\pm 90°$, the deflection of the CPDMS membrane is 0 µm and the variable capacitance ΔC_{TP} [in Eq. (1)] is zero. The sensor capacitance (C_S) has a value of 0.834 pF at ±90. With an increase of tilt, the deflection along the positive Z-axis increases, with other words the distance between T and the CPDMS membrane increases. This decreases the sensor capacitance. Similarly, a tilt angle below 90° increases the deflection of the CPDMS membrane into the direction of the negative Z-axis and this increases the sensor capacitance. The variable capacities (ΔC_{TP}) simulated for different tilt angles are shown in Fig. 9. For comparison, ΔC_{TP} calculated from the analytical model is also plotted in the Fig. 9. The results of both models are similar. Slight variations are possibly caused by the resistance of the CPDMS-membrane R_p and the deflection of the membrane in directions perpendicular to the sensitive axis are neglected in the analytical model.

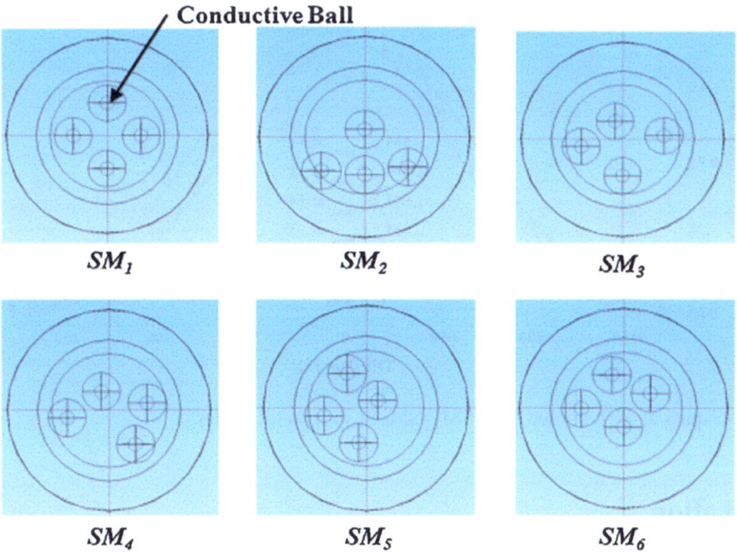

Fig. 10 Ball arrangements in six sensor models (SMs)

The effect of random ball distribution on the capacitance of the sensors was verified, too. Six sensor models SM_1–SM_6, with different conductive ball arrangements, were considered. The arrangements of steel balls are illustrated in Fig. 10. For a tilt from $-180°$ to $+180°$ the capacitance behavior was simulated. Simulated values of total sensor capacitance (C_S) from SM_1 to SM_6 for various tilt angles are plotted in Fig. 11. Owing to the different mass distributions, the deflection characteristic of each membrane is different, and so the different models showed unique output characteristics. In Fig. 9, as the mass on the membrane is evenly distributed, the maximum deflection of the membrane was at 0° and $\pm 180°$. The sensors showed maximum and minimum capacitance at 0° and $\pm 180°$, respectively. In SM_1–SM_6, the random arrangement of the balls modified the weight distribution of the membrane. Therefore, the angle at which the sensors showed maximum (and minimum) capacitance varied, which actually improves the uniqueness of the sensors. The tilt at which each SM showed maximum capacitance was evaluated (using a curve fitting technique) and the value is shown in Fig. 11. The simulation results show that the random distribution of conductive balls alone can introduce a high uniqueness in the sensor properties.

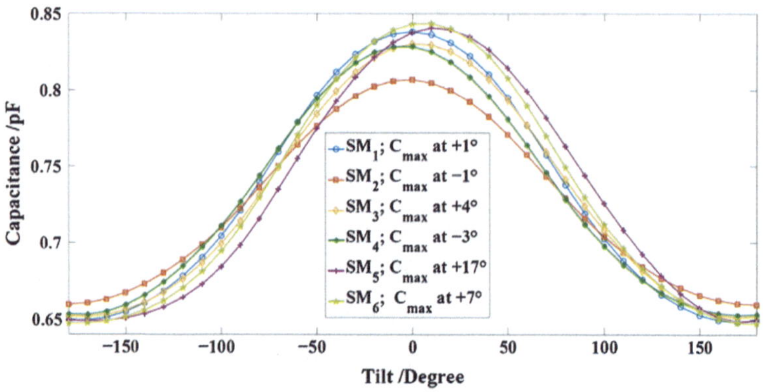

Fig. 11 Capacitances recorded from SM_1–SM_6 at various tilt angles

Fabrication

Conductive PDMS (CPDMS) Membranes

The sensor shown above requires a flexible and conductive structure. Silicon based flexible elastomer Poly-Di-Methyl-Siloxane (PDMS) is a suitable choice. However, intrinsically PDMS is not electrically conductive. The conductivity of PDMS can be achieved by adding conductive fillers such as carbon black (CB) particles, silver (Ag) particles or multi-walled carbon nanotubes (MWCNTs) [24, 25]. Compared to Ag and MWCNTs, carbon black (CB) filler is a low-cost choice.

Conductive PDMS (CPDMS) composites for the fabrication of membranes were prepared by mixing CB fillers in a PDMS matrix. Therefore, a soft PDMS Alpa-Sil Classic from Alpina was used [26]. Alpa-Sil Classic has a Young's modulus of 160 kPa, which is much lower than that of the most commonly used Sylgard 184 (about 2 MPa) [27]. Carbon black, Vulcan XC72 from Cabot [28], was used as filler. It has a particle size of about 50 nm. In order to improve the dispersion of CB particles inside PDMS, Methanol was used as solvent.

The steps for the preparation of CB-Methanol-PDMS composite were as follows:

1. The required amount of CB fillers was placed in a vessel. Methanol (three times the volume of CB) was added to the CB fillers. The mixture was ultra-sonicated for 15 min and then stirred for 1 h with the help of a magnetic stirrer. The CB fillers were (partially) dissolved in methanol.
2. The PDMS was immediately added to the CB-Methanol mixture. The composite was stirred for 30 min. As the composite was getting highly viscous, hand-mixing was carried out. During the process, the CB-PDMS base mixture started to settle on the bottom of the vessel and the methanol floated above the mixture. Excess methanol was poured-off the CB-PDMS mixture.

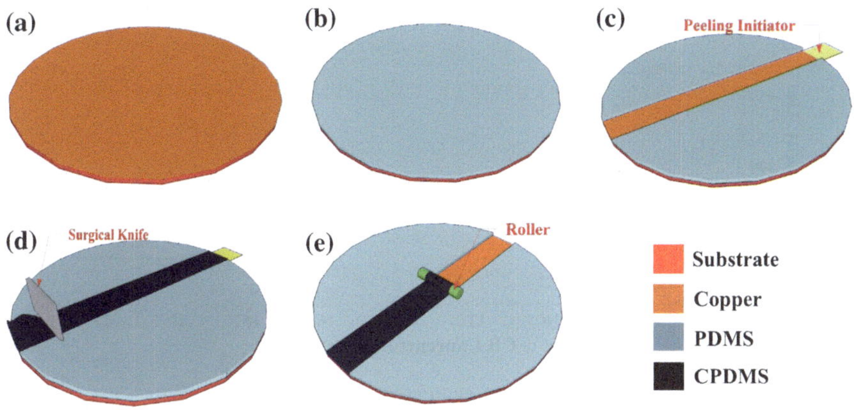

Fig. 12 Fabrication process for soft, thin CPDMS membranes. **a** Template for membrane fabrication. **b** and **c** Patterning of the master mold. **d** Fabrication of CPDMS membranes. **e** Peeling off the fabricated membrane with the help of a peeling initiator and a roller

3. The required amount of Curing Agent (CA) (PDMS:CA in a ratio of 10:1) was added to the CB-Methanol-PDMS mixture and mixed for 5 min.

CPDMS composites prepared using the methods discussed above were highly viscous. Hence, they could not be spin-coated to fabricate thin membranes. Steps for the fabricating of thin CPDMS membrane are illustrated in Fig. 12. A printed circuit board (PCB), cut in the shape of a 4 in. wafer shown in Fig. 12a, was used as substrate for the fabrication of CPDMS membranes. As result of the poor adhesion of PDMS to metal, finally fabricated CPDMS membranes can be easily peeled off from PCB's. Initially, a pure PDMS (Sylgard 184) was prepared by mixing the base component and the CA at a ratio of 10:1 and kept for 30 min to remove air bubbles. Next, the Sylgard mixture was deposited on the PCB substrate (Fig. 12b) by spin coating (800 rpm, 15 s). That PDMS coated substrate was placed in an oven for curing (80 °C, 30 min). After curing, a rectangular section was removed, to create a master mold for the fabrication of CPDMS membranes (Fig. 12c). A piece of rectangular Scotch Tape was attached to the edge of the substrate, as helping tool for the peel off the CPDMS membrane after fabrication. The CPDMS composite was deposited into the mold and spread across (Fig. 12d). The whole arrangement was kept in an oven for curing (70 °C, 30 min's). During this process, methanol evaporated and the CB filled PDMS cross-linked. After curing, the CPDMS membrane was removed with the help of a roller of PDMS (Fig. 12d). The roller reduces the force on the membrane and prevents the membrane from tearing during peeling [29]. Additionally, it prevents wrinkling and helps to store the membrane without contamination.

From the electrical and analytical models, it is known the sensitivity of such sensors depends on the Young's modulus (E_Y) and the resistivity (ρ) of the CPDMS membrane. In CPDMS, these properties vary with the CB concentration. Membranes with CB concentrations from 3-wt% to 17-wt% were fabricated and characterized.

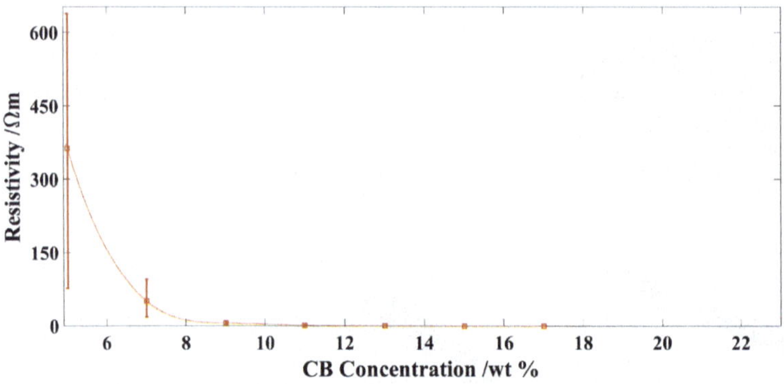

Fig. 13 Resistivity of CPDMS membranes at different CB concentrations

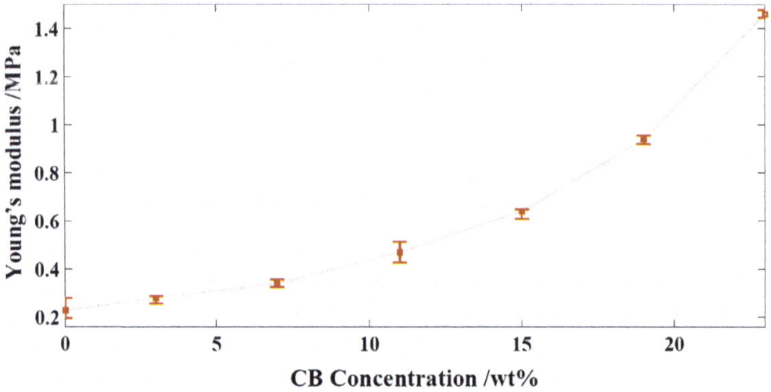

Fig. 14 Young's modulus of CPDMS membranes at different CB concentrations

(ρ)- and (E_Y)-values depending on the CB-concentration are shown in Figs. 13 and 14, respectively. The desirable characteristics of the sensors are minimal values of E_Y and ρ. Even though, ρ decreases with an increase of CB, E_Y increases. After substituting E_Y and ρ values in the sensor model, 5-wt% was found as optimum value for the sensors.

Sensor Structure

The fabrication of the sensor structure is illustrated in Fig. 15. Each sensor requires two CPDMS membranes—one as the active element and the other one as the distance holder. On the active element, four conductive balls are randomly distributed. Prior to distribution, the balls were dipped in PDMS, acting as adhesive to attach the balls to the CPDMS membrane. A mask was used to limit the ball distribution region within the active area of the membrane. A circular hole was punched through the distance

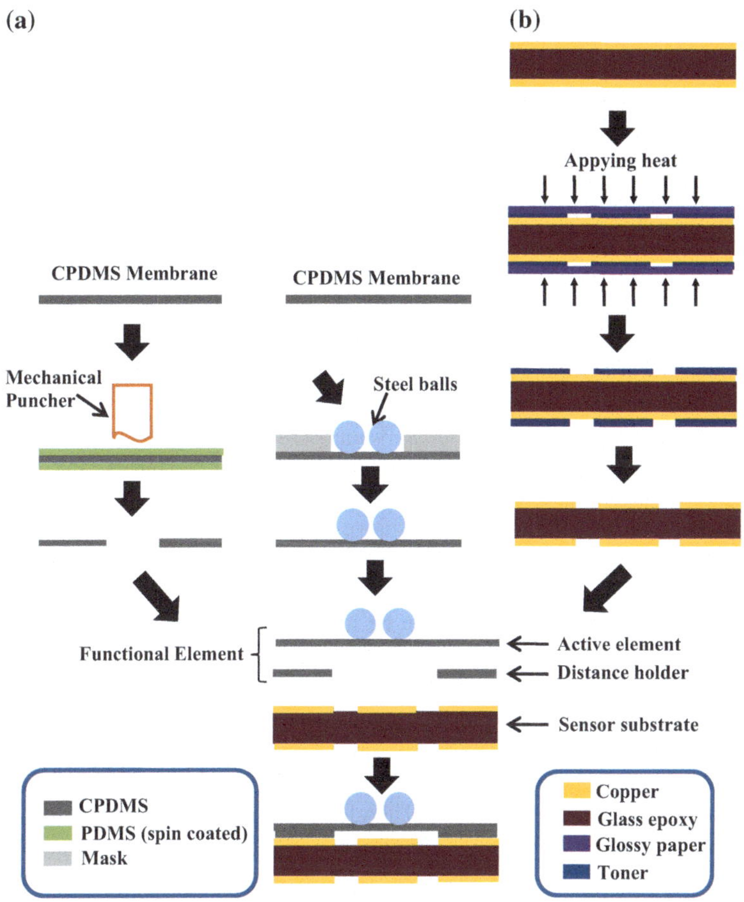

Fig. 15 Sensor fabrication **a** flexible element, **b** substrate

holder, using a mechanical puncher. During punching, the CPDMS membrane was sandwiched between two protective layers to prevent breakage, contamination and stretching. This is of importance as stretching beyond a certain limit could disturb the electromechanical properties of the CPDMS membrane. The sensor was fabricated on a two-sided PCB. Electrode structures were initially printed on glossy paper and transferred to the PCB by a laminator. Etching the copper layer of the PCB shaped the electrode structure. The active element, the distance holder and the sensor substrate layer, were aligned manually and bonded together by a CPDMS composite.

Sensor Parameters

All important parameters for evaluation and comparison of the capacitance-based unique and unclonable sensors are discussed in this section

(1) *Offset capacitance* (C_{OS}): The offset capacitance (C_{OS}) represents the fixed capacitance of the electrode structure when the membrane is at the zero deflection position. In case of tilt, the membrane deflection has a minimum at $\pm 90°$. However, there is a slight difference in the capacitance value at +90° and −90°, due to the random ball distribution. The capacitance at +90° is considered as C_{OS}. This fixed offset value does not contain any measurement information and reduces the sensitivity of the entire sensing system. Hence, it is desirable to have a minimum C_{OS}.

(2) *Maximum capacitance variation* (ΔC_{max}): The sensor shows maximum capacitance at a tilt of 0° (C_0) and minimum capacitance at 180° (C_{180}). ΔC_{max} is the difference between C_0 and C_{180}. In addition, ΔC_{max} is proportional to the sensor sensitivity. The goal is to achieve a maximum ΔC_{max}.

(3) *Concordance correlation coefficient* (ρ_c): The concordance correlation coefficient (ρ_c) defines the uniqueness and reliability of the sensor. Uniqueness means how effective a sensor can be identified among a group of other sensors. Reliability defines the ability of a sensor to reproduce the same output for a similar excitation. In devices such as PUFs, uniqueness and reliability are commonly measured in terms of Hamming distances [30], i.e. the minimum number of substitutions required to transform one string to another. Techniques more suitable for measuring the uniqueness of static digital PUFs cannot be applied to analog sensor outputs. Therefore, methods such as paired T-test [31], Bland-Altman plot [32] and correlation coefficients [33], are considered for deriving the uniqueness and reliability parameters. While the paired T-test and the Bland-Altman plot have severe disadvantages the estimation of correlation coefficients is the best choice to compare similar devices. The correlation coefficient represents a single value of similarity. The coefficient varies between −1 and 1. The higher the value, the higher is the similarity. Nonetheless, an ordinary correlation coefficient is independent of scale and bias. In order to overcome this disadvantage, Lin's proposed the concordance correlation coefficient [34]. It is used to measure the uniqueness and reliability of sensors with unique and unclonable characteristics. The concordance correlation coefficient between two variables can be calculated as:

$$\rho_c = \frac{2\rho\sigma_x\sigma_y}{\sigma_x^2 + \sigma_y^2 + \left(\mu_x - \mu_y\right)^2} \tag{6}$$

where μ_x and μ_y are the means and and are the variances of the variables. ρ is the correlation coefficient between the variables

In order to identify a sensor from its properties, for the same input (or measurand) the output characteristic of each sensor should be measurably different from others. On the other hand, each sensor should show exactly the same characteristic when the same input is applied multiple times. In practical use, different environmental factors such as changes in temperature or humidity affect the reliability of sensors. For evaluation of uniqueness and reliability, the parameters 'inter-concordance correlation coefficient' (inter$-\rho_c$) and 'intra-concordance correlation coefficient' (intra$-\rho_c$) have been introduced. The "inter$-\rho_c$" measures any (dis)similarity in the sensor characteristic of different sensors for the same excitation. The "intra$-\rho_c$" measures the sensor characteristic when the same sensor is excited multiple times. "Inter$-\rho_c$" represents therefore uniqueness while "intra$-\rho_c$" shows reliability. The goal is to maximize "intra$-\rho_c$" (best value is 1) and to minimize "inter$-\rho_c$" (best value is 0).

(4) *C_{OS} Spread*: The C_{OS} spread presents the extent of C_{OS} variations among a group of sensors, due to random ball distributions. C_{OS} Spread is the difference between the maximum and minimum values of C_{OS} among the sensors.

Experimental Set-up and Results

Six different sensors (SM_1–SM_6) were fabricated. The ball distribution was random in all these sensors. With exception of this ball arrangement, all other dimensions were similar. The electrode capacitances were measured with the evaluation board (AD7746) from Analog Devices [35]. The evaluation board builds around an IC AD7746 a capacitance-to-digital converter which converts the capacitance into a digital voltage. The IC AD7746 has an accuracy of 4 fF, a resolution down to 4 aF and can measure changes of the capacitance up to ± 4 pF. These specifications are enough to evaluate sensor prototypes. A shielded cable was used to connect sensor electrodes to the evaluation board. The shielded cable removes the effect of external interference on the sensor output. Capacitances were recorded using an in-built software suite available with the evaluation board.

The performance of the sensors was evaluated for tilt. The sensors, along with the measurement circuit, were placed on a tiltable platform and tilted from $-180°$ to $+180°$. The capacitances were recorded gradually every 10°. All experiments were carried out in a normal laboratory environment (23 °C). The output signals of the prototypes (SM_1–SM_6) are plotted in Fig. 16. Each single sensor showed a unique output characteristic. The reliability of each sensor was also verified. To do so, each sensor was tested six times. The output recorded from SM_1 is shown in the Fig. 17. In order to include the effect of environmental variations, measurements were repeated after 1 h. All SMs showed high reliability.

For all 6 sensor models (SM_1–SM_6), 15 different combinations of inter$-\rho_c$-values can be estimated. Similarly, as each sensor was tested 6 times, there are also 15 different combinations of intra$-\rho_c$. All these values are plotted in Fig. 18. The minimum value of the intra$-\rho_c$ (worst-case reliability) and maximum value of the inter$-\rho_c$ (worst-case uniqueness) are important. When considering a large number of sensors they offer a high probability when their inter$-\rho_c$-values is less than the

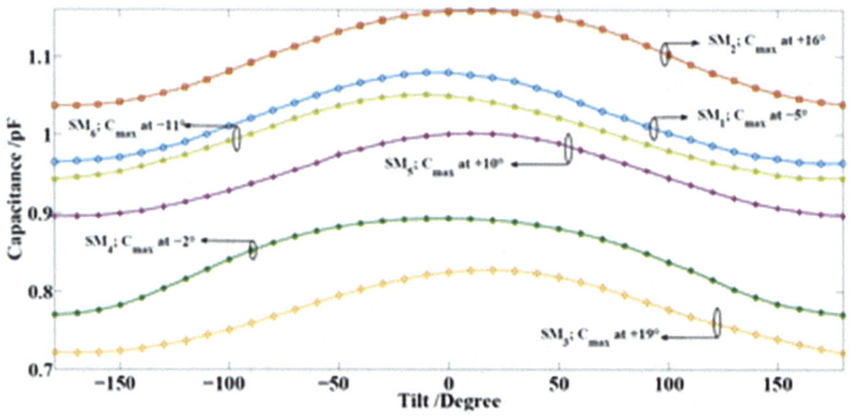

Fig. 16 Sensor capacitances recorded from SM_1 to SM_6 at different tilt angles

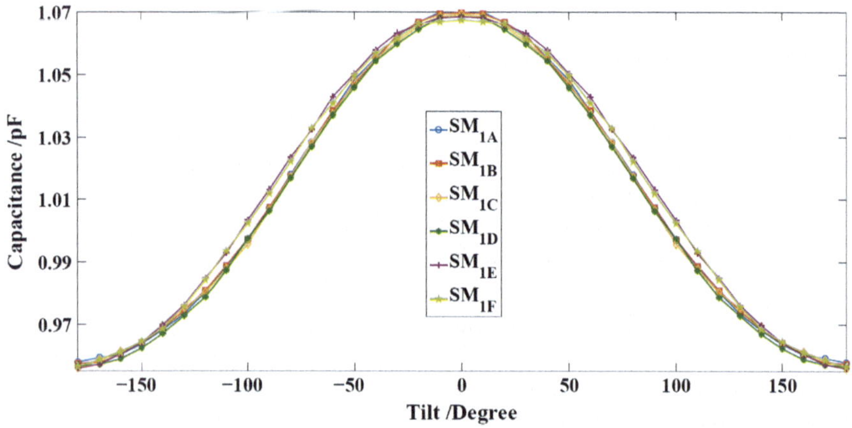

Fig. 17 Sensor capacitance recorded when sensor SM1 was tilted six times

Max(inter$-\rho_c$) and when their intra$-\rho_c$ is greater than the Min(intra$-\rho_c$) Therefore, in order to uniquely identify a sensor from its properties, one has to pay attention to the following condition.

$$\text{Min}(\text{intra} - \rho_c) > \text{Max}(\text{inter} - \rho_c) \quad (7)$$

This implies that the worst-case reliability of the SMs should be greater than the worst-case uniqueness observed among the SMs. The higher the difference, the better the uniqueness. In Fig. 18, a clear difference between the Max(inter$-\rho_c$) and the Min(intra$-\rho_c$) can be seen. Max(inter$-\rho_c$) and Min(intra$-\rho_c$) together with other parameters evaluated from the sensor parameters are given in Table 2.

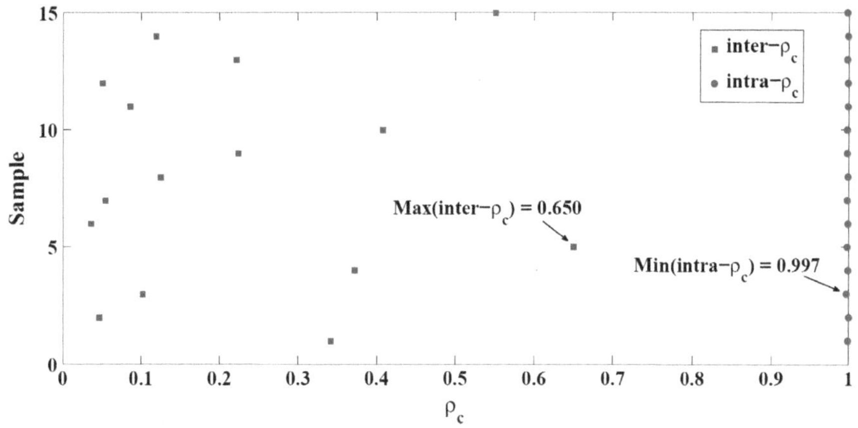

Fig. 18 The uniqueness parameters inter–ρ_c and the reliability parameter intra−ρ_c values evaluated from prototype sensor output

Table 2 Sensor parameters

Sensor parameters	Values
C_{OS} (pF)	0.9505
ΔC_{max} (pF)	0.1232
C_{OS} spread (pF)	0.3273
Max(inter−ρ_c)	0.6500
Max(intra−ρ_c)	0.9970

In comparison with FEA models, fabricated sensors show a much higher uniqueness. This is caused mainly due to the random structural variations in the CPDMS membrane. Tolerances as result of the CPDMS membrane fabrication alter the electromechanical characteristic (Young's modulus) and thickness of the membranes. The thickness variation of the distance holder varies the capacitance C_{TP} between T and the CPDMS membrane. Variations of the Young's modulus and the thickness of the membrane affect mainly the deflection characteristic. The random variations in the membrane characteristic improve the degree of uniqueness and unclonablity of the sensors.

Application

The main motivation of the development of unique and unclonable sensor is the security of sensor modules. The sensor described above is useful for different applications. One is briefly discussed in this section.

The sensor is interesting for applications where tilt needs to be detected with trust, for instance a tilt gesture based access control. The integration of a unique and

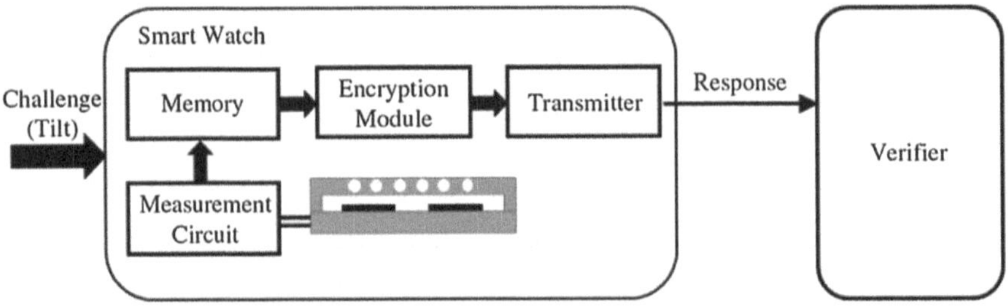

Fig. 19 Block representation of tilt gesture based access control using proposed unique and unclonable sensors

unclonable sensor with associated circuitry into a smart wrist watch can be considered as an identifier to get access to buildings, cars etc. A block diagram of the system is shown in Fig. 19. The tilt gesture in a predefined pattern generates a unique sensor output. The encryption module encrypts the output and transmits it to the verifier (e.g. a vehicle). Upon receiving the information, the verifier decrypts and compares it with the value stored in its database. The unique sensor output, associated with the particular tilt gesture, was previously enrolled in the verifier database, during the training phase. It is also possible to associate a particular gesture with a particular action. For instance, tilting in a clockwise direction corresponds to a door unlocking and tilting in an anticlockwise direction is for boot opening. For each pattern, the sensor output is different. From this difference, the verifier can determine the required action. If the transmitted information matches with the data stored in the memory, a desired action is carried out.

A tilt gesture-based keyless system has many advantages. The identifier in a watch gives the flexibility of using it as personal identifier for different applications, including access control as well as authorization of a person in vehicles, buildings, garage doors, etc. There is no need to search for keys inside a pocket or a handbag, because, for most people, the watch is always on the wrist, and a wrist-tilt gesture can perform the required action. Wrist tilting is generally considered as one of the easiest gestures. It is possible (to some extent) even when both hands are occupied (e.g. with shopping bags). Furthermore, wrist tilting requires only limited space and can be performed without much physical or mental effort. In the gesture-based keyless system, a user interface only in the form of a tilt is necessary to activate the key and generate a valid response. Therefore, it is difficult to implement a relay-based attack. Moreover, such a system has all the advantages of PUF-based key/identifier generations such as high security. It is difficult to duplicate such a device which offers additionally low resource requirements.

Conclusion

Capacitive sensors with unique and unclonable characteristics for developing nature inspired secure sensing modules have been presented. The sensors consist of a CPDMS membrane with randomly distributed steel balls on top. A planar electrode structure converts the deflection of the membrane into a capacitance output signal. Due to the random distribution of the balls and random variations of the CPDMS membranes, each sensor has a unique capacitance characteristic and cannot be duplicated. Each sensor can be identified by its own properties. This uniqueness can be further improved by the intentional variation of several parameters such as the number of conductive balls, ball sizes CB-concentration and distribution. These properties make the sensors attractive for several security solutions such as access control.

References

1. J. Jackson, Ready, aim, record: armys prototype system uses RFID, GCN Government Computer News (2018)
2. Y. Shoukry, P. Martin, P. Tabuada, M. Srivastava, Non-invasive spoofing attacks for anti-lock braking systems, in *International Workshop on Cryptographic Hardware and Embedded Systems*, Santa Barbara, USA (2013)
3. I. Haide, M. H¨oberl, B. Rinner, Trusted sensors for participatory sensing and IoT applications based on physically unclonable functions, in *2nd ACM International Workshop on IoT Privacy, Trust, and Security*, New York, USA (2016)
4. R. Maes, *Physically Unclonable Functions* (Springer, Berlin, Germany, 2013)
5. S. Morozov, A. Maiti, P. Schaumont, An analysis of delay based PUF implementations on FPGA, in *International Symposium on Applied Reconfigurable Computing*, Bangkok, Thailand (2010)
6. J. Guajardo, S. S. Kumar, G.-J. Schrijen, P. Tuyls, FPGA intrinsic PUFs and their use for IP protection, in *International workshop on Cryptographic Hardware and Embedded Systems*, Vienna, Austria (2007)
7. J.T. Ficke, W.E. Hall, T.B. Hook, M.A. Sperling, L. Wissel, Physically unclonable function implemented through threshold voltage comparison, US Patent 8,619,979, 31 Dec 2013
8. R. Pappu, B. Recht, J. Taylor, N. Gershenfeld, Physical one-way functions. Science **297**(5589), 2026–2030 (2002)
9. D. Roy, J.H. Klootwijk, N.A. Verhaegh, H.H. Roosen, R.A. Wolters, Comb capacitor structures for on-chip physical uncloneable function. IEEE Trans. Semicond. Manuf. **22**(1), 96–102 (2009)
10. D.F. Kune, J. Backes, S.S. Clark, D. Kramer, M.Reynolds, K. Fu, Y. Kim, W. Xu, Ghost talk: mitigating EMI signal injection attacks against analog sensors. Secur. Priv. (SP), Washington, USA (2013)
11. M. Burge, W. Burger, Ear Biometrics, in *Biometric* (Springer, Berlin, Germany, 1996), pp. 273–285
12. D. Zhang, Z. Liu, J. Yan, P. Shi, Tongue-print: a novel biometrics pattern, in *International Conference on Biometrics*, Crystal City, USA (2007)
13. P.M. Hofman, J.G.V. Riswick, A.J.V. Opstal, Relearning sound localization with new ears. Nat. Neurosci. **1**(5), 417–421 (1998)

14. A.N.F.G. Aysu, Z. Franklin, M.P. Yali, P. Schaumont, Digital fingerprints for low-cost platforms using MEMS sensors, in *Proceedings of the Workshop on Embedded Systems Security* (2013)
15. O. Willers, C. Huth, J. Guajardo, H. Seidel, MEMS-based gyroscopes as physical unclonable functions, in *IACR Cryptology ePrint Archive* (2016)
16. U. Rührmair, J. Martinez-Hurtado, X. Xu, C. Kraeh, C. Hilgers, D. Kononchuk, J. Finley, W. Burleson, Virtual proofs of reality and their physical implementation, in *IEEE Symposium on Security and Privacy (SP)*, San Jose, USA (2015)
17. Y. Gao, H. Ma, D.C. Ranasinghe, S.F. Al-Sarawi, D. Abbott, Exploiting unreliability of the PUF to secure remote wireless sensing, in *Cryptology ePrint Archive* (2015)
18. K. Shimizu, T. Sugawara, A.D. Suzuki, PUF as a sensor, in *IEEE 4th Global Conference on Consumer Electronics (GCCE)*, Osaka, Japan (2015)
19. J. Rajendran, J. Tang, R. Karri, Securing pressure measurements using sensor PUFs, in *IEEE International Symposium on Circuits and Systems (ISCAS)*, Montreal, Canada (2016)
20. K. Rosenfeld, E. Gavas, R. Karri, Sensor physical unclonable functions, in *IEEE International Symposium on Hardware-Oriented Security and Trust (HOST)*, Anaheim, USA (2010)
21. C.B. Karuthedath, N. Schwesinger, An improved capacitance-based intentionally imperfect sensor for measuring mechanical parameters, in *8th International Conference on Sensing and Technology (ICST)*, Liverpool, UK (2014)
22. C.B. Karuthedath, S. Aung, N. Schwesinger, Finite element analysis of differential capacitive PUF sensors, in *IEEE Sensors Applications Symposium (SAS)*, Catania, Italy (2016)
23. W.K. Schomburg, Membranes, in *Introduction to Microsystem Design* (Springer, Berlin, Germany, 2015), pp. 29–52
24. A. Khosla, Nanoparticle-doped electrically-conducting polymers for flexible nanomicro systems. Electrochem. Soc. Interface **21**(1), 67–70 (2012)
25. J. Ruhhammer, M. Zens, F. Goldschmidtboeing, A. Seifert, P. Woias, Highly elastic conductive polymeric MEMS. Sci. Technol. Adv. Mater. **16**(1–10) (2015)
26. AlpaSil Classic-Safety Data Sheet, Alpina (2016)
27. Sylgard 184-Safety Data Sheet, Sigma-Aldrich (2016)
28. Vulcan XC72-Safety Datasheet, Cabot Corporation (2016)
29. S. Krishnan, On the manufacture of very thin elastomeric films by spin-coating, Ph.D. dissertation, Massachusetts Institute of Technology, Massachusetts, USA (2007)
30. M. Abhranil, V. Gunreddy, P. Schaumont, A systematic method to evaluate and compare the performance of physical unclonable functions, in *Embedded systems design with FPGAs*, New York, USA (2013)
31. J.H. McDonald, *Handbook of Biological Statistics*, (Sparky House, Baltimore, USA, 2009)
32. J.M. Bland, D. Altman, Statistical methods for assessing agreement between two methods of clinical measurement. The Lancet **327**(8476), 307–310 (1986)
33. J.L. Rodgers, W.A. Nicewander, Thirteen ways to look at the correlation coefficient. Am. Stat. **42**(1), 59–66 (1988)
34. I. Lawrence, K. Lin, A concordance correlation coefficient to evaluate reproducibility. Biometrics **45**(1), 255–268 (1989)
35. AD7746 evaluation board: EVAL-AD7746 EB datasheet, USA: Analog Devices (2015)

Internet of Things Scalability: Communications and Data Management

Laith Farhan and Rupak Kharel

Abstract Internet of Things (IoT) is becoming more and more pervasive in everyday life and connecting an array of diverse physical objects. It is fast growing and receiving a tremendous amount of research focus. Billions of objects communicate each other with or without human intervention to achieve smart applications. Most of the connected devices are constrained nodes to its ecosystem which have limited memories, CPU capabilities and power sources. Therefore, for implementing autonomous smart systems, efficient energy consumption is imperative. This chapter introduces a novel scheduling algorithm called Long Hop (LH) first to optimize energy usage on a wireless sensor network (WSN) that enables IoT systems. The selected algorithm proposes an optimized solution to the energy efficient for scalable IoT networks. LH technique assigns high priority for packets coming with more hops and longer distances to be served first at the cluster head (CH) nodes. Since these packets require more links and nodes (thus increased energy and bandwidth usage) to reach the ultimate destination if not prioritized. The proposed technique reduces the overall energy usage and minimizes the total number of packets re-transmission and the effective data transmission distances. Thus, this improves the overall system performance and elongates the network lifetime.

The Future Network Era Begins

Recently, the desire for connectivity has caused an exponential growth in wireless communication. In particular, wireless sensor networks (WSNs) technology has led the trend due to increase in applications exchanging data utlizing in internet services. WSNs are spatially distributed autonomous sensors in a sensing field to collect/exchange data and to monitor specific phenomenon [1]. It interacts between

L. Farhan (✉) · R. Kharel
Manchester Metropolitan University, Manchester, UK
e-mail: l.al-bayati@mmu.ac.uk

L. Farhan
University of Diyala, Baqubah, Iraq

S. C. Mukhopadhyay et al. (eds.), *Modern Sensing Technologies*,
Smart Sensors, Measurement and Instrumentation 29,
https://doi.org/10.1007/978-3-319-99540-3_16

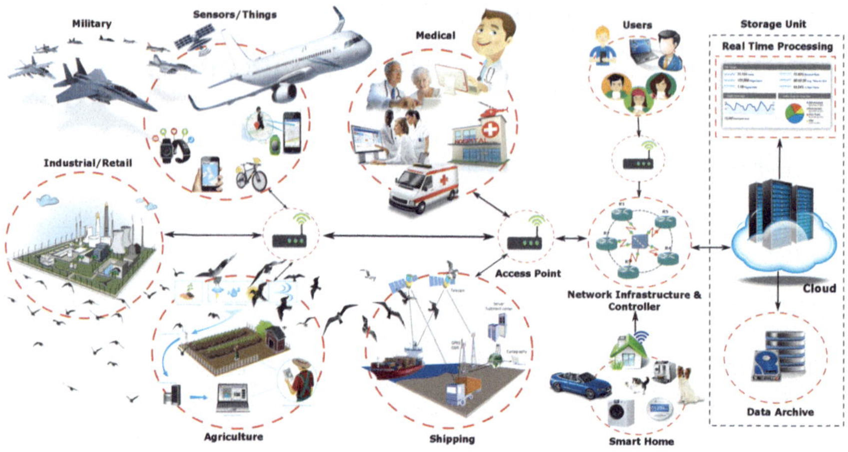

Fig. 1 Network of the future

persons or computers, and the surrounding environments and thus it is one of the key enabling technologies for the internet of things (IoT) system.

The term of internet of things has coined by Kevin Ashton in 1999 [2]. It is a rapidly growing technology with wide range of applications in various fields. It has unified a plethora of devices and infrastructure under the same umbrella and is considered by many technologies leaders as the network of the future. Internet of thing is expected to increase in popularity and become even more pervasive in the next few years. IoT-applications are growing at an exponential pace with wearable devices, kitchen appliances, connected cars, healthcare devices, etc. becoming more and more commonplace [3]. Soon, every device we own will be connected to the internet. In Fig. 1, some of the IoT applications have been described.

The connectivity of these devices will significantly increase over the next few years, according to numbers forecasted by Cisco Systems, from 10 billion in 2014 to 50 billion by 2020 [2]. This will provide a new paradigm of communication between digital contents, computing, applications and services within the physical world. This creates unprecedented opportunities for governments, education, and industries in a wide variety of sectors [4, 5]. The communication of IoT networks combine three categories based on their technology elements that can be summarized below and seen in Fig. 2.

- **People-to-People (P2P) connection**: is the data transfer from one person to the other. It occurs through video call, telephone call, and social communications. It is usually called collaboration connection.
- **Machine-to-People (M2P) connection**: is the data transfer from machines such as computing devices, sensor nodes or others to the users for analysis purposes. For example: weather forecasting uses smart devices to gather the data from the environment and send it back to the administrators in the control centre for further analysis.

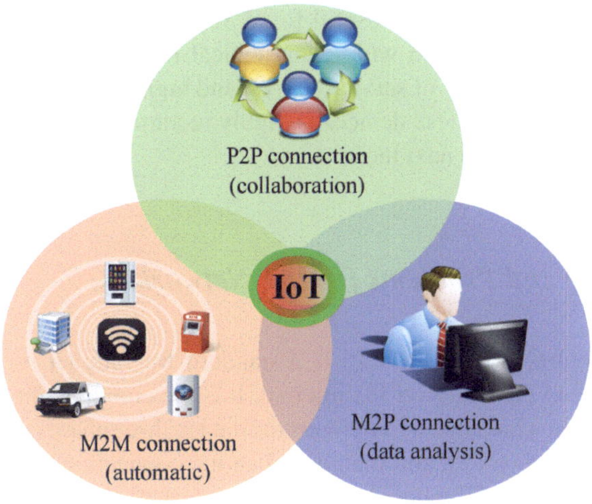

Fig. 2 Internet of things (IoT) elements

- **Machine-to-Machine (M2M) connection**: is the data transfer between devices without human interactions. For instance, a car talking to another car about its speed, lane change or breaking intentions, etc.

Therefore, we summarize the IoT components into the simple equation below:

$$\boldsymbol{IoT} = \boldsymbol{Human} + \boldsymbol{Physical\ Objects}(\boldsymbol{sensors}, \boldsymbol{controllers}, \boldsymbol{devices}, \boldsymbol{storages}) + \boldsymbol{Internet}$$

Smart city concept is another powerful application of IoT generating curiosity among world's population. Recent statistics observe that 70% of the world population will be living in cities and neighboring regions by 2025 [6]. This leads to billions of devices communicating with each other and massive amount of data being generated every day. The amount of data gathering from these connected devices is 30 times more than that from human society [7]. In this sense, scalability and big data applications require collection of data from millions of sensor nodes and thus it is a challenge to implement IoT architecture that is energy optimized such that the network lasts longer without running out of power.

Most of IoT-devices run on batteries or by energy harvesting, which have limited shelf life [8]. This means it will not be rational to waste the energy on the transmission of unneeded data or protocol overheads like existing protocols do such as Hypertext Transfer Protocol (HTTP) and Transmission Control Protocol (TCP), etc. Therefore, efficient energy consumption is the key issue for longer network and system lifetime. It is related to the sustainability of the network and energy consumption by each single device. The connected devices introduce several services and therefore send heterogeneous packets. These packets access multiple devices and links (mostly wireless communication) to reach the ultimate receiver using multi-hop technology.

Multi-hop topology is practically useful for WSN and IoT networks because both techniques are deployed over a wide geographical region [9]. It also provides an opportunity for better quality of services (QoS) and higher network capacity. Therefore, management data of these devices is highly recommended to optimize energy uses and success IoT as the next technology.

TCP Overview and Problems for IoT Networks

Transmission control protocol (TCP) is a standard for internet protocol (IP) that defines how to send the data over the internet. It is the embodiment of reliable end-to-end transmission which means it delivers the data in three phases: Request-to-Send (RTS), Clear-to-Send (CTS), and then establish the connection between the source and destination as seen in the Fig. 3.

Assuming sensor1 needs to send a packet to the destination, a packet transfer would proceed as follows:

- Sensor1 starts to transmit a series of RTS beacons to the destination node via intermediate nodes and waits for an CTS respond from it.
- Destination node gets the RTS message that been sent from the sensor1 and now transmits a CTS packet to sensor1.
- After acknowledgment of message receipt, sensor1 returns to periodic transmission of query beacon packets.

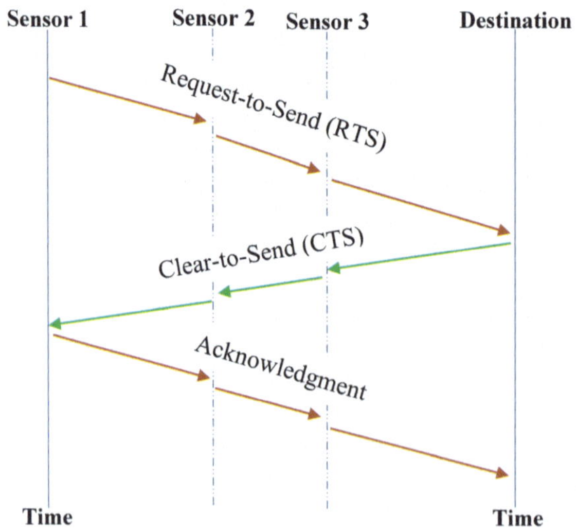

Fig. 3 Data transaction and aggregation

Sensor nodes use the TCP protocol to communicate to the neighbor nodes or may connect to another network without a gateway using multi-hop technique. During multi hop of packets amongst the nodes especially on a large network, the probability of the packet drops increase. This is because of various factors such as packet arrival rate, timeout for message expiry or simply due to the limitations of node because of its constrained nature (low processing, memory and bandwidth resources). Therefore, to avoid packet loss in the network, receipt acknowledgment of transmitted packets or otherwise retransmission of the lost data packets must happen. Internet control message protocol (ICMP) generated by the network devices is an error-reporting protocol sent to the original source whose IP address is encapsulated in the IP packet. The minimum size of ICMP packet would be 8 bytes of ICMP header + 20 bytes of IPv4 or 8 bytes of ICMP header + 40 bytes of IPv6 for ICMPv4 and ICMPv6 respectively [10]. Each retransmission will require transmission of these headers and overheads as well. This can significantly impact system performance if the retransmission is not optimized properly. It can add more load on the already constrained network and contribute further to the power depletion of the nodes. This affects the quality of service (QoS) and throughput of the network [9] which ultimately drains the node battery leaving the area uncovered and thus reducing the network lifetime. The problem begins serious when more and more devices/things are being linked and communicated to each other. It is a matter of utmost concern to consider scalability and adaptability of these devices to change the environment and need of the people. There are innumerable solutions proposed for reducing energy consumption of WSN and IoT applications. However, most of previous works focused in a small area with less number of nodes which is not consistent with IoT applications. Network of the future will deploy numerous devices that cover larger areas. Therefore, the solutions should work on larger areas with higher number of nodes.

To this end, the focus and scope of this chapter is on the scalability and energy aspects of the IoT networks. Thus, we explain an energy saving scheme by applying a novel dynamic scheduling algorithm called long hop (LH) first. LH algorithm operates at the cluster heads (CHs) of a WSN and schedules high priority for messages coming with more hops and from longer distances to be routed first to the ultimate receiver. We elucidate more information for this lovely technique in sect. "Implementation of Efficient Scheduling Technique and Procedures".

A Practical Data Management of IoT Applications

Scheduling algorithms have been widely investigated by the IoT research community in recent years. In such schemes can achieve energy efficiency in data collecting, processing, managing and thus increasing the sustainability of the WSN and IoT networks. As already mentioned, WSNs technology is part of IoT networks and it is a similar field as well. Therefore, we take into account some practical related works from the WSNs aspect.

The work in [11] describes the earliest deadline first algorithm (EDF) which is a dynamic scheduling algorithm used for real-time operating systems. EDF requires to sort all tasks based on their deadlines time and then it gives high priority for packets closest to their exceeded time. However, it does not consider managing time redundancy and packets spend high energy. Therefore, scheduling tasks will complete within expire times even in the presence of faults. The proposed work in [12] uses an idea to cluster IoT networks into sub-groups and within each group a message broker was placed. The broker job is to gather the information from the sensors around it and forward it to the last destination. The short process time (SPT) algorithm implemented at broker level to select and deliver packets based on their arrival time. Each message is presented as Mess (R_{time}, T_{trans}), where R_{time} and T_{trans} are request time and successful transmission time period respectively. It is applied when the network is unstable (traffic intensity > 1). The proposed method promotes IoT system efficiency by improving service response time and reducing the overall energy consumption. Another study [13] schedules packets into three classes priority for large-scale wireless sensor networks. The higher priority assigns for emergency real-time packets and preempt the processing of packets at other queues. Lowest priority packets wait a certain number of time slots after processing higher priority packets. The proposed work improves the system in term of end-to-end data transmission delay. Authors in [14] intend the border nodes of wireless sensor networks. These nodes located between two sensing area consume a large amount of energy due to the listening mode. Thus, S-MAC protocol minimizes the listening time of each node. The proposed scheme extends the lifetime and reduces the energy consumption for the system. Heterogeneous dual-core processor for IoT applications were carried out by [15]. Authors show that up to 2.62x energy efficiency improvements can be realised without deadline messages expiry. Another approach [16] investigates energy saving scheme by implementing dynamic voltage and frequency scaling (DVFS) scheduling algorithm. DVFS uses low time complexity to avoid the deadlines of the real-time tasks and shows that it can minimize up to 64% energy used for each task on a separate core.

Although the previous studies improve on the energy usage, there is no considerable work done on dealing the optimized scheduling for nodes deployed in different locations (thus still being valid for larger areas and hence scalable networks). Normally, in multi-hop networks, devices send their packets to the destination through intermediate nodes. In small IoT networks, objects collect the necessary information and then send it to the receiver directly through the CHs within maximum single or two hops [17]. However, in large networks, the data accesses many links and devices using multi-hop technique to reach the intend destination. Therefore, the data coming from nodes located in longer distances is passed through multiple intermediate nodes to reach the ultimate receiver. Accessing multiple nodes lead to use the bandwidth, throughput and consumption of previous node energy. Therefore, while designing and implementing the efficient energy algorithms, the factor of distance (and hence number of hops) should also be taken into consideration to minimize the power spent and prolong the network lifetime.

This chapter introduces one of the perfect solutions technique to meet the demands of IoT scalable networks. Long hop first scheduling algorithm focuses on real-time tasks to place processes in a priority queue. LH algorithm assigns high priority for messages coming from far distances and thus accessing multiple nodes to get served first at CHs. It is beneficial for large-scale networks that can manages data communication and reduces the energy consumption for sensor devices and thus elongate the lifetime of the network.

Implementation of Efficient Scheduling Technique and Procedures

On a larger scale, internet of things connects a huge number of devices/objects and this provides and shares a large amount of applications via internet. This has created a complex interoperability and integration challenge to realize large-scale heterogeneous IoT ecosystems. Scalability means the flexibility that allows us to better address, communication, management, and achieves the specific needs as they arise.

Therefore, a huge number of nodes have been assumed to be deployed in an outdoor area and classified into: (i) Sensors and Actuators (SNs) that gather all the necessary information from the environment and send it to CH nodes within single or multiple hops. (ii) cluster heads receive the heterogeneous data from the SNs around it, compress it, and then send to the intend destination. (iii) Base station (BS) is responsible for collecting the information from all sensors via CHs. The proposed system architecture is clustered into sub-groups, as showed in Fig. 4. Each group is composed number of nodes deployed randomly in the sensing field and connected to a single CH node. CH becomes overloaded due to the number of nodes connected to it. Furthermore, memory of CHs become full. Therefore, software or hardware flow control must prevent overflow and thus the data loss, otherwise, the node must re-send it again [18].

Long Hop (LH) First Scheduling Strategy

Multi-hop technology helps the devices to connect and communicate with each other even if sensors are located out of the transmission range. The nodes should depend on some intermediate nodes for relaying packets. These packets come from nodes located on the border use high number of hops to reach the ultimate receiver. Thus, these nodes consume larger amount of energy and bandwidth during the transmitting and receiving of the packets by other nodes.

Figure 5 shows the energy consumption for individual data packet at different nodes as a function of number of hops and distances. The plot was generated by

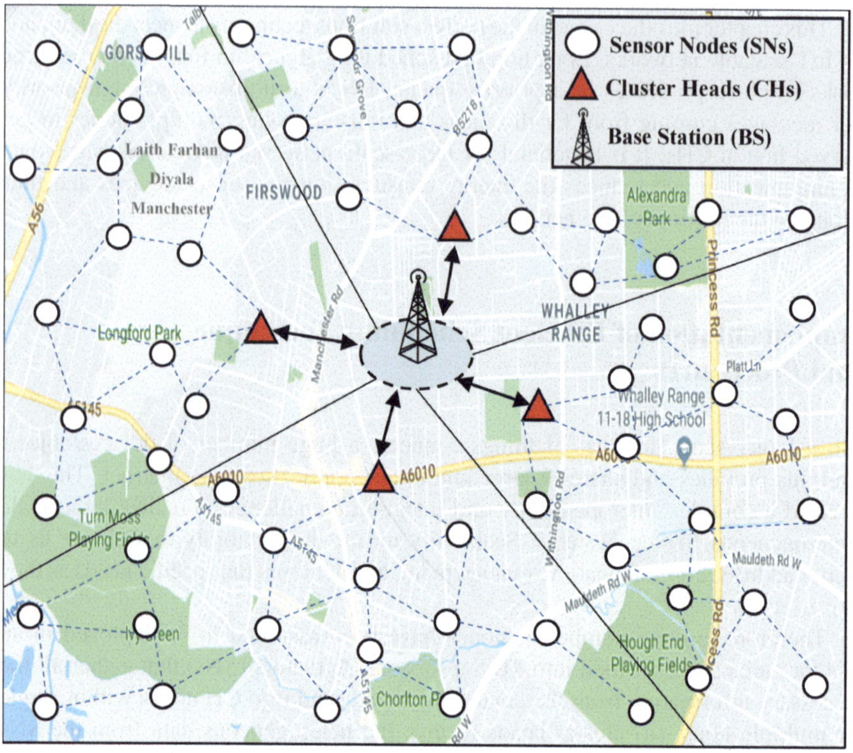

Fig. 4 Typical IoT system architecture

randomly picking fifteenth nodes from the sensing field. These nodes located in different locations and used some intermediate nodes to deliver their packets. It clearly indicates that the packets come with higher hops "i.e. 13" uses maximum energy. When multiple data packets have same number of hops "i.e. 4, 7, and 11", the packet comes from far distance consumes more energy. Due to this reason, it is reasonable to assign high priority for these data packets during scheduling algorithm to conserve energy. This is the key idea behind the LH algorithm and gives high priority to data based on sensors locations and number of sensors accessed.

Meanwhile, the main task is to optimize by power control, the communication towards the destination in terms of interference, power consumption, capacity, etc. Therefore, Fig. 6 proves the relation between number of hops, distances and transmitting power. It observes that higher number of hops and longer distances used have the maximum energy, interference, and capacity usage. The ideal energy is used when a packet sends with a transmission distance and number of hops as less as possible.

Figure 7 shows set of sensor nodes connect each other using multi-hop technique. We assume that these packets are transmitted in different time from each sensor and arrive at the same time at CH where they are placed within a queue at the cluster head node before delivering to the BS. Each task has different number of hops and

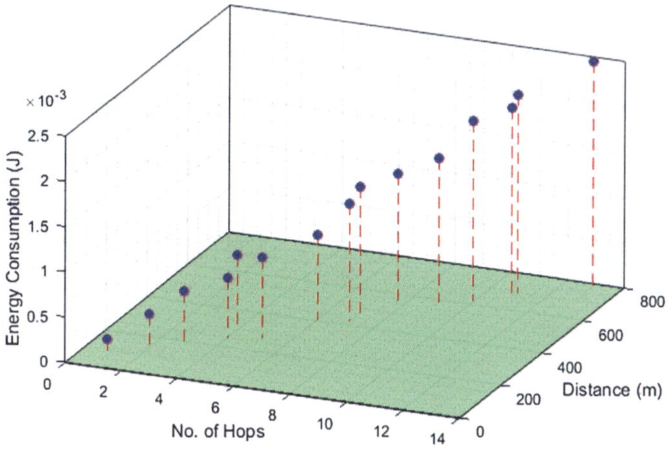

Fig. 5 Energy consumption by number of hops versus distances

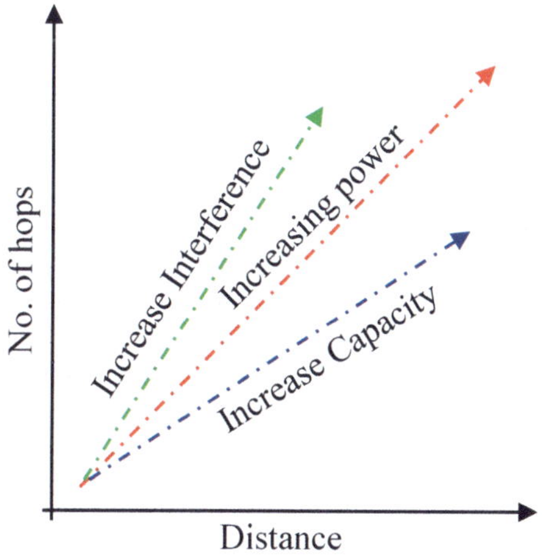

Fig. 6 Relation between number of hops, distance and transmitting power

distance. LH protocol re-arranges the packets at CH buffer based on higher number of hops and longer distances to forward it first to the ultimate receiver. There must be at least a single task execution through CH to be forwarded to the exchange centre within one spin. During queuing time, some of these packets discard due to time exceeded message, buffer is full, or quench source "i.e. yellow packet". Also, the orange and green packets reached the RF antenna at the same time. In this case, the antenna can serve one packet only at each time and discard other. Therefore, LH policy takes the packet that used higher energy "i.e. green packet" and drops other.

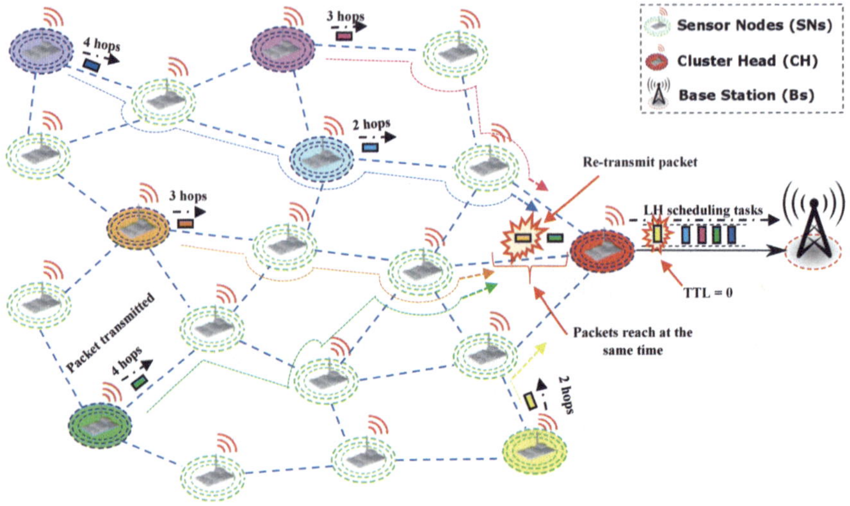

Fig. 7 The partial schedule of four tasks under LH algorithm

Queuing Model

M/M/1 queuing model has been used to calculate service rate and arrival rate for all messages coming from devices. M/M/1 is a queuing theory within the mathematical theory of probability that shows the queue length of a single server in the system. Service times have an exponential distribution and exponential arrivals are determined by a Poisson process [19]. Each packet has Successful transmission time of the request and request time period that can be represented as Mess (T_{trans}, R_{time}). The service rate and arrival rate for $\boldsymbol{m}$ messages are introduced by $\boldsymbol{\mu}$ and $\boldsymbol{\lambda}$ respectively.

$$\lambda = \frac{1}{R_{time}} \tag{1}$$

$$\mu = \frac{1}{T_{trans}} \tag{2}$$

$$P = \frac{\lambda}{\mu} \tag{3}$$

$$P = \frac{T_{trans}}{R_{time}} \tag{4}$$

Then, the total traffic intensity ($\boldsymbol{P_i}$) for the overall system in each IoT sub-group becomes as follows:

$$P_i = \sum_{1}^{n} \frac{\lambda}{\mu} = \sum_{1}^{n} \frac{T_{trans}}{R_{time}} < 1 \tag{5}$$

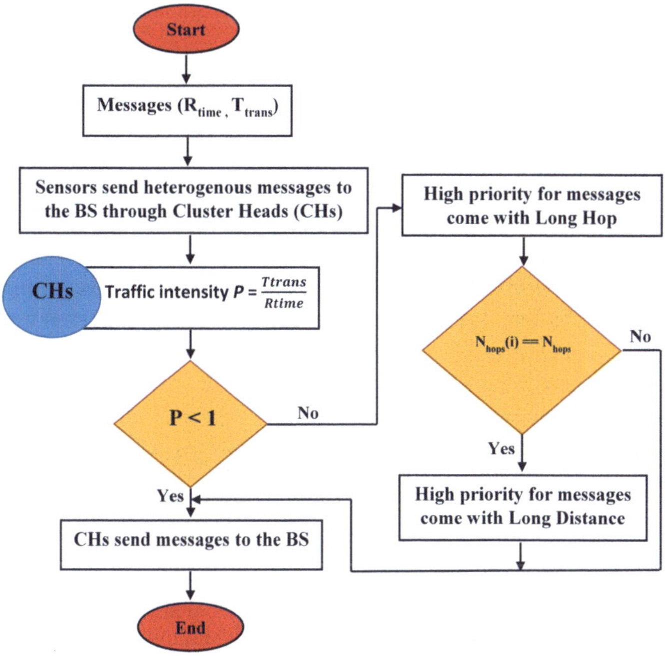

Fig. 8 Flow-chart of LH procedure

After evaluating the system traffic at the cluster heads, traffic intensity (***P***) should be less than 1. However, if ***P*** is larger than 1, the proposed algorithm avoids packets accessing many devices and links from retransmission. The proposed algorithm is shown in Fig. 8 and explained the steps of LH scheduling packets to the intend destination.

Nodes Placement

Let N be the number of sensor nodes in the system model, and $loc = (x, y)$ is the location of each node. The distance (***d***) between two nodes is given Euclidean mathematical method [20] as:

$$d_i = \sqrt{((x_i - x) + (y_i - y))^2},\ \mathrm{i} = 1, 2, 3, \ldots, N \tag{6}$$

Direct communication, whenever possible, is certainly the best way for data dissemination. Therefore, sensors follow shortest path route to deliver their packets to the next hop. Dijkstra algorithm [21, 22] has been applied to find the shortest path ($\boldsymbol{d_i}$) between these sensor nodes to reach the ultimate receiver, i.e.

$$\sum_{i=1}^{N} d_i \rightarrow min \tag{7}$$

Network Buffer Sizing

Congestion may happen in the network specially at CHs. They carry their data and other forwarded data from various sensors. Sensor devices have a very limited buffer (or as known data buffer). It is a block of physical memory that temporarily stores packets until it is being moved [19]. All network devices (i.e. sensors, gateway, routers, etc.) normally contain buffers to hold packets during transmitting and receiving packets. As the network load increases, some packets drop due to excessive incoming traffic. Therefore, buffer plays a very crucial aspect on the performance of the network.

The rule-of-thumb and Standford rule are most important rules for dimensioning network queues [23, 24]. Rule-of-thumb states that each link requires a buffer of size $B = RTT * C$, where C is the bottleneck capacity and RTT is the average round-trip time of the flow passing across the link. This rule is often applied at the edge or cluster devices of the network when the bandwidth capacity and number of flows are small. While the Stanford rule is used for large number of TCP flows and higher speed links. The recommended router requires a buffer of size $(RTT\ *\ C)/\sqrt{n}$, where n is the number of TCP flows sharing the bottleneck link [23]. The rule-of-thumb has been used for this study since the flows at each CH is relatively small.

Energy Consumption Model

The rapid growth in the population density in the cities reflects the phenomenal growth in the use of smart devices. Therefore, they demand tolerable provision of infrastructures and services. In such systems, a large number of connected devices are deployed in a large geographical area and generally configured in a mesh network, ultimately sending a large volume of data to the base station or gateway. The packets to be transmitted are forwarded with multiple hops to reach the intend destination. This causes lots of redundant data, and hence their transmitting brings in a large waste of bandwidth and limited power, resulting in low energy efficiency and short network lifetime. So, energy efficiency and big data are two sides of the same coin.

Therefore, the aim of proposed algorithm is to implement message scheduling algorithm that minimizes energy consumption and avoids the transmissions of redundant information and thus elongates lifetime of the network. Most of the energy is consumed in listening, transmitting and receiving of packets in the network. A common power model [18] is used and shown in Fig. 9. The total energy [25] spent by the system can be represented as follows:

- To transmit a number of bits:

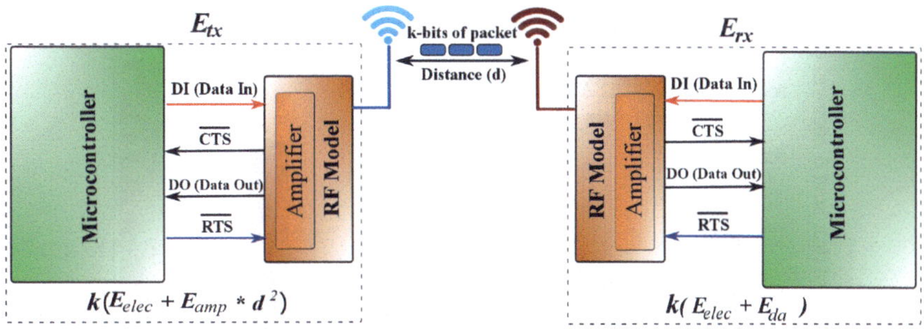

Fig. 9 The wireless communication power model

$$E_{tx} = k\left(E_{elec} + E_{amp} * d^2\right) \tag{8}$$

- To receive a number of bits:

$$E_{rx} = (k * E_{elec}) \tag{9}$$

- Total energy used by each sensor:

$$E_{total} = L(E_{tx}) + M(E_{rx}) \tag{10}$$

where $\boldsymbol{E_{tx}}$ is the energy dissipated to disseminate number of bits ($\boldsymbol{k}$) from the source to the next object. $\boldsymbol{E_{rx}}$ is the energy depleted to receive chunk of bits. $\boldsymbol{E_{elec}}$ is presented the energy depleted to run the receiver or transmitter circuitry, and $\boldsymbol{E_{amp}}$ is the energy dissipated in transmission to amplify the signal enough to reach the next target. $\boldsymbol{L}$ and $\boldsymbol{M}$ are number of transmitted and received packets from a sensor node respectively. The distance between two transmitters is denoted as $\boldsymbol{d}$.

Complexity Analysis

The complexity of the selected algorithm can be analysed in terms of storage and computation complexity. WSNs have small CPU that carries out the instructions of a computer program to send and receive packets. It is important to reduce the burden on this processor unit to prevent the fault. Therefore, the computational complexity is the major components in the analysis of the proposed protocol. The time complexity of the LH algorithm is (n^2 +8n). Therefore, the time complexity is obtained to be O(n^2), which is similar or better than other protocols which have complexity in order of O(n^2) and O(n^3).

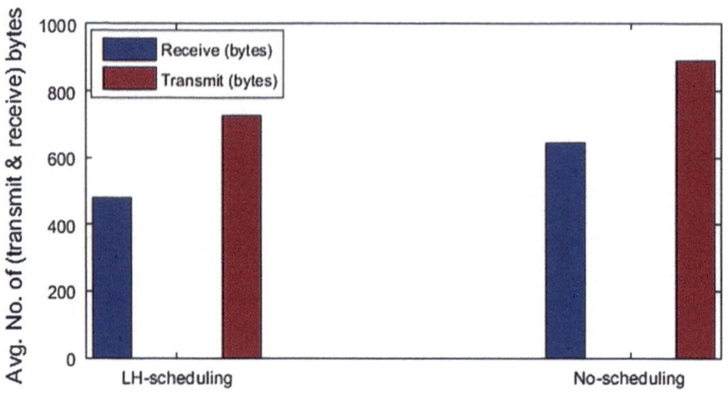

Fig. 10 Avg. number of (transmitting and receiving) packets

Performance Evaluation and Discussion

In this section, we evaluate and analyse the performance of the proposed scheme by using *Matlab* simulation. A performance comparison is done when the proposed scheduling algorithm and no scheduling is used, that was reported in [25]. It also compared with first-come-first-serve (FCFS) and nearest-job-next (NJN) techniques was notified in [26].

Internet of things consists of number of devices/things deploy randomly in a large scale. Therefore, the simulation is performed with 100 and 200 nodes that are distributed randomly in an area of (500×500) m^2, using mesh topology consisting multi-hop network scenario in [25, 26] respectively. We also assume that the sensor nodes are deployed outdoor to monitor the environment that send the information to the central application. CHs job is to collect the data coming from the sensors, process and deliver it to the BS. In every round, each sensor sends the data environment to the BS via the CHs. The BS is set at the centre of the square field. The assumptions and parameters that been used in the simulation model are shown and elucidated in the table 1 of [25, 26].

Some of the performance comparison results in terms of scheduling and no scheduling system are applied. It is important to minimize the number of packets transmit and receive in each individual sensor that can further reduce the processing time, capacity used, and energy consumption of the network. A transmit/receive happens when the packet is passed to the next target and receive it successfully. Figure 10 shows that the proposed algorithm has lower number of transmitting and receiving packets in the network. This means that the use of LH algorithm has led to the reduction in total transmitting and receiving power thus and ultimately extending the network lifetime.

Number of hops is the summation of data relays on multi-hop communication to reach the intended destination. A hop occurs when the packet is passed to the next network device. Next hop is a routing term that refers to the next device based on the

Fig. 11 Total number of hops

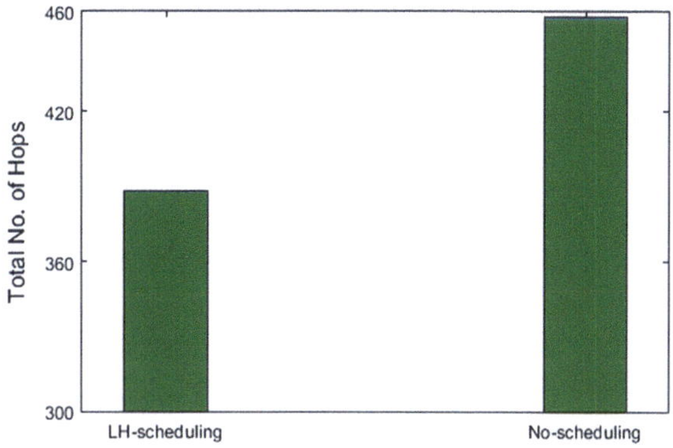

Fig. 12 Avg. transmission distances (m)

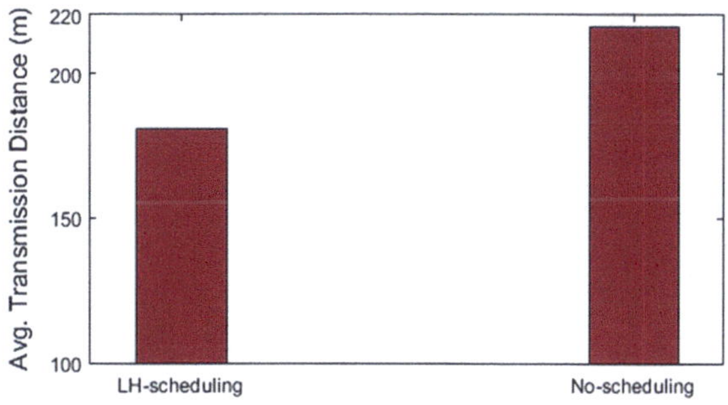

type of routing algorithm is used. The purpose is to minimize the power by reducing the relay packets from the sender to the receiver. Figure 11 shows that proposed algorithm has minimum number of hops compared to when the algorithm is not implemented.

Transmission distance is the physical path (wire/wireless) between transmitter and receiver within single or multi-hop communication. It is only plausible to say that longer distances from sources to the intended destination will consume more power. Therefore, reducing the average transmission distance for the packets will have positive impact on the energy consumption and delay time. Figure 12 depicts the average transmission distances from each single node to the BS. It shows that when the algorithm is implemented, the average transmission distance is reduced. Figure 13 shows the average energy consumption for each round. According to the Figs. 11 and 12, the proposed algorithm reveals less number of hops and transmission distances. As a result, the energy consumption of the proposed scheme is lower.

As already mentioned, the proposed technique is compared with FCFS and NJN algorithms. FCFS is a popular technique uses to forward the packets according to their interval time on the queue. First packet arrives in the queue then first packet delivers to the destination. While NJN is a scheduling scheme that always selects the nearest device as the first job to collect the data and then disseminates it to the ultimate receiver. Both techniques are unsuitable for data accessing multiple devices

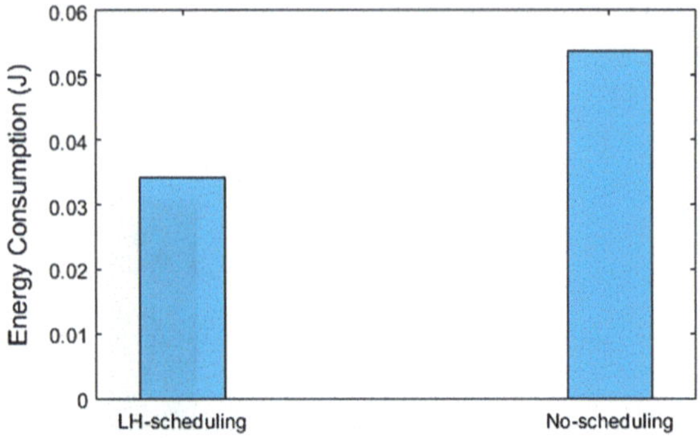

Fig. 13 Avg. energy consumption (J)

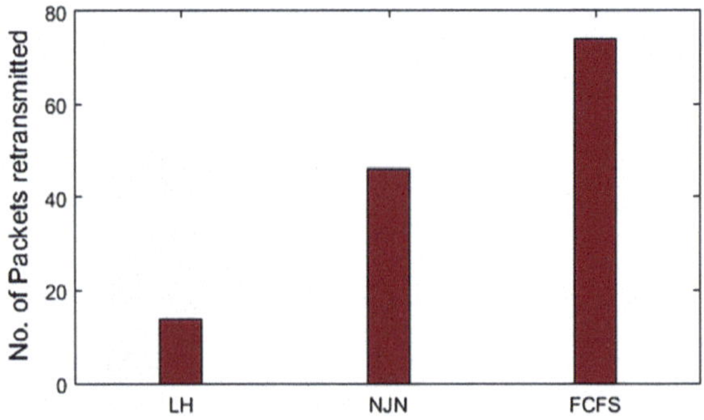

Fig. 14 Number of packets loss

and links and thus uses high energy waiting in the system. Therefore, the probability of retransmitting these packets will increase due to waiting time. Below, the main crucial results are introduced based on [26].

The proposed technique takes into consideration to calculate the number of packets loss. It is important to reduce the retransmission packets that can further minimise the memory used, energy consumption, processing time of the network. Figure 14 shows the proposed algorithm has lower number of retransmission packets than FCFS and NJN schemes. This finding evidence that higher number of retransmission packets means resend the data over the entire path, thus it incurs wasting bandwidth and higher latency.

Figure 15 reveals the throughput of the network. It shows our method gives 7.2% and 4.3% increase in throughput over FCFS and NJN respectively. This is because the number of hops and packets error probability decrease thus, causing the network throughput to maximise.

Simulation setting has been adopted as in [20] where it takes 2 ms for a sensor node to make a transmission. The length of an interval period to update packets is 200 ms. End-to-End delay [27] is the time taken by bits to travel through the wire or wireless communication from the source to the intend destination. Delay time depends on congestion in the network and number of hops. Basically, end-to-end delay categories into four types in the network:

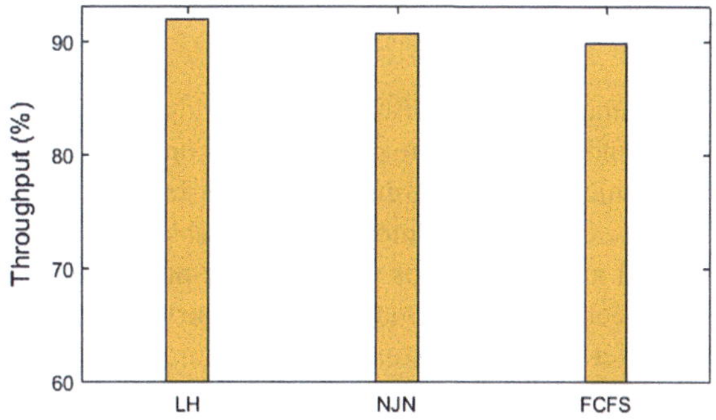

Fig. 15 Throughput (%)

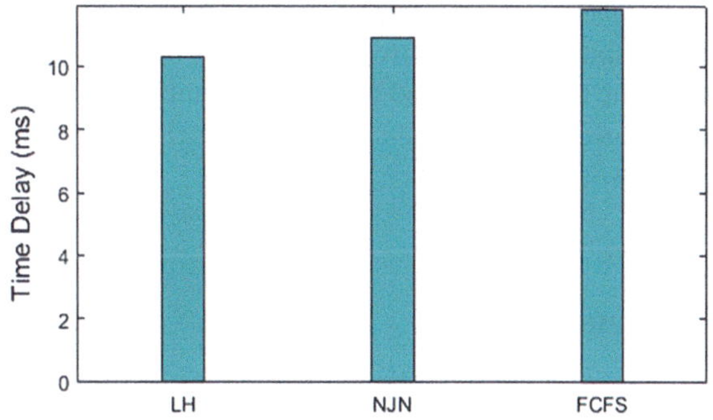

Fig. 16 Delay time (ms)

- **Propagation delay (p_d)**: is the time it takes for single bit to travel over a link from the sender to the receiver at propagation speed over the specific media as shown below in Eq. (11).
- **Transmission delay or (store-and-forward delay) (t_p)**: is the time required to push all bits into wire/wireless as revealed in Eq. (12).
- **Processing Delay (p_s)**: is the time it takes to process all the bits in a network device.
- **Queuing Delay (Q(t))**: is the time for each packet waits in a queue until it can be forwarded to the next hop. Therefore, the total end-to-end delay (t_{total}) measured by Eq. (13).

Figure 16 shows the superiority of proposed scheme over two methods in reducing the end-to-end delay due to minimize the retransmission packets.

$$P_d = \frac{Length}{Speed} \tag{11}$$

$$t_p = \frac{Packet\ size}{rate} \tag{12}$$

$$t_{total} = \sum P_d + t_p + p_s + Q(t) \tag{13}$$

Conclusion

Data management in WSNs and IoT fields is a critical especially for large-scale networks and should optimize the power consumption without compromising reliable communication. Therefore, this chapter introduces a novel scheduling algorithm to optimize energy consumption for scalable IoT networks. The proposed technique reduced number of hops of the sensor nodes and thus minimise the packets loss. Since packet loss occurs at CH nodes due to quench, time to live exceeding for packets, buffer is full, or destination being unreachable, application of the proposed long hop (LH) first algorithm on the CH nodes shows promising results for energy optimization. The proposed algorithm assigns high priority for messages coming from far distances to be scheduled first thus preventing re-transmission. The preliminary results showed the effectiveness and the efficiency of the proposed algorithm to maximize the throughput and reduce the number of re-transmission packets, end-to-end delay, energy consumption, transmission distance, number of hops, and thus elongate the network lifetime.

References

1. M. Hammoudeh, F. Al-fayez, H. Lloyd, R. Newman, B. Adebisi, System: deployment issues and routing protocols. IEEE Sens. **17**(8), 2572–2582 (2017)
2. W. Stallings, *Foundations of Modern Networking: SDN, NFV, QoE, IoT, and Cloud* (2015)
3. H. Ghayvat, S. Mukhopadhyay, X. Gui, N. Suryadevara, WSN- and IOT-based smart homes and their extension to smart buildings. Sensors (Switzerland) **15**(5), 10350–10379 (2015)
4. D.R. Dandekar, P.R. Deshmukh, Energy balancing multiple sink optimal deployment in multi-hop wireless sensor networks, in *2013 3rd IEEE International Advance Computing Conference (IACC)* (2013), pp. 408–412
5. L. Farhan, A.E. Alissa, S.T. Shukur, M. Alrweg, U. Raza, R. Kharel, A survey on the challenges and opportunities of the internet of things (IoT), in *International Conference on Sensing Technology* (2017)
6. M.M. Rathore, A. Ahmad, A. Paul, S. Rho, Urban planning and building smart cities based on the internet of things using big data analytics. Comput. Netw. **101**(2016), 63–80 (2016)
7. Project team IEC, Internet of things : wireless sensor networks executive summary, in *International Electrotechnical Commission* (2014), p. 78
8. M. Hammoudeh, R. Newman, Adaptive routing in wireless sensor networks: QoS optimisation for enhanced application performance. Inf. Fusion **22**, 3–15 (2015)
9. A. Abuarqoub, M. Hammoudeh, B. Adebisi, S. Jabbar, A. Bounceur, H. Al-Bashar, Dynamic clustering and management of mobile wireless sensor networks. Comput. Netw. **117**, 62–72 (2017)
10. S. Hagen, IPV6 essentials, in *Grundlagen—Funktionalität, Integr.* 2nd edn. (Sunny ED, 2009)
11. H. Wang, J. Jin, Z. Wang, L. Shu, On a novel property of the earliest deadline first algorithm, in *2011 Eighth International Conference on Fuzzy Systems and Knowledge Discovery (FSKD)* (2011), pp. 197–201
12. S. Abdullah, K. Yang, An energy-efficient message scheduling algorithm in internet of things environment, in *2013 9th International Wireless Communications and Mobile Computing Conference (IWCMC)* (2013), pp. 311–316

13. L. Karim, N. Nasser, T. Taleb, A. Alqallaf, An efficient priority packet scheduling algorithm for wireless sensor network, in *2012 IEEE International Conference on Communications (ICC)* (2012), pp. 334–338
14. D. Saha, M.R. Yousuf, M.A. Matin, Energy efficient scheduling algorithm for S-mac protocol in wireless sensor network. Int. J. Wirel. Mob. Netw. **3**(6), 129–140 (2011)
15. Z. Wang, Y. Liu, Y. Sun, Y. Li, D. Zhang, H. Yang, An energy-efficient heterogeneous dual-core processor for internet of things, in *2015 IEEE International Symposium on Circuits and Systems (ISCAS)* (2015), pp. 2301–2304
16. W.Y. Lee, Energy-saving DVFS scheduling of multiple periodic real-time tasks on multi-core processors, in *2009 13th IEEE/ACM International Symposium on Distributed Simulation and Real Time Applications* (2009), pp. 216–223
17. Z. Vincze, R. Vida, A. Vidacs, Deploying multiple sinks in multi-hop wireless sensor networks, in *IEEE International Conference on Pervasive Services* (2007), pp. 55–63
18. XBee Modules, XBee®/XBee-PRO® RF Modules, in *Product Manual v1. xEx-802.15.4 Protocol* (2009)
19. L. Farhan, R. Kharel, O. Kaiwartya, M. Hammoudeh, B. Adebisi, Towards green computing for internet of things: energy oriented path and message scheduling approach. Sustain. Cities Soc. **38**(July 2017), 195–204 (2018)
20. Y. Zheng, L. Wan, Z. Sun, S. Mei, A long range DV-Hop localization algorithm with placement strategy in wireless sensor networks, in *2008 4th International Conference on Wireless Communications, Networking and Mobile Computing* (2008), pp. 1–5
21. S. Broumi, A. Bakal, M. Talea, F. Smarandache, L. Vladareanu, Applying Dijkstra algorithm for solving neutrosophic shortest path problem, in *International Conference on Advanced Mechatronic Systems (ICAMechS)* (2016), pp. 412–416
22. A.M. Ahmed, S.H. Ahmed, O.H. Ahmed, Dijkstra algorithm applied: design and implementation of a framework to find nearest hotels and booking systems in Iraqi, in *International Conference on Current Research in Computer Science and Information Technology (ICCIT 2017)* (2017), pp. 126–132
23. N.M.G. Appenzeller, I. Keslassy, Sizing router buffers. SIGCOMM Comput. Commun. **34**(4), 281–292 (2004)
24. D. Raca, A.H. Zahran, C.J. Sreenan, Sizing network buffers : a HTTP adaptive streaming perspective, in *The IEEE 6th International Conference on Future Internet of Things and Cloud Work* (2016)
25. L. Farhan, A.E. Alissa, S.T. Shukur, M. Hammoudeh, R. Kharel, An energy efficient long hop (LH) first scheduling algorithm for scalable internet of things (IoT) networks, in *11th International Conference on Sensing Technology* (2017)
26. L. Farhan, L. Alzubaidi, M. Abdulsalam, A.J. Abboud, M. Hammoudeh, R. Kharel, An efficient data packet scheduling scheme for internet of things networks, in *Diyala Third International Scientific Conference of Engineering Sciences 2018, 1st Diyala International Scientific Conference of Engineering Sciences 2018*
27. Y.S. Uddin, F. Saremi, T. Abdelzaher, End-to-end delay bound for prioritized data flows in disruption-tolerant networks. In *IEEE Real-Time Systems Symposium* (2010) https://doi.org/10.1109/RTSS.2010.39

Internet of Things: Vision, Future Directions and Opportunities

Laith Farhan and Rupak Kharel

Abstract Internet of things (IoT) is considered to revolutionize the way internet works and bring together the concepts such as machine to machine (M2M) communication, big data, artificial intelligence, etc. to work under a same umbrella such that cyber space and human (physical systems) are more intertwined and thus ubiquitous giving rise to cyber physical systems. This will involve billions of connections and smart products communicating with each other mostly without human intervention to achieve smart objectives. The idea of IoT has enticed significant research attentions since the massive connectivity bring varieties of challenges and obstacles including heterogeneity, scalability, security, big data, energy requirements, etc. The chapter looks into providing a concise review of the concepts on IoT and applications describing the main features, vision, and future directions. Furthermore, open issues and challenges that need addressing by the research community and some potential solutions are discussed.

Introduction to Internet of Things

During the last decade, Internet of Things (IoT) has attracted intensive attention due to a wide range of applications in industrial, biomedical observation, agriculture, smart cities, environmental monitoring and other fields (Fig. 1) [1]. IoT is the internetworking of physical devices used in our daily lives that use standard communications architectures to provide new services to end users [2].

It is envisioned that by 2020 the future Internet will include tens of billions of smart objects/devices [3]. IoT technology provides better services to end users via real-time data processing, communications and visualization. IoT can be extended to almost everything from refrigerator to washing machine, wristwatches to smartphones, home

L. Farhan (✉) · R. Kharel
Manchester Metropolitan University, Manchester, UK
e-mail: l.al-bayati@mmu.ac.uk

L. Farhan
University of Diyala, Baqubah, Iraq

S. C. Mukhopadhyay et al. (eds.), *Modern Sensing Technologies* ,
Smart Sensors, Measurement and Instrumentation 29,
https://doi.org/10.1007/978-3-319-99540-3_17

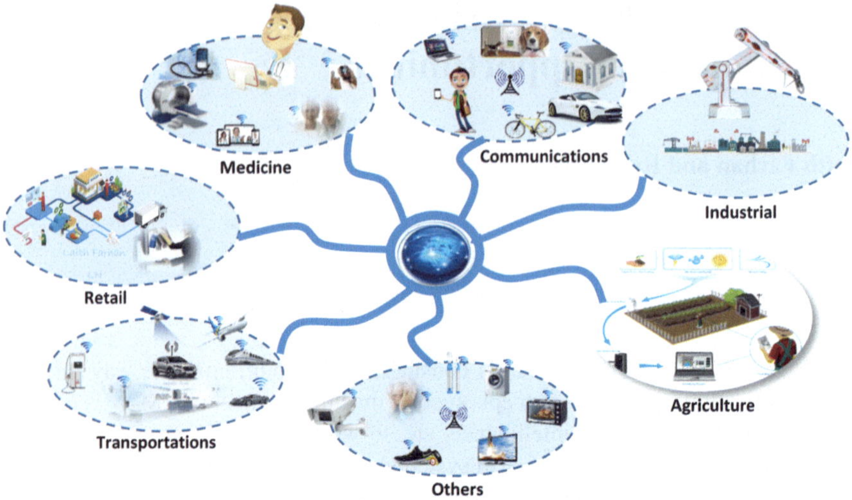

Fig. 1 Network of the future

security to alarm system, etc. [4]. For example, smart refrigerators can tell us the end of the validity of food using bar-codes or which items to buy during our shopping in the market. On the other hand, imagine that we can control our house from anywhere. By using smart phones or tablets with just simple touch we can set a desired temperature or turn lights on or off before getting home. These are examples of just a few applications out of thousands being currently developed every day in the field of IoT.

The massive growth in the number of devices connected to the internet (up to 100 billion devices), poses a huge range of challenges. In the future IoT will not be islands of isolated systems, but will be an integration of many islands of connected systems, applications, services and underlying devices. At the moment, each of these devices and services work on their own architectures, data format, and own existing protocol stacks. They are all still at early stages of development. Hence, the communication between these objects is insecure, suffers from interoperability and integration issues [5]. Furthermore, the sources of energy required to power these devices are very precious due to the fact that most of them are powered by battery or by means of energy harvesting. Therefore, there is a need for comprehensive review of existing unconstrained and constrained devices protocols with the view of developing unified, dynamic, standardized, energy efficient and intelligent protocol stacks with recourse to node identity (both capacity and capability). So far, most of these new challenges and concerns have started to attract the attention of academic researchers and companies.

IoT Applications, Future Directions and Challenges

Many researchers and early adopters have come up with promising solutions to overcome the problems and challenges in realizing IoT applications. However, IoT opens up new horizontal challenges that demand latest research and capacities to address them. In this section, we briefly highlight the key challenges for the future IoT systems (see Fig. 2). Wireless Sensor Networks (WSN) are considered as one of the fundamental underlying technologies for IoT. Therefore, the study has also considered some previous works in the WSN field.

Fault Tolerance

Fault tolerance is one of the most important issues in the area of WSN and IoT applications. Both technologies involve large number of heterogeneous sensor nodes spread over a large geographic area to perform a specific task. The explosive growth of the connected devices demands higher reliability and performance from modern

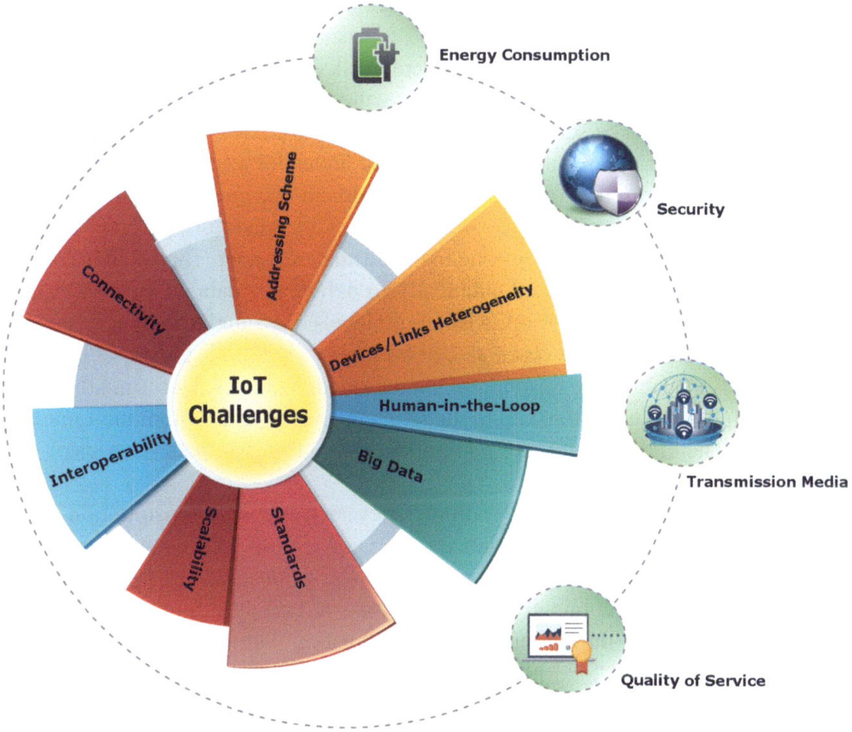

Fig. 2 IoT challenges and opportunities

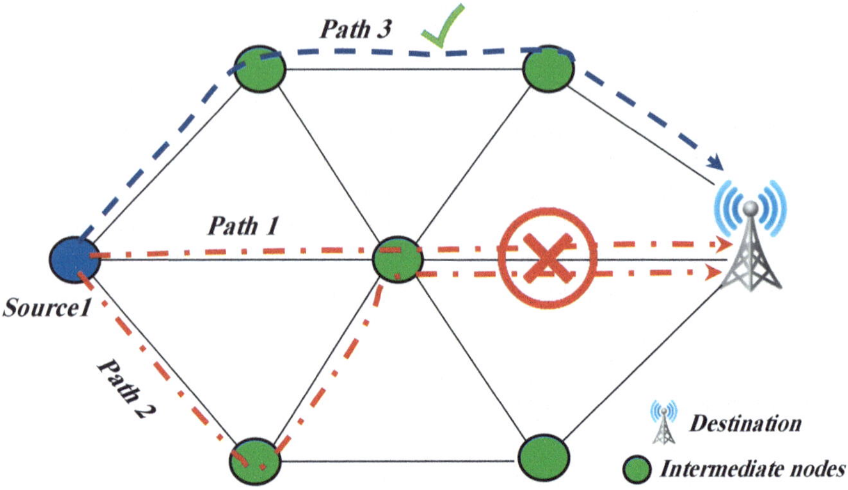

Fig. 3 Multi-level paths between source and destination

networks. Some sensor nodes may be blocked or fail due to exposure to harsh environments, interference, physical damage or lack of power [6]. Probability of sensor node failure increases with rise in number of sensors and increasing sensing field. Such nodes failure should not affect the overall task of the sensor network [7]. Most of WSNs and IoT routing protocols should have the ability to recover from the failure of a sensor node [8]. The sensor is reported as a failure node when a node cannot receive or communicate messages with neighbor nodes for a specific period of time and thus excluded from the routing path. The problem becomes worse when two or more sensor nodes fail in the same area. The network might cripple very quickly because other nodes cannot find a route to the ultimate receiver. Therefore, a routing protocol must follow new links dynamically to deliver the data collected by other devices to the intended destination. Also, multi levels of redundancy may be needed in a fault tolerant sensor networks as can be seen in Fig. 3. It clearly shows that source 1 sends its data to the destination via intermediate nodes. Unfortunately, path 1 and 2 are failing to carry the data to the final destination due to malfunctioning of some nodes. Therefore, the topology might require to find new path for packets and reorganize the network.

Many design goals are related to routing policy such as energy consumption, small delay, high throughput, limited variance of the connection quality, etc. Authors in [9] present expected transmission count (ETX) algorithm that performs up to two times better than minimal hop-count for long links in term of throughput. The ETX strategy is to find high throughput paths on multi-hop wireless communication. It minimizes the expected total number of packets transmission (including retransmissions) that requires to successfully deliver data to the final destination. Another study [10] implements new routing technique called balanced energy adaptive routing (BEAR) for IoT networks. The proposed method operates in three phases: (i)

Sensor nodes share the information related to their locations and residual energy. (ii) BEAR protocol elects neighbor nodes and selects the facilitator and successor nodes based on the node locations. (iii) Data transmission: The results of BEAR show improvement of network lifetime by up to 55%. In [11], authors investigate a novel context-aware routing (CAR) method for IoT applications. The proposed scheme improves the current request response model, during the exchange in peer to peer fashion. The simulation results of CAR show reduction in the total wasted time in network delay and improvement of network service and bandwidth.

Quality of Service (QoS)

The concept of QoS is relatively new in the area of IoT. There are a few number of researches currently dealing with this feature. In many applications, gathering data needs to be delivered within a certain time to the ultimate receiver otherwise the data will be of less value. The QoS requirements are met with differentiated services and delay management, packet loss, and bandwidth parameters in a network. These requirements become the secret to a successful end-to-end service. We review the most critical challenges faced in IoT infrastructures while implementing QoS requirements and are listed below and seen in Fig. 4.

- **Scalability**: IoT field is widely supported by an increasing number of devices that should not affect the QoS.
- **Communication range**: Most of these devices are limited in the transmission range that lead to a major source of QoS degradation in IoT nodes problem. Therefore, a node should depend on other intermediate nodes to be able to communicate

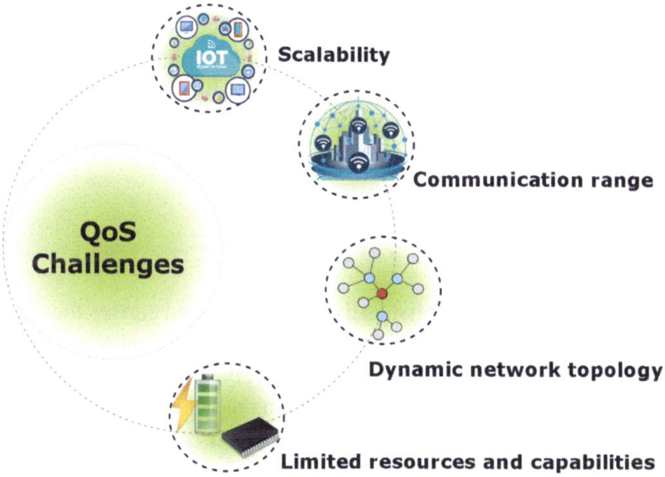

Fig. 4 Quality of service challenges

with others out of transmission range. These intermediate nodes act as relays for packets.

- **Dynamic network topology**: IoT network should have dynamic topology due to route changes, node power failures, link failure, etc. Therefore, QoS of these applications should not be affected with dynamic changes.
- **Limited resources and capabilities**: As already mentioned, most of the connected devices are constrained nodes to its ecosystem which have limited memories, CPU capabilities, power sources and bandwidth. Therefore, the QoS should be aware of limited resources and capabilities of these IoT-objects.

As a result, quality of service needs research and stabilization for implementation, optimization and management. In Ref. [12], authors present a general model to support the QoS aware deployment of multi-components IoT cloud infrastructures. The proposed model introduces suitable operational systemic qualities of fog facilities. Another study, authors [13] show a hybrid push-pull traffic scheme for data exchange in IoT environment. The proposed algorithm reduced 50% of network load and throughput compared with traditional IPv6 and showed minimal packet drop.

Humans in the Loop (HITL)

As IoT technology proliferates and things become more sophisticated, many of these new applications will require some form of human interaction. Human-in-the-loop allows the user to change the outcome of an event or process. For example, self-driving cars (also known as auto cars) are a great example when we mention HITL. The car mostly drives itself, but it still needs human to be alert in the case of emergency. When the sensor system sense something unusual on the road (e.g., there is snow, construction, fire, possible collision, etc.), it probably has to hand the control of the car to human. For example, consider a controversial, unfortunate and ethical scenario, such as a self-driving car going to crash inflicting either significant third-party damages or serious injuries to the driver. Should human make the final decision rather than an artificial intelligence algorithm in such circumstances (Fig. 5)?

Transmission Media (TM)

TM is the physical path that establishes the connection and carries the data from the sender to the receiver. IoT networks use different types of technology to transmit and receive the data such as RFID, Bluetooth, LiFi, Zigbee, LoraWAN, Sigfox, etc. The traditional problems associated with transmission media (e.g., bandwidth, high error rate, fading, inference, etc.) exist for IoT applications as well. Each transmission medium requires specialized energy, network hardware, bandwidth that has to be

Fig. 5 Human in the loop challenges

Fig. 6 Different technologies of IoT

compatible with that medium. Therefore, optimizing the TM is a challenge in IoT environment to sustain and prolong the lifetime of networks.

According to study in [14], authors discuss and analyse the capacity and coverage of LoRaWAN and Sigfox in a large-scale area. They measured the interference in the European frequency 868 MHz. The study illustrated both technologies provided uplink and downlink failure rates of less than 1%. Furthermore, improvement of indoor coverage up to 99% is also shown. In a different study [15], authors design full-duplex wireless information transfer and use backscatter antennas to reduce latency and energy consumption from reader to the target. The proposed scheme suppressed interference from coexisting links. It also enabled simultaneous backward/forward information transfer (Fig. 6).

Big Data

Internet of things leverage massive amount of data aggregated via smart objects and is one of the most striking features of this new technology. It will be necessary to develop techniques that convert this data into usable knowledge. Every two years,

Fig. 7 The "5V" challenges

data is doubling in size and is expected to reach 44 Zettabytes in the next four years [16].

The "5V" (Value, Velocity, Volume, Variety, and Veracity) are important challenges in IoT applications as can be seen in Fig. 7 and explained below:

- **Velocity**: refers to the speed at which the data is being collected, transferred and processed. The speed of processing data is different based on the type of application. For some applications, the arrival of data can be handled within a short time while in other applications, real-time processing is required such as analytics programs.
- **Variety**: refers to the different types of data collected by end devices such as smart-phones, machines, sensors, etc. The data content is unstructured of different types such as audio, video, images, XML format, plain text, CSV format, etc. The variety of data should be organized and processed in a meaningful and consistent way.
- **Veracity**: means making sure that the data gathered and stored are accurate. This might mean filtering out any unwanted or corrupted data to enhance quality of applications.
- **Volume**: is the amount of all types of data that is collected, stored, retrieved, updated from different sources. IoT is creating enormous amount of data that is rising exponentially. The question is can we incorporate volume and velocity together?
- **Value**: Once the massive data is gathered accurately, the next step is to get the value out of the data. Therefore, various algorithms such as feature extraction,

trend analysis using artificial intelligence that enables informed decision, within the required time frame, is another challenge.

There are several contemporary big data management and analytics applications that can be applied in the area of IoT. Authors in [17] focus on the real-time data coming from IoT devices in a smart building. The framework presents new technique on analysis and storage of high-speed data generated by sensors. The system shows the monitoring and control of a large amount of data without human intervention and improved users experience. Previous study in [18] reported an integrated IoT architecture for gas and water smart meter. The intelligent model provides the information to the utility and customers for a large amount of data. The proposed system shows benefits for utility companies and customers such as reduce energy consumption, automatic and real-time meter reading that saved physical meter reading and thus maintaining environmental sustainability.

Security

The security problem is one of the major challenges of networks over the years. Thus, security, privacy and trust are critical factors for IoT applications as well. When packets are routed through different links and devices to reach ultimate receiver on the internet, measures should be taken so that the confidentiality and integrity of the data is maintained. Moreover, most of the IoT devices are low power constrained devices, therefore, already established cryptographic solutions cannot be directly applied in the IoT scene [16]. Also, currently, the integration of application in the network infrastructure is focused on only achieving the functionality rather than holistically considering the security requirements when the application is designed. This is leaving door open for attacks and hacking attempts. Cybersecurity experts have warned that IoT is one of the most vulnerable technology and they expect more targeted attacks on existing and emerging infrastructures, e.g., data theft, physical injury, DDoS attack, ransomware for smart homes or smart cars, etc. Four key IoT security challenges can be seen in Fig. 8:

- **Trust and data integrity**: is to ensure the data has not changed from the moment it is sensed until it reaches the final destination. It also involves verifying the data and validating the verification certificate.
- **Trillion points of vulnerability**: with each device getting connected to IoT represents a potential risk. This leads to questions: how confident can an organization be of the data gathered and the integrity of the data sent? How to make sure data has not been interfered or compromised with?
- **Data protection**: is the law required to be designed to protect and control individual and organization data gathered by sensors or applications and stored to be part of a filing system.

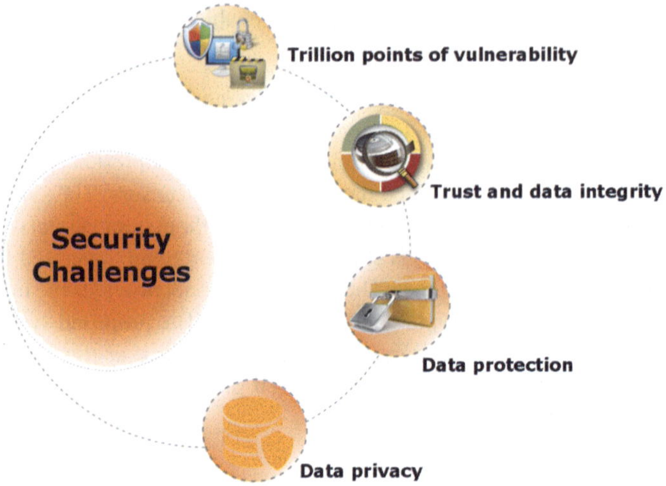

Fig. 8 Security challenges

- **Data privacy**: is to protect the data from exposure in the IoT environment. For instance, any logical or physical entity can be given a unique address and the ability to communicate automatically over the network.

A novel dynamic defence frame for IoT security has been implemented in [19]. The proposed method is divided into two phases: (i) The first stage applies based on recognition of security threats. (ii) The second stage uses real data provided by first stage. The authors provide a great solution method to ensure IoT security. With the same objective, authors in [20] propose a lightweight trust design to identify and isolate common routing attacks for IoT applications. In this protocol, the SecTrust framework mitigate routing attacks by enhancing the integrity and confidentiality of the IoT routing protocols. A new lightweight privacy-preserving data aggregation protocol has been investigated by [21] to enhance IoT security. The LPDA scheme supports fault-tolerance and efficiently aggregate IoT devices. It also early filters false data infected by attackers.

Addressing Schemes

Uniquely identifying objects is a critical issue for the operation and success of IoT applications. IoT-objects require to uniquely classify thousands of devices and to manage and control them remotely through the internet. A few most critical features of creating unique address are reliability, uniqueness, scalability and persistence [3]. These smart devices require a suitable and unique address that will make them able to communicate each other and become part of the internet. Internet protocol version 4 (IPv4) uses 32-bit addresses and provides capacity for only 4.3 billion IP addresses which are almost out of addresses. The next generation is IPv6 and it uses 128-bit addresses and has massive address abundance 3.4×10^{38} or (340 trillion trillion trillion) [22, 23]. A group of researchers [24] design a lightweight addressing scheme to solve IoT heterogeneity and scalability problems. In this model, virtual domain and multi encoding have been used to implement the nodes addressing. The proposed scheme shows suitable and fixable interconnection between WSNs and internet based on IPv6 over 6LowPAN. As highlighted by [25], authors propose identification and addressing scheme for IoT devices where they used distributed address allocation algorithm to implement automatic ID for IoT nodes. This work also presents addressing scheme combining cluster tree algorithm with AODV routing protocol. It implements nodes addressing scheme in the local and wide area of IoT networks. To this end, a unified addressing scheme for IoT is a very popular research field and a big challenge.

Devices/Links Heterogeneity

Another important feature of the vision of IoT is the variety of devices and links since it will be working on different sets of protocol suite, data formats, operating systems, etc. In WSNs, most of the sensors are homogeneous, i.e., having the same power, communication, capacity, and processor in term of computation. While IoT technology implements a wide variety of networks, links, and devices connectivity to provide different services [1]. Thus, heterogeneous nature of links and objects play a critical role in the interconnection of the IoT devices and thus add a unique challenge to address. Therefore, the question might be, is it possible to have a unified architectural model that can be deployed that it is able to support these wide ranges of devices and applications? In [26] an architecture is presented for heterogeneity of devices and networks based on SDN-Docker techniques. The proposed work reveals the feasible architecture and communication established between IoT devices through an SDN-based network. The DIAT scheme is a simple, scalable distributed architecture for large-scale IoT networks. It is specially designed to overcome the interoperability among various devices and deployments [27].

Mobility, Geography

Internet of things with mobile sink are expected to increase the flexibility of provide services and gathering data in large-scale sensing fields and detecting environments. The use of geographical position of nodes is necessary to detect nodes locations and simplify routing for WSNs and IoT networks. In such systems, it requires each node to determine its own location as well as the positions of its neighbour nodes. This information can be obtained with

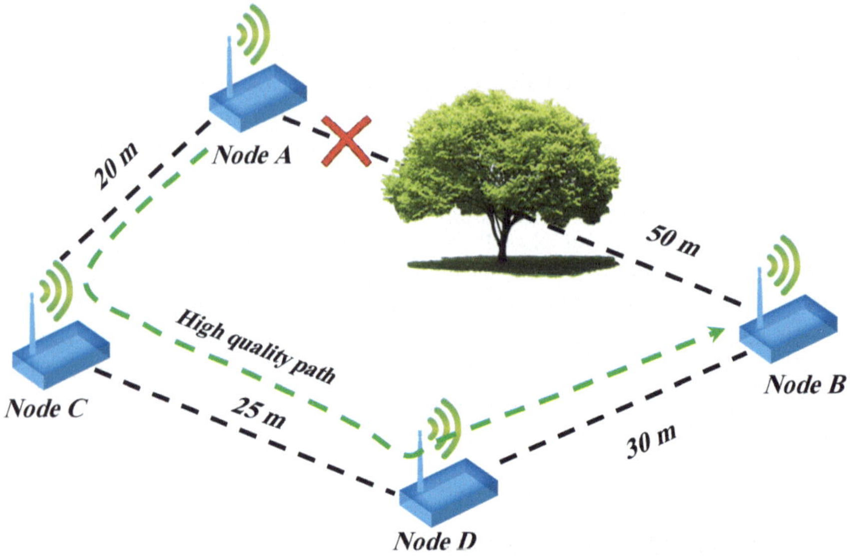

Fig. 9 Although the direct link from node A to B is less distance, the quality of path is more efficient to route traffic over node C and D that accepts a geographically longer path

the help of global positioning system (GPS) [28]. With this location information, a message can be routed to the ultimate receiver after including these information into routing algorithm in various ways [29]. Direct communication, whenever possible, is certainly the best way for data dissemination [22]. However, the quality of a link can be significantly decreased by obstacles such as buildings, trees, walls (see Fig. 9). Therefore, the researchers should take into consideration a signal strength metric than by using the geographical distance.

The 5G and 4G Technologies Enabled IoT

The fourth generation (4G) technology has been widely used in the IoT and continuously evolving to match the needs of the future networks [30]. The main generations evolved from 1G to 4G, recently work is progressing on 5G. With each technology, new features are added and issues are resolved. The 4G technology with long term evolution (LTE) provide high quality video stream and audio over end-to-end user with high bandwidth speed that could reach up to 1 Gbps. While the fifth generation (5G) is the next technology of mobile internet networks and is expected to be operational in next two years [31]. It is the extension of 4G LTE however with 10 times faster, higher data rates, more secure, lower latency, long battery lifetime, and reliable connections on smartphones and other devices than ever before (see Fig. 10). It expects to handle about

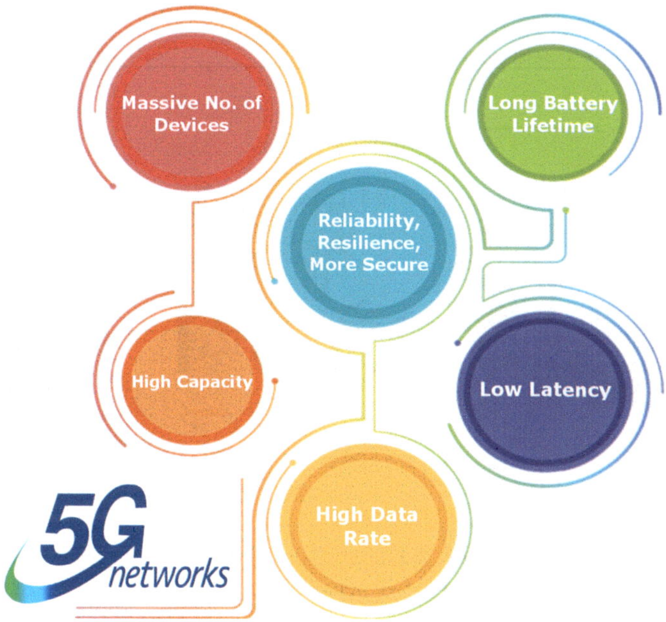

Fig. 10 5G enabled IoT innovation

more than 1000 times of mobile data than recent cellular systems. These features will certainly be expected to make the next generation is the optimal network and solution for IoT applications. This finding provides evidence that the 5G platform will help the IoT technology to meet the requirements of applications and market demands. In additional, it will become the solution of IoT applications such as non-orthogonal multiple access, non-orthogonal waveforms, massive MIMO systems, machine to machine (M2M) communications, etc.

Internet of Things Enabled by LiFi Technology

Light Fidelity (LiFi) is the wireless technology that uses the visible light communication networks instead of radio waves for data transmission. It is low cost and efficient technology due to it uses only LED lamp to transmit data. This technology provides highspeed, lightweight, secure and fully networked wireless via light. As we are aware that the light travels faster than the radio waves. Therefore, the data could be transferred 250 times faster than the high-speed broadband [32]. In additional, the technology involves visible light wavelength rather than radio waves that can lead to human disorders.

As the market for IoT-devices grows and tiny sensors are embedded to more and more things, buildings. Thus, these things produce huge amount of data and require

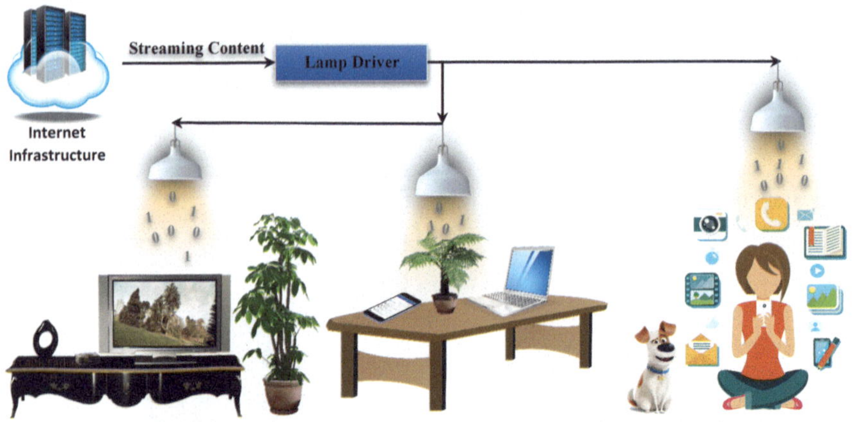

Fig. 11 LiFi-technology enabled IoT innovation

faster data transmission to reach the target. Therefore, LiFi technology offers many key benefits and best solutions for IoT networks in recent decade (Fig. 11).

Power Line Communication (PLC) for IoT Applications

PLC is a communication technology that carries data on a conductor and sending it over existing power cables. Nowadays, power supply lines not only use for power transmission but it also for data communication as a second job. It offers great ecosystems advantages and opens up a wide field for different applications [33]. A wide range of PLC technologies are used for different purposes such as smart grid, home automation, IoT, etc. Internet of Things is recently popular field and an innovation buzzword. It has several applications that will enable a smarter world such as smart home, smart cities, intelligent enterprise. At the concept of smart home, there is always an electronic network linked each other. So, with the PLC platform make it possible to create communications between a variety of devices/things by using any existing cable network. This finding provides a great opportunity that minimizes the cost to deploy and further expansion of the network [34] (Fig. 12).

Conclusion

In summary, with the emergence of IoT, new regulatory approaches to ensure its energy, scalability, Heterogeneity, human-in-the-loop, big data, etc. become necessary. The IoT revolution is expanding connectivity via the internet and a wide range of applications (e.g., actuators, sensors and other embedded systems). This will have

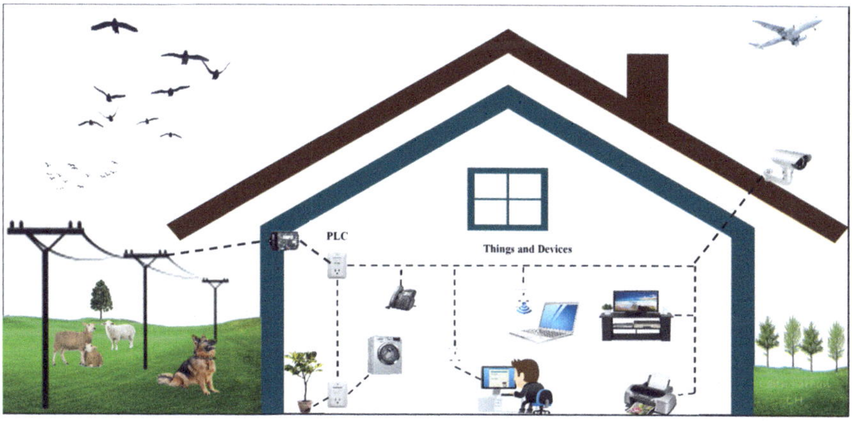

Fig. 12 PLC-technology enabled IoT innovation

an effect on the quality, different life styles and the way we behave and interact with humans, machines and devices in the future. Therefore, new research challenges and problems will emerge due to the largescale device proliferation and their intercommunication. This chapter gives an overview of the key issues related to the IoT services and technologies. A number of researcher challenges have been described, which are expected to become a major research trends in the next decade. A number of previous works and new technologies have been analysed, and most relevant WSN and IoT applications were presented.

Acknowledgements The authors would like to thank Ministry of Higher Education and Scientific Research (Iraq) and University of Diyala for the funding to conduct the research. Also, many thanks to Manchester Metropolitan University for their help and support.

References

1. S. Li, L. Da Xu, S. Zhao, The internet of things : a survey, Inf. Syst. Front. **17** (April 2014), 243–259 (2015)
2. W. Stallings, *Foundations of Modern Networking: SDN, NFV, QoE, IoT, and Cloud* (2015)
3. J. Gubbi, R. Buyya, S. Marusic, M. Palaniswami, Internet of Things (IoT): A vision, architectural elements, and future directions. Futur. Gener. Comput. Syst. **29**(7), 1645–1660 (2013)
4. A. Abuarqoub, M. Hammoudeh, B. Adebisi, S. Jabbar, A. Bounceur, H. Al-Bashar, Dynamic clustering and management of mobile wireless sensor networks. Comput. Netw. **117**, 62–72 (2017)
5. L. Farhan, R. Kharel, O. Kaiwartya, M. Quiroz-castellanos, A. Alissa, A concise review on Internet of Things (IoT)—problems, challenges and opportunities, in*11th International Symposium Communication System Networks, Digital Signal Processing, Hungary* (2018)
6. E. Pignaton de Freitas et al., Handling failures of static sensor nodes in wireless sensor network by use of mobile sensors, in *2011 IEEE Working International Conference* on *Advanced Information Networking* and *Applications* (2011)

7. L. Farhan, R. Kharel, O. Kaiwartya, M. Quiroz-castellanos, LQOR: link quality-oriented route selection on Internet of Things networks for green computing, in *11th International Symposium on Communication Systems Networks, Digital Signal Processing Hungary* (2018)
8. K.Y. Bendigeri, J.D. Mallapur, S.B. Kumbalavati, Recovery based fault tolerance in wireless sensor networks, in *International Conference* on *Electrical, Electronics Communication* and *Computational Optimum Technologies* (2017)
9. D. Couto, D. Aguayo, J. Bicket, R.M.A, High throughput path metric for multi-hop wireless routing. Wirel. Netw. **11**(4), 419–434 (2005)
10. N. Javaid, S. Cheema, M. Akbar, N. Alrajeh, M.S. Alabed, N. Guizani, Balanced energy consumption based adaptive routing for IoT enabling underwater WSNs. IEEE Access **5**, 10040–10051 (2017)
11. F. Al-turjman, M. Gunay, CAR approach for the Internet of Things approche de la CAR pour l' internet des objets. Can. J. Electr. Comput. Eng. **39**(1), 11–18 (2016)
12. A. Brogi, S. Forti, QoS-aware deployment of IoT applications through the fog. IEEE Internet Things J. **4662**(c), 1–1 (2016)
13. S. Muralidharan, B.J.R. Sahu, N. Saxena, A. Roy, PPT: a push pull traffic algorithm to improve QoS provisioning in IoT-NDN environment. IEEE Commun. Lett. **21**(6), 1417–1420 (2017)
14. B. Vejlgaard, M. Lauridsen, H. Nguyen, I.Z. Kovacs, P. Mogensen, M. Sørensen, *Interference impact on coverage and capacity for low power wide area IoT networks* (IEEE Wirel. Commun. Netw. Conf, WCNC, 2017)
15. W. Liu, K. Huang, X. Zhou, S. Durrani, Full-duplex backscatter interference networks based on time-hopping spreading spectrum. IEEE Trans. Wirel. Commun. **1276**(c), 1–16 (2017)
16. L. Farhan, A.E. Alissa, S.T. Shukur, M. Alrweg, U. Raza, R. Kharel, A survey on the challenges and opportunities of the Internet of Things (IoT), in *11th International Conference* on *Sensor Technologies* (2017)
17. M.R. Bashir, A.Q. Gill, A. Iot, Towards an IoT big data analytics framework : smart buildings systems, in *14th International Conference on Smart City* (2016)
18. J. Lloret, J. Tomas, A. Canovas, L. Parra, An integrated IoT architecture for smart metering. IEEE Commun. Mag. **54**(12), 50–57 (2016)
19. C. Liu, Y. Zhang, H. Zhang, A novel approach to IoT security based on immunology, in *Proceedings—9th International Conference* on *Computational Intelligence* in *Security, CIS 2013* (2013), pp. 771–775
20. D. Airehrour, J. Gutierrez, S.K. Ray, A lightweight trust design for IoT routing, in *2016 IEEE 14th Intl Conf on Dependable, Autonomic and Secure Computing, 14th International Conference on Pervasive Intelligence and Computing, 2nd International Conference on Big Data Intelligence and Computing and Cyber Science and Technology Congress (DASC/PiCom/DataCom/CyberSciTech)* (2016), pp. 552–557
21. R. Lu, K. Heung, A.H. Lashkari, A.A. Ghorbani, A lightweight privacy-preserving data aggregation scheme for fog computing-enhanced IoT, in *Spec. Sect. Secur. Priv. Appl. Serv. Futur. Internet Things*, vol. 5 (2017), pp. 3302–3312
22. L. Farhan, R. Kharel, O. Kaiwartya, M. Hammoudeh, B. Adebisi, Towards green computing for Internet of things: energy oriented path and message scheduling approach. Sustain. Cities Soc. **38**(July 2017), 195–204 (2018)
23. L. Farhan, A.E. Alissa, S.T. Shukur, M. Hammoudeh, R. Kharel, An energy efficient long hop (LH) first scheduling algorithm for scalable Internet of Things (IoT) networks, in *11th International Conference* on *Sensor Technologies* (2017)
24. B. Luo, Z. Sun, Research on the model of a lightweight resource addressing. Chin. J. Electron. **24**(4), 832–836 (2015)
25. R. Ma, Y. Liu, C. Shan, X.L. Zhao, X.A. Wang, Research on identification and addressing of the Internet of Things, in *Proceedings—2015 10th International Conference on P2P, Parallel, Grid, Cloud Internet Computing 3PGCIC 2015* (2016), pp. 810–814
26. I. Bedhief, M. Kassar, T. Aguili, SDN-based architecture challenging the IoT heterogeneity, in *2016 3rd Smart Cloud Networks System SCNS 2016* (2017), pp. 14–16

27. C. Sarkar et al., DIAT: a scalable distributed architecture for IoT. IEEE Internet Things **2**(3), 230–239 (2015)
28. L. Farhan, L. Alzubaidi, M. Abdulsalam, A.J. Abboud, M. Hammoudeh, R. Kharel, An efficient data packet scheduling scheme for Internet of Things networks, in *Diyala Third* Scientific *Conference* of *Engineering Science, 1st Diyala Inernational. Conference* of *Engineering Science 2018* (2018)
29. R. Baumann, S. Heimlicher, M. Strasser, A. Weibel, A survey on routing metrics, in *TIK Report 262, ETH-Zentrum, Computer Engineering* and *Networks* Laboratory (2007)
30. S. Li, L. Da Xu, S. Zhao, 5G Internet of Things: a survey. J. Ind. Inf. Integr., (January), 1–10 (2018)
31. M.B. Yassein, S. Aljawarneh, Challenges and features of IoT communications in 5G networks, in *2017 International Electrical* and *Computing Technologies* and *Applications* (2017)
32. M. Aswin, A.R.G, L.V.S.M, Iot enabled by Li-Fi technology, in *Proceedings of Nattional Conference on Communications and Informatics-2016* (2018)
33. G. Bernhard, A.M. Lehmann, J.B. Huber, A power line communication topology module for NS-3 and DCE, in *IEEE International Conference* on *Smart Grid Communications*, no. October, pp. 295–301 (2017)
34. A.A. Zhilenkov, D.D. Gilyazov, I.I. Matveev, Y.V. Krishtal, Power line communication in IoT-systems, in *2017 IEEE Conference* of Russian *Young* Researchers in *Electrical* and *Electronic Engineering* (2017)

A Novel Unobtrusive Vibration Sensing System for Machine Inspection

Dibyendu Roy, Prasant Misra, Tapas Chakravarty, Arijit Sinharay, Raj Rakshit and Arpan Pal

Abstract Spectral vibration signatures of a machine offer one of the earliest indication of a potential failure. The journey from potential failure point to full functional failure is often seen to get traversed in weeks. Thus, it becomes imperative that any condition monitoring system is able to immediately detect anomalous vibration signatures, as they begin to build up. The machines with rotating parts often experience anomalies due to defects like looseness in joints, misalignments, imbalance, mechanical wear and tear etc. Such incidents result in vibrations induced on non-rotating body parts. The preferred mode of vibration sensing is through in situ sensors like accelerometers attached on specific locations of the machine; however a significant benefit (with respect to legacy machines) will be derived if an accurate measurement can be done from a distance, without attaching the sensor on the machine parts. In this regard, we have designed an affordable, unobtrusive and autonomous vibration sensing system, called "ShakeMeter", where we combine two cost-effective measurement principles namely optical stroboscope with a standard low-frame rate camera and a Doppler sensor. The optical stroboscope detects the vibration frequency with a high degree of precision by capturing the modulo (of division) between the difference between the object's vibration frequency and the optical frequency. The

D. Roy · T. Chakravarty (✉) · A. Sinharay · R. Rakshit · A. Pal
TCS Research and Innovation, Kolkata, India
e-mail: tapas.chakravarty@tcs.com

D. Roy
e-mail: roy.dibyendu@tcs.com

A. Sinharay
e-mail: arijit.sinharay@tcs.com

R. Rakshit
e-mail: raj.rakshit@tcs.com

A. Pal
e-mail: arpan.pal@tcs.com

P. Misra
TCS Research and Innovation, Bangalore, India
e-mail: prasant.misra@tcs.com

S. C. Mukhopadhyay et al. (eds.), *Modern Sensing Technologies*,
Smart Sensors, Measurement and Instrumentation 29,
https://doi.org/10.1007/978-3-319-99540-3_18

vibration signal may consist of multiple independent signals. For incorporating such effects, multiple mutually co-prime strobing frequencies are generated and utilized to efficiently estimate the multiple frequency components via Chinese Remainder Theorem (CRT). Experimental results show that our proposed technique can effectively estimate multiple frequencies with a frequency detection error of 0.5% for vibrations occurring up to 1 kHz. While functionalizing this system is an important milestone, its deployment feasibility in real world scenarios is subject to various factors like; sensing distance from the object, luminance conditions, duty cycle of the optical array etc. We study these operational challenges and analyze the performance of this system under those constraints. Based on these empirical studies, we endorse that the system should be operated at a distance of 1–2 m from the vibration scene, and at a strobing duty-cycle rate that does not exceed 1%. Finally, to estimate the vibration signatures induced on a large surface (at multiple points), a Doppler radar is fused with the mentioned technique. Experimental result illustrates that this method can be very beneficial for quickly capturing the spatial vibration information.

Introduction

Machine condition monitoring is an *established* industrial practice [1] of finding out the health of machines, in order to *identify* changes that may be indicative of a *developing fault, damage, or degradation*. An early detection of such warning signals help to prevent unscheduled outages, optimize machine performance, and reduce the repair time and maintenance costs thereof. Vibration, noise and temperature are often used as progressive *external* indicators to obtain better insights about the internal state of machines. With regards to rotating and reciprocating machines, *vibration* is typically the *first* warning sign of a possible fault under manifestation. Vibration is a function of the machine dynamics [2, 3], and provides valuable insights into various operational aspects of machines such as condition of bearings or gears, alignment efficacy of shafts and balance, impact of structural resonance, etc. Therefore, vibration-based condition monitoring is widely used for assessing machine wellness.

Machine vibrations are generally measured using accelerometers or displacement sensors [4], both of which are *contact-based* and need to be directly connected to the physical scene-of-interest. Hence, it is an *arduous* task to deploy such sensor types and can be exceptionally challenging in hazardous machining environments, such as structures exposed to high temperature and pressure. *Stand-off* vibration sensing, on the other hand, is an alternative to alleviate such operational risks. The two most promising techniques, in this regard (although with certain impediments), are the laser Doppler vibrometer (LDV) [5] and the near-field acoustic holography (NAH) [6]; which are based on direct laser and indirect acoustic pressure measurements, respectively, in the near field of the vibrating surface. While the sensing precision of LDVT is very high, the system is exorbitantly costly. On the other hand, NAH is capable of completely describing the acoustic field (including vibrations) generated by a structure; but has severe restrictions with respect to boundary conditions compared to traditional intensity measurement systems [7]. Furthermore, depending on the vibration span, high frame rate camera systems can also be an elegant means of measurements, but they also tend to be very expensive [8].

In this chapter, we present the design of ShakeMeter: a low-cost, reliable stand-off vibration sensing system consisting of two sensors connected in a manner so as to offer much enhanced features as compared to a traditional industry grade stroboscope. However, each sensor can also be independently considered for specific application scenarios or where cost of deployment is a critical requirement. To begin with, we describe how an optical stroboscope using a low FPS camera can search for multiple vibration frequencies [9–11]. In the case of standalone operation of optical strobing unit, a unique feature is the manner in which it is able to measure multiple vibration frequencies with a low frame rate camera, whose limited sampling rate does not support this direct measurement. Here, the principle of optical strobing is used to make the vibrating surface appear stationary or slow moving when imaged with a camera, and this affect is noted when the strobing frequency is in very close range of the actual vibrating frequency of the measurement structure [12]. We build on this basic sensing premise, and augment a new estimation method to accurately obtain the multiple vibration frequencies prevalent in the machine structure. We further present an optimization strategy to reduce the response time of the ShakeMeter system by elegantly fusing it with a Doppler radar.

Principle of Strobing

The basic principle of strobing plays an important role in the study of mechanical motion like vibrations. Authors have demonstrated its utilization in vibration detection [3] and condition monitoring of machines [1].

Let us assume that a vibrating surface generates a signal $x(t)$ with a center frequency of f, and is strobed with a signal $s(t)$ with a centre frequency of f_s.

The resultant sampled signal can be written as:

$$w_s(f) = \frac{2\pi}{T_s} \sum_{n=-\infty}^{\infty} X(f - n \times f_s) \tag{1}$$

where, $\frac{2\pi}{f_s}$ and $n \in Z^+$.

The resultant signal is then captured by a camera, having a sampling rate of f_c frames per second (FPS). In order to get meaningful insights from the consecutive camera images, the following theorem should be followed.

Theorem 1 (Condition of Perfect Strobing): *From (1), the aliased images can be efficiently recovered, if f_c is greater than or equal to twice of $(f - n \times f_s)$ (as per the Nyquist criterion). In addition, the frequency folding phenomena observed in aliased images is used for better recovery. The perfect strobing conditions are observed; if:*

$$|f - n \times f_s| \leq \frac{f_c}{2}$$
$$||f - n \times f_s| - f_c| \leq \frac{f_c}{2} \qquad (2)$$

The unknown vibration frequency f can be obtained from the aliased images, if n can be is accurately estimated.

In the next section, we describe the entire "*Stroboscope*" system design for validating Theorem 1.

Stroboscope: System Design and Implementation

System architecture—The system consists of *three* units as shown in Fig. 1. It consists of (i) sensing unit, comprising of a low frame rate camera, (ii) actuation unit, comprising of a custom designed optical strobing system, and (iii) control and processing unit, comprising of a computing station. The *sensing unit* is used to record the vibration scene. The actuation unit uses the stroboscopic effect to create the optical illusion of slowdown of the vibrating scene at the sensing unit, when flashed at a frequency that is close to the vibration frequency (or, is submultiples of it). The stroboscopic conditions ensure that the visibility of the vibrating surface is broken into a series of brief episodes in which its vibration motion appears to be minimal. The control and processing unit is a computing station that progressively tunes the strobing frequency of the actuation unit so as to bring it down to the sampling speed of the sensing unit; and after obtaining the correct spectral span, estimates the frequency of the vibrating scene.

Hardware platform details—The vibrating surface is emulated by using the diaphragm of a speaker, which is made to oscillate at a pre-configured frequency. A visual marker is attached to this moving surface for the camera unit (recording at 30 FPS or frames per second) to sense these vibrations. The stroboscopic effect

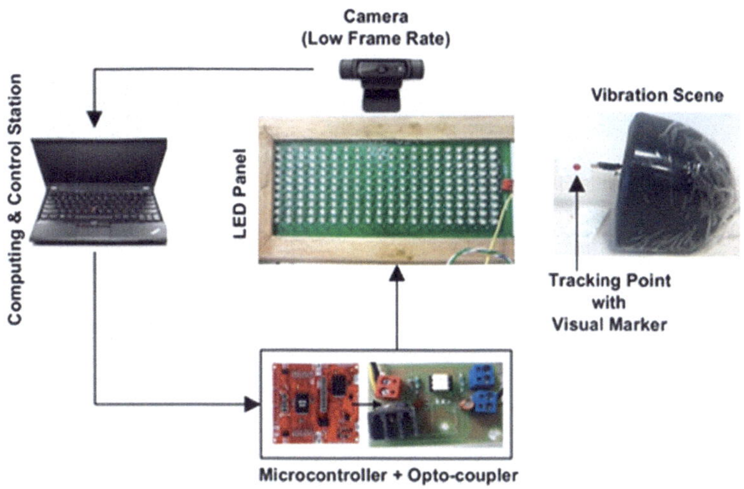

Fig. 1 The Stroboscope system and emulation of vibration scene

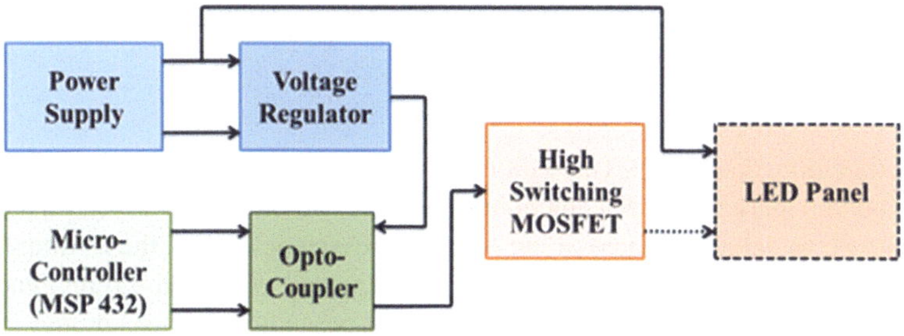

Fig. 2 Optical sampler hardware

is ingested into the vibration scene by the optical strobing system. It is a custom designed unit that consists of: (i) a 200 element light emitting diode (LED) panel, and (ii) a MSP432 microcontroller [from M/s Texas Instruments] interfaced with an opto-coupler (for high frequency switching). The schematic of the optical sampler is shown in Fig. 2

Software details—In the software part, the impact of the optical sampler on the vibrating object is captured by the camera. For easy tracking of the resultant rotational movement, a visual marker is attached to the vibrating surface-of-interest of the machine [10, 11]. The captured images are transformed into their respective gray-scale representation, and then an image complement is applied to make the

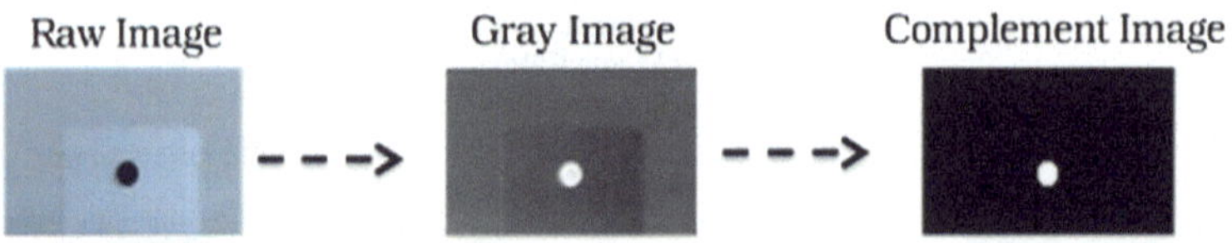

Fig. 3 Processing pipeline for maker tracking

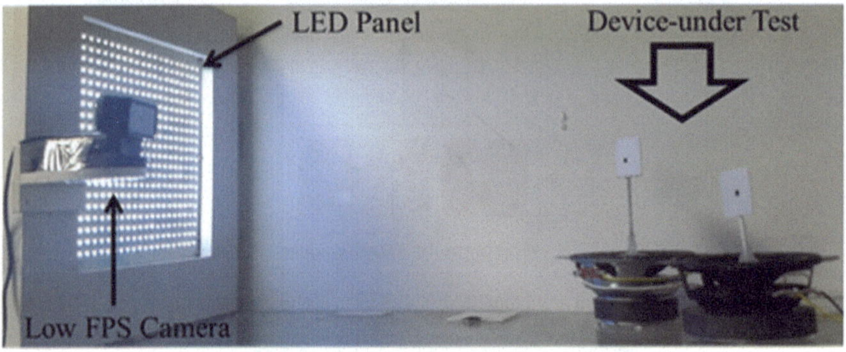

Fig. 4 Optical Stroboscope prototype

marker spot more distinctive. Thereafter, the centroid of the visual scene is tracked for finding out the marker's displacement in each frame. At this stage, a spectral analysis is performed to get an estimate of the dominant frequencies, and the resultant SNR indicates the need to increase or decrease the frequency of the pulse train generated by the optical strobe. The overall image tracking procedure of the captured images are shown in Fig. 3.

Figure 4 displays a desktop version of the optical stroboscope designed using the principles mentioned before. Here, two independent speakers are emulating two vibrating bodies, a standard camera is placed in the same plane as the LED panel. A more compact version of the said unit had been built in later design phase.

Stroboscope: Single/Dual Frequency Vibration Measurement

Our first set of empirical studies to validate the techniques was performed on surfaces emitting singular frequency components. In this regard, we tuned the speaker to generate a single frequency, and this served as the ground truth for the experimental validation.

Let us assume that the optical unit is strobing with a frequency f_s such that condition given by Eq. (2) holds true (i.e., resultant images are within the aliased zone). In such a case, the observed frequency (f_o) is denoted as:

$$f_o = f - n \times f_s \tag{3}$$

In order to estimate n with high accuracy, the strobing frequency is increased by a small amount Δ and hence takes the new form—($f_s + \Delta$). As a result, in the subsequent parse, the new observed frequency denoted by f_{o1} is given as:

$$f_{o1} = f - n \times (f_s + \Delta) \tag{4}$$

From (3) and (4), the value of n can be estimated as:

$$n = \frac{f_{o1} - f_o}{\Delta} \tag{5}$$

Once the value of n is calculated, the unknown vibrating frequency f can be estimated from (3).

It is important to note that the same technique can be used to measure dual frequency components, if the frequency range is known a priori. The corresponding results are shown in Fig. 5.

Figure 5a represents the observed frequency spectrum of the aliased images time-series when $f_s = 250$ Hz and $f = 755$ Hz. The peak frequency is observed at $f_o =$ 5.09 Hz; which validates our strobing principle (i.e. (2)) experimentally.

For the dual frequency estimation, the result is shown in Fig. 5b. Here, $f_s = 755$ Hz and $f = 328$ and 668 Hz. The corresponding observed frequencies are: $f_o = 2.05$ and 8.11 Hz.

Limitation: The aforementioned strategy has two drawbacks

1. It is assumed that the vibrational frequency range is known a priori; and that the strobing frequency, starting from a default value, will be incremented linearly. However, in practice, such a range is indefinite; and hence, the entire process will be very time consuming.
2. Machines generate multiple frequencies as opposed to single or dual frequencies' and hence, the proposed technique will not suffice.

To overcome these limitations, we have proposed an improvement on our searching strategy along with multi-frequency detection algorithm so that convergence timing will be much less than the previous technique.

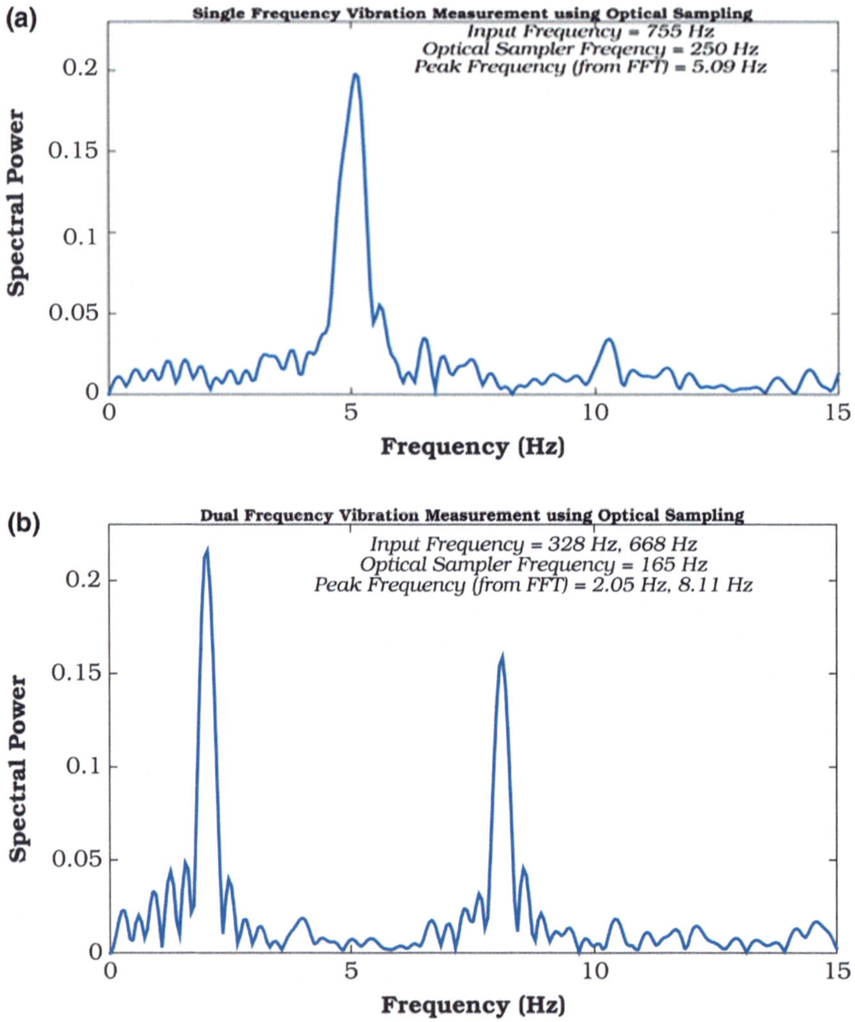

Fig. 5 Spectrum of vibration measurement. **a** Single frequency, **b** dual frequency

Stroboscope: Multiple Frequency Vibration Measurement

We now extend our analysis to the case where the vibrating signal $x(t)$ consists of multiple frequency components—$[f_1 f_2 \ldots f_N]$. As explained in the previous section, $x(t)$ is sampled with a strobing signal $s(t)$ with a center frequency—f_s. The resultant sampled signal can be denoted as:

$$w_s(f) = \frac{2\pi}{T_s} \sum_{n=-\infty}^{\infty} \sum_{i=1}^{N} X(f - n \times f_s) \tag{6}$$

Therefore, for multi-frequency vibration detection scenario, the condition of strobing can be given as:

$$\sum_{i=1}^{N} |f_i - n \times f_s| = f_o$$

$$\sum_{i=1}^{N} ||f_i - n \times f_s| - f_c| = f_o \tag{7}$$

where, $f_o \leq \frac{f_c}{2}$ is the frequency observed by the camera.

For each sampling instance f_{si} (where, i ∈ [1, 2, …, p]), a set of f_o will be obtained. Hence, for each r (where, $1 \leq r \leq p$), the residue set $S_r(f_1, f_2, \ldots, f_N)$ can be estimated, assuming $n \in Z^+$ and $f_c = 30$. The final residue set is—[S_1, S_2, …, S_N]. On this residue set- S, an improved Chinese Remainder Theorem (CRT) based algorithm has been applied for detection of the multi-frequency components [13]. The detection algorithm is given below:

Detection Algorithm

The detail steps for the multiple frequency vibration detection are given below:

Step 1 Take an arbitrary vector set: $(k_1, k_2, \ldots, k_p) \in S$.

Step 2 For each $1 \leq r \leq p$, define a set;
$\Gamma_r = \{k_r + n \times f_s; k_r + n \times f_s\}$
Where, n is an integer. All numbers in Γ_r, have same residue modulo f_s, which is called as coset of k_r.

Step 3 There exist integers $(r_1, r_2, \ldots, r_\alpha)$ with $1 \leq r_1 < r_2 < \cdots < r_\alpha \leq p$, such that the residues $(k_{r1}, k_{r2}, \ldots, k_{r\alpha})$ are from a common frequency (i.e., $\tau = \Gamma_{r_1} \cap \Gamma_{r_2} \cap \ldots \cap \Gamma_{r_\alpha} \neq \varnothing$). So, $\tau = \{\bar{N}\}$.
Here, it is important to check whether $\bar{N}$ is a valid frequency by verifying its residue vector $(k_1, k_2, \ldots, k_p)\, mod\, (f_{s_1}, f_{s_2}, \ldots, f_{s_p})$ belongs to the set S. If not, find another set of $1 \leq r_1 < r_2 < \cdots < r_\alpha \leq p$, such that $\tau = \Gamma_{r_1} \cap \Gamma_{r_2} \cap \ldots \cap \Gamma_{r_\alpha} \neq \varnothing$. Repeat the step, until N is a valid frequency denoted by $N_N \in \Gamma$.

Step 4 For, r = 1, 2, …, p, remove $k_{p,r} = N_N\, mod\, f_s$ from S_r.

Step 5 Go to step 1 by replacing p with $p - 1$ and replace $S(N_1, N_2, \ldots, N_p)$ with $S(N_1, N_2, \ldots, N_{p-1})$. Repeat the steps until all N are determined.

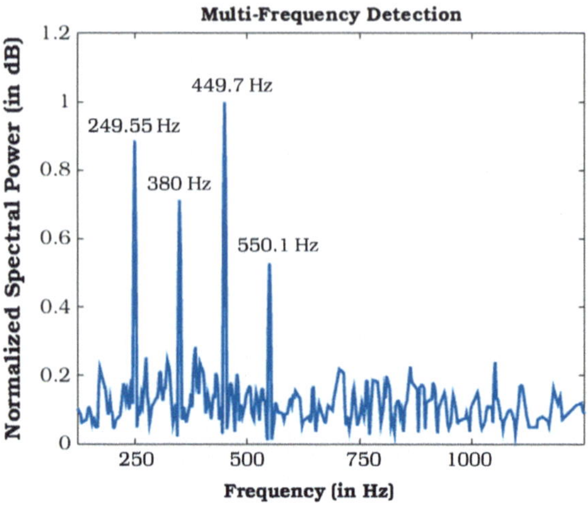

Fig. 6 Multiple frequency vibration measurement {*Ground Truth* = 250, 380, 450 and 550 Hz}

Table 1 Comparison results (up to 1 kHz)

	Traditional technique (in s)	Co-prime sampling technique (in s)
Single frequency measurement	150–200	50–75
Dual frequency measurement	225–500	75–180
Multi-frequency measurement	>2000	75–500

We experimentally study the accuracy of our proposed algorithm multi frequency detection algorithm; for which, multiple frequencies (250, 380, 450 and 550 Hz) are generated through the speaker membrane.

Figure 6 shows the result of our proposed CRT-based method. It is observed that the detection error of our proposed technique is within the range of 0.1–0.2%. It is important to note that sequential search methods can also obtain an accurately similar to the reported results; but take a significantly longer time (approx. 2000 s.) in contrast to around 350 s reported on this case. The comparison results are given in Table 1.

Stroboscope: Performance Analysis

Stroboscope is an optical sampling technique for unobtrusive vibration sensing. As a consequence, factors such as: camera and strobe separation distance from the vibration scene and its angular displacement, luminance conditions, marker size for optical tracking, and duty cycle of the strobing unit; have a substantial influence on the resultant signal-to-noise ratio (SNR) of the captured scene. In this section, we conduct a detailed study of these performance aspects and report out empirical findings [14].

Case 1: *Impact of strobing distance on detection accuracy under ambient light and dark conditions.* We conduct a study to quantify the system performance under varying sensing distances and luminance levels. It is performed by fixing the camera at a distance of 50 cm, and progressively changing the strobing distance from the vibration scene under two different lightning (ambient and dark) conditions. The result is depicted in Fig. 7a, b, where we observe that the detection accuracy exceeds 95% for distances up to 90 cm under ambient light conditions; and is noted to extend up to 110 cm under no background illuminance.

Case 2: *Impact of camera distance on detection accuracy under ambient light conditions.* Figure 7c shows the outcome of this study on the impact of camera distance on detection accuracy. In this setup, the strobing unit was placed at a distance of 50 cm and the distance of the camera module from the vibration scene was progressively increased. We observe that the detection accuracy exceeds 95% for distances up to 110 cm; with a sharp decrease in detection accuracy, if sensed from a farther distance.

Case 3: *Impact of angular displacement on detection accuracy under ambient light conditions.* We observe that the system is resilient to an angular displacement of 10 degrees, beyond which the detection fails. Hence the best position of the camera to track the marker perfectly is when the camera is directly looking at the scene with the strobing system lighting it up. The corresponding result is shown in Fig. 7d.

Case 4: *Impact of marker size on detection accuracy under ambient light and dark conditions.* Here, we perform some more detailed experiments to verify this aspect with three different marker sizes, which we refer to as: large (5×5 mm), medium (2×2 mm), and small (1×1 mm). The results are shown in Fig. 7e, f.

Case 5: *Impact of duty cycle on detection accuracy under ambient light conditions.* As stated before, the necessary condition for observing the strobing effect is that the sampling frequency of the strobe should be equal to the frequency of motion or multiples thereof. The tracking point (i.e., the marker) appears to be blurry upon increasing the duty cycle as the LED stays on for a longer time, until eventually at 100% duty cycle the strobe becomes a regular light. This is an artifact of strobes that give far more useful lighting power than continuous lightning systems. Hence, as shown in Fig. 7g, the best performance is obtained at 1% duty cycle compared to lower switching rates.

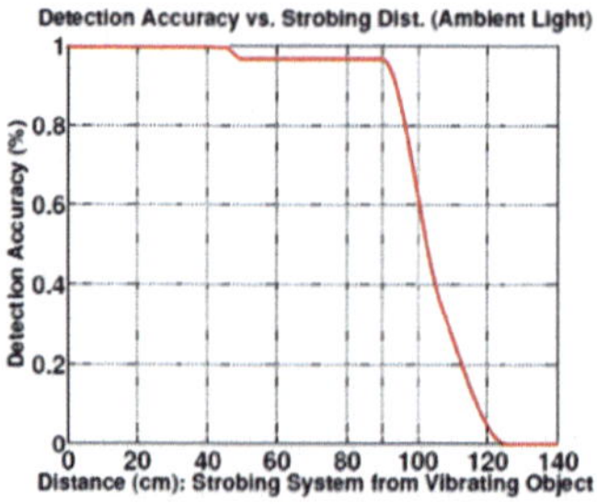

(a) Case-1

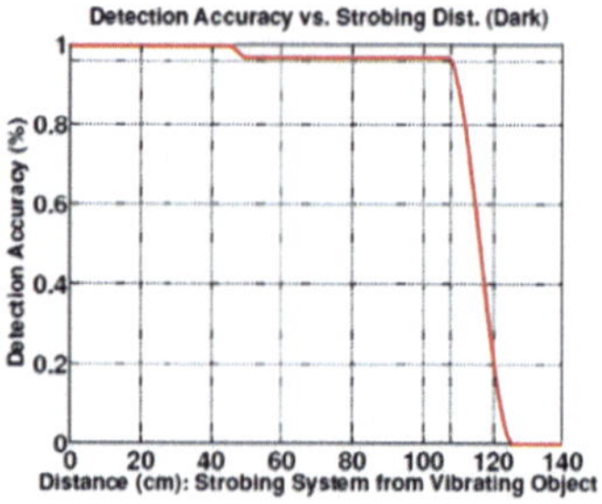

(b) Case-1

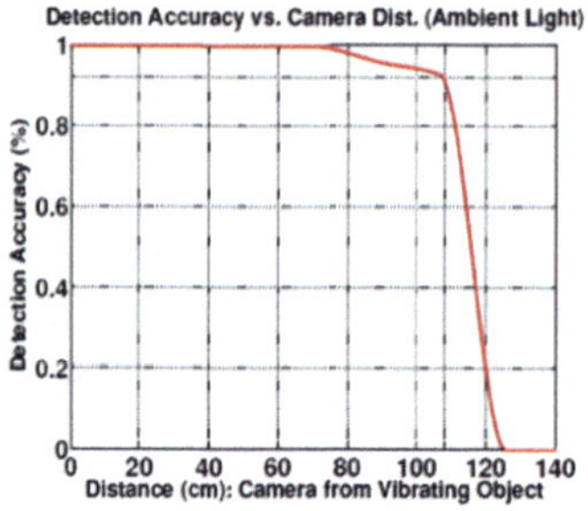

(c) Case-2

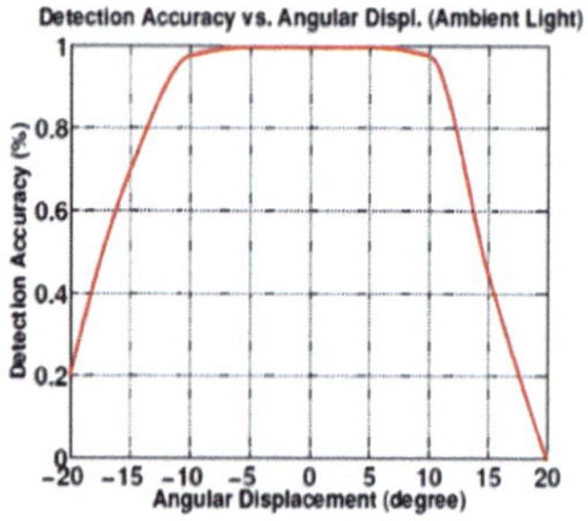

(d) Case-3

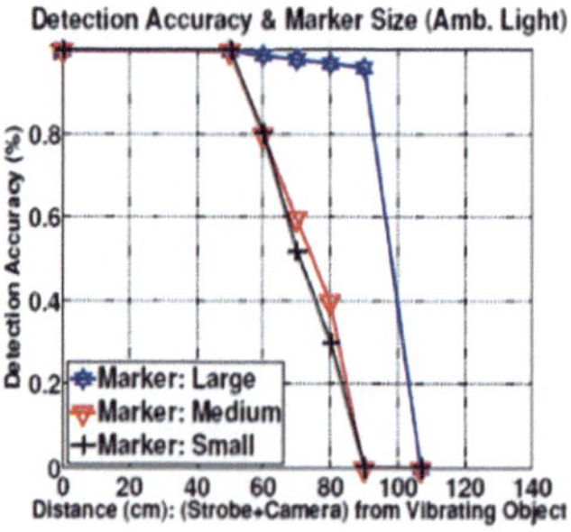

(e) Case-4

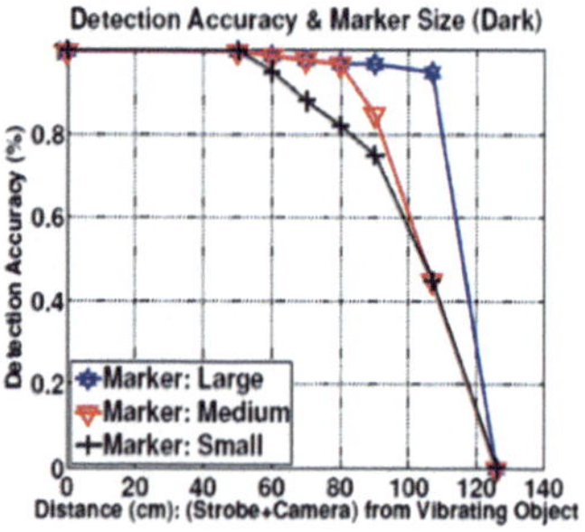

(f) Case-4

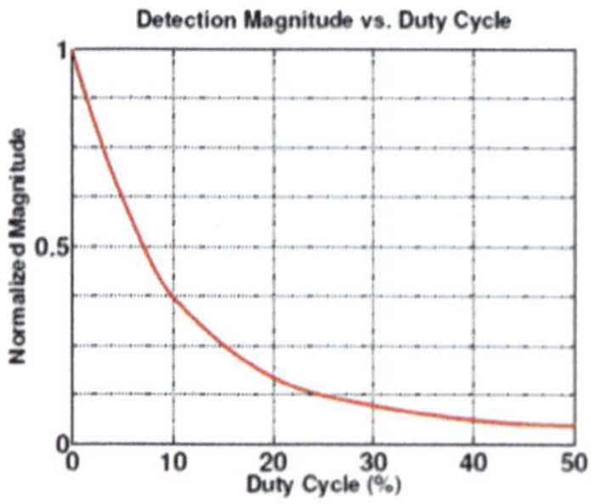

(g) Case-5

Fig. 7 Performance analysis of 'Stroboscope' [14]

Table 2 Performance analysis

Factor	Performance limit
Strobing distance (AL)	<90 cm
Strobing distance (D)	<110 cm
Camera distance (AL)	<110 cm
Angular displacement (AL)	±10°
Maker size (AL)	<95 cm
Maker size (D)	<110 cm
Duty cycle (AL)	<1%

AL: Ambient Light, D: Dark

The optimum performance measures are shown in Table 2.

In a realistic scenario, a machine will vibrate with numerous frequency components in different locations due to multiple connected rotating parts. Our proposed CRT-based frequency detection technique is an efficient method of detecting frequencies of a solitary location. In order to measure vibration of a large surface, the proposed method is very time-consuming. Hence, to reduce the overall detection timing over a huge vibrating surface, a Doppler radar is fused with optical stroboscope: together we call it as "ShakeMeter". In the next section, we offer the details of this new proposition.

Shakemeter: RF Guided Stroboscope

From the discussions presented in previous sections, it is to be noted that the proposed optical stroboscope will face challenges from deployment considerations. These are:

a. Speed of measurement for a machine with multiple target parts is a major concern. Each target needs to be sequentially monitored; one at a time.
b. The harmonic signatures occurring on 1X vibration will not get captured since the focus lies on detecting fundamental resonance with precision. The nature of harmonic signatures carries significant insights into the type of faults being generated.

In order to mitigate the above issues we designed an elegant combination of best of both worlds namely Stroboscope and Radar. The schematic is presented in Fig. 8.

This system combines two separate sensing methodologies described as above, in a unified approach for inference. Initially, the Doppler radar captures the spectrum of all vibrations in its field of view (FOV). The set of deduced frequencies are then messaged to Stroboscope. Stroboscope, in turn, illuminates the FOV using a LED panel, one frequency at a time; thereby enabling the identification of all the parts. This system is completely automated and significantly quick as compared to a standard Stroboscope which needs to search for the unknown frequency. Coupled with a

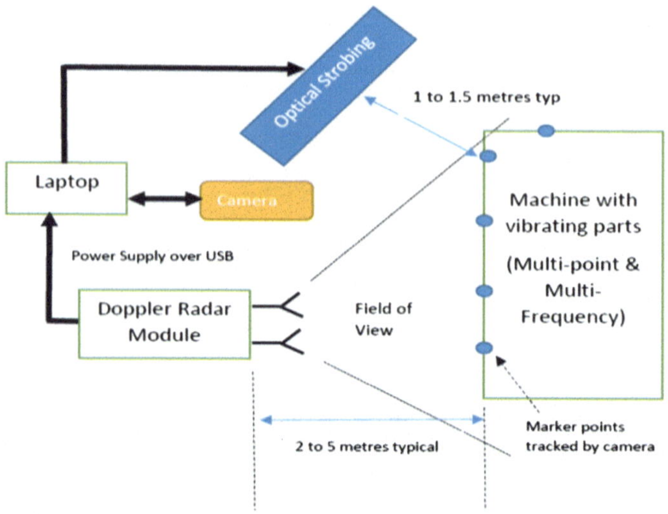

Fig. 8 Schematic of Shakemeter: RF guided Stroboscope for machine inspection

detailed spectral analysis of vibration (amplitude and frequency of harmonics), this low-cost and non-contact system offers elegant solution for legacy machines as well as for extreme environments like high temperature operations.

Utilizing the best features of two techniques, the following objectives have been achieved;

1. Identify all the interconnected moving parts of a machine i.e. detection is done without the need of pointing the sensors towards each vibrating part.
2. The entire process gets over a very short time. Significant improvement has been achieved over any other method which requires one to direct sensing beam to each vibrating part.

Moreover, resolving two closely spaced vibrating bodies in space requires pointed beam like manually rotated Laser beams. This requirement is completely obviated by this combination. Radar collects all the frequencies (in its wide FOV) and feeds the set to stroboscope. Stroboscope, then strobes the FOV selecting one frequency at a time; the zone which becomes visually still is identified and assigns the corresponding frequency to this specific location.

Doppler Measurement

The mathematical foundation for Doppler measurements of vibrating plates is drawn from the work of Li et al. [15]. Figure 9 displays the measurement principle.

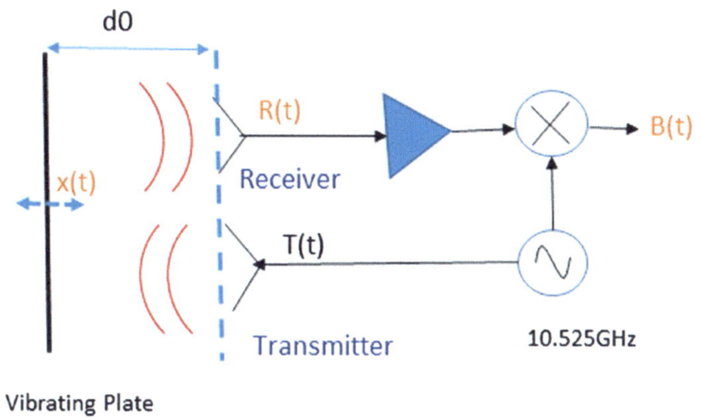

Fig. 9 Doppler measurement set up for a vibrating plate

Let us assume that a metallic plate is vibrating at a frequency f and the vibration is defined as

$$\mathrm{x(t)} = A sin\ \omega t \tag{8}$$

where A is the vibration amplitude.

Here, the transmitted signal $T(t)$ gets reflected from the plate and is received at receiver antenna. The received signal $R(t)$ is mixed with the same transmitted signal resulting in a baseband signal—$B(t)$. The vibration signature of the target plate is deduced from $B(t)$;

$$B(t) = \cos(\theta_0 + \frac{4\pi x(t)}{\lambda} + \Delta\theta(t)) \tag{9}$$

where,

$\theta_0 = \frac{4\pi d_0}{\lambda}$ is the phase shift due to the range and $\Delta\theta(t) = \theta(t) - \theta(t - \frac{2d_0}{c})$ is defined as residual phase noise.

The spectral analysis of the baseband signal $B(t)$ gives us the vibrating frequencies and their harmonics. In the proposed system, the detection error for Doppler radar is in the range of 0.2–0.3%. However, when taken together with stroboscope, the detection error for Shakemeter is brought down to the range of 0.02–0.03%.

The overall performance indices are reported in Table 3.

From Table 3, it is seen that the detection time has significantly improved in all respect for Shakemeter.

Table 3 Comparison results (up to 1 kHz)

	Traditional technique (in s)	Co-prime sampling technique (in s)	RF guided stroboscope (in s)
Single frequency single point	150–200	50–75	10–12
Dual frequency single point	225–500	75–180	15–20
Multi-frequency single point	Very large time to converge	75–500	15–30
Multi-frequency multiple point	Very large time to converge	Very large time to converge	15–50

Conclusions

The motivation behind the work is to enable condition monitoring of machines by detecting anomalous vibrations at an early stage, through the design of non-contact sensing mechanism. Speed of detection and autonomy of the detection process is the given objective. Towards such objective, we describe a new sensing concept—named as "Shakemeter-RF guided stroboscope" where two measurement principles namely optical stroboscope using standard camera and a RF based Doppler sensor is combined in an elegant manner. This chapter has briefly described about the affordable, unobtrusive vibration measurement technique which can efficiently detect vibrations induced on multiple surfaces in one single sweep, with quick detection time. We describe the mechanism where a low FPS camera can be used to detect unknown vibration frequency. Extending it further, an improved Chinese Remainder Theorem based algorithm has been proposed for estimating multiple frequency vibration information. After that the performance analysis has been analysed. Finally we have outlined the functional flow for RF guided Stroboscope and demonstrated its efficacy. Based on the empirical studies, we recommend that the ShakeMeter should be operated at a sensing distance in the range of 1–2 m from the vibration scene, and at a strobing duty cycle rate within 1%.

References

1. A.R. Mohanty, *Machinery Condition Monitoring: Principles and Practices* (CRC Press, 2014). ISBN 9781466593046
2. R. Yan, R.X. Gao, Hilbert–Huang transform-based vibration signal analysis for machine health monitoring. IEEE Trans. Instrum. Meas. **55**(6), 2320–2329 (2006)
3. C.W. de Silva, *Vibration: Fundamentals and Practice*, 2nd edn. (CRC press, 2006). ISBN 978084931987
4. http://www.ni.com/white-paper/3807/en/

5. P. Castellini, M. Martarelli, E.P. Tomasini, Laser doppler vibrometry: development of advanced solutions answering to technology's needs. Mech. Syst. Signal Process. **20**(6), 1265–1285 (2006). Special Issue: Laser Doppler Vibrometry
6. M. Martarelli, G.M. Revel, Laser Doppler vibrometry and near-field acoustic holography: Different approaches for surface velocity distribution measurements. Mech. Syst. Signal Process. **20**(6), 1312–1321 (2006). Special Issue: Laser Doppler Vibrometry
7. J.M. Robichaud, P. Eng. Reference standards for vibration monitoring and analysis. Bretech Engineering Ltd , 70 (2009)
8. J.G. Chen, Video camera-based vibration measurement of infrastructure, Ph.D. thesis, Massachusetts Institute of Technology (2016)
9. A. Veeraraghavan, D. Reddy, R. Raskar, Coded strobing photography: compressive sensing of high speed periodic videos. IEEE Trans. Pattern Anal. Mach. Intell. **33**(4), 671–686 (2011)
10. D. Roy, S. Mukherjee, B. Bhowmik, A. Sinharay, R. Dasgupta, A. Pal, An autonomous, non-invasive vibration measurement system using stroboscope, in *Proceedings of the 10th International Conference on Sensing Technology, ICST'16*, Nov 2016, pp. 1–6
11. D. Roy, S. Mukherjee, T. Chakravarty, A. Sinharay, A. Ghose, A. Pal, Demo abstract: ShakeMeter: an autonomous vibration measurement system using optical strobing, in *Proceedings of the 14th ACM Conference on Embedded Network Sensor Systems, SenSys'16* (ACM, New York, NY, USA, 2016), pp. 294–295
12. L. L´evesque, Nyquist sampling theorem: understanding the illusion of a spinning wheel captured with a video camera. Phys. Educ. **49**(6), 697 (2014)
13. H. Xiao, G. Xiao, Notes on CRT-based robust frequency estimation. Sig. Process. **133**, 13–17 (2017)
14. P. Misra, Roy D, T. Chakravarty, A. Pal, Unobtrusive vibration sensing using optical strobing: performance analysis, in *2017 Eleventh International Conference on Sensing Technology (ICST)* (IEEE, 4 Dec 2017), pp. 1–4
15. C. Li, Y. Xiao, J. Lin, Experiment and spectral analysis of a low-power Ka-Band heartbeat detector measuring from four sides of a human body. IEEE Trans. Microw Theory Tech. **54**(12), 4464–4471 (2006)

Compensation Techniques for Vibration Sensors with Application in Structural Health Monitoring

Navid Hakimitoroghi, Rabin Raut, Ashutosh Bagchi and Mehrdad Mirshafiei

Abstract With the prolific usage of vibration analysis to extract information about different structures (buildings and bridges etc.), vibration sensors have become an important device in Structural Health Monitoring (SHM). We tried to modify the sensor systems with a concentration on the analog subsystem sensors. Different types of seismic sensors with different analog characteristics are available to use in vibration sensors. Regarding the analog characteristics, different analog stages are needed to shape the signal appropriately. For the required level of sensitivity in SHM, the optimum choice for our case is a geophone. Mass-spring velocity meters or geophones are available in different sizes and characteristics. The main drawback of the geophones is their high-pass frequency response. As a result, vibrations below a certain frequency which is called corner frequency, is hard to detect. In this study, along with the adequate explanation on the filter and amplifier selection, several analog methods have been proposed to modify the lower frequency response of the geophone. In this chapter, after an introduction to the vibration sensors, we briefly described the geophone. Then we proposed some compensation techniques along with other analog stages. The experimental results verify the correctness of the techniques.

N. Hakimitoroghi (✉) · R. Raut
Department of Electrical and Computer Engineering,
Concordia University, Montreal, Canada
e-mail: navid.hakimi66@gmail.com

A. Bagchi
Department of Building, Civil & Environmental Engineering,
Concordia University, Montreal, Canada

M. Mirshafiei
Sensequake, Montreal, Canada

S. C. Mukhopadhyay et al. (eds.), *Modern Sensing Technologies*,
Smart Sensors, Measurement and Instrumentation 29,
https://doi.org/10.1007/978-3-319-99540-3_19

Introduction

The Advances in seismology started with the availability of precise seismometer from 1900 and afterward. Using primary seismometers, the scientists could measure the magnitude of earthquakes. Using seismometers scattered over a large area and with accurate record of timing, they could locate the position of the earthquake by the evaluation of delays in earthquake vibrations. At this point, the seismometers were not sensitive enough to measure smaller vibrations. For more precise applications in geological sciences (i.e., evaluation of the natural movement of the earth's crust) one needed to have information about the natural background vibrations. With this demand on detecting and processing of vibrations of the earth, knowledge on vibration sensors, and applying these for processing vibration signals have become vital in geology and seismology.

Besides the value of a vibration signal, the changes in the vibration signal with time has a vital role in seismic investigations. This item is defined as the frequency of the vibration. The frequency band of the vibrations in seismology ranges from 10^{-5} Hz to1000 Hz [1].

It is practically impossible to design one sensor to cover the wide range of magnitudes and frequencies. Therefore, sensor designers focus on building sensors that meet the requirements for a specific application [1].

One of the specific application of the vibration sensors is in structural and civil engineering. Civil engineers could extract important information about the structure by analyzing the vibration profile of the structure. The extracted information is used to model the structural behavior in earthquakes. This model is used to predict the behavior of the structure during an earthquake. Based on the predictions, they can reinforce the structure to prevent disaster in case of severe earthquakes. Evaluation of the seismic performance of a structure is also possible using numerical methods. These methods are based on linear or non-linear dynamic analysis and have different results due to using different approaches adopted to reach the result. On the other hand, the data used for such method is mainly gathered from the technical drawings of the building which are not accurate enough (especially for older buildings) [2].

One of the methods for investigation that is extensively discussed in [2] is ambient vibration testing (AVT). This method utilizes ambient vibrations that gathered by an array of sensors to evaluate the seismic performance of the buildings. Figure 1 shows a building under test with four different set of sensors (red spot) to measure the vibration. Location of each sensor is predefined by the structural specialists.

The long-term goal of this study is to develop a vibration sensor system that could mimic the seismic performance evaluation similar to AVT.

Seismic Sensor System

A seismic sensor system is an instrument to measure and record the vibration of the ground or any other moving structure. Regarding the application, the sensor may be used as a single unit (i.e. bearing vibration sensor) or as an array of sensors (i.e. seismic evaluation sensor in SHM). The sensor system must be equipped with adequate arrangement to manage (i) the time synchronization among the sensors, (ii) communication of each sensor with a host, data storage, etc. Later, the gathered data from all the sensors can be processed by software routines for more study by the structural engineering specialists. Figure 2 shows one the results of analyzing the building that is shown in Fig. 1. Noted that in the visualization, the vibrations are always exaggerated to have more visibility.

Seismic Sensor

The main component of each vibration sensor system is a seismic sensor. A seismic sensor generates a physical signal (i.e. voltage or current) regarding the motions of the body of the sensor. Seismic sensors, traditionally work based on a mass-spring structure.

Regarding the physical parameter that the seismic sensor measures, it can be a velocity meter or an accelerometer. The output of these two types of sensors can be converted easily. Therefore, it is possible to use both accelerometer and velocity meter in a seismic sensor system. In this work, a geophone, which is a velocity meter

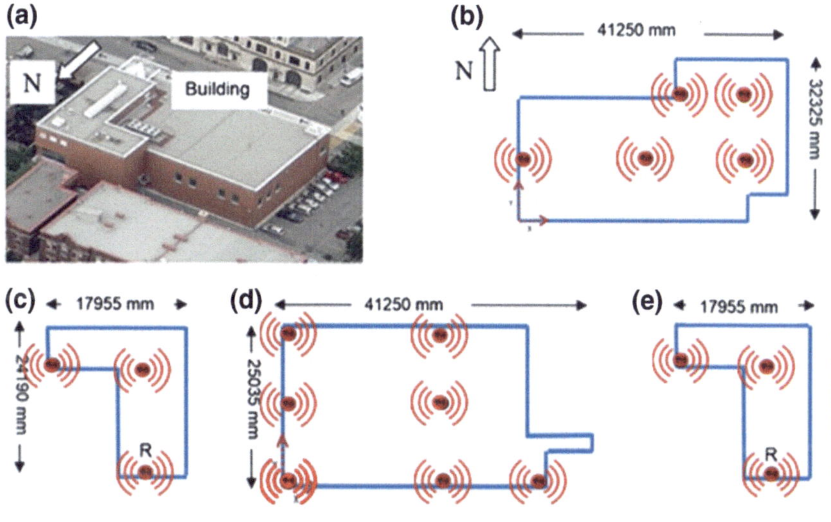

Fig. 1 **a** A sample building, **b–e** layout of sensor array (sensequake.com)

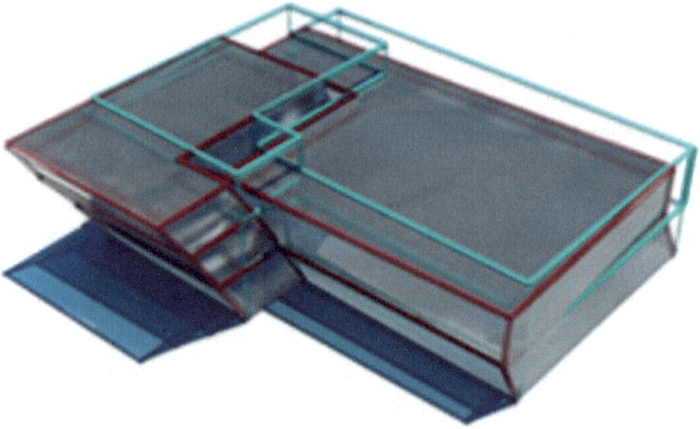

Fig. 2 Visualized model of a building

has been used and tested. In the following, the velocity meter that has been used frequently for the tests and experiments are introduced.

Geophone

A geophone is a sensor that converts the ground velocity to a voltage signal. The structure of a basic geophone is shown in Fig. 3.

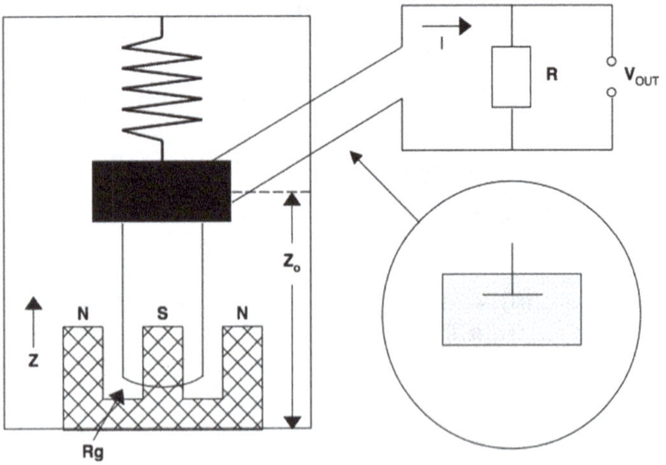

Fig. 3 Geophone structure

The mass connected to the spring fluctuates because of the vibrations of the body of the sensor. The spring moves accordingly. The moving spring is surrounded by a magnet is subjected to an induced electric current by the principle of electromagnetism. The generated induced current in the geophone is proportional to the velocity of the mass. GS-11D from Geospace [3] is the geophone that has been used frequently in this study. Advantages and disadvantages of the geophone are as follow:

Advantages:

- Regarding the applications, geophones can be designed and built to operate over a variety of frequencies and signal magnitudes. Note that geophones with higher sensitivity are bigger and heavier.
- On the economic side, for the same price, geophones afford to more sensitivity to other types of sensors.
- Geophones are passive sensors and they do not need an external power supply.

Disadvantages:

- The main disadvantages of the geophones are their responses to vibrations at low frequencies, especially for frequencies of interest in SHM applications. Figure 4 shows the frequency response of GS11-D with three different damping resistors. As we can see, the response is severely attenuated before the corner frequency.

In the following we describe the analog circuits that have been used with the geophone on hand in our seismic sensor.

Analog Subsystem

In this section, the analog subsystem for the sensor system has been explained. Regarding the seismic sensor that has been used, the analog subsystem may have different structures (see Fig. 5). As an example, we do not need a frequency response compensator for a MEMS sensor. Because the response of the MEMS accelerometers is flat in our frequencies of interest.

The response compensator, amplifier, and filters are the modules that have been used in this study. One of them, all of them or a combination can be used for the analog subsystem in the seismic sensor system. In Fig. 5 two high-level diagrams of the possible analog subsystem are shown. Analog subsystem (a) shows a setup using inverse transfer function technique (this technique will be introduced later in this chapter) and analog subsystem (b) shows the setup that has been used for MEMS. In the following, the amplifiers and the filters that have been used in this study are briefly introduced. Then, several compensation techniques which are the main contributions of the authors have been discussed.

Amplifiers

The amplifiers are required to meet the total gain budget of the chosen analog subsystem (see Fig. 5). In other words, we need to cascade several stages to amplify and

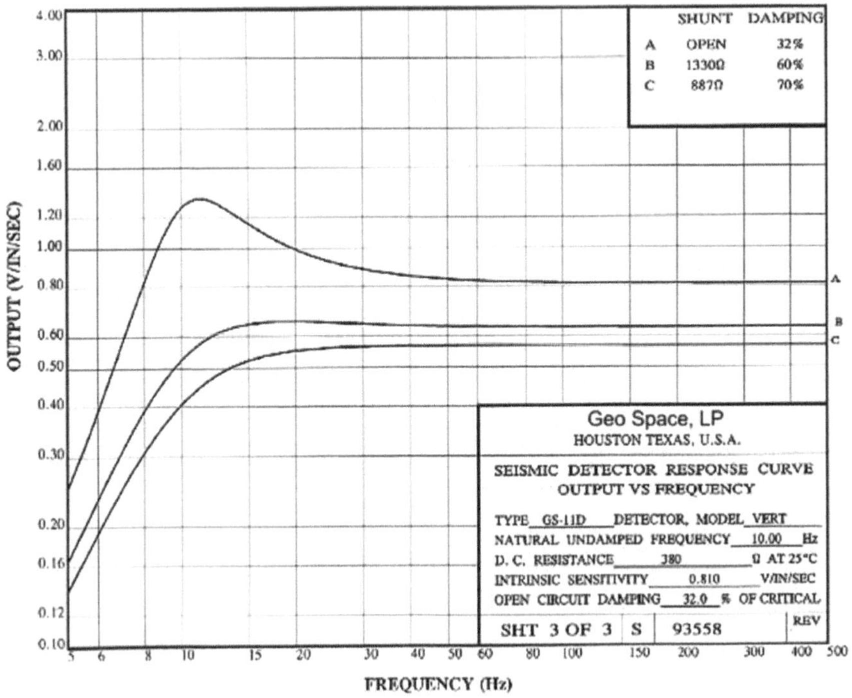

Fig. 4 Frequency response of GS11-D for three different loads [3]

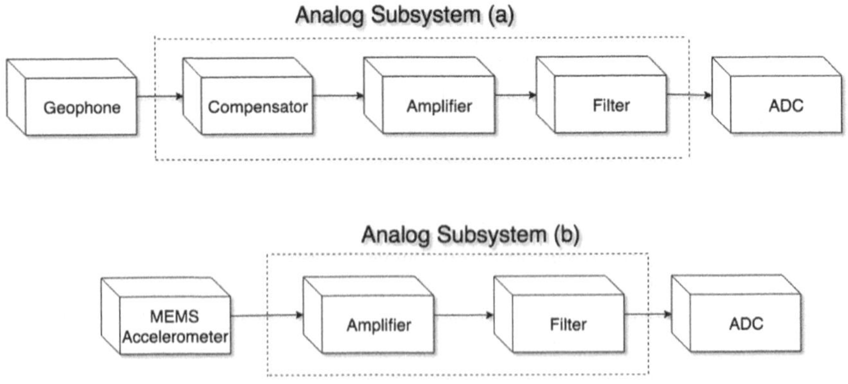

Fig. 5 High level diagrams for the analog subsystem

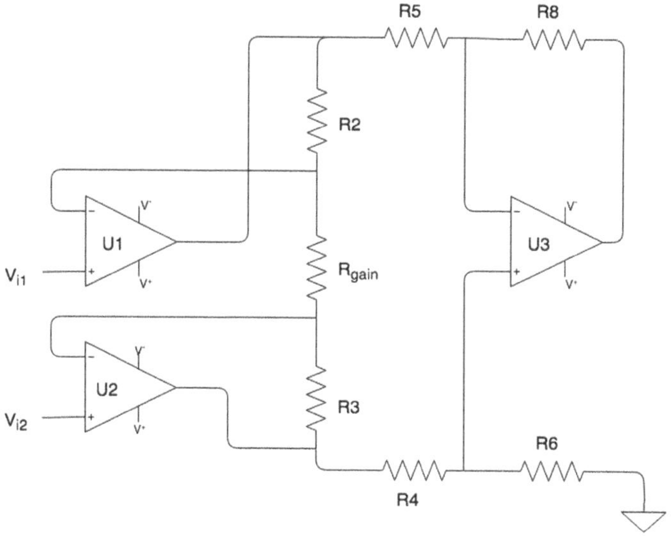

Fig. 6 Instrumentation amplifier [5]

shape the analog signal before interfacing with the ADC. Each of these stages has its own gain value. Multiplication of all these gains is the total gain of the analog subsystem. The total gain is important because we want the output signal of the analog subsystem to cover the whole range of operation of the ADC. The procedure of determining the required gain of the amplifier chain is explained in the following.

In [3], the sensitivity of the geophone has been reported as 32 mV per one unit of velocity in meters per second (m/s). For the ADC on hand [4], the input range is 5 volts. Therefore, a gain of ~150 is required for the analog sub-system. The same procedure can be used for different combinations of ADCs and sensors.

To achieve the above order of gain, we have tried the following alternatives:

- **Instrumentation amplifier**: Instrumentation amplifiers are useful in cases when high gain is required for an input voltage signal of the order of milli-volts [5]. We have tried both integrated instrumentation amplifiers (INA133 from Texas Instruments) and discrete instrumentation amplifier using three op-amps (see Fig. 6).
- **Non-inverting amplifier**: The work in [5] recommends using a non-inverting amplifier. as the first stage of the analog conditioning system. The reason behind the recommendation is that in case of using an inverting amplifier, the feedback resistor would load the internal coil of the geophone. This current will reduce the geophone sensitivity significantly.

As we will see in the experimental result section, we need to use amplifiers to adjust the overall gain while utilizing compensator in the design. Therefore, we abandoned the differential amplifiers for the analog MEMS and used non-inverting amplifiers to reduce the noise level (as well as the complexity) for the analog subsystem.

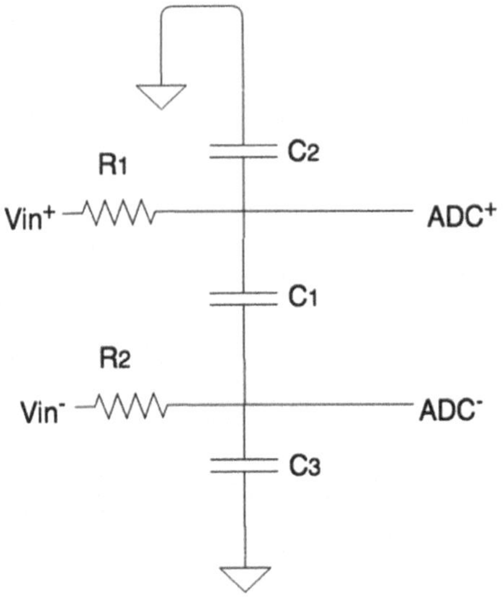

Fig. 7 Passive RC filter for differential ADC

Filters

We have tested and evaluated up to 4 different ADCs. In all the models, an anti-aliasing filter is recommended to use by the manufacturer. The anti-aliasing filter is the last stage of the analog system before the ADC. Figure 7 shows the RC filter that has been used for the ADC in differential mode. The details about the ADC and data gathering module are provided in the experimental results section.

Compensator

In the previous subsections, we explained the basic electronics that we have been used for the sensor system except for the compensator which plays a vital role in sensor system quality. In the following subsections, several compensators that have been proposed in this study are explained. To use every numerical or circuit simulation software, it is mandatory to have a valid model for the geophone.

Geophone Model

First, to use circuit simulation tools, we need to define an equivalent circuit model that represents the electrical behavior of the geophone. In [5], the author has modeled the geophone based on its electrical and mechanical characteristics. Therefore, in the

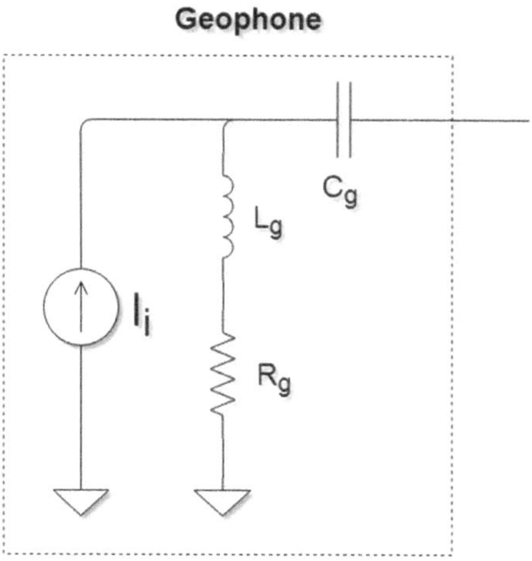

Fig. 8 Geophone equivalent circuit with current source

circuit model presented in [5] R_{eq}, L_{eq} and C_{eq} are used to model the mechanical characteristics and L_c and R_c are defined for the resistance and inductance of the coil.

Clearly, as we can see in Fig. 4, the frequency response of the geophone can be modeled as a second-order high-pass filter. So, we used curve fitting techniques to extract electrical and characteristics of the geophone in the coefficients of a second-order high-pass filter using curve fitting techniques. The achieved transfer function is:

$$H_g(s) = \frac{s^2 + \frac{R_g}{L_g}s}{s^2 + s\left(\frac{R_g}{L_g}\right) + \frac{1}{L_g C_g}}$$

For GS11-D, the geophone that has been used in our study, the values of R_g, L_g and C_g are 33 Ω, 33 μF and 33 H. The transfer function in above equation has two alternative equivalent circuits as shown in Figs. 8 and 9. The circuit in Fig. 8 is more compatible with the physical nature of the geophone as a current is induced to the center coil by the ambient vibrations.

After deriving the components of the equivalent circuit model we can embark on realization of several compensator circuits as explained in the following subsections.

Passive Compensator

Generally, the output signal for the seismic sensors is of the order of milli volts. Using many components in the compensator and analog subsystem will add more

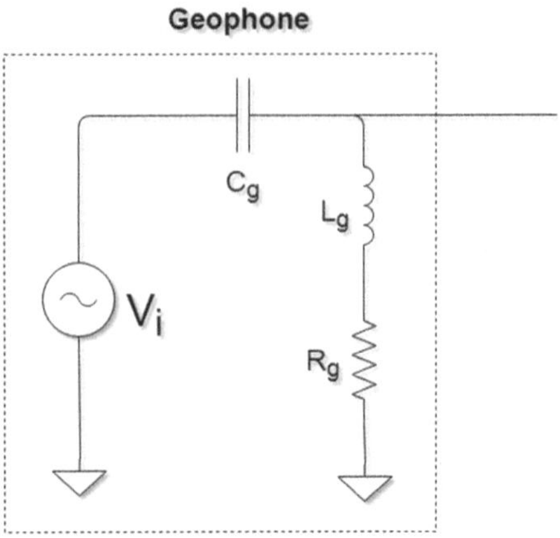

Fig. 9 Geophone equivalent circuit with voltage source

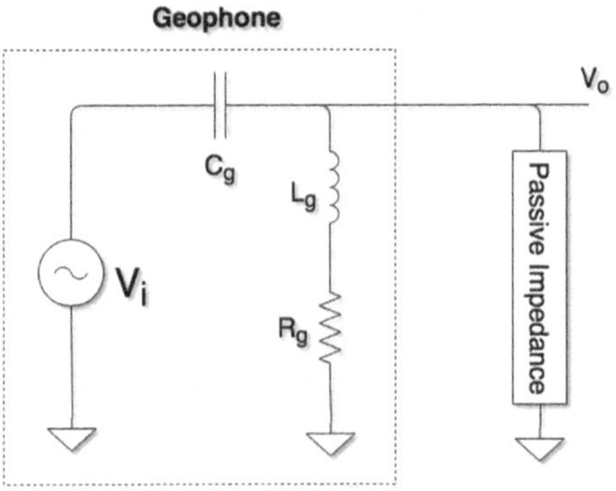

Fig. 10 Passive compensator structure

noise to the system. Therefore, the first step was to design a compensator using minimal passive components.

Figure 10 shows the structure of a passive compensator added to a geophone voltage model. To keep the design simple and effective, we started experimenting with different RLC, RL, and RC circuits in place of the Passive Impedance box of Fig. 10. Each time the overall frequency response graph was obtained. The best result was obtained with a series RC circuit representing the Passive Impedance. Figure 11 shows the circuit for the proposed passive compensator.

Using numerical methods, we tried to find the optimum values for the R_1 and C_1 (in Fig. 11), which extends the corner frequency backward toward DC (i.e., zero frequency). In the simulations, we considered that values of R_1, and C_1 as the integer multiples of R_g and C_g.

Results of the simulations showed that the maximum backward stretch occurs for $R_1 = -1 * R_g$ and $C_2 = -2 * C_g$.

Figure 12 show the comparison between the response of the geophone (blue) and the response of the compensated geophone (orange) using passive elements.

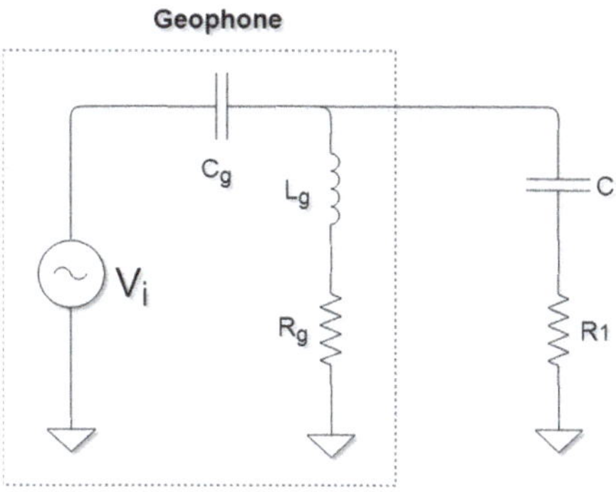

Fig. 11 Passive compensator circuit

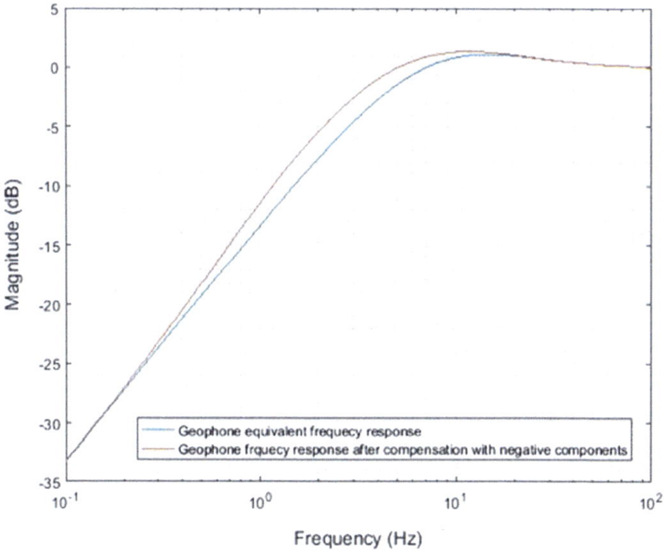

Fig. 12 Numerical simulation result of passive compensator

Two drawbacks of this technique for compensation lead us to design the next version. First, the negative values for the resistor and capacitor are only possible using active components. Second, as we presented in experimental results, the circuit could compensate the effect of the geophone in low frequency, however, the amount of backward stretching is not enough for our application. It could be concluded that use of active components to realize an optimum compensator is imperative. Therefore, we embarked on finding a frequency response compensator system using active circuits.

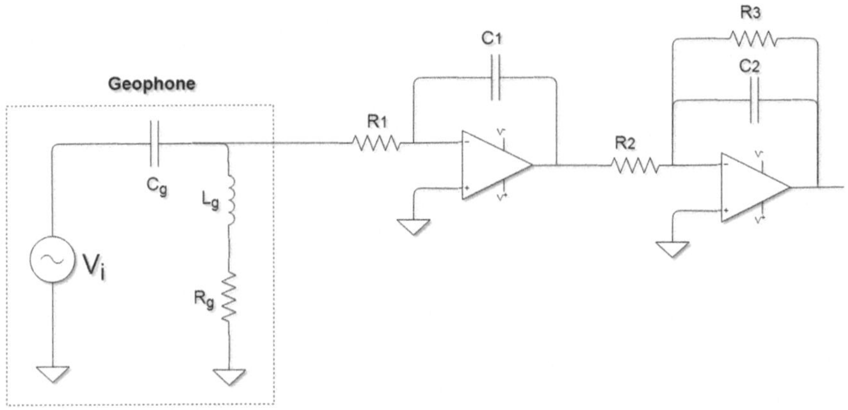

Fig. 13 Inverse transfer function circuit

Inverse Transfer Function Method

Re-writing the numerator of the geophone transfer function as:

$$H_g(s) = \frac{(s)\left(s + \frac{R_g}{L_g}\right)}{s^2 + s\left(\frac{R_g}{L_g}\right) + \frac{1}{L_g C_g}}$$

We can see that there are two zeros in $H_g(s)$ which are clearly responsible for the high-pass characteristics of the transfer function. Therefore, use of an integrator (i.e., $1/s$), and a lossy integrator (i.e., $\frac{1}{s+a}$ can cancel out the zeros and convert the high-pass characteristics to a low-pass function of frequency.

$$H_e(s) = \frac{(s)\left(s + \frac{R_g}{L_g}\right)}{s^2 + s\left(\frac{R_g}{L_g}\right) + \frac{1}{L_g C_g}} * \frac{K_1}{s + \frac{1}{C_1 R_1}} * \frac{K_2}{C_2 s}$$

Figure 13 shows the circuit that has been used in circuit simulations and experiments. Note that the values for the K_1 and K_2 are to be adjusted to meet the total gain requirement.

Figure 14 shows the numerical simulation of the $H_e(g)$ in MATLAB. Clearly, the geophone transfer has been converted to a second order low-pass transfer function.

In the inverse transfer function method, after we canceled the zeros, the result is the low-pass transfer function. While the total gain of the transfer function can be adjusted by the values of resistors and capacitors in the integrator and lossy integrator, the −3 dB frequency still remains dependent on the geophone parameters (R_g, L_g, and C_g). In order to realize the −3 dB frequency independent of the natural frequency of the geophone device, the overall transfer function of the geophone plus

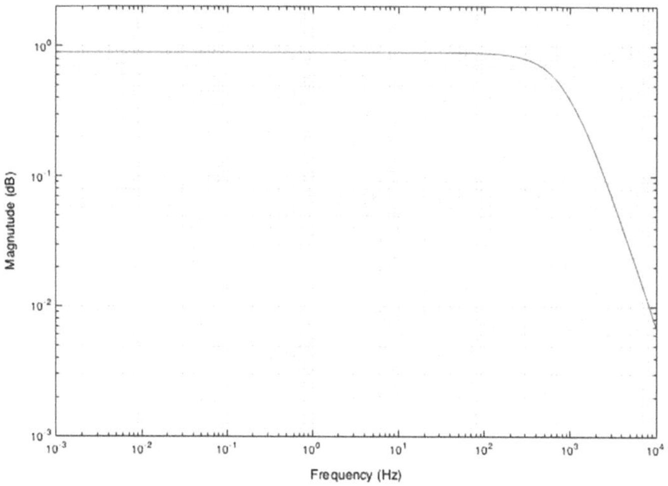

Fig. 14 Numerical simulation result for inverse transfer function method

the compensator need to include additional poles. This can be achieved by using multi-loop feedback principle [ ]. This is discussed in the following.

Multi-loop Feedback Method

Our objective is to create a low pass filter with adjustable cut-off frequency. The simplest transfer function that represents these characteristics is:

$$H_t(s) = \frac{K_0}{s + a}$$

$H_t(s)$ is the overall transfer function of the geophone cascaded plus the compensator. To achieve this form, an unknown impedance box is considered in series with geophone. Unlike the passive compensator design, this time the impedance box can also include active components (Fig. 15).

Mathematical form of the circuit with the target function can be represented as:

$$H_t(s) = \frac{R_g + sL_g}{R_g + sL_g + \frac{1}{sC_g} + Z(s)} = \frac{K_0}{s + a}$$

To achieve the unknown impedance, we solved the above function for $Z(s)$. leading to:

$$Z(s) = \frac{s^3 L_g C_g + s^2 C_g (R_g - aL_g - K_0 L_g) + sC_g R_g (a - K_0) - K_0}{sC_g K_0}$$

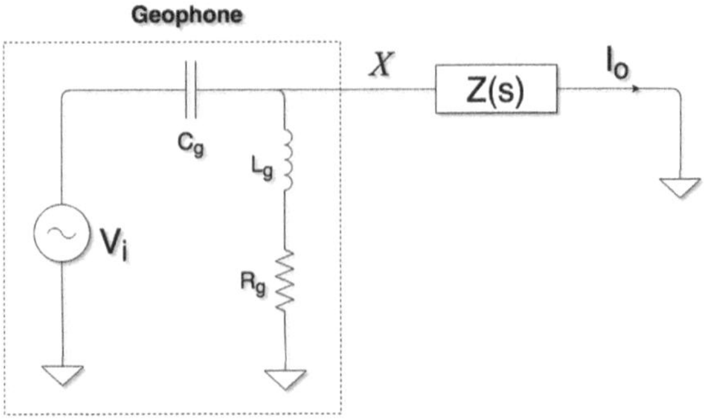

Fig. 15 Geophone cascaded with an unknown impedance Z(s)

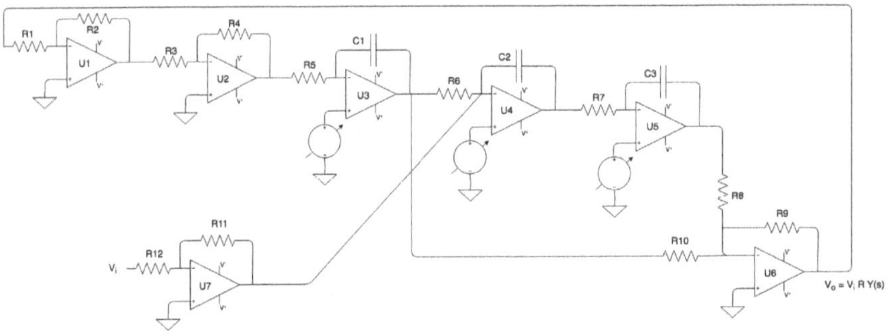

Fig. 16 Schematics of Y(s)

Therefore, according to the Fig. 15 if we use impedance $Z(s)$ in series with the geophone, it will result in a first-order low-pass transfer function. As an alternative to this problem, we can implement the admittance instead:

$$Y(s) = Z^{-1}(s) = \frac{sC_gK_0}{s^3L_gC_g + s^2C_g(R_g - aL_g - K_0L_g) + sC_gR_g(a - K_0) - K_0}$$

The admittance function could be inverted in a feedback configuration to produce the desired impedance function. Figure 16, shows the circuit that has been designed based on the procedures provided in [6].

Note that in the arrangement in Fig. 16, Y(s) leads to a voltage transfer function, to be later converted to an admittance by passing the voltage transfer function through a transconductance amplifier. Further, to obtain an impedance from an admittance function, negative feedback is to be applied. This is displayed in Fig. 17.

The overall realized function can be easily understood through numerical simulations. For the numerical simulation, as we expect, the result should be a first-order low-pass filter. This is shown in Fig. 18.

In the next section, we presented three experiments to verify the correctness of the proposed circuit.

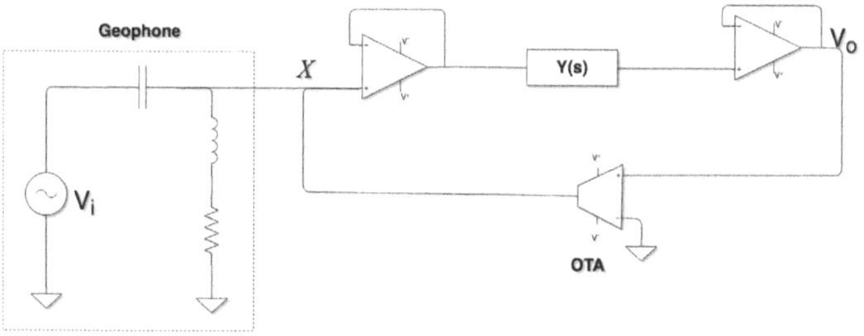

Fig. 17 Admittance convertor

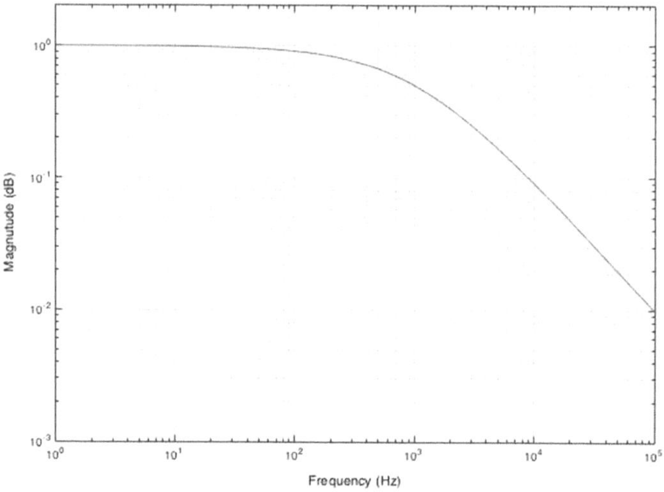

Fig. 18 Numerical simulation of the multi-feedback method

Experimental Results

Using laboratory setups, we verified the correctness of the proposed inverse transfer function and multi-loop feedback methods for the geophone compensation in this section.

In all the tests, we have used a shaker to excite the geophone at certain frequencies. The shaker was specially designed and built in our lab. The working mechanism is as follows. A square wave signal from a signal generator is amplified with a power amplifier and applied to drive a loud speaker. The speaker is positioned in a plastic box. The unit under test (geophone) placed on top of the box detects the hits and generates a voltage related to the intensity of the hits. Then the voltage is read by the data gathering module and the results are record on a personal computer (PC) for more processing. Figure 19 shows the setup that has been used for the tests.

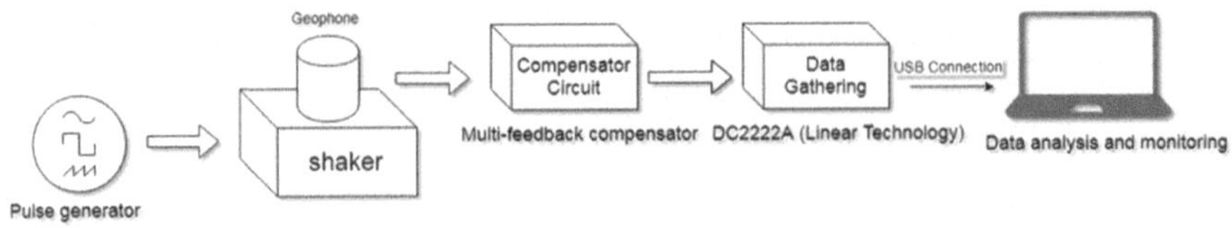

Fig. 19 Test setup

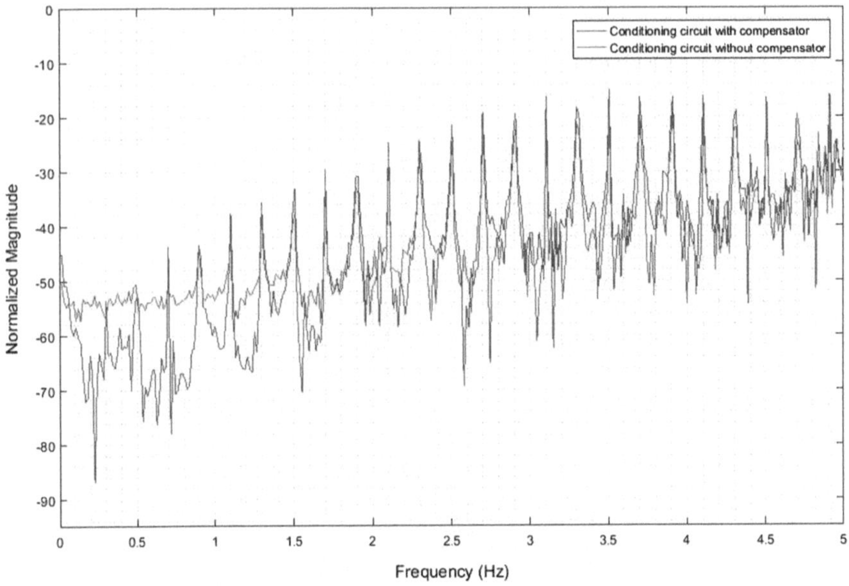

Fig. 20 Geophone response compensated by inverse transfer function method (blue) versus geophone (orange). Shaker on 0.1 Hz

Inverse Transfer Function Method

For the inverse transfer function method, first, we measured the response of the geophone without the compensator. Then, using exactly the same setup, we measured the compensated geophone response

We repeated the experiment for two different shaking profiles. Once that shaker hits the geophone each second and the other time the shaker hits the geophone at every 10 s. The results of this profile are shown in Figs. 20 and 21.

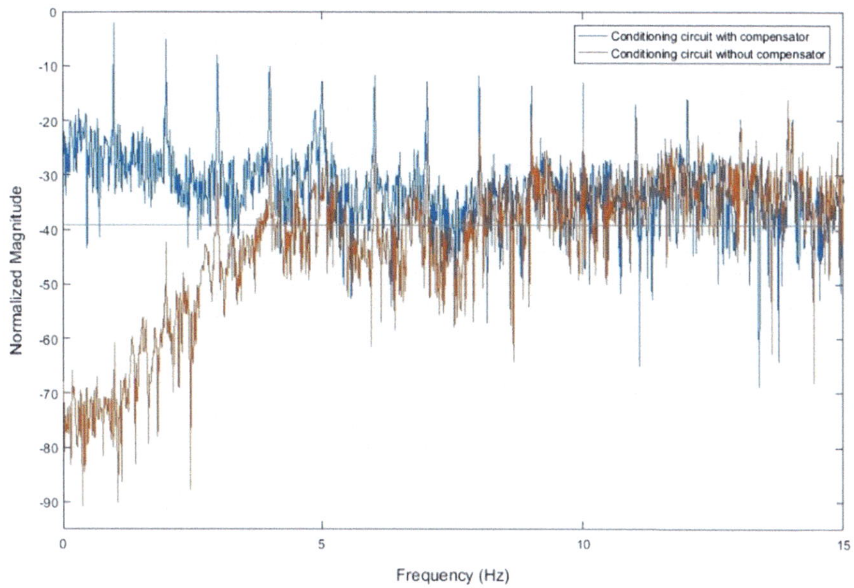

Fig. 21 Geophone response compensated by inverse transfer function method (blue) versus geophone (orange). Shaker on 1 Hz

Multi-loop Feedback Method

For the Multi-feedback method, we have to check both the capability of the circuit on compensating the low-frequency response and tailoring the high-frequency noises to make the response appropriate for the ADC.

For the benchmark sensor in the following tests, we have used a piezoelectric sensor. Piezoelectric sensors are generally more sensitive than the geophones and have a flat response from 0 to 1 KHz [7].

For the first experiment, we verified the correctness of the circuit presented in Fig. 16. We measured the value extracted from the experiment and compared them with the circuit simulation result of the circuit. The results are shown in Fig. 22. Note that the dots are the results of measurements and the solid line is the result of the simulation software.

We next used the system in Fig. 17 to convert the admittance and connected with a geophone device (GS-11D) for vibration tests. Figure 23 shows the responses of the compensated geophone system and of the piezoelectric sensor over a frequency span of DC to 20 Hz. Clearly, the odd harmonics of the main frequency of the shaker is also visible in Fig. 23 as we used a square wave to feed the shaker.

In all the experiments, practically we have two different sources of vibration. The shaker and the ground (building in our cases). The ground vibrations constitute an artifact signal.

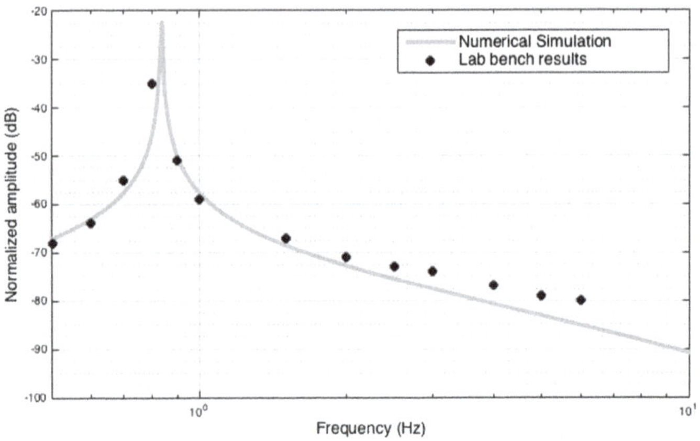

Fig. 22 Experiment to verify the circuit in Fig. 16

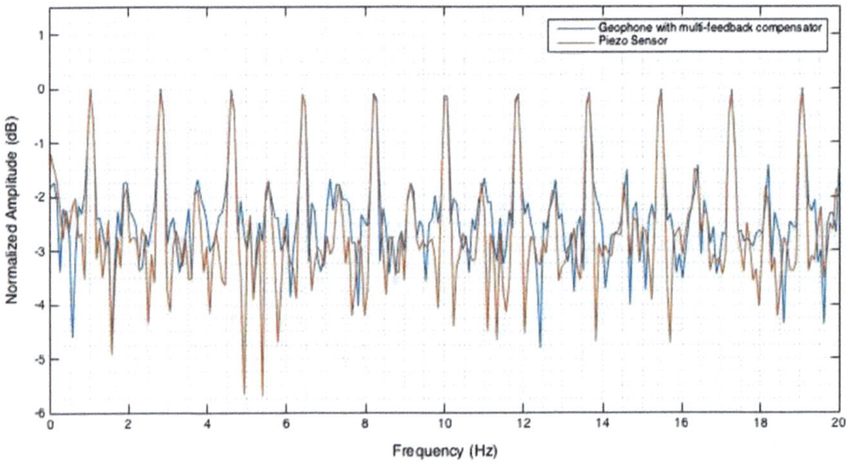

Fig. 23 Geophone response compensated by the multi-feedback method (blue) versus piezoelectric sensor. Shaker on 1 Hz

To gain an estimation of the ground vibrations, we measured the response of the geophone on the test location and subtract it from our result to have a virtually flat response. Figure 23 shows the smoothed result of the measurement from compensated geophone and the piezoelectric sensor.

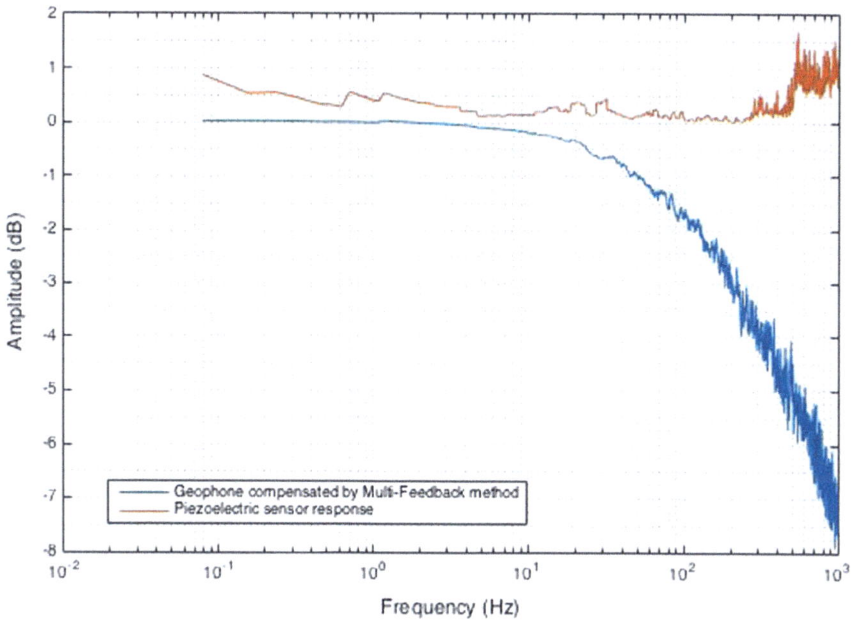

Fig. 24 Smoothed response of the geophone compensated by multi-loop feedback method (blue) versus smoothed response of the piezoelectric sensor

Discussion

Figures 20 and 21 proved the effectiveness of the inverse transfer function method. For this method of compensation, we need an extra level of anti-aliasing filter for the ADC. To eliminate this filter, we applied the multi-loop feedback method. Figure 23 shows the comparison between compensated geophone and a piezoelectric sensor. In Fig. 23, the clock source for the data gathering module has been chosen from an inaccurate source. This clock source determines the sampling period of the ADC. As a result, we can see a small amount of shift in the response of both sensors.

In Fig. 24 we show that the response of compensated geophone is similar to the target function that we had defined. Noted that, the −3 dB frequency of the first order target function with the multi-loop feedback technique was set to 100 Hz. In Fig. 24 the −3 dB is about 120 Hz. As we could not choose circuit components for any arbitrary values, the above disparity can be considered acceptable.

Conclusions

In this study, we targeted to develop the task of analog signal conditioning of the output of a geophone used as vibration sensor. As mentioned, the frequency response of the geophone attenuates the output signals below its own −3 dB corner frequency. As a result, the information carried by the analog signal for lower frequency is hard to detect. We proposed several techniques to reconstruct the attenuated output signal below the corner frequency of the geophone. The circuit simulation and models have been verified by a set of experiments with a real geophone (GS-11D) and a shaker.

References

1. J. Havskov, G. Alguacil, *Instrumentation in Earthquake Seismology* (Springer International Publishing, 2015)
2. F. Mirshafei, M. Mirshafei, G. McClure, A new three-dimensional seismic assessment method (3d-sam) for buildings based on experimental modal analysis. Comput. Struct. **180**, 125–137 (2017)
3. GS-11D geophone specifications, http://www.geospace.com/geophones-gs-11d/
4. LTC2508-32 ADC specifications, http://www.analog.com/media/en/technical-documentation/data-sheets/250832fc.pdf
5. P.W. Rodgers, Maximizing the signal-to-noise ratio of the electromagnetic seismometer: the optimum coil resistance, amplifier characteristics, and circuit. Bull. Seismol. Soc. Am. **2** (1993)
6. R. Raut, M.N.S. Swamy, *Modern Analog Filter Analysis and Design: A Practical Approach* (Wiley, 2010)
7. PCB Piezotronic product specifications, http://www.pcb.com/ Products/model/393B31

Considerations Needed for Sensing Mineral Nutrient Levels in Pasture Using a Benchtop Laser-Induced Breakdown Spectroscopy System

H. Jull, R. Künnemeyer and P. Schaare

Abstract Precision agriculture aims to increase yield and profits while reducing costs, waste, and environmental side-effects. This is achieved through a process of measuring, modelling and acting; for example, laser-induced breakdown spectroscopy (LIBS) can be used to measure macro and micro nutrients in crops to determine nutrient requirements. The limiting factor with quantitative LIBS analysis of plant nutrient levels is the variation between shots on the same sample. Following a review of current literature relevant to LIBS for agriculture, this work investigates whether different chemometric methods can mitigate these variations and can create quantitative calibrations for nutrient levels in fresh and dried pelletised pasture under laboratory conditions. The methods explored were Savitzky Golay filtering, multiple linear regression, principal component regression, partial least-squares regression, gaussian process regression, and artificial neural networks. The algorithms that performed best were partial least-squares with gaussian process regression (R^2 of 0.93, 0.95, and 0.92 for K, Na, and Mn, respectively), principal components analysis with artificial neural networks (R^2 of 0.94, 0.83, and 0.80 for Fe, Ca, and Mg, respectively), and partial least-squares with artificial neural networks (R^2 of 0.77 for B). Removing the moisture from the pasture improved model R^2 values by 4–5% on average. Acquiring spectra under an argon purge produced a small reduction in accuracy for some nutrients compared to models acquired in air. Including categorical data in the principal component regression and the artificial neural networks produced negligible improvements in prediction. This chapter will give an introduction of using different types of chemometric analyses on spectra generated by LIBS to measure micro and macro nutrients in pasture under laboratory conditions. It discusses the challenges faced when building models for each nutrient.

H. Jull (✉) · R. Künnemeyer
School of Engineering, The Dodd Walls Centre for Photonic and Quantum Technologies, University of Waikato, Hamilton, New Zealand
e-mail: harrisson.jull@gmail.com

P. Schaare
Plant and Food Research, Hamilton, New Zealand

S. C. Mukhopadhyay et al. (eds.), *Modern Sensing Technologies*, Smart Sensors, Measurement and Instrumentation 29,
https://doi.org/10.1007/978-3-319-99540-3_20

Introduction

Precision agriculture is a farming technique that measures and reacts to site-specific crop variability. One method is to optimise crop yield through precisely timed and tailored fertiliser application and application rates. This not only creates financial benefits, through increased yield and reducing the amount of fertiliser applied, but also reduces the environmental side effects caused by over fertilisation, i.e., leaching and runoff into streams and rivers. Accurate sensor technologies able to give real-time information on the nutritional status of crops are needed to realise precision agriculture [1]. Laser-induced breakdown spectroscopy (LIBS) is a technique that has become popular for its ability to make fast elemental analyses. LIBS uses a high powered pulsed laser to create a plasma on a target material's surface through inverse Bremsstrahlung [2]. As the plasma cools, electrons transition from high energy levels to lower levels and emit photons. Photons are emitted at specific wavelengths corresponding to the particular elements that constitute the plasma and represent the elemental makeup of the sample. The concentration of these elements directly affects the intensities of the emission lines in the spectrum [3]. LIBS is a sensitive technique that needs no sample preparation and has the potential to produce reliable calibrations that determine the nutrient levels in pasture. The essential elements needed for plant growth are the macro (N, P, K, Ca, S, and Mg) and micro nutrients (B, Cl, Mn, Fe, Zn, Cu, Mo, and Ni).

LIBS spectra exhibit strong within- and between-sample variation which reduces the repeatability of measurements [4]. The shot-to-shot variations originate from multiple sources, like sample inhomogeneity, surface roughness, matrix effects, sample moisture, and differences in experimental parameters. The main cause of variability in spectra is the difference in plasma temperatures. The exponential Boltzmann factor and the partition function in the emission line intensity equation both have strong temperature dependence. Averaging multiple spectra from the same sample, using internal standards, normalisation, and chemometric algorithms can reduce the shot-to-shot variation. Some of the first quantitative chemometric methods that have been used for LIBS include linear or rank correlation [5], multiple linear regression (MLR) [6], principal component regression (PCR) [7], partial least squares regression (PLS) [8], and artificial neural networks (ANN) [9]. Increasing the signal to noise ratio by increasing the emission line intensities can produce additional lines that contain valuable information. This is achieved through increasing the temperature in the plasma by increasing the laser pulse energy or using an argon atmosphere.

Literature Review

There have been many studies using LIBS on agriculture. A summary of the majority of these studies is presented in Table 1. This table contains studies performed on agricultural material, including all those which are relevant for the following discussion. Various wavelengths, pulse durations, and repetition rates have been used for the lasers in these studies. Nd:YAG lasers are typically used for LIBS because they are reliable, compact, and are easy to use [10]. Most of the studies mentioned below use a 1064 nm Nd:YAG laser. Unfortunately, there are a few studies where there is no information on the spectrometer settings.

The extensive literature list of Table 1 shows that there have been many studies on biological material and agricultural products, but there are very few related to pasture. Studying fresh herbage using LIBS has not been explored in-depth. Chauhan et al. [17] used fresh Bermuda grass in their study of Si distribution within leaves. Spectra were averaged and the Si I 288.15 nm emission line was used to analyse the amount of silicon at different locations on the leaves. Jull et al. [137] investigated fresh and dried, pelletized pasture in air and under an argon atmosphere using PLS. The root mean square error of cross-validation (RMSECV) of the models were better for pellets, but the fresh pasture models had better precision. The pellet spectra exhibited less between-sample variation because the moisture was removed from the samples. The moisture reduced the intensity of the fresh pasture emission lines. An explanation of the limiting factors inhibiting calibrations were given for each macro and micro nutrient. The best results were obtained for K, Na, and Mn. Jull et al. [138] studied temperature correction on pasture spectra using Boltzmann plots, Saha-Boltzmann plots, and internal standards. The Boltzmann plot and Saha-Boltzmann plot methods needed additional emission lines, confirmation of local thermodynamic equilibrium, and the degree of instrument broadening to produce better accuracies. PLS models were created using spectra normalised to carbon lines. Violet band and swan band heads produced better results. The best results were generated by taking the ratio of two emission lines with similar upper level electron energies. Boyain-Goitia et al. [67] investigated similarities between fresh grass fragments and pollen using LIBS spectra. The main difference was that the pollen spectra had Cr and Fe emission lines. Concentrations for Ca and Al were similar for the samples. Differentiating between the samples was hard due to the limited number of samples. Si was found in the grass spectra. This was attributed to soil dust on the samples. It was found that normalising the spectra with the CN violet band produced a reduction in matrix effects.

There have been other studies on grass that do not use fresh samples. Bermuda grass (*Cynodon dactylon*) was investigated for a possible treatment for diabetics by Rai et al. [18]. The sap in the samples was removed, and the resulting plant material was crushed and lyophilized into powders which were dissolved in distilled water. The C III 229.6 nm line was used to normalise the emission lines of interest. The resulting ratios identified that certain amounts of K, Na, and Mg in Bermuda grass have antidiabetic effects. Tall Fescue Grass (*Festuca arundinacea*) was formed into pellets and investigated for endophyte infection by Martin et al. [107]. The small number of samples made it difficult to discriminate between healthy and infected grass. Braga et al. [14] used pelletized grass (*Axonopus obtusifolius*) to validate micronutrient calibration models. The study compared multivariate (PLS) and univariate calibrations. The spectra were normalised by the C I 193.09 nm line to reduce variations. The PLS models performed better than the univariate models. The physical and chemical makeup of the samples used in the calibration should be similar

Table 1 Laser parameters and spectrometer settings for LIBS on various agricultural samples (λ: wavelength, t_p: pulse duration, f: pulse frequency, E: pulse energy, t_d: gate delay, Δt: integration time)

Sample	Laser				Spectrometer		Refs.
	λ (nm)	t_p (ns)	f (Hz)	E (mJ)	t_d (μs)	Δt (μs)	
Algae	1,064	7	2	100	6	0.6	[11]
Algae	1,064	5–7	10	30	–	2,000,000	[12]
Apple leaves, pine needles, citrus leaves, tea leaves, rice flour, Cannabis plant leaves and flower tops	1,064	7	10	200	–	–	[13]
Auatic plant, aquatic moss, bush branches and leaves, cabbage, soya flour, rice flour, wheat flour, spinach leaves, brachiaria, banana leaves coffee leaves, maize leaves, mango leaves, pepper leaves, soya leaves, olive leaves, apple leaves, guava leaves, grass and jackfruit leaves	532	12	10	71	9	1.1	[14]
Bean leaves	1,064	5	10	200	2	5	[15]
Bean leaves, bush branches and leaves, cabbage, soya flour, rice flour, apple leaves, peach leaves, wheat flour and spinach leaves	1,064	5	10	360	2	5	[16]
Bermuda grass	532	4	–	10	–	–	[17]
Bermuda Grass and *Musa paradisiaca*	–	–	1	175	–	–	[18]
Bitter gourd (*Momordica charantia L.*)	1,064	–	–	200	–	–	[19]
Bitter Melon	–	–	10	40	–	–	[20]

(continued)

Table 1 (continued)

Sample	Laser				Spectrometer		Refs.
	λ (nm)	t_p (ns)	f (Hz)	E (mJ)	t_d (μs)	Δt (μs)	
Black tea (*Anxi, Fujian*), Huangya tea (*Ya'an, Sichuan*), Longjing tea and Pu'er tea (*Pu'er, Yunnan*)	1,064	8	20	20–100	2–30	2	[21]
Boldo leaves (*Peumus boldus Molina*)	1,064	5	10	110 and 220	2	5	[22]
Bran tissues and wheat grain	193	15	1	–	–	2,000	[23]
Broad bean (*Vicia faba*)	266	–	1	5	1 (delay)	10	[24]
	1,064	–	1	100	0.5 (interpulse delay)		
Cabbage leaf	532	4	4	20	1.5	–	[25]
Capsicum leaves	532	5	–	10	1	10	[26]
Carrot Root (*Daucus carota*)	532	8	10	–	–	–	[27]
Cauliflower and broccoli	532	7	2	96	1	5	[28]
Chinese cabbage	1,064	8	2	120	2	2,000	[29]
Chinese tea leaves, Longjing green tea, Mengding Huangya, White tea, Tie Guanyin, Wuyi black tea, and Pu'er tea	1,064	5.82	–	50	1.2	4	[30]

(continued)

Table 1 (continued)

Sample	Laser				Spectrometer		Refs.
	λ (nm)	t_p (ns)	f (Hz)	E (mJ)	t_d (μs)	Δt (μs)	
Citrus leaf	355	5	10	5	1	50	[31]
	1,064	5	10	110			
Citrus leaves	1,064	–	10	50	48.5	–	[32]
Citrus leaves	1,064	8	10	50	2.5	2,000	[33]
Coffee	1,064	8	10	50	–	–	[34]
Coffee	266	8	20	31	0.4	–	[35]
Coffee beans	1,064	8	–	50	11	2,100	[36]
Cotton	1,064	5	0.67	–	1.3	4.5	[37]
Cucurbita maxima seeds	532	4	1	100	–	–	[38]
Dallisgrass (*Paspalum dilatatum*); wheat (*Triticum aestivum*), soybean (*Glycine max*); bell pepper (*Capsicum annuum*)	800	0.000035	1,000	0.3	–	500,000	[39]
Duckweed (*Lemna minor*)	266	5	–	10	1	10	[40]
	1,064	5	–	100	1	10	
Emblica officinalis	–	–	2	175	–	–	[41]
Ficus bengalensis	–	–	–	–	–	–	[42]

(continued)

Table 1 (continued)

Sample	Laser				Spectrometer		Refs.
	λ (nm)	t_p (ns)	f (Hz)	E (mJ)	t_d (μs)	Δt (μs)	
Folium lycii	1,064	8	–	–	0.8	2,000	[43]
Gannan navel orange	–	–	2	20	1.2	2,000	[44]
Gannan navel orange	1,064	8	0.1–10	110	2	–	[45]
Green herb, tomato leaves and coffee beans	1,064	5	–	–	–	–	[46]
Guava	–	–	2	175	–	–	[47]
Holly leaves	1,064	10	10	45	0.5	2,000	[48]
	266	–	–	45	0.5	2,000	
Holly leaves	1,064	–	–	10–80	–	–	[49]
	266	–	–	10–80	–	–	
Juncus efiusus L.	1,064	3–5	2	340	1	1,000	[50]
Lettuce	532	5	–	10	1	10	[51]
Ligusticum wallichii	1,064	5.82	20	26.8	1.5	20	[52]
Maize (*Zea mays L.*)	–	–	–	–	–	–	[53]
Maize (*Zea mays*), sunflower (*Helianthus annuus*) and lettuce (*Lactuca sativa*)	532	5	–	10	1	10	[54]
Maize and holly (*Ilex chinensis Sims*)	1,064	–	–	90	–	–	[55]
Maize leaf and Red osier dogwood leaves	795	0.00016	10	0.1	–	–	[56]

(continued)

Table 1 (continued)

Sample	Laser				Spectrometer		Refs.
	λ (nm)	t_p (ns)	f (Hz)	E (mJ)	t_d (μs)	Δt (μs)	
Mustard grass	1,064	–	10	300	–	–	[57]
Navel oranges	1,064	8	10	200	1.28	2,000	[58]
Ocimum sanctum, Ocimum americanum, Ocimum gratissimum, and *Ocimum basilicum*	532	4	10	–	–	–	[59]
Orange leaves	1,064	10	10	200	1.28	–	[60]
Orange peel	1,064	8	10	200	1.28	2,000	[61]
Orange tree leaves	1,064	8	10	–	1.28	2,000	[62]
Peach, apple, and spinach	1,064	–	–	10–30	–	–	[63]
Perennial ryegrass (*Lolium perenne*) and white clover (*Trifolium repens*)	1,064	–	–	200	1	–	[64]
Phaleria Macrocarpa leaves	1,064	10	–	100	–	–	[65]
Pollen	–	10	–	30	1.5	2,000	[66]
Pollen and fresh grass fragments	1,064	–	5	20–30	1	4	[67]
Poplar tree leaves	800	0.000,1	10	25	–	–	[68]
Potato skin	266	5	10	5–20	Various	Various	[69]
Potato skin	1,064	5	–	10	1	1	[70]

(continued)

Table 1 (continued)

Sample	Laser				Spectrometer		Refs.
	λ (nm)	t_p (ns)	f (Hz)	E (mJ)	t_d (μs)	Δt (μs)	
Potato skin and flesh	1,064	4	20	10	1	5	[71]
Potato, carrot, celery and aubergine	266	–	10	10	0.1	5	[72]
Powdered rice, starch and seaweed	10,600	200	10	1,500	1	100	[73]
Prickle pears	1064	20	–	300	–	–	[74]
Psoralea corylifolia seed	–	–	2	175	–	–	[75]
Psoralea corylifolia seed	–	–	2	175	–	–	[76]
Red Fuji apples and Hosui pears	1,064	–	20	160	2	–	[77]
Rice	1,064	5	1	–	–	–	[78]
Rice Leaves (*Oryza sativa L.*)	532	8	1	60	4	20	[79]
Rice seed and milk powder	10,600	200	20	1,500	–	2,000,000	[80]
Saffron (*C. sativus L.*), safflower (*Carthamus tinctorius L.*), marigold flower (*Calendula officinalis L.*), turmeric (*Curcuma longa L.*)	1,064	–	4	38	0.3	20	[81]
Scented geranium	1,064	8	–	80	5	15	[82]
Scented geranium	1,064	12	1	–	2	2	[83]

(continued)

Table 1 (continued)

Sample	Laser				Spectrometer		Refs.
	λ (nm)	t_p (ns)	f (Hz)	E (mJ)	t_d (μs)	Δt (μs)	
Seed kernels of pumpkin (*Cucurbita maxima*), ash gourd (*Benincasa hispida*), watermelon (*Citrullus lanatus*) and muskmelon (*Cucumis melo*)	532	4	2	40	1	5	[84]
Sophora leaves	800	0.00015	10	5	–	–	[85]
Soybean leaves	532	4	–	1, 4, 8, 16 and 32	1	–	[86]
	1,064	6	–	1, 4, 8, 16 and 32	1	–	
Spinach leaf powder and unpolished rice flour	1,064	7	10	80–140	–	–	[87]
Spinach leaves	532	12	10	70	1.1	9	[88]
spinach, mustard, chenopodium, fenugreek, and chickpea leaves	532	4	10	10	–	–	[89]
Spirulina and chlorella	10,600	100	1.3	Various	–	30,000,000	[90]
Starch powders and rice flour	1,064	7	10	30	–	–	[91]
Sugarcane	1,064	5	10	220	2	4.5	[92]
Sugarcane	–	–	–	50, 75	0.5, 1		[93]
Sugarcane (*Saccharum offcinarum*)	532	4	10	10	–	–	[94]

(continued)

Table 1 (continued)

Sample	Laser				Spectrometer		Refs.
	λ (nm)	t_p (ns)	f (Hz)	E (mJ)	t_d (μs)	Δt (μs)	
Sugar cane and Boldo leaves	1,064	5	10	365	2	5	[95]
Sugar cane leaves	1,064	5	10	–	2	5	[96]
Sugar cane leaves	1,064	5	10	110	2	4.5	[97]
Sugar cane leaves	1,064	5	10	110	2	5	[98]
Sugar cane, orange tree leaves and soy leaves	1,064	5	10	110	2	4.5	[99]
Sugarcane (*Saccharum officinarum*), soy (*Glycine max*), citrus (*Citrus sinensis*), coffee (*Coffea arabica*), maize (*Zea mays*), eucalyptus (*Eucalyptus sp.*), mango (*Mangifera indica*), bean (*Phaseolus vulgaris*), banana (*Musa paradisiaca*), lettuce (*Lactuca sativa*), brachiaria (*Brachiaria decumbens*), pearl millet (*Pennisetum americanum*), grape (*Vitis sp.*), rubber tree (*Hevea brasiliensis*), tomato (*Solanum lycopersicum*), apple leaves, peach leaves, spinach leaves, tomato leaves, and pine needles	1,064	6	3.3	70	0.75	3	[100]
	532	6	3	7	0.75	3	
	266	6	3.3	70	0.75	3	
	800	0.0006	1,000	1.65	0.35	0.25	
Sunflower (*Helianthus annuus*), leaves/stem	532	5	–	10	1	10	[101]

(continued)

Table 1 (continued)

Sample	Laser				Spectrometer		Refs.
	λ (nm)	t_p (ns)	f (Hz)	E (mJ)	t_d (μs)	Δt (μs)	
Sunflower (*Helianthus annuus*), leaves/stem	795	0.000,16	10	0.1	–	–	[102]
Sunflower (*Helianthus annuus*), leaves/stem	532	5	–	10	1	10	[103]
Sunflower (*Helianthus annuus*), leaves/stem	532	5	–	10	1	10	[104]
Sunflower (*Helianthus annuus*), leaves/stem	532	5	–	10	1	10	[105]
Sunflower, leaves/stem	355	6	20	–	–	0.50	[106]
	790	0.00003	20	–	–	0.50	
Tall fescue and apple leaf	532	–	–	23	0.5	10	[107]
Tangerine leaves and Rhododendron leaves	1,064	20	100	1.1	–	100,000	[108]
Tea leaves	1,064	8	4	30	1.5	2	[109]
Tea plants (*Sambucus nigra L., Hypericum perforatum L., Crataegus oxyacantha auct. non L., Rubus idaeus L. and Betula species L.*)	1,064	8	10	70	1	1,050	[110]
Tobacco	266	6	10	25	0.25	1,050	[111]
Tobacco leaves	532	8	1	60	1.5	10	[112]
Tomato leaves	1,064	6	20	100	2	5	[113]
Tomato leaves, spinach leaves, apple leaves, peach leaves Spanish moss and pine needles	1,064	5	1	100	1	10	[114]

(continued)

Table 1 (continued)

Sample	Laser				Spectrometer		Refs.
	λ (nm)	t_p (ns)	f (Hz)	E (mJ)	t_d (μs)	Δt (μs)	
Trichosanthes dioica fruit	532	3–4	2	175	–	–	[115]
Tsumura kackontou	10,600	200	–	1,500	10	100	[116]
Tsumura kackontou	10,600	200	–	1,000–1,500	–	–	[117]
Turmeric	532	4	10	18	–	–	[118]
Wheat and gardenia	800	0.000,035	1,000	0.3	–	–	[119]
Wheat flour	1,064	5	–	–	1–3	1–10	[120]
Wheat flour	1,064	–	4	38	0.3	1,050	[121]
Wheat grain	193	15	1	35	–	–	[122]
Wheat grain	532	5	20	80	0.5	2,000	[123]
	193	15	1	–			
Wheat leaves, poppy leaves, barley leaves and rape leaves	1,064	12	–	80–150	–	1	[124]
Wheat seedlings (*Triticum aestivum L.*)	532	4	4	10	–	–	[125]
Wheat seedlings (*Triticum aestivum L.*)	532	4	4	10	–	–	[126]
White chickpea	1,064	–	8	40	0.650	1,050	[127]
Withania coagulans fruit	–	–	–	–	–	–	[128]

(continued)

Table 1 (continued)

Sample	Laser				Spectrometer		Refs.
	λ (nm)	t_p (ns)	f (Hz)	E (mJ)	t_d (μs)	Δt (μs)	
Withania coagulans fruit	–	–	–	–	–	–	[129]
Wood	1,064	5	10	50–70	3	10	[130]
Wood	1,064	8	2	–	17	5,000	[131]
Wood	1,064	10	2	–	–	5,000	[132]
Wood	532	–	–	45	–	10	[133]
Wood	–	10	10	200	–	–	[134]
Wood	532	5	–	–	–	–	[135]
Wood and *babassu mesocarp*	266	–	5–20	5–25	0–0.4	–	[136]

to the intended plant species to be studied. This will produce superior calibrations. Pelletized pasture samples were used to create calibration models for macro and micronutrients by Devey et al. [64]. LIBS was compared to inductively coupled plasma—optical emission spectroscopy (ICP-OES). The spectra were processed using Savitzky Golay (SG) smoothing, vector normalisation, and wavelet smoothing before PLS models were generated on different spectral regions using selected wavelengths only. The best results were found for Na, K, Ca, and P and were similar to those for ICP-OES.

Experimental Setup

A LIBS-6 (Applied Photonics, UK) system with a Nd:YAG laser (Big Sky Ultra, Quantel, France) operating at 1064 nm with a pulse width of 7 ns, and pulse energy of 100 mJ was focused perpendicular to the sample surface to generate the plasma. The setup had a fixed distance to the sample. Each spectrum was acquired with six (Avantes, The Netherlands) spectrometers in a LIBS-6 unit covering the range 182.26–908.07 nm. All spectrometers were set to start recording after a delay time of 1.27 μs with respect to the laser pulse and an integration time of 1 ms. A 3-axis translation stage was employed to move the sample so that each LIBS pulse was on a fresh surface but at the same height.

Fresh pasture (a mix of ryegrass and clover) from 20 different plots was harvested over a 13 month period creating a total of 280 samples. The fresh pasture was pressed flat in a holder. An accumulation of 100 shots under an air atmosphere, and 100 shots under an argon purge were taken for each batch of harvested pasture. Each shot was taken from a new location on the sample. The pasture samples were sent to a commercial analytical laboratory where they were dried at 62 °C overnight and ground to pass through a 1 mm screen. Nitrogen was estimated by Dumas combustion calibrated NIR. All other elements were determined by nitric acid/hydrogen peroxide digestion followed by ICP-OES. Table 2 contains the limits of detection for the qualitative analysis performed by the laboratory. A portion of the powder was returned and pellets were pressed for each batch of harvested pasture. An accumulation of 100 shots under an air atmosphere, and 100 shots under an argon purge were taken for each pellet. Each shot was taken from a new location on the sample. All spectra were corrected for dark current and background.

Data Treatment

Building quantitative models on raw spectroscopic data can lead to low correlations. Pre-processing can enhance the data by removing noise and recovering peaks. Regression techniques can then build models on improved data, producing better results. Pre-processing of data includes smoothing, removal of dark current and

Table 2 Description of the macro and micro nutrients in the dataset determined by a commercial analytical laboratory

Element	Range	Mean	Detection limit
N (wt%)	1.8–4.0	2.8 ± 0.49	0.1
P (wt%)	0.25–0.59	0.36 ± 0.054	0.02
K (wt%)	1.0–4.0	2.7 ± 0.68	0.1
S (wt%)	0.23–0.46	0.33 ± 0.042	0.02
Ca (wt%)	0.39–1.31	0.68 ± 0.18	0.02
Mg (wt%)	0.12–0.28	0.19 ± 0.03	0.02
Na (wt%)	0.0730–0.608	0.26 ± 0.12	0.002
Fe (mg/kg)	48–644	100 ± 82	5
Mn (mg/kg)	30–144	61 ± 25	3
Zn (mg/kg)	16–82	44 ± 57	2
Cu (mg/kg)	5–12	7.5 ± 1.8	1
B (mg/kg)	3–16	6.2 ± 3.0	1

background, and transformations. After pre-processing, calibrations are produced. Simple univariate linear regression on LIBS spectra can produce poor results due to the variations in the data. These variations include nonlinearities, interferences, and noise. The advantages of using multivariate over univariate models include data compression, reduction of noise, increased tolerance of interferences, instrument selectivity becomes less important, and outliers are easily detected. The multivariate models also provide information on which variables are important in the calibration [139]. Multivariate chemometric methods can build models on the entire spectrum captured by the spectrometers whereas univariate methods only use one wavelength, or more in the case of internal normalisation. The methods used in this study are briefly explained. A detailed explanation can be found in the references quoted.

Savitzky Golay

A SG filter is a type of filter that is used for reducing noise, thus smoothing the data [140]. The result is an increase to the signal-to-noise ratio without overly distorting the data. Filtering is done by applying a polynomial function to a windowed set of the data and then adjusting the coefficients of the polynomial to minimise the mean-squared approximation error for the window. The central data point in the window is then replaced with the new value obtained from the polynomial function. The window is propagated through the entire data set until all data points have been processed. SG filtering can be used to calculate the derivatives of a signal.

Multiple Linear Regression

MLR is an extended version of simple linear regression which builds a model between the variables in the predictors and the output responses [141]. The predictor variables are regressed to find constant regression coefficients that minimises the sum of squares on the responses. Problems occur when some of the predictors are strongly correlated with each other (multicollinearity), or with a linear combination of several predictors.

Principal Component Regression

PCR compresses data by creating new variables called Principal Components (PCs) [141]. The PCs are linear combinations of the predictors and are uncorrelated. The first PC contains as much of the variability in the predictors as possible. Each successive PC accounts for the highest amount of variance in an orthogonal direction to the preceding PCs. With a small number of PCs the data set can be reduced but still retains a large portion of the variability. The drawback of PCR is that the variance caught by some of the PCs does not correspond to the responses. Thus there are various ways to determine which PCs to retain.

Partial Least-Squares Regression

PLS is a chemometric technique that has been employed to overcome the drawbacks of between spectra variability, which is not related to the responses [141]. PLS is used when there are many variables, compared to observations, and when there is high collinearity between variables. This is perfectly suited to LIBS spectra that have thousands of variables, many of which are collinear. PLS maximises the covariance between the predictors and responses by projecting the responses and the predictors to a new space. An underlying relationship can be found that is not easily detected by inspecting the spectra. This relationship is conveyed through the Latent Variables (LVs) which, like PCs, are a compression of the original data. PLS has found great success building calibration models that perform better than simple linear regression of a single emission line [97, 142].

Gaussian Process Regression

Gaussian process regression (GP) generates nonparametric kernel-based probabilistic models [143]. They take a collection of random variables which defines a distribution over functions. This is done by including only those functions that agree with the observations reducing the uncertainty around those data points. A noise function is added to the model that explains how much deviation there is between the response and function values. The function used in this study is the squared exponential covariance function, also known as the radial basis function. Tuning of the noise and length-scale parameters is needed to produce an accurate model.

Artificial Neural Networks

ANN are a group of techniques that based on the workings of biological systems. A feed-forward network is one of the types of ANN that handles nonlinearities in data [141]. The predictors are called the input layer and the responses are called the output layer. Between these layers are hidden layers that are comprised of elements called neurons. Each input to a neuron has an associated weighting which it is multiplied by before being summed and passed on to all the neurons in the next layer. The output layer and neurons are non-linear functions of linear combinations from the input layer and are commonly a sigmoid function. To train the network back-propagation is commonly used to update the regression coefficients as each observation is processed by the network. This continues until the prediction error is as small as possible.

Chemometric methods can fail when there are a large number of predictors and not enough data in the responses. Dimension reduction is a method of overcoming this problem. Techniques such as PCR and PLS compress the data into a few new variables (PCs and LVs). The scores from theses can then be used as inputs for methods like ANN and GP significantly improving results [144, 145].

Figures of Merit

To ensure that statistical results generalise for prediction of an independent dataset a model validation technique known as cross-validation was used [146]. Cross-validation splits the data up into training and test sets. A model is created on the training set and then evaluated on the test set. The process is repeated choosing a new test set which was not a part of the previous test set. This continues until each observation has been included in a test set. The number of iterations and the size of the test and training sets are determined by the split in the dataset. The percentage split is determined by how many folds are specified, this is known as k-fold cross-validation. A moderate number for k reduces the variance, increasing the performance of the method [147]. In this study 10-fold cross-validation is used. The accuracy of a model is determined by the difference between the predicted results and the actual responses.

One measure of accuracy for a cross-validated model is the RMSECV, which is calculated as

$$\mathrm{RMSECV} = \sqrt{\frac{\sum_{i=1}^{n} \left(\hat{y}_i - y_i\right)^2}{n}} \tag{1}$$

where $\hat{y}$ are the predictions from the cross-validation for the i-th sample which was not used in building the model for that fold, y are the actual responses, and n is the number of samples. Normalizing the RMSECV provides a better comparison between the models created for the different nutrients. The normalized root mean squared error of cross-validation (NRMSECV) is

$$\mathrm{NRMSECV} = \frac{\mathrm{RMSECV}}{y_{max} - y_{min}} \tag{2}$$

where y_{max} and y_{min} refer to the maximum and minimum concentrations of the particular nutrient the model was built for. The result is expressed as a percentage.

Calibration models for N, P, K, S, Ca, Mg, Na, Fe, Mn, Zn, Cu, and B were generated using multiple chemometric methods. No standardisation was done on the spectra before analyses. SG filtering, SG first derivative, and SG second derivative were performed on the spectra followed by PLS to find out if this form of pre-processing would benefit this particular type of data. The harvest site and date were investigated with PCR to determine whether this information would explain some variations in the data. PCR performed on the spectra alone was used for a comparison. MLR, PCR, PLS, PLS + GP, PCR + ANN, and PLS + ANN performed on the spectra were compared to determine which method would be best for pasture. The results were used to investigate the effects that argon and moisture had on the models. All methods used were investigated using 10-fold cross-validation to tune the parameters reducing the RMSECV. The software packages that were used were Matlab (MathWorks, US) and Weka (The University of Waikato, NZ) (for a description of the software refer to [148]).

Results and Discussion

Savitzky Golay Filtering

SG filtering was investigated to see if improvements could be made through pre-processing the spectra. SG, SG first derivative, and SG second derivative filtering followed by PLS were performed on the four datasets (fresh pasture in air, fresh pasture under argon, pellets in air, and pellets under argon) and compared to PLS with no filtering. A 6th order polynomial fit with a window length of 11 were used for filtering with the number of LVs depending on the element for the model.

SG filtering of LIBS spectra from nitrogen in pellets sampled in air resulted in a reduction in NRMSECV of 1.7% which was the only model to show a decrease of NRMSECV greater than 1%. Overall SG was considered ineffective since there was little to no reduction in NRMSECV. For this reason, it was not used in any further analyses.

Categorical Data

There may be perturbations between samples caused by site specific environmental effects and date specific events. These are hard to identify through visual inspection, but the influence can be detected and compensated for through chemometric techniques such as PCR and ANN. PCR and PCR + ANN were used to investigate whether including categorical data (date and site of harvest) would increase predictions in the models. There were six neurons in the first hidden layer and three in the second hidden layer for each model. The number of PCs depended on the element analysed.

There was very little change in the results using the categorical data. The biggest change recorded was a 0.2% decrease in NRMSECV when including the categorical data for potassium in pellet samples in argon. Performing ANN after PCR did not significantly increase the accuracy of the models. The model that had the largest improvement in NRMSECV, using PCR + ANN, was sulphur (0.4%) in pellets under argon.

Argon and Moisture Effects

Argon is commonly used in plasma spectroscopy to increase the electron density and temperature in plasmas. The chemometric methods used to investigate the effects of an argon purge were MLR, PCR, PLS, PLS + GP, PCR + ANN, and PLS + ANN. The noise and length-scale parameters were optimised with values of 0.01 and 5.77 respectively for GP and 6 neurons in the first hidden layer and 3 in the second hidden layer for ANN. The number of PCs and LVs were dependant on the element of interest.

Using an argon purge produced small differences. The changes can be seen in Fig. 1. The NRMSECV for these models were averaged to see the overall effect that argon had on fresh and pelletized pasture. Table 3 displays the results. Models created from fresh pasture under argon for N, K, Fe, Mn, Zn, and B produced NRMSECV over 1% better than models created in air, Mn having the largest increase of 2.1%. Pelletized pasture models created under argon only produced 1.2% better NRMSECV for N and Cu. NRMSECV for Ca and B were worse, a decrease of 1.2% and 1.6%, respectively.

There is a marked increase in accuracy for models built on pellets compared to models built on fresh samples, due to the reduction in moisture content in the samples. The moisture changes the breakdown threshold of a material which increases both within-sample and between-sample variations [79]. Figure 1 and Table 3 show that

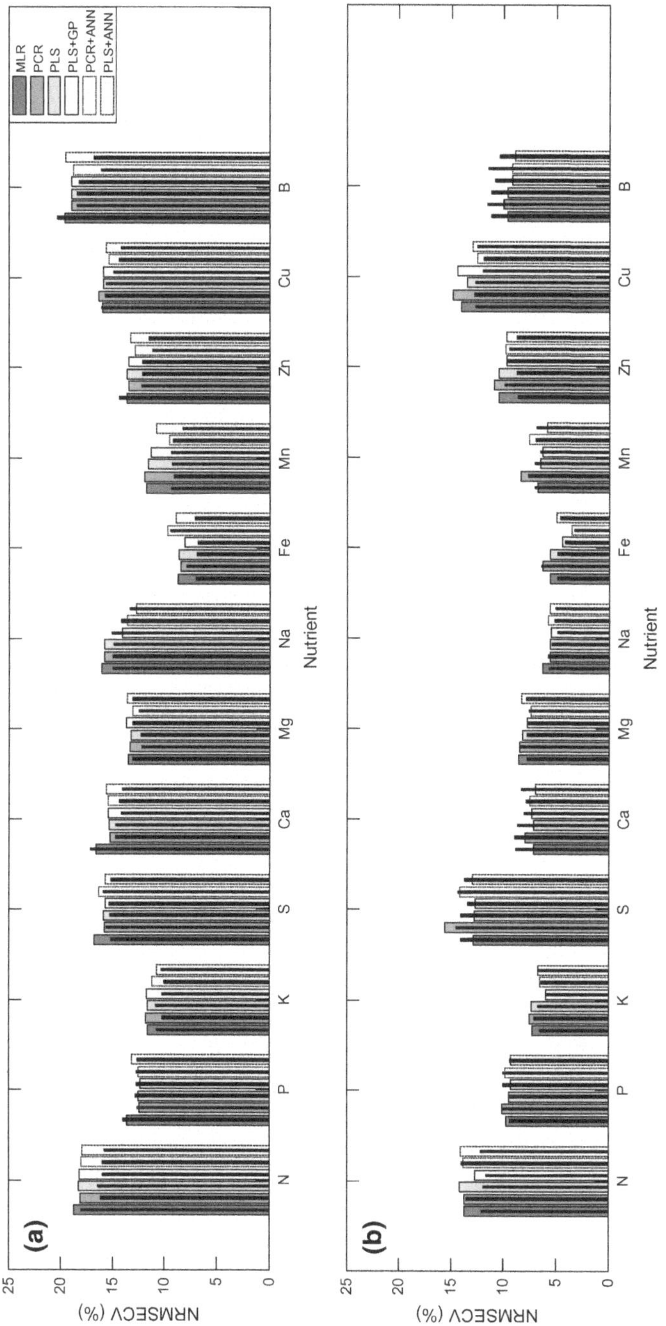

Fig. 1 Accuracy of models created for **a** fresh and **b** pelletized pasture. The bars indicate NRMSECV produced from spectra recorded in air and the solid black lines within the bars represent values created under argon

Table 3 Average NRMSECV values calculated from MLR, PCR, PLS, PLS + GP, PCR + ANN, and PLS + ANN. The uncertainty is one standard deviation

Element	Fresh pasture		Pelletized pasture	
	In air	Under argon	In air	Under argon
N (%)	18.2 ± 0.3	16.4 ± 0.8	13.7 ± 0.5	12.5 ± 0.9
P (%)	12.8 ± 0.5	12.9 ± 0.5	9.6 ± 0.3	9.7 ± 0.4
K (%)	11.5 ± 0.4	10.4 ± 0.3	6.9 ± 0.6	6.5 ± 0.4
S (%)	16.0 ± 0.4	15.4 ± 0.3	13.5 ± 1.1	14.0 ± 0.4
Ca (%)	15.6 ± 0.5	14.9 ± 1.2	7.3 ± 0.3	8.5 ± 0.4
Mg (%)	13.4 ± 0.2	12.7 ± 0.5	8.1 ± 0.5	7.9 ± 0.3
Na (%)	14.7 ± 1.4	14.6 ± 0.7	5.8 ± 0.3	5.4 ± 0.4
Fe (%)	8.7 ± 0.5	7.5 ± 1.0	5.1 ± 1.0	4.7 ± 1.0
Mn (%)	11.1 ± 0.9	9.0 ± 0.4	6.9 ± 0.9	7.0 ± 0.4
Zn (%)	13.4 ± 0.3	12.2 ± 1.1	10.2 ± 0.5	9.2 ± 0.6
Cu (%)	15.8 ± 0.3	15.1 ± 0.8	13.7 ± 0.9	12.5 ± 0.4
B (%)	19.1 ± 0.3	18.0 ± 1.5	9.5 ± 0.4	11.1 ± 0.5

removing the moisture in the pasture leads to an average increase in accuracy of 4–5% with the highest increase recorded for B, that of 9.6%. The increase may also be partially attributed to the increased homogenisation produced by grinding the samples. Fresh pasture will have different nutrient distributions in each blade of grass.

Comparison of Models

Comparing the different chemometric algorithms, there were mixed results with no single method outperforming the rest. Different techniques worked better on different elements. Figures 2 and 3 display the best calibration curves generated for fresh pasture in air and pelletized pasture under argon. Looking at which methods produced the best models for each element in the different datasets (fresh pasture in air, fresh pasture under argon, pelletized pasture in air, and pelletized pasture under argon) can give a general overview of which algorithm performed best overall.

Looking at the best performing models for the 12 nutrients from the 4 data sets showed that PLS + GP, PCR + ANN, and PLS + ANN were the best algorithms producing 19, 15, and 7 of the 48 models respectively. MLR, PCR, and PLS only produced only one, four, and two of the best models. Looking specifically at elements that produce models with R^2 values over 0.8 and NRMSECV under 10%, the more complex methods, PLS + GP, PCR + ANN, and PLS + ANN produced models with higher accuracy for K, Na, Ca under argon, Mg and Fe in pelletized samples, and Mn in air. The largest improvement was a reduction in NRMSECV of 3.3% when

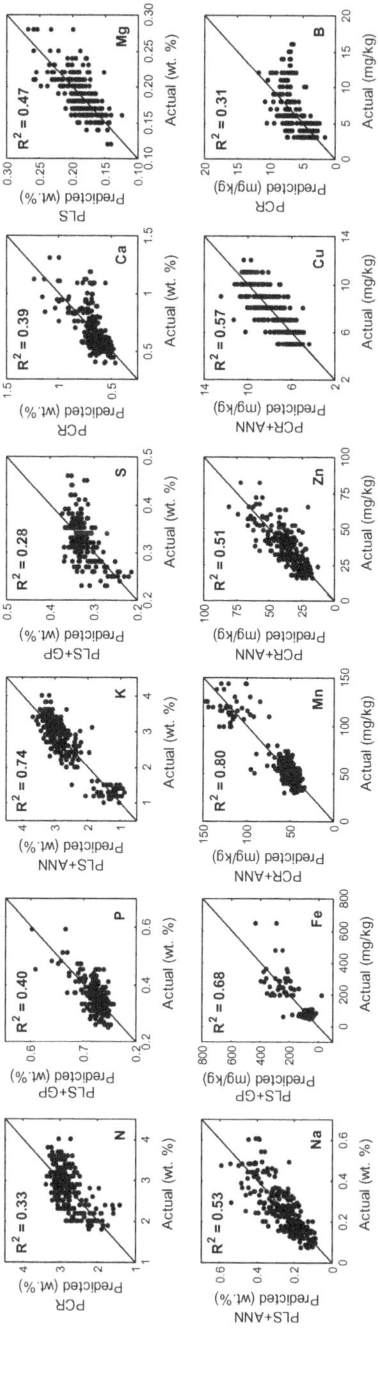

Fig. 2 The best calibrations created for fresh pasture in air

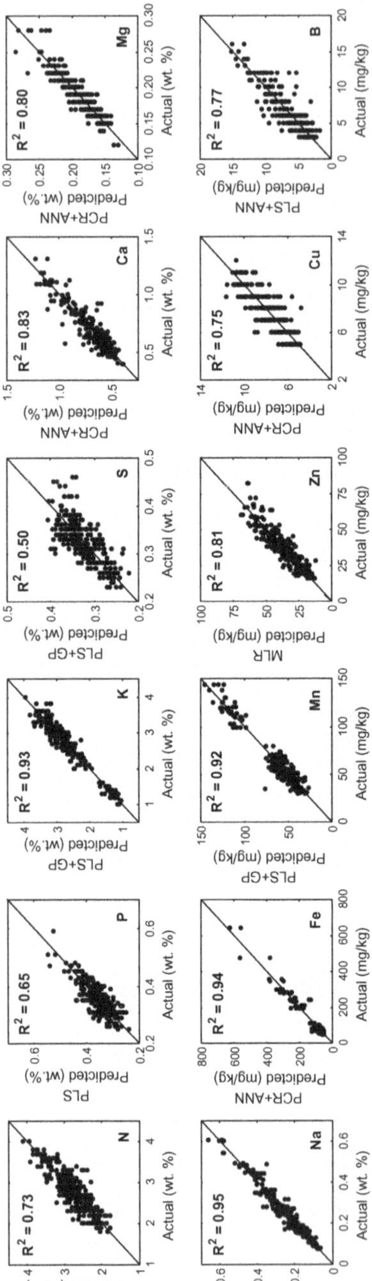

Fig. 3 The best calibrations created for pelletized pasture under argon

comparing the PLS + ANN model with the MLR model for Na in fresh pasture. The time taken to perform ANN is the compromise for increasing the accuracy of the model. Inspection of Figs. 2 and 3 shows that there are two groupings of concentrations for K and Mn. The low K and high Mn concentrations harvested in March 2014 were caused by a drought in the region. The non-Normal range of concentrations provides added leverage for R^2 and NRMSECV values making them look similar to other elements.

Table 4 displays the RMSECV values to provide a better idea of how the predictions are distributed. The data in Table 4 was made by choosing the best RMSECV with the best R^2 value from all algorithms. K has the second worse RMSECV out of the macronutrients, but because of the range of concentrations in the samples it has one of the best correlation coefficients even without the drought data.

Another way to increase the predictive ability of the models is to build them on better data with stronger, consistent emissions. The spectra in this work had weak emission lines for N, P, Mg, Fe, and Cu. Zn lines only appeared under an argon atmosphere, and B lines were only visible on a few occasions. The major problem with N comes from the N contribution in the atmosphere. Using argon did increase the accuracy of the N models. Emissions for S could not be obtained with the spectrometers used in this work, so the models would have been created on inferred data. The strongest line for P had interferences from Fe which would limit quantitative analysis. Ca and Mg produced some of the better models even though some of the emissions from these elements were not from ionised Ca and Mg. The ratio of neutral calcium to ionised calcium changes from shot to shot producing varying spectra for the same sample. This is the same for magnesium.

In Field Application

As discussed above, the major issue with LIBS measurements on pasture is variability. The chemometric methods used in this study provide reasonable models to estimate semi-quantitative levels of nutrients. These laboratory results indicate how an in field system would perform minus the added in field variability. For in field use to be realised, improved models from better quality spectra with less shot-to-shot variation are needed [149]. The variations in experimental parameters, surface roughness, and local inhomogeneity may be mitigated by increasing the spot size and ablating more material. Finding a way to normalise the spectra to disregard moisture would help increase predictability of quantitative calibration models. Accurate chemometric models can be produced once these factors are minimised. Infield implementation can then be realised for fresh pasture using a LIBS sensor to determine fertilizer application and rate in real time.

Table 4 The best algorithms for the nutrients in each dataset

Element	Pasture in air		Pasture under argon	
	Method	RMSECV	Method	RMSECV
N (wt%)	PCR	0.40	PLS+GP	0.35
P (wt%)	PLS+GP	0.042	PLS+GP	0.043
K (wt%)	PLS+ANN	0.32	PCR+ANN	0.30
S (wt%)	PLS+GP	0.036	PLS+GP	0.035
Ca (wt%)	PCR	0.14	PLS+GP	0.13
Mg (wt%)	PLS	0.021	PCR	0.020
Na (wt%)	PLS+ANN	0.068	PCR+ANN	0.076
Fe (mg/kg)	PLS+GP	48	PLS+GP	41
Mn (mg/kg)	PCR+ANN	11	PLS+ANN	9.4
Zn (mg/kg)	PCR+ANN	8.5	PCR+ANN	7.3
Cu (mg/kg)	PCR+ANN	1.1	PCR+ANN	1.0
B (mg/kg)	PCR	2.5	PCR+ANN	2.1
Element	Pellets in air		Pellets under argon	
	Method	RMSECV	Method	RMSECV
N (wt%)	PLS+GP	0.28	PLS+GP	0.26
P (wt%)	PLS+GP	0.032	PLS	0.032
K (wt%)	PLS+GP	0.18	PLS+GP	0.18
S (wt%)	PLS+GP	0.029	PLS+GP	0.031
Ca (wt%)	PLS+ANN	0.064	PCR+ANN	0.073
Mg (wt%)	PCR+ANN	0.012	PCR+ANN	0.012
Na (wt%)	PLS+GP	0.030	PLS+GP	0.026
Fe (mg/kg)	PCR+ANN	21	PCR+ANN	20
Mn (mg/kg)	PLS+ANN	6.7	PLS+GP	7.4
Zn (mg/kg)	PLS+GP	6.5	MLR	5.7
Cu (mg/kg)	PCR+ANN	0.88	PCR+ANN	0.84
B (mg/kg)	PLS+ANN	1.2	PLS+ANN	1.3

Conclusions

SG filtering was ineffective, there was little to no reduction in NRMSECV since the main source of disparity is shot-to-shot variation, not noise in the spectra. Including categorical data for PCR and PCR+ANN did not produce a significant increase in prediction. The use of an argon purge increased the predictive capability in fresh samples with a reduction of over 1% in NRMSECV for N, K, Fe, Mn, Zn, and B. For pelletised samples only N, Cu, and Zn had an improvement over 1% in NRMSECV with Ca and B producing a decrease of 1.2% and 1.6%, respectively. The highest increase in predictive ability was produced by using dried pellets instead of fresh samples. The reduction in moisture lead to models having an average increase in accuracy of 4–5%. No single chemometric method provided improvement for all nutrients. The algorithms that performed the best were PLS + GP, PCR + ANN, and PLS + ANN.

To increase the predictive ability of chemometric models improved spectra are needed and the shot-to-shot variation needs to be reduced. Temperature differences between plasmas is a main factor contributing to the variations. Ca and Mg suffer the most from the temperature difference between plasmas because the temperature changes the amount of Ca and Mg being ionised. This significantly changes the intensities of the neutral and ionised emission lines. Other issues that need resolving are that some of the nutrients displayed weak lines or interferences from other lines and the surrounding atmosphere. The chemometric methods used in this study provide reasonable models. Before infield use is viable, a moisture normalisation method needs to be developed to produce accuracies similar to that of dried pasture.

Acknowledgements We would like to acknowledge and offer continuing respect and admiration to deceased colleague Dr. Sadhana Talele, who was not able to see this work completed. This work was supported by New Zealand's Ministry of Business, Innovation and Employment under contract C11X1209.

References

1. M. van Maarschalkerweerd, S. Husted, Recent developments in fast spectroscopy for plant mineral analysis. Front. Plant Sci. **6**, 169 (2015)
2. D. Cremers, L. Radziemski, Introduction, in *Handbook of Laser-Induced Breakdown Spectroscopy*, 2nd edn. (Wiley, West Sussex, UK, 2013), p. 3
3. S.N. Thakur, Atomic emission spectroscopy, in *Laser-Induced Breakdown Spectroscopy*, ed. by J.P. Singh, S.N. Thakur (Elsevier, Oxford, UK, 2007), pp. 29–30
4. D.W. Hahn, N. Omenetto, Laser-induced breakdown spectroscopy (LIBS), part II: review of instrumental and methodological approaches to material analysis and applications to different fields. Appl. Spectrosc. **66**, 347–419 (2012)
5. G. Galbács, I.B. Gornushkin, B.W. Smith, J.D. Winefordner, Semi-quantitative analysis of binary alloys using laser-induced breakdown spectroscopy and a new calibration approach based on linear correlation. Spectrochim. Acta Part B Atomic Spectrosc. **56**, 1159–1173 (2001)
6. S. Palanco, J.J. Laserna, Full automation of a laser-induced breakdown spectrometer for quality assessment in the steel industry with sample handling, surface preparation and quantitative analysis capabilities. J. Anal. Atomic Spectrom. **15**, 1321–1327 (2000)
7. R. Wisbrun, I. Schechter, R. Niessner, H. Schröder, K.L. Kompa, Detector for trace elemental analysis of solid environmental samples by laser plasma spectroscopy. Anal. Chem. **66**, 2964–2975 (1994)
8. U. Panne, M. Clara, C. Haisch, R. Niessner, Analysis of glass and glass melts during the vitrification of fly and bottom ashes by laser-induced plasma spectroscopy. Part II. Process analysis. Spectrochim. Acta Part B: Atomic Spectros. **53**, 1969–1981 (1998)
9. J.B. Sirven, B. Bousquet, L. Canioni, L. Sarger, Laser-induced breakdown spectroscopy of composite samples: comparison of advanced chemometrics methods. Anal. Chem. **78**, 1462–1469 (2006)

10. D. Cremers, L. Radziemski, LIBS apparatus fundamentals, in *Handbook of Laser-Induced Breakdown Spectroscopy*, ed. by D. Cremers, L. Radziemski, 2nd edn. (Wiley, West Sussex, UK, 2013), pp. 69–121
11. M. Garcimuño, D.M. Díaz Pace, G. Bertuccelli, Laser-induced breakdown spectroscopy for quantitative analysis of copper in algae. Opt. Laser Technol. **47**, 26–30 (2013)
12. L. Niu, H.H. Cho, K.S. Song, H. Cha, Y. Kim, Y.I. Lee, Direct determination of strontium in marine algae samples by laser-induced breakdown spectrometry. Appl. Spectrosc. **56**, 1511–1514 (2002)
13. M.M. El-Deftar, J. Robertson, S. Foster, C. Lennard, Evaluation of elemental profiling methods, including laser-induced breakdown spectroscopy (LIBS), for the differentiation of Cannabis plant material grown in different nutrient solutions. Forensic Sci. Int. **251**, 95–106 (2015)
14. J.W.B. Braga, L.C. Trevizan, L.C. Nunes, I.A. Rufini, D. Santos Jr., F.J. Krug, Comparison of univariate and multivariate calibration for the determination of micronutrients in pellets of plant materials by laser induced breakdown spectrometry. Spectrochim. Acta Part B Atomic Spectrosc. **65**, 66–74 (2010)
15. L.C. Trevizan, D. Santos Jr., R.E. Samad, N.D. Vieira Jr., L.C. Nunes, I.A. Rufini, F.J. Krug, Evaluation of laser induced breakdown spectroscopy for the determination of micronutrients in plant materials. Spectrochim. Acta Part B Atomic Spectrosc. **64**, 369–377 (2009)
16. L.C. Trevizan, D. Santos Jr., R.E. Samad, N.D. Vieira Jr., C.S. Nomura, L.C. Nunes, I.A. Rufini, F.J. Krug, Evaluation of laser induced breakdown spectroscopy for the determination of macronutrients in plant materials. Spectrochim. Acta Part B Atomic Spectrosc. **63**, 1151–1158 (2008)
17. D.K. Chauhan, D.K. Tripathi, N.K. Rai, A.K. Rai, Detection of biogenic silica in leaf blade, leaf sheath, and stem of Bermuda grass (Cynodon dactylon) using LIBS and phytolith analysis. Food Biophys. **6**, 416–423 (2011)
18. P.K. Rai, D. Jaiswal, N.K. Rai, S. Pandhija, A.K. Rai, G. Watal, Role of glycemic elements of Cynodon dactylon and Musa paradisiaca in diabetes management. Lasers Med. Sci. **24**, 761–768 (2009)
19. M. Iqbal, Z. ul Haq, A. Malik, C.M. Ayoub, Y. Jamil, J. Nisar, Pre-sowing seed magnetic field stimulation: a good option to enhance bitter gourd germination, seedling growth and yield characteristics. Biocatal. Agric. Biotechnol. **5**, 30–37 (2016)
20. N.K. Rai, P.K. Rai, S. Pandhija, G. Watal, A.K. Rai, D. Bicanic, Application of LIBS in detection of antihyperglycemic trace elements in Momordica charantia. Food Biophys. **4**, 167–171 (2009)
21. P. Zheng, M. Shi, J. Wang, H. Liu, The spectral emission characteristics of laser induced plasma on tea samples. Plasma Sci. Technol **17**, 664–670 (2015)
22. G.G.A. de Carvalho, D. Santos Jr., M. da Silva Gomes, L.C. Nunes, M.B.B. Guerra, F.J. Krug, Influence of particle size distribution on the analysis of pellets of plant materials by laser-induced breakdown spectroscopy. Spectrochim. Acta Part B Atomic Spectrosc. **105**, 130–135 (2015)
23. M.R. Martelli, F. Brygo, P. Delaporte, X. Rouau, C. Barron, Estimation of wheat grain tissue cohesion via laser induced breakdown spectroscopy. Food Biophys. **6**, 433–439 (2011)
24. L. Krajcarová, K. Novotný, M. Kummerová, J. Dubová, V. Gloser, J. Kaiser, Mapping of the spatial distribution of silver nanoparticles in root tissues of Vicia faba by laser-induced breakdown spectroscopy (LIBS). Talanta **173**, 28–35 (2017)
25. S. Awasthi, R. Kumar, A. Devanathan, R. Acharya, A.K. Rai, Multivariate methods for analysis of environmental reference materials using laser-induced breakdown spectroscopy. Anal. Chem. Res. **12**, 10–16 (2017)
26. M. Galiová, J. Kaiser, K. Novotný, M. Hartl, R. Kizek, P. Babula, Utilization of laser-assisted analytical methods for monitoring of lead and nutrition elements distribution in fresh and dried Capsicum annuum l. leaves. Microsc. Res. Tech. **74**, 845–852 (2011)
27. N. Shukla, A.S. Bharti, S. Srivastava, K.N. Uttam, Determination of elements in carrot root by laser induced breakdown spectroscopy. Nat. Acad. Sci. Lett. **40**, 47–51 (2017)

28. C.R. Bhatt, B. Alfarraj, C.T. Ghany, F.Y. Yueh, J.P. Singh, Comparative study of elemental nutrients in organic and conventional vegetables using laser-induced breakdown spectroscopy (LIBS). Appl. Spectrosc. **71**, 686–698 (2017)
29. M. Yao, H. Yang, L. Huang, T. Chen, G. Rao, M. Liu, Detection of heavy metal Cd in polluted fresh leafy vegetables by laser-induced breakdown spectroscopy. Appl. Opt. **56**, 4070–4075 (2017)
30. J. Wang, P. Zheng, H. Liu, L. Fang, Classification of Chinese tea leaves using laser-induced breakdown spectroscopy combined with the discriminant analysis method. Anal. Methods **8**, 3204–3209 (2016)
31. M.M. Suliyanti, M. Pardede, T.J. Lie, K.H. Kurniawan, A. Khumaeni, K. Kagawa, M.O. Tjia, Y.I. Lee, Direct powder analysis by laser-induced breakdown spectroscopy utilizing laser-controlled dust production in a small chamber. J. Korean Phys. Soc. **58**, 1129–1134 (2011)
32. F.M.V. Pereira, D.M.B.P. Milori, A.L. Venâncio, M.D.S.T. Russo, P.K. Martins, J. Freitas-Astúa, Evaluation of the effects of Candidatus Liberibacter asiaticus on inoculated citrus plants using laser-induced breakdown spectroscopy (LIBS) and chemometrics tools. Talanta **83**, 351–356 (2010)
33. A.C. Ranulfi, R.A. Romano, A. Bebeachibuli Magalhães, E.J. Ferreira, P. Ribeiro Villas-Boas, D. Marcondes Bastos Pereira Milori, Evaluation of the nutritional changes caused by Huanglongbing (HLB) to citrus plants using laser-induced breakdown spectroscopy. Appl. Spectrosc. **71**, 1471–1480 (2017)
34. E.J. Ferreira, E.C. Ferreira, A.C.B. Delbem, D.M.B.P. Milori, Ensemble of predictors and laser induced breakdown spectroscopy for certifying coffee. Electron. Lett. **47**, 967–969 (2011)
35. M.A. Gondal, U. Baig, M.A. Dastageer, M. Sarwar, Determination of elemental composition of coffee using UV-pulsed laser induced breakdown spectroscopy, in *AIP Conference Proceedings*, Gizan, Saudi Arabia (2016), p. 030007
36. T.V. Silva, S.Z. Hubinger, J.A. Gomes Neto, D.M.B.P. Milori, E.J. Ferreira, E.C. Ferreira, Potential of laser induced breakdown spectroscopy for analyzing the quality of unroasted and ground coffee. Spectrochim. Acta Part B Atomic Spectrosc. **135**, 29–33 (2017)
37. E.R. Schenk, J.R. Almirall, Elemental analysis of cotton by laser-induced breakdown spectroscopy. Appl. Opt. **49**, C153–C160 (2010)
38. D.K. Kushawaha, M. Yadav, S. Chatterji, A.K. Srivastava, G. Watal, α-amylase and α-glucosidase inhibitory activity assessment of Cucurbita maxima seeds—a LIBS based study. Int. J. Phytomed. **8**, 312–318 (2016)
39. J.N. Kunz, D.V. Voronine, B.A. Ko, H.W.H. Lee, A. Rana, M.V. Bagavathiannan, A.V. Sokolov, M.O. Scully, Interaction of femtosecond laser pulses with plants: towards distinguishing weeds and crops using plasma temperature. J. Mod. Opt. **64**, 942–947 (2017)
40. P. Modlitbová, K. Novotný, P. Pořízka, J. Klus, P. Lubal, H. Zlámalová-Gargošová, J. Kaiser, Comparative investigation of toxicity and bioaccumulation of Cd-based quantum dots and Cd salt in freshwater plant Lemna minor L. Ecotoxicol. Environ. Safety **147**, 334–341 (2018)
41. S. Mehta, P.K. Rai, D.K. Rai, N.K. Rai, A.K. Rai, D. Bicanic, B. Sharma, G. Watal, LIBS-based detection of antioxidant elements in seeds of Emblica officinalis. Food Biophys. **5**, 186–192 (2010)
42. R.K. Singh, S. Mehta, D. Jaiswal, P.K. Rai, G. Watal, Antidiabetic effect of Ficus bengalensis aerial roots in experimental animals. J. Ethnopharmacol. **123**, 110–114 (2009)
43. D. Sun, M. Su, C. Dong, D. Zhang, X. Ma, A semi-quantitative analysis of essential micronutrient in folium lycii using laser-induced breakdown spectroscopy technique. Plasma Sci. Technol **12**, 478–481 (2010)
44. H. Hu, L. Huang, M. Liu, T. Chen, P. Yang, M. Yao, Nondestructive determination of cu residue in orange peel by laser induced breakdown spectroscopy. Plasma Sci. Technol **17**, 711–715 (2015)
45. M. Yao, L. Huang, J. Zheng, S. Fan, M. Liu, Assessment of feasibility in determining of Cr in Gannan Navel Orange treated in controlled conditions by laser induced breakdown spectroscopy. Opt. Laser Technol. **52**, 70–74 (2013)

46. A.A. Bol'shakov, J.H. Yoo, C. Liu, J.R. Plumer, R.E. Russo, Laser-induced breakdown spectroscopy in industrial and security applications. Appl. Opt. **49**, C132–C142 (2010)
47. P.K. Rai, N.K. Rai, A.K. Rai, G. Watal, Role of LIBS in elemental analysis of Psidium guajava responsible for glycemic potential. Instrum Sci. Technol. **35**, 507–522 (2007)
48. D.C. Zhang, X. Ma, W.Q. Wen, H.P. Liu, P.J. Zhang, Studies of laser induced-breakdown spectroscopy of holly leaves. J. Phys: Conf. Ser. **185**, 012058 (2009)
49. D.-C. Zhang, X.-W. Ma, W.-Q. Wen, P.-J. Zhang, X.-L. Zhu, B. Li, H.-P. Liu, Influence of laser wavelength on laser-induced breakdown spectroscopy applied to semi-quantitative analysis of trace-elements in a plant sample. Chin. Phys. Lett. **27**, 063202 (4 pp.) (2010)
50. X. Liu, J. Huang, Z. Wu, Q. Zhang, X. Shi, N. Zhao, S. Jia, Y. Qiao, Microanalysis of multi-element in Juncus effusus L. by LIBS technique. Plasma Sci. Technol **17**, 904–908 (2015)
51. V. Diopan, V. Shestivska, O. Zitka, M. Galiova, V. Adam, J. Kaiser, A. Horna, K. Novotny, M. Liska, L. Havel, J. Zehnalek, R. Kizek, Determination of plant thiols by liquid chromatography coupled with coulometric and amperometric detection in lettuce treated by lead(II) ions. Electroanalysis **22**, 1248–1259 (2010)
52. J. Wang, S. Xue, P. Zheng, Y. Chen, R. Peng, Determination of lead and copper in ligusticum wallichii by laser-induced breakdown spectroscopy. Anal. Lett. **50**, 2000–2011 (2017)
53. S. Qamar, M. Aslam, M.A. Javed, Determination of proximate chemical composition and detection of inorganic nutrients in maize (Zea mays L.), in *Materials Today: Proceedings*, New York, USA (2016), pp. 715–718
54. O. Krystofova, V. Shestivska, M. Galiova, K. Novotny, J. Kaiser, J. Zehnalek, P. Babula, R. Opatrilova, V. Adam, R. Kizek, Sunflower plants as bioindicators of environmental pollution with lead (II) ions. Sensors **9**, 5040–5058 (2009)
55. C. Zhao, D. Dong, X. Du, W. Zheng, In-field, in situ, and in vivo 3-dimensional elemental mapping for plant tissue and soil analysis using laser-induced breakdown spectroscopy, *Sensors (Basel, Switzerland)*, **16**, 1764 (2016)
56. O. Samek, J. Lambert, R. Hergenroder, M. Liska, J. Kaiser, K. Novotny, S. Kukhlevsky, Femtosecond laser spectrochemical analysis of plant samples. Laser Phys. Lett. **3**, 21–25 (2006)
57. M. Barbafieri, R. Pini, A. Ciucci, E. Tassi, Field assessment of Pb in contaminated soils and in leaf mustard (Brassica juncea): the LIBS technique. Chem. Ecol. **27**, 161–169 (2011)
58. Y. Mingyin, L. Jinlong, L. Muhua, L. Qiulian, L. Zejian, Discrimination of Ca, Cu, Fe, and Na in Gannan Navel orange by laser induced breakdown spectroscopy, in *4th Conference on Computer and Computing Technologies in Agriculture (CCTA)*, Nanchang, China (2011), pp. 608–613
59. D.K. Tripathi, A.K. Pathak, D.K. Chauhan, N.K. Dubey, A.K. Rai, R. Prasad, An efficient approach of laser induced breakdown spectroscopy (LIBS) and ICAP-AES to detect the elemental profile of Ocimum L. species. Biocatal. Agric. Biotechnol. **4**, 471–479 (2015)
60. M. Yao, M. Liu, J. Zhao, L. Huang, Identification of nutrition elements in orange leaves by laser induced breakdown spectroscopy, in *2010 Third International Symposium on Intelligent Information Technology and Security Informatics (IITSI)*, Jinggangshan, China (2010), pp. 398–401
61. Z. Lei, M. Yao, M. Liu, Q. Peng, X. Zhang, T. Chen, Y. Xu, Analysis the macronutrients in Gannan navel oranges for three kinds of sample state by laser induced breakdown spectroscopy, in *2011 4th International Congress on Image and Signal Processing (CISP)*, Shanghai, China (2011), pp. 2464–2466
62. X. Zhang, M. Yao, M. Liu, Z. Lei, Analysis of trace elements in leaves using laser-induced breakdown spectroscopy," in *5th International Conference on Computer and Computing Technologies in Agriculture, CCTA*, Beijing, China (2011), pp. 334–339
63. S.I. Gornushkin, I.B. Gornushkin, J.M. Anzano, B.W. Smith, J.D. Winefordner, Effective normalization technique for correction of matrix effects in laser-induced breakdown spectroscopy detection of magnesium in powdered samples. Appl. Spectrosc. **56**, 433–436 (2002)
64. K. Devey, M. Mucalo, G. Rajendram, J. Lane, Pasture vegetation elemental analysis by laser-induced breakdown spectroscopy. Commun. Soil Sci. Plant Anal. **46**, 72–80 (2015)

65. Z. Haider, J. Ali, M. Arab, Y.B. Munajat, S. Roslan, R. Kamarulzman, N. Bidin, Plasma diagnostics and determination of lead in soil and phaleria macrocarpa leaves by ungated laser induced breakdown spectroscopy. Anal. Lett. **49**, 808–817 (2016)
66. A.C. Samuels, F.C. DeLucia Jr., K.L. McNesby, A.W. Miziolek, Laser-induced breakdown spectroscopy of bacterial spores, molds, pollens, and protein: Initial studies of discrimination potential. Appl. Opt. **42**, 6205–6209 (2003)
67. A.R. Boyain-Goitia, D.C.S. Beddows, B.C. Griffiths, H.H. Telle, Single-pollen analysis by laser-induced breakdown spectroscopy and raman microscopy. Appl. Opt. **42**, 6119–6132 (2003)
68. S. Ma, X. Gao, K. Guo, M. Kahsay, J. Lin, Analysis of the element content in poplar tree leaves by femtosecond laser-induced breakdown spectroscopy. Sci. China Phys. Mech. Astron. **54**, 1953–1957 (2011)
69. W. Lei, V. Motto-Ros, M. Boueri, Q. Ma, D. Zhang, L. Zheng, H. Zeng, J. Yu, Time-resolved characterization of laser-induced plasma from fresh potatoes. Spectrochim. Acta Part B Atomic Spectrosc. **64**, 891–898 (2009)
70. S. Beldjilali, W.L. Yip, J. Hermann, T. Baba-Hamed, A. Belasri, Investigation of plasmas produced by laser ablation using single and double pulses for food analysis demonstrated by probing potato skins. Anal. Bioanal. Chem. **400**, 2173–2183 (2011)
71. S. Beldjilali, D. Borivent, L. Mercadier, E. Mothe, G. Clair, J. Hermann, Evaluation of minor element concentrations in potatoes using laser-induced breakdown spectroscopy. Spectrochim. Acta Part B Atomic Spectrosc. **65**, 727–733 (2010)
72. V. Juvé, R. Portelli, M. Boueri, M. Baudelet, J. Yu, Space-resolved analysis of trace elements in fresh vegetables using ultraviolet nanosecond laser-induced breakdown spectroscopy. Spectrochim. Acta Part B Atomic Spectrosc. **63**, 1047–1053 (2008)
73. A. Khumaeni, Z.S. Lie, H. Niki, K.H. Kurniawan, E. Tjoeng, Y.I. Lee, K. Kurihara, Y. Deguchi, K. Kagawa, Direct analysis of powder samples using transversely excited atmospheric CO 2 laser-induced gas plasma at 1 atm. Anal. Bioanal. Chem. **400**, 3279–3287 (2011)
74. T. Flores, L. Ponce, M. Arronte, E. de Posada, Free-running and Q: switched LIBS measurements during the laser ablation of prickle pears spines. Opt. Lasers Eng. **47**, 578–583 (2009)
75. P. Dhar, I. Gembitsky, P. Rai, N. Rai, A.K. Rai, G. Watal, A possible connection between antidiabetic & antilipemic properties of psoralea corylifolia seeds and the trace elements present: a LIBS based study. Food Biophys. **8**, 95–103 (2013)
76. P. Dhar, I. Gembitsky, P.K. Rai, N.K. Rai, A.K. Rai, G. Watal, A possible connection between antidiabetic & antilipemic properties of psoralea corylifolia seeds and the trace elements present: a LIBS based study, Food Biophys. 1–9 (2012)
77. X. Du, D. Dong, X. Zhao, L. Jiao, P. Han, Y. Lang, Detection of pesticide residues on fruit surfaces using laser induced breakdown spectroscopy. RSC Adv. **5**, 79956–79963 (2015)
78. T.J. Jiang, Z. Guo, M.J. Ma, L. Fang, M. Yang, S.S. Li, J.H. Liu, N.J. Zhao, X.J. Huang, W.Q. Liu, Electrochemical laser induced breakdown spectroscopy for enhanced detection of Cd(II) without interference in rice on layer-by-layer assembly of graphene oxides. Electrochim. Acta **216**, 188–195 (2016)
79. J. Peng, Y. He, L. Ye, T. Shen, F. Liu, W. Kong, X. Liu, Y. Zhao, Moisture influence reducing method for heavy metals detection in plant materials using laser-induced breakdown spectroscopy: a case study for chromium content detection in rice leaves. Anal. Chem. **89**, 7593–7600 (2017)
80. A. Khumaeni, M. Ramli, Y. Deguchi, Y. Lee, N. Idris, K.H. Kurniawan, T.J.J. Lie, K. Kagawa, New technique for the direct analysis of food powders confined in a small hole using transversely excited atmospheric CO2 laser-induced gas plasma. Appl. Spectrosc. **62**, 1344–1348 (2008)
81. S. Varliklioz Er, H. Eksi-Kocak, H. Yetim, I.H. Boyaci, Novel spectroscopic method for determination and quantification of saffron adulteration. Food Anal. Methods **10**, 1547–1555 (2017)

82. M. Hassan, M. Sighicelli, A. Lai, F. Colao, A.H.H. Ahmed, R. Fantoni, M.A. Harith, Studying the enhanced phytoremediation of lead contaminated soils via laser induced breakdown spectroscopy. Spectrochim. Acta Part B Atomic Spectrosc. **63**, 1225–1229 (2008)
83. M. Hassan, M. Abdelhamied, A.H. Hanafy, R. Fantoni, M.A. Harith, Laser monitoring of phytoextraction enhancement of lead contaminated soil adopting EDTA and EDDS, in *AIP Conference Proceedings*, Cairo, Egypt (2011), pp. 93–100
84. J. Singh, R. Kumar, S. Awasthi, V. Singh, A.K. Rai, Laser Induced breakdown spectroscopy: a rapid tool for the identification and quantification of minerals in cucurbit seeds. Food Chem. **221**, 1778–1783 (2017)
85. M. Bossu, Z.Q. Hao, M. Baudelet, J. Yu, Z. Zhang, J. Zhang, Femtosecond laser-induced breakdown spectroscopy for detection of trace elements in sophora leaves. Chin. Phys. Lett. **24**, 3466–3468 (2007)
86. G. Nicolodelli, G.S. Senesi, A.C. Ranulfi, B.S. Marangoni, A. Watanabe, V. de Melo Benites, P.P.A. de Oliveira, P. Villas-Boas, D.M.B.P. Milori, Double-pulse laser induced breakdown spectroscopy in orthogonal beam geometry to enhance line emission intensity from agricultural samples. Microchem. J. **133**, 272–278 (2017)
87. G. Kim, J. Kwak, J. Choi, K. Park, Detection of nutrient elements and contamination by pesticides in spinach and rice samples using laser-induced breakdown spectroscopy (LIBS). J. Agric. Food Chem. **60**, 718–724 (2012)
88. L.C. Nunes, G.A. da Silva, L.C. Trevizan, D. Santos Júnior, R.J. Poppi, F.J. Krug, Simultaneous optimization by neuro-genetic approach for analysis of plant materials by laser induced breakdown spectroscopy. Spectrochim. Acta Part B Atomic Spectrosc. **64**, 565–572 (2009)
89. P. Shukla, R. Kumar, A.K. Raib, Detection of minerals in green leafy vegetables using laser induced breakdown spectroscopy. J. Appl. Spectrosc. **83**, 872–877 (2016)
90. S. Zivkovic, M. Momcilovic, A. Staicu, J. Mutic, M. Trtica, J. Savovic, Spectrochemical analysis of powdered biological samples using transversely excited atmospheric carbon dioxide laser plasma excitation. Spectrochim. Acta Part B Atomic Spectrosc. **128**, 22–29 (2017)
91. H.H. Cho, Y.J. Kim, Y.S. Jo, K. Kitagawa, N. Arai, Y.I. Lee, Application of laser-induced breakdown spectrometry for direct determination of trace elements in starch-based flours. J. Anal. Atomic Spectrom. **16**, 622–627 (2001)
92. P.F. de Souza, D. Santos, G.G.A. de Carvalho, L.C. Nunes, M. da Silva Gomes, M.B.B. Guerra, F.J. Krug, Determination of silicon in plant materials by laser-induced breakdown spectroscopy. Spectrochim. Acta Part B Atomic Spectrosc. (2013)
93. J.P.R. Romera, P.L. Barsanelli, F.M.V. Pereira, Expeditious prediction of fiber content in sugar cane: An analytical possibility with LIBS and chemometrics. Fuel **166**, 473–476 (2016)
94. D.K. Tripathi, R. Kumar, D.K. Chauhan, A.K. Rai, D. Bicanic, Laser-induced breakdown spectroscopy for the study of the pattern of silicon deposition in leaves of saccharum species. Instrum Sci. Technol. **39**, 510–521 (2011)
95. G.G.A. De Carvalho, D. Santos Jr., L.C. Nunes, M.D.S. Gomes, F.D.O. Leme, F.J. Krug, Effects of laser focusing and fluence on the analysis of pellets of plant materials by laser-induced breakdown spectroscopy. Spectrochim. Acta Part B Atomic Spectrosc. **74–75**, 162–168 (2012)
96. M.B. Bueno Guerra, A. Adame, E. De Almeida, G.G. Arantes De Carvalho, M.A. Stolf Brasil, D. Santos, Jr., F.J. Krug, Direct analysis of plant leaves by EDXRF and LIBS: microsampling strategies and cross-validation. J. Anal. Atomic Spectrom. **30**, 1646–1654 (2015)
97. L.C. Nunes, J.W. Batista Braga, L.C. Trevizan, P. Florêncio De Souza, G.G. Arantes De Carvalho, D.S. Júnior, R.J. Poppi, F.J. Krug, Optimization and validation of a LIBS method for the determination of macro and micronutrients in sugar cane leaves. J. Anal. Atomic Spectrom. **25**, 1453–1460 (2010)
98. M. da Silva Gomes, G.G.A. de Carvalho, D. Santos Jr., F.J. Krug, A novel strategy for preparing calibration standards for the analysis of plant materials by laser-induced breakdown spectroscopy: a case study with pellets of sugar cane leaves. Spectrochim. Acta Part B Atomic Spectrosc. **86**, 137–141 (2013)

99. M.D.S. Gomes, D. Santos Jr., L.C. Nunes, G.G.A. De Carvalho, F. De Oliveira Leme, F.J. Krug, Evaluation of grinding methods for pellets preparation aiming at the analysis of plant materials by laser induced breakdown spectrometry. *Talanta* **85**, 1744–1750 (2011)
100. G.G. Arantes de Carvalho, J. Moros, D. Santos Jr., F.J. Krug, J.J. Laserna, Direct determination of the nutrient profile in plant materials by femtosecond laser-induced breakdown spectroscopy. Anal. Chim. Acta **876**, 26–38 (2015)
101. M. Galiova, J. Kaiser, K. Novotny, O. Samek, L. Reale, R. Malina, K. Palenikova, M. Lika, V. Cudek, V. Kanicky, V. Otruba, A. Poma, A. Tucci, Utilization of laser induced breakdown spectroscopy for investigation of the metal accumulation in vegetal tissues. Spectrochim. Acta Part B Atomic Spectrosc. **62**, 1597–1605 (2007)
102. J. Kaiser, O. Samek, L. Reale, M. Liška, R. Malina, A. Ritucci, A. Poma, A. Tucci, F. Flora, A. Lai, L. Mancini, G. Tromba, F. Zanini, A. Faenov, T. Pikuz, G. Cinque, Monitoring of the heavy-metal hyperaccumulation in vegetal tissues by X-ray radiography and by femto-second laser induced breakdown spectroscopy. Microsc. Res. Tech. **70**, 147–153 (2007)
103. M. Galiova, J. Kaiser, K. Novotny, J. Novotny, T. Vaculovic, M. Lika, R. Malina, K. Stejskal, V. Adam, R. Kizek, Investigation of heavy-metal accumulation in selected plant samples using laser induced breakdown spectroscopy and laser ablation inductively coupled plasma mass spectrometry. Appl. Phys. A Mater. Sci. Process. **93**, 917–922 (2008)
104. S. Krizkova, P. Ryant, O. Krystofova, V. Adam, M. Galiova, M. Beklova, P. Babula, J. Kaiser, K. Novotny, J. Novotny, M. Liska, R. Malina, J. Zehnalek, J. Hubalek, L. Havel, R. Kizek, Multi-instrumental analysis of tissues of sunflower plants treated with silver(I) ions—plants as bioindicators of environmental pollution. Sensors **8**, 445–463 (2008)
105. J. Kaiser, M. Galiova, K. Novotny, R. Cervenka, L. Reale, J. Novotny, M. Lika, O. Samek, V. Kanicky, A. Hrdlicka, K. Stejskal, V. Adam, R. Kizek, Mapping of lead, magnesium and copper accumulation in plant tissues by laser-induced breakdown spectroscopy and laser-ablation inductively coupled plasma mass spectrometry. Spectrochim. Acta Part B Atomic Spectrosc. **64**, 67–73 (2009)
106. A. Assion, M. Wollenhaupt, L. Haag, F. Mayorov, C. Sarpe-Tudoran, M. Winter, U. Kutschera, T. Baumert, Femtosecond laser-induced-breakdown spectrometry for Ca 2+ analysis of biological samples with high spatial resolution. Appl. Phys. B Lasers Opt. **77**, 391–397 (2003)
107. M.Z. Martin, A.J. Stewart, K.D. Gwinn, J.C. Waller, Laser-induced breakdown spectroscopy used to detect endophyte-mediated accumulation of metals by tall fescue. Appl. Opt. **49**, C161–C167 (2010)
108. T. Ohta, M. Ito, T. Kotani, T. Hattori, Emission enhancement of laser-induced breakdown spectroscopy by localized surface plasmon resonance for analyzing plant nutrients. Appl. Spectrosc. **63**, 555–558 (2009)
109. J. Wang, M. Shi, P. Zheng, S. Xue, Quantitative analysis of lead in tea samples by laser-induced breakdown spectroscopy. J. Appl. Spectrosc. **84**, 188–193 (2017)
110. D.F. Andrade, E.R. Pereira-Filho, P. Konieczynski, Comparison of ICP OES and LIBS analysis of medicinal herbs rich in flavonoids from Eastern Europe. J. Braz. Chem. Soc. **28**, 838–847 (2017)
111. D.M. Silvestre, F. de Oliveira Leme, C.S. Nomura, A.N. do Nascimento, Direct analysis of barium, calcium, potassium, and manganese concentrations in tobacco by laser-induced breakdown spectroscopy. Microchem. J. **126**, 545–550 (2016)
112. J. Peng, K. Song, H. Zhu, W. Kong, F. Liu, T. Shen, Y. He, Fast detection of tobacco mosaic virus infected tobacco using laser-induced breakdown spectroscopy. Sci. Rep. **7**, 44551 (2017)
113. X. Fang, S.R. Ahmad, Elemental analysis in environmental land samples by stand-off laser-induced breakdown spectroscopy. Appl. Phys. B, 1–7 (2013)
114. Q. Sun, M. Tran, B.W. Smith, J.D. Winefordner, Direct determination of P, Al, Ca, Cu, Mn, Zn, Mg and Fe in plant materials by laser-induced plasma spectroscopy. Can. J. Anal. Sci. Spectrosc. **44**, 164–170 (1999)
115. G. Watal, B. Sharma, P. K. Rai, D. Jaiswal, D.K. Rai, N.K. Rai, A.K. Rai, LIBS-based detection of antioxidant elements: a new strategy. Methods Mol. Biol. (Clifton, N.J.) **594**, 275–285 (2010)

116. A. Khumaeni, H. Niki, K.I. Fukumoto, Y. Deguchi, K. Kurihara, K. Kagawa, Y.I. Lee, A unique technique of laser-induced breakdown spectroscopy using transversely excited atmospheric CO2 laser for the sensitive analysis of powder samples. Curr. Appl. Phys. **11**, 423–427 (2011)
117. A. Khumaeni, H. Niki, Y. Deguchi, K. Kurihara, K. Kagawa, Y.I. Lee, Analysis of organic powder samples by using the metal-assisted subtarget effect in a Transversely-Excited Atmospheric (TEA) CO2 laser-induced he gas plasma at 1 atm. J. Korean Phys. Soc. **55**, 2441–2446 (2009)
118. M. Tiwari, R. Agrawal, A.K. Pathak, A.K. Rai, G.K. Rai, Laser-induced breakdown spectroscopy: An approach to detect adulteration in turmeric. Spectrosc. Lett. **46**, 155–159 (2013)
119. J.N. Kunz, D.V. Voronine, H.W.H. Lee, A.V. Sokolov, M.O. Scully, Rapid detection of drought stress in plants using femtosecond laser-induced breakdown spectroscopy. Opt. Express **25**, 7251–7262 (2017)
120. L.C. Peruchi, L.C. Nunes, G.G.A. de Carvalho, M.B.B. Guerra, E. de Almeida, I.A. Rufini, D. Santos Jr., F.J. Krug, Determination of inorganic nutrients in wheat flour by laser-induced breakdown spectroscopy and energy dispersive X-ray fluorescence spectrometry. Spectrochim. Acta Part B Atomic Spectrosc. **100**, 129–136 (2014)
121. G. Bilge, B.Sezer, K.E. Eseller, H. Berberoğlu, H. Köksel, İ.H. Boyacı, Determination of Ca addition to the wheat flour by using laser-induced breakdown spectroscopy (LIBS). Eur. Food Res. Technol. **242**, 1685–1692 (2016)
122. M.R. Martelli, C. Barron, P. Delaporte, G. Viennois, X. Rouau, A. Sadoudi, Pulsed laser ablation: a new approach to reveal wheat outer layer properties. J. Cereal Sci. **49**, 354–362 (2009)
123. M.R. Martelli, F. Brygo, A. Sadoudi, P. Delaporte, C. Barron, Laser-induced breakdown spectroscopy and chemometrics: a novel potential method to analyze wheat grains. J. Agric. Food Chem. **58**, 7126–7134 (2010)
124. M. Pouzar, T. Cernohorsky, M. Prusova, P. Prokopcakova, A. Krejcova, LIBS analysis of crop plants. J. Anal. Atomic Spectrom. **24**, 953–957 (2009)
125. D.K. Tripathi, V.P. Singh, S.M. Prasad, D.K. Chauhan, N. Kishore Dubey, A.K. Rai, Silicon-mediated alleviation of Cr(VI) toxicity in wheat seedlings as evidenced by chlorophyll florescence, laser induced breakdown spectroscopy and anatomical changes. Ecotoxicol. Environ. Safety **113**, 133–144 (2015)
126. D.K. Tripathi, V.P. Singh, S.M. Prasad, N.K. Dubey, D.K. Chauhan, A.K. Rai, LIB spectroscopic and biochemical analysis to characterize lead toxicity alleviative nature of silicon in wheat (Triticum aestivum L.) seedlings. J. Photochem. Photobiol. B Biol. **154**, 89–98 (2016)
127. B. Sezer, G. Bilge, A. Berkkan, U. Tamer, I. Hakki Boyaci, A rapid tool for determination of titanium dioxide content in white chickpea samples. Food Chem. **240**, 84–89 (2018)
128. D. Jaiswal, P.K. Rai, G. Watal, Hypoglycemic and antidiabetic effects of withania coagulans fruit ethanolic extract in normal and streptozotocin-induced diabetic rats. J. Food Biochem. **34**, 764–778 (2010)
129. D. Jaiswal, P.K. Rai, G. Watal, Antidiabetic effect of Withania coagulans in experimental rats. Indian J. Clin. Biochem. **24**, 88–93 (2009)
130. A. Uhl, K. Loebe, L. Kreuchwig, Fast analysis of wood preservers using laser induced breakdown spectroscopy. Spectrochim. Acta Part B Atomic Spectrosc. **56**, 795–806 (2001)
131. T.M. Moskal, D.W. Hahn, On-line sorting of wood treated with chromated copper arsenate using laser-induced breakdown spectroscopy. Appl. Spectrosc. **56**, 1337–1344 (2002)
132. H.M. Solo-Gabriele, T.G. Townsend, D.W. Hahn, T.M. Moskal, N. Hosein, J. Jambeck, G. Jacobi, Evaluation of XRF and LIBS technologies for on-line sorting of CCA-treated wood waste. Waste Manag **24**, 413–424 (2004)
133. M.Z. Martin, N. Labbé, N. André, R. Harris, M. Ebinger, S.D. Wullschleger, A.A. Vass, High resolution applications of laser-induced breakdown spectroscopy for environmental and forensic applications. Spectrochim. Acta Part B Atomic Spectrosc. **62**, 1426–1432 (2007)
134. B.A. Gething, J.J. Janowiak, R.H. Falk, Assessment of laser induced breakdown spectroscopy (LIBS) for classification of preservative in CCA-treated lumber. Forest Prod. J. **59**, 67–74 (2009)

135. M.Z. Martin, N. Labbé, T.G. Rials, S.D. Wullschleger, Analysis of preservative-treated wood by multivariate analysis of laser-induced breakdown spectroscopy spectra. Spectrochim. Acta Part B Atomic Spectrosc. **60**, 1179–1185 (2005)
136. D.M. Silvestre, F.M. Barbosa, B.T. Aguiar, F.O. Leme, C.S. Nomura, Feasibility study of calibration strategy for direct quantitative measurement of K and Mg in plant material by laser-induced breakdown spectrometry. Anal. Chem. Res. **5**, 28–33 (2015)
137. H. Jull, R. Künnemeyer, P. Schaare, Nutrient detection in fresh and dried pasture using laser-induced breakdown spectroscopy. Precis. Agric. (2018)
138. H. Jull, R. Künnemeyer, S. Talele, P. Schaare, M. Seelye, Laser-induced breakdown spectroscopy analysis of sodium in pelletised pasture samples, in *2015 6th International Conference on Automation, Robotics and Applications (ICARA)* (2015), pp. 262–268
139. R. Bro, Multivariate calibration: what is in chemometrics for the analytical chemist? Anal. Chim. Acta **500**, 185–194 (2003)
140. A. Savitzky, M.J.E. Golay, Smoothing and differentiation of data by simplified least squares procedures. Anal. Chem. **36**, 1627–1639 (1964)
141. T. Næs, T. Isaksson, T. Fearn, T. Davies, *A User-Friendly Guide to Multivariate Calibration and Classification* (NIR publications, Chichester, UK, 2002)
142. H. Fink, U. Panne, R. Niessner, Process analysis of recycled thermoplasts from consumer electronics by laser-induced plasma spectroscopy. Anal. Chem. **74**, 4334–4342 (2002)
143. C.E. Rasmussen, C.K.I. Williams, *Gaussian Processes for Machine Learning* (University Press Group Limited, 2006)
144. C. Borggaard, H.H. Thodberg, Optimal minimal neural interpretation of spectra. Anal. Chem. **64**, 545–551 (1992)
145. P.J. Gemperline, J.R. Long, V.G. Gregoriou, Nonlinear multivariate calibration using principal components regression and artificial neural networks. Anal. Chem. **63**, 2313–2323 (1991)
146. I.H. Witten, E. Frank, M.A. Hall, Chapter 5—credibility: evaluating what's been learned, in *Data Mining: Practical Machine Learning Tools and Techniques (Third Edition)* (Morgan Kaufmann, Boston, 2011), pp. 147–187
147. R. Kohavi, A study of cross-validation and bootstrap for accuracy estimation and model selection, in *Proceedings of the 14th International Joint Conference on Artificial Intelligence*, vol. 2, Montreal, Quebec, Canada (1995)
148. I.H. Witten, E. Frank, M.A. Hall, *Data Mining: Practical Machine Learning Tools and Techniques*, 3rd edn. (Morgan Kaufmann, Boston, USA, 2011)
149. D. Cremers, L. Radziemski, Chemometric analysis in LIBS, in *Handbook of Laser-Induced Breakdown Spectroscopy*, 2nd edn. (Wiley, West Sussex, UK, 2013), pp. 223–255

The manufacturer's authorised representative in the EU is Springer Nature Customer Service Centre GmbH, Europaplatz 3, 69115 Heidelberg, Germany. If you have any concerns regarding our products, please contact ProductSafety@springernature.com

Printed and bound by CPI Group (UK) Ltd, Croydon, CR0 4YY
06/10/2026
02228973-0006